Two-Dimensional Materials for Nonlinear Optics

Two-Dimensional Materials for Nonlinear Optics

Fundamentals, Preparation Methods, and Applications

Edited by Qiang Wang and Hao-Li Zhang

WILEY-VCH

Editors

Prof. Qiang Wang
Lanzhou University
College of Chemistry and Chemical Engineering
No. 222 Tianshui South Road
730000 Lanzhou
China

Prof. Hao-Li Zhang
Lanzhou University
College of Chemistry and Chemical Engineering
No. 222 Tianshui South Road
730000 Lanzhou
China

Cover Images: © Serg Myshkovsky/Getty Images; © Marco_de_Benedictis/ Getty Images

Library of Congress Card No.: applied for

British Library Cataloguing-in-Publication Data
A catalogue record for this book is available from the British Library.

Bibliographic information published by the Deutsche Nationalbibliothek The Deutsche Nationalbibliothek lists this publication in the Deutsche Nationalbibliografie; detailed bibliographic data are available on the Internet at <http://dnb.d-nb.de>.

Print ISBN: 978-3-527-35059-9
ePDF ISBN: 978-3-527-83826-4
ePub ISBN: 978-3-527-83827-1
oBook ISBN: 978-3-527-83828-8

Typesetting Straive, Chennai, India
Printing CPI Group (UK) Ltd, Croydon, CR0 4YY
C9783527350599_270923

Contents

Preface *xiii*
List of Abbreviations *xv*

1 Preparation of 2D Materials *1*
Yue Tang and Hua Xu
1.1 Mechanical Exfoliation of 2D Materials *2*
1.2 Liquid-Phase Exfoliation of 2D Materials *4*
1.3 Chemical Vapor Deposition Growth of 2D Materials *6*
1.4 CVD Growth of Wafer-Scale Single Crystal 2D Materials *8*
1.5 Thickness Control in CVD Growth of 2D Materials *10*
1.6 Phase Control in CVD Growth of 2D Materials *12*
1.7 Summary and Prospect *14*
References *15*

2 An Introduction to the Nonlinear Optical Properties of 2D Materials *21*
Bolong Wang and Hao-Li Zhang
2.1 Introduction *21*
2.2 Nonlinear Optics of 2D Materials *22*
2.2.1 SHG, THG, and HHG Setups *24*
2.2.2 Four-Wave Mixing *26*
2.2.3 Z-Scan Techniques *27*
2.2.4 Nonlinear Optical Imaging *29*
2.2.5 Pump–Probe Techniques *32*
2.3 Application of 2D Nonlinear Materials *35*
2.3.1 Optical Limiting *35*
2.3.2 Q-Switched and Mode-Locked Lasers *38*
2.3.3 Optical Switch and Modulation *40*
2.3.4 Other Nonlinear Optical Phenomena *41*
2.4 Prospect *43*
2.4.1 Precise Fabrication and Functionalization of 2D Materials *43*
2.4.2 High Resolution Spatiotemporal Characterizing Techniques *44*

2.4.3 New Physics of 2D Materials *44*
2.4.4 Ever-Lasting Expansion of Applications *44*
Acknowledgment *45*
References *45*

3 Modulation and Enhancement of Optical Nonlinearity in 2D Materials *55*
Xinglin Wen and Qihua Xiong
3.1 Introduction *55*
3.2 Nonlinear Optics in 2D Materials *57*
3.2.1 Basics of Nonlinear Optics *57*
3.2.2 Parametric Nonlinear Optics *58*
3.2.3 Nonparametric Nonlinear Optics *59*
3.3 Nonlinearity Modulation in 2D Materials *60*
3.3.1 Structural Engineering *60*
3.3.1.1 Layer Number *60*
3.3.1.2 Twisted Angle *62*
3.3.1.3 Stacking Order *64*
3.3.1.4 Edge States *65*
3.3.2 Optical Modulation *66*
3.3.2.1 Excitonic Effects Enhanced Nonlinearity *66*
3.3.2.2 Valley-Dependent Nonlinearity *69*
3.3.2.3 Optical Cavity-Enhanced Nonlinearity *71*
3.3.2.4 All-Optical Modulation *73*
3.3.3 Electrical Modulation *75*
3.3.4 Magnetic Modulation *78*
3.3.5 Perspectives *78*
Author Contributions *78*
Notes *79*
Acknowledgment *79*
References *79*

4 Characterizing the Nonlinear Optical Properties of 2D Materials by Double 4f Nonlinear Imaging System with Phase Object and Four-Wave-Mixing Microscopy *87*
Zhongguo Li and Yinglin Song
4.1 Introduction *87*
4.2 Principle of NLO Measurement Technique *88*
4.2.1 Nonlinear Optical Interaction in 2D Materials *88*
4.2.2 Typical NLO Measurement Method in 2D Material Research *89*
4.2.2.1 Self-Phase Modulation (SPM) *89*
4.2.2.2 Z-Scan *89*
4.2.2.3 Third-Harmonic Generation (THG) *90*
4.2.2.4 Two-Photon-Induced Photoluminescence (TPL) *90*

4.3 Characterizing NLO Response of 2D Materials via Double 4f Nonlinear Imaging System with Phase Object (NIT-PO) Microscopy *91*
4.4 Characterizing NLO Response of 2D Materials via Four-Wave-Mixing Microscopy *95*
4.5 Outlook and Perspective *97*
References *98*

5 Ultrafast Carrier Dynamics in Emerging 2D Materials *103*
Jiawei Huang, Ningning Dong, and Jun Wang
5.1 Introduction *103*
5.2 Ultrafast Time-Resolved Spectroscopy *104*
5.2.1 Principle of Pump–Probe Measurement *104*
5.2.2 Experiment for Ultrafast Spectroscopy *107*
5.2.3 Fundamental Mechanism of Recombination Dynamics *107*
5.3 Ultrafast Optics in van der Waals 2D Materials *109*
5.3.1 Graphene *109*
5.3.2 Transition Metal Dichalcogenides (TMDC) *109*
5.3.3 Black Phosphorus *115*
5.4 Ultrafast Optics in Emerging Quasi-2D Materials *118*
5.4.1 2D Perovskite *118*
5.4.2 Non-Layered 2D Platinum Sulfide *120*
5.5 Perspectives on Ultrafast Optics for 2D Material *124*
Author Contributions *125*
Notes *125*
Acknowledgment *125*
References *125*

6 Transient Terahertz Spectroscopy for 2D Materials *131*
Jingyin Xu, Kai Zhang, Hong Li, and Tianwu Wang
6.1 Introduction *131*
6.2 Generation and Detection of THz Radiation *132*
6.2.1 Generation of THz Radiation Based on Nonlinear Optical Processes *132*
6.2.1.1 Photoconductive Antennas (PCAs) *133*
6.2.1.2 Electro-optic (EO) Crystals *135*
6.2.1.3 Air Plasma *136*
6.2.1.4 THz Emission of 2D Semiconductor Materials *140*
6.2.2 Typical THz Time-Domain Spectroscopy Setup *141*
6.2.2.1 Time-Domain THz Spectroscopy Under Pressure *143*
6.2.3 Photoinduced Changes of the Dielectric Function in 2D via THz Spectroscopy *147*
6.3 Nanoscale THz Scanning Probe Microscopy of 2D Materials *149*
6.3.1 Scanning Near-Field THz Microscopy for Polariton in 2D Materials *150*
6.3.2 THz-STM *151*
6.4 Perspectives *154*
6.4.1 Transient THz Spectroscopy *154*

6.4.2 High-Resolution Spatial and Temporal THz Spectroscopies *155*
Acknowledgment *155*
References *155*

7 Graphene Glass for Nonlinear Optics *163*
Qi Xiao, Jingyu Sun, and Hao-Li Zhang
7.1 Light Absorption of Graphene *163*
7.2 Nonlinear Optical Properties of Graphene *164*
7.2.1 Third-Order Nonlinear Optical Properties of Graphene *164*
7.2.2 Saturable Absorption of Graphene *165*
7.2.3 High-Order Harmonic Generation of Graphene *165*
7.3 Nonlinear Optical Properties of Graphene Glass *167*
7.3.1 Graphene-Doped Glass *167*
7.3.2 Graphene Glass Fabricated by Spin Coating *167*
7.3.3 Graphene Glass Fabricated by Transfer Method *168*
7.3.4 Directly Growth Graphene on Glass *168*
7.3.4.1 Graphene Growth on High-Temperature-Resistant Glass *168*
7.3.4.2 Graphene Growth on Molten Glass *171*
7.3.4.3 Metal-Catalyst-Assisted Graphene Growth *173*
7.3.4.4 Plasma-Enhanced Low-Temperature Graphene Growth *176*
7.4 Perspectives *179*
Acknowledgment *180*
References *180*

8 2D Materials for Nonlinear Optical Limiting *185*
Wen Shang, Bolong Wang, and Qiang Wang
8.1 Introduction *185*
8.2 Nonlinear Optical Limiting Mechanism *186*
8.2.1 Reverse Saturable Absorption *186*
8.2.2 Two-Photon/Multiphoton Absorption *187*
8.2.3 Free Carrier Absorption *188*
8.2.4 Nonlinear Refraction *188*
8.2.5 Nonlinear Scattering *189*
8.3 2D Materials for Optical Limiting *189*
8.3.1 Graphene and Analogs *190*
8.3.1.1 Graphene *190*
8.3.1.2 Transition Metal Dichalcogenides *194*
8.3.1.3 Black Phosphorus *196*
8.3.1.4 Hexagonal Boron Nitride (h-BNS) *197*
8.3.2 Novel 2D Materials *197*
8.3.2.1 Antimonene *197*
8.3.2.2 Tellurene *198*
8.3.2.3 2D Perovskites *200*
8.3.2.4 2D Metal–Organic Frameworks (2D-MOFs) *201*
8.3.3 Materials with New Structures and Mechanism *203*

8.3.3.1 Nonlinear Photonic Metamaterials *203*
8.3.3.2 Plasmonic Effect-Enhanced Nonlinearity *205*
8.3.4 Multifunctional Optical Limiting Devices *206*
8.4 Conclusions and Prospects *209*
Acknowledgments *210*
References *210*

9 The Saturable Absorbers Based on 2D Materials *221*
Xin-Hai Yan, Lei Zhang, and Kai-Ge Zhou
9.1 Introduction *221*
9.2 The Fundaments in the 2D Materials-Based Saturable Absorbers *222*
9.2.1 SA Theory *222*
9.2.2 Slow- and Fast-Saturable Absorber *223*
9.2.2.1 Slow-Saturable Absorber *224*
9.2.2.2 Fast-Saturable Absorber *225*
9.3 The Family of 2D Material-Based Saturable Absorbers *225*
9.3.1 Mono-Elemental 2D Materials *226*
9.3.1.1 Group IVA-Elemental 2D Materials *226*
9.3.1.2 Group VA-Elemental 2D Materials *230*
9.3.2 Dual-Elemental 2D Materials *234*
9.3.2.1 TMDs *234*
9.3.2.2 Topological Insulators (TIs) *235*
9.3.3 Multi-Elemental 2D Materials *237*
9.3.3.1 MOFs *237*
9.3.3.2 MXenes *237*
9.3.3.3 2D Perovskites *239*
9.4 Applications *239*
9.4.1 Fabrication of SA Device Based on 2D Materials *239*
9.4.1.1 Scotch Tape Exfoliation *240*
9.4.1.2 CVD, PVD *240*
9.4.1.3 LPE *240*
9.4.2 Q-Switched Laser *242*
9.4.2.1 Q-Switched Fiber Laser *243*
9.4.3 Mode-Locked Laser *246*
9.5 Perspectives and Outlook *248*
Acknowledgment *249*
References *249*

10 Second-Harmonic and Third-Harmonic Generations in 2D Layered Materials *257*
Xudong Jin and Min Zhao
10.1 Introduction *257*
10.1.1 Fundamentals of SHG and THG *258*
10.1.2 2DLM-Based SHG *259*
10.1.3 Graphene *260*

10.1.4 Transition Metal Dichalcogenides *262*
10.1.5 Other 2DLMs *266*
10.1.6 2DLM-Based THG *268*
10.1.6.1 Graphene *268*
10.1.7 Transition Metal Dichalcogenides *268*
10.1.7.1 Other 2DLMs *270*
10.2 Conclusions and Outlook *274*
References *274*

11 2D Perovskites for Nanolasers *281*
Wei Yuan and Chuanjiang Qin
11.1 Introduction *281*
11.2 Laser Formation Principle and Performance Parameters *283*
11.2.1 Basic Optical Physical Process *283*
11.2.2 Laser Formation Condition *283*
11.2.3 Performance Parameters *284*
11.2.4 Optical Cavities *285*
11.3 The Application of 2D Perovskites in Nanolasers *286*
11.3.1 Microplates *287*
11.3.2 Microrods *287*
11.3.3 Microcrystals *291*
11.3.4 Nanowires *291*
11.3.5 Microrings *293*
11.3.6 Thin Films *294*
11.3.7 Thin Flakes *298*
11.4 Prospect *299*
References *300*

12 2D Materials for Space Use *303*
Shuyan Wang and Qiang Wang
12.1 Introduction *303*
12.2 Space Radiation and Two-Dimensional Materials *305*
12.2.1 Space Radiation Environment *305*
12.2.2 2D Materials Under Irradiation *307*
12.2.2.1 Electron Irradiation *307*
12.2.2.2 Ion Irradiation *307*
12.2.2.3 Proton Irradiation *308*
12.2.2.4 γ-Ray Irradiation *308*
12.3 2D Materials for Space Use *309*
12.3.1 Emerging Nonlinear Optical Applications *309*
12.3.1.1 Laser Protection and Mode Locking *310*
12.3.1.2 Other Nonlinear Phenomena *313*
12.3.2 Space X-ray Detection and Imaging *314*
12.3.3 Perovskite Solar Cells (PSCs) *316*
12.3.4 2D Single-Photon Sources *317*

12.3.5 Integrated Optoelectronic Platform *318*
12.3.5.1 Field-Effect Transistor and Photodetectors *318*
12.3.5.2 Sensors *320*
12.3.6 Other Applications: Lubricants *320*
12.4 Perspectives *321*
Acknowledgment *322*
References *322*

Index *333*

Preface

The development of two-dimensional (2D) materials has garnered increasing interest due to their unique structure and versatile applications. For instance, in the optoelectronic and photonic fields, 2D materials have been broadly employed for optical modulation through optical limiting/mode locking, photodetectors, optical communications, miniaturized all-optical devices, and so on. The ever-expanding 2D family particularly offers abundant sources to the emerging new physical insights and phenomena in traditional nonlinear optics. Although previously there have been many excellent reviews on the applications of 2D materials in one specific field, systematic introduction integrating all of the key elements ranging from material fabrications to novel characterizing techniques and myriad nonlinear optical applications is still rare.

Hence, in this book, collective efforts through researchers from interdisciplinary fields focus on 2D materials for nonlinear optical applications to further the exciting territory toward a new horizon. Preparation methods for various 2D materials including graphene glass, transition metal dichalcogenides (TMDs), single elemental 2D materials from Group 15, and perovskites have been thoroughly summarized. Moreover, powerful nonlinear optical characterizing techniques, such as Z-scan measurement, femtosecond transient absorption spectroscopy and microscopy are introduced. The origin of the unique optoelectronic properties of 2D materials is elucidated from various perspectives like excited-state dynamics.

Specifically, Chapter 1 provides a comprehensive perspective on the fundamentals and preparation methods of 2D materials with nonlinear optical properties. Chapter 2 is an overall introduction to the nonlinear optical properties of 2D materials, and Chapter 3 elaborates on the modulation and enhancement of optical nonlinearity in 2D materials. Furthermore, Chapters 4–6 discuss the nonlinear optical characterization techniques for 2D materials and their applications in the specific field, including novel nonlinear optical imaging systems, ultrafast time-resolved spectroscopy for investigating carrier dynamics in emerging 2D materials, transient terahertz spectroscopy, etc. Chapter 7 expounds on graphene glass for nonlinear optics while Chapters 8–12 emphasize the applications of 2D materials for optical limiting, saturable absorber, SHG and THG, nanolasers, and space use, respectively. The topics covered in this book are of vital importance in understanding the unique optoelectronic properties and applications of 2D materials and the physical insights behind them. We hope the book will be a handy toolkit for professionals, researchers, and graduate students working or interested in the field of nonlinear optics and materials.

The editors thank all the contributors of this book for their insightful chapters. We are grateful to the production editors at Wiley, Ms. Lifen Yang and Katherine Wong, for their diligent work and expertise in managing the process.

Lanzhou, People's Republic of China
Qiang Wang
Hao-Li Zhang

List of Abbreviations

Number

2D	Two dimensional
2DLMs	2D layered materials
2DCS	Two-dimensional coherent spectroscopy
1D	One dimensional
0D	Zero dimensional
2PA	Two-photon absorption
3PA	Three-photon absorption
3PM	Three-photon microscopy

A

ABCD	Air-biased coherent detection
AC-STEM	Aberration-corrected scanning transmission election microscopy
AIE	Aggregation-induced emission
AFM	Atomic force microscope
ASE	Amplified spontaneous emission

B

Bi	Bismuthene
BC	Birefringence crystals
BP	Black phosphorus
BS	Beam splitter
BPDs	Balanced photodiodes
BBO	Beta barium borate
BN	Boron nitride
BLG	Bilayer graphene
BGR	Bandgap renormalization

C

CARS	Anti-Stokes Raman scattering
CMOS	Complementary metal-oxide semiconductor
CVD	Chemical vapor deposition
CSS	Chinese Space Station
COFs	Covalent organic frameworks
CB	Conduction bands
CW	Continuous wave

D

DAC	Diamond anvil cell
DFG	Difference-frequency generation
DLC	Diamond-like carbon
D-A	Donor–acceptor
DBR	Distributed Bragg reflector
DSM	Dirac semimetal
DFWM	Degenerate FWM
DT	Differential transmission
DFB	Distributed feedback
DWP	Double-wavelength wave plate

E

ESA	Excited-state absorption
EDF	Erbium-doped fiber
EBL	Electron beam lithography
ENZ	Epsilon near zero
ED	Electric dipole
EQ	Electric quadrupole
EO	Electro-optic
EFISH	Electric-field-induced second harmonics

F

FWM	Four-wave mixing
FWR	Four-wave rectification
fs	Femtosecond
FCA	Free carrier absorption

FET	Field-effect transistor
FTO	Fluorine-doped tin oxide
FW	Fundament wavelength
F-P	Fabry–Pérot cavity
FWHM	Full width at half maxima

G

Ge	Germanium
GO	Graphene oxide
GD	Graphdiyne
GEOs	Geostationary orbits
GSAM	Graphene saturable absorber mirror

H

HG	Harmonic generation
HHG	High-harmonic generation
h-BN	Hexagonal boron nitride
HEOs	Highly elliptical orbits
HAADF	High-angle annular dark field
HT	High temperature
HP	Half-wave plate
HEG	Hydrogen-induced exfoliated graphene
h-MO	Hexagonal metal oxides

I

ISS	International Space Station
IPA	Isopropylammonium

L

LEED	Low-energy electron diffraction
LD	Laser diode
LEOs	Low-Earth orbits
LiDAR	Light detection and ranging
LDOS	Local density of states
LPE	Liquid-phase exfoliation
LAPCAs	Large-aperture photoconductive antennas

M

M	Magnetization
MPL	Multiphoton lithography
MPEA	Chiral β-methylphenethylammonium
MPA	Multiphoton absorption
MRRs	Micro-ring resonators
MZI	Mach–Zehnder interferometer
MEOs	Medium-Earth orbits
MOFs	Metal–organic frameworks
MOCVD	Metal–organic chemical vapor deposition
MRA	Metamaterial resonant absorber
MD	Magnetic dipole
MI	Multiphoton ionization
MSHG	Magnetization-induced SHG
MAEF	Multiphoton-absorption-excited fluorescence
MXenes	Transition metal carbides and nitrides

N

NLO	Nonlinear optics
NLA	Nonlinear absorption
NLR	Nonlinear refraction
NLS	Nonlinear scattering
NWs	Nanowires
NRs	Nanorods
NDs	Nanodots
NSs	Nanosheets
NIT-PO	Nonlinear imaging system with phase object
NMP	N-methylpyrrolidone

O

OPA	Optical parametric amplifier
OM	Optical microscope
OL	Optical limiting
OC	Output coupler
OA	Open aperture
OR	Optical rectification

P

PMMA	Polymethyl methacrylate
PDMS	Polydimethylsiloxane

PEEM	Photoemission electron microscopy
Pt	Platinum
PC	Polarization controller
PD	Photodiode
PCAs	Photoconductive antennas
PO	Phase object
PI-ISO	Polarization-independent isolator
PSE	Polymer solid electrolyte
PVD	Physical vapor deposition
PVT	Physical vapor transport
PB	Photoinduced bleaching
PM	Parabolic mirrors
PMT	Photomultiplier tube
PIA	Photoinduced absorption
PtS	Platinum sulfide
PMT	Photomultiplier tube

Q

QWs	Quantum wells
QDs	Quantum dots

R

RSA	Reverse saturable absorption
rGO	Reduced graphene oxide
RIE	Reactive ion etching

S

SHG	Second-harmonic generation
SRS	Stimulated Raman scattering
SA	Saturable absorption
SAE	Sonication-assisted exfoliation
SEEs	Single-event effects
SEU	Single-event upset
SEFI	Single-event functional interruption
SET	Single-event transient
SEL	Single-event latch-up
SNSPD	Superconducting nanowire single-photon detectors
SDI	spectral-domain interferometry
SOI	Silicon on insulator
SEM	Scanning electron microscopy
SFM	Sum-frequency mixing

SPM	Self-phase modulation
SNOM	Scanning near-field optical microscopy
STE	Self-trapped exciton
SVEA	Slowly varying envelope approximation
SESAM	Semiconductor saturable absorber mirror
STM	Scanning tunneling microscopy
STS	Scanning tunneling spectroscopy
SECT	Slow evaporation at constant temperature
SNR	Signal-to-noise ratio

T

TPA	Two-photon absorption
THG	Third-harmonic generation
TPF	Two-photon fluorescence
TPF	Tilted pulse front
THz	Terahertz
TFISH	THz-field-induced second harmonic
TPL	Two-photon luminescence
TPEF	Two-photon excitation fluorescence
TAM	Transient absorption microscopy
TAS	Transient absorption spectroscopy
TAC	Thermally assisted conversion
TCR	Transition-Cherenkov radiation
TMDs	Transition metal chalcogenides
TMDCs	Transition metal dichalcogenides
TI	Tunnel ionization
TID	Total ionizing dose
tr-SNOM	Time-resolved scanning near-field optical microscope
TRTS	Time-resolved THz time-domain spectroscopy

V

vdW	van der Waals
VB	Valence bands

W

W	Tungsten
WP	Wollaston prism
WDM	Wavelength division multiplexer
WGM	Whispering-gallery mode

X

XUV	Extreme ultraviolet

1

Preparation of 2D Materials

Yue Tang and Hua Xu

Shaanxi Normal University, Key Laboratory of Applied Surface and Colloid Chemistry of Ministry of Education, Shaanxi Key Laboratory for Advanced Energy Devices, School of Materials Science and Engineering, Xi'an 710119, P. R. China

Two-dimensional (2D) materials, an emerging new class of nanomaterials with rich structures and remarkable properties, would bring many transformative technologies and applications [1]. Since the discovery of graphene for the first time in 2004, the 2D material family has expanded dramatically to include insulators (hexagonal boron nitride [h-BN]), semiconductors (most transition metal dichalcogenides [TMDCs], black phosphorus [BP], and tellurium [Te]), semi-metals (some TMDCs and graphene), metals (transition metal carbides and nitrides [MXenes]), superconductors ($NbSe_2$), and topological insulators (Bi_2Se_3 and Bi_2Te_3) [2, 3]. The atomic thickness and dangling free surface of 2D materials, together with their superior optical, electrical, magnetism, thermal, and mechanical properties, endow them with great promise for applications in optical communication, electronics, optoelectronics, spintronics, memory, thermoelectric, and energy conversation and storage devices [4, 5].

As highlighted by the famous nanomaterial scientist Zhongfan Liu, "preparation determines the future" is an inexorable law for all materials. In the past decade, a series of preparation technologies have been developed to fabricate 2D materials for satisfying the requirements of their fundamental studies and various applications. In view of the layered structure of 2D materials, the primary preparation technologies can be divided into two major types: top-down and bottom-up approaches. In this chapter, we will introduce the recently developed preparation technologies for 2D materials, including two top-down approaches (mechanical exfoliation and liquid exfoliation) and one bottom-up approach (vapor phase growth). Here, we give more space to introduce single crystal growth, thickness control, and phase control in the vapor phase growth of 2D materials.

Two-Dimensional Materials for Nonlinear Optics: Fundamentals, Preparation Methods, and Applications, First Edition. Edited by Qiang Wang and Hao-Li Zhang.

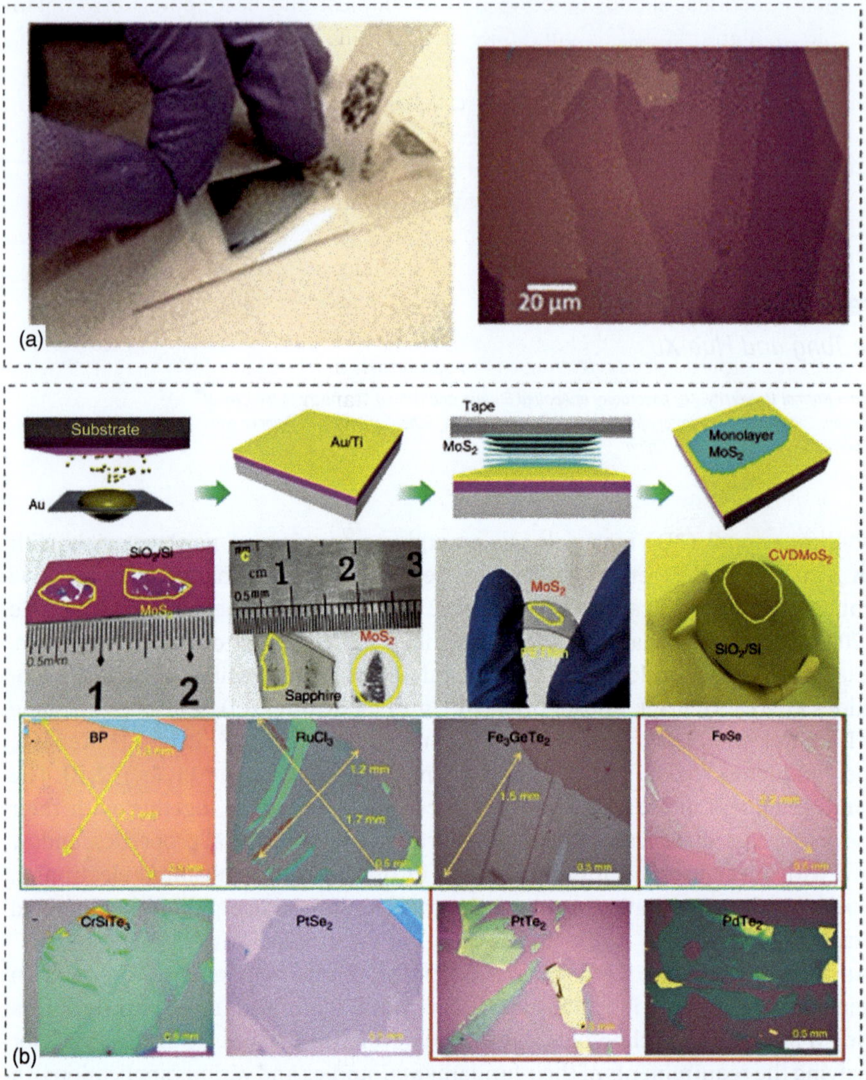

Figure 1.1 Mechanical exfoliation preparation of 2D materials. (a) Schematic diagram of mechanical cleavage process and optical micrograph of one of the graphene flakes on the SiO_2/Si substrate with different thicknesses. Source: Reproduced with permission from Yuan Huang et al. [7]/American Chemical Society. (b) Schematic diagram of the Au-assisted mechanical exfoliation process and corresponding optical microscope (OM) images of obtained vdW-layered 2D materials. Source: Yuan Huang et al. [8]/Springer Nature/CC BY 4.0.

1.1 Mechanical Exfoliation of 2D Materials

In 2004, Geim and Novosolov firstly prepared monolayer graphene by exfoliating bulk graphite using Scotch tape [6]. The schematic exfoliation method and the obtained 1-4-layer graphene are shown in Figure 1.1a [7]. Since then, mechanical cleavage, commonly referred to as the Scotch tape method and involving no chemical reactions, is considered to be the simplest and best approach to obtain large-area, high-quality 2D materials that retain their pristine structures and properties. Until 2010, Heinz's group firstly extended the mechanical cleavage method to prepare monolayer MoS_2 and discovered the indirect to direct bandgap transition with the thickness changing from bulk to monolayer [9]. Then, A. Kis's group fabricated the field effect transistor based on the exfoliated monolayer MoS_2 and achieved room-temperature current on/off ratios of 10^8 and ultralow standby power dissipation [10]. Later, mechanical cleavage techniques were widely used to produce dozens of 2D transition metal dichalcogenides (TMDCs: WS_2, WSe_2, $MoSe_2$, ReS_2, etc.) [11]. However, with the increasing requirements on thickness and domain size of 2D materials for device fabrication and property exploration, cleavage technology pursues efficient preparation of 2D materials with large areas and high quality. Hence, many optimized mechanical cleavage techniques have been developed in recent years. For example, Huang *et al.* developed a universal Au-assisted mechanical cleavage technique, as shown in Figure 1.1b [8]. Theoretical calculation indicates that Au and many 2D materials can form quasi-covalent bonds, which are larger than van der Waals (vdW) interactions but smaller than covalent bonds. In the experiment, a thin Au layer was firstly deposited onto a substrate covered with a thin Ti or Cr adhesion layer, and a freshly cleaved bulk crystal on tape was brought into contact with the Au layer. Then the adhesive tape was placed on the outward side of the crystal, and gentle pressure was applied to establish a good crystal/Au contact. Finally, peel off the tape to remove the major portion of the crystal, leaving monolayer or few-layer flakes on the Au surface. Using this approach, they obtained large-area monolayer flakes on the Au surface, including MoS_2, FeSe, $PtSe_2$, $PtTe_2$, $PdTe_2$, and $CrSiTe_3$. Using the above exfoliation methods, a series of new 2D-layered materials have been successfully prepared, such as Fe_3GeTe_2, $MnBi_2Te_4$, CrOCl, $NbSe_2$, and $NbOCl_2$, [12–16] which exhibit superior ferromagnetic, superconduction and nonlinear optic properties. In a word, mechanical cleavage technology has been widely utilized to prepare 2D-layered materials for studying their fundamental properties.

However, the methods mentioned above are restricted to 2D-layered materials in which the interlayer interactions are dominated by weak vdW force, and thus these methods are not applicable to materials in which interlayer interactions are dominated by non-vdW force. In view of this, Zhang et al. developed a new mechanical cleavage strategy [17]. The polished metallic surface is oxidized under a controlled environment to enable the growth of hexagonal metal oxides (h-MO). These high-crystalline h-MO with layered structures, without ionic dopants or vacancies, can be easily exfoliated by stamping them onto the target substrates. This cleavage strategy was firstly applied to prepare the three-dimensional (3D) transition metal-based h-MO (TiO_2, Fe_2O_3, and Ni_2O_3), and it could be readily extended to prepare a variety of other metal oxides for exploring their novel 2D quantum properties. Most recently, Fengxia Geng's group reported a calendaring pretreatment mechanically exfoliate approach to prepare 2D materials from non-vdW structures [18]. On the basis of the traditional scotch tape method, an external mechanical force was applied to the non-vdW materials. This approach involves laterally sliding in the closely packed neighboring layers to transform the structure from a stable to a metastable phase, weakening the interlayer binding. Using this approach, a variety of 2D materials have been prepared, including metals (Bi, Sb), semiconductors (SnO, V_2O_5, Bi_2O_2Se), and superconducting compounds (KV_3Sb_5). This method for mechanically exfoliating non-vdW materials increases the availability of 2D materials for the exploration of their physical characteristics and potential applications. To sum up, using the mechanical cleavage method can prepare most of the target 2D materials with high crystal quality, but the samples acquired via this approach also possess several problems, such as irregular morphology, uncontrollable thickness, small domain size, and low yield. Therefore, the present mechanical cleavage technology for 2D material preparation still faces great challenges, which limits its basic research to practical application. It is worth to further optimize the present cleavage methods and explore novel cleavage technologies to achieve 2D materials with controllable layers and sizes.

1.2 Liquid-Phase Exfoliation of 2D Materials

As demonstrated above, the preparation of 2D materials via mechanical exfoliation is limited by the inevitable low yield and uncontrollable thickness, domain size, and morphology. To resolve these issues, liquid-phase exfoliation (LPE) was developed to prepare 2D materials, which has the advantages of high efficiency, large scale, and better controllability [19–23]. The LPE of 2D materials can be divided into two primary approaches: direct exfoliation and chemical exfoliation [24]. Direct exfoliation methods include sonication-assisted exfoliation [SAE] and shear exfoliation. Chemical exfoliation methods include chemical intercalation and electrochemical intercalation [25, 26].

For the SAE method, layered bulk crystal was firstly dispersed in a solvent, followed by material exfoliation via ultrasonic energy and removal of the non-exfoliated material via centrifugation [27]. The dispersion is then centrifuged and purified to achieve 2D nanosheets with uniform domain size and thickness. Moreover, a shear exfoliation method has been widely used to disperse the pretreated layered crystals, which enables the production of high-quality 2D nanosheets while avoiding SAE's fragile issues. However, the exfoliation is only suitable for 2D materials with weak interlayer force, while it is difficult to effectively exfoliate 2D materials with strong interlayer coupling.

Huang et al. developed an efficient and stable three-stage sonication-assisted liquid-phase exfoliation (TS-LPE) method for mass preparation of high-structural-integrity few- and single-layer Fe_3GeTe_2 nanoflakes [28]. The three stages include slicing crystals, weakening interlayer vdW forces, and using ultrasonic cavitation (Figure 1.2a). At the first stage, ball milling is used to reduce the thickness and size of Fe_3GeTe_2 crystals. At the second stage, the shear of the solvent via stirring in *N*-methylpyrrolidone (NMP) and the shock waves caused by the collision events in the container lead to expanding layer spacing and weakening interlayer vdW forces to facilitate the subsequent shedding of the nanoflakes. The third stage is ultrasonic cavitation to facilitate the exfoliation of nanoflakes. Finally, Fe_3GeTe_2 nanoflakes

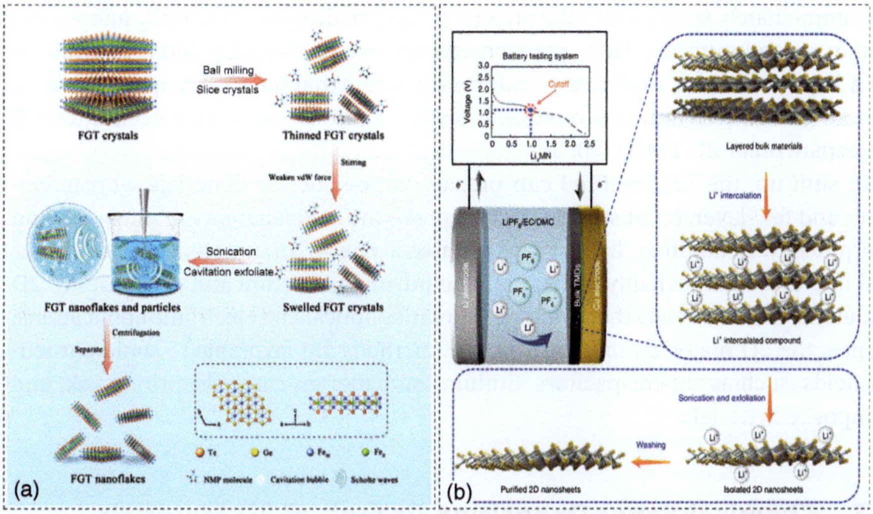

Figure 1.2 Liquid-phase exfoliation of 2D materials. (a) Schematic diagram of the TS-LPE method. Source: Reproduced with permission from Ma et al. [28]/American chemical society. (b) Schematic illustrations of the electrochemical Li^+ intercalation-based exfoliation process. Source: Reproduced with permission from Yang et al. [29]/Springer Nature.

were extracted by centrifugation. The highest yield of Fe_3GeTe_2 nanoflakes is 22.3 wt%, with single layers accounting for 6%, and a maximum of 103 μm is available. The 200 mg level of output has overcome the limitations of mechanical exfoliation and molecular beam epitaxy in economically amplificated production.

Zeng et al. developed an electrochemical lithium-ion intercalation-based exfoliation method that can be easily conducted at room temperature within 26 hours (Figure 1.2b) [29]. This exfoliation method is relatively simple and straightforward. It offers a higher degree of control under mild conditions, and thus it was widely utilized to prepare various 2D materials on a large scale. The obtained 2D nanosheets possess high crystal quality, uniform thickness, and a relatively large domain size. However, during the lithium-ion intercalation process, the insertion of each Li^+ ion involves the injection of one electron into the host crystal. Thus, the intercalation of a large number of Li^+ ions (one per formula unit in $LiMoS_2$) would lead to massive electron injection into the MoS_2 crystal, which would induce an undesired phase transition from the semiconducting 2H phase to the metallic 1T phase [30–32]. Theoretical studies suggest that this phase transition occurs only when the electron injection exceeds a certain threshold (0.29 electrons per MoS_2 formula unit) [33, 34]. A possible way of reducing electron injection into the host 2D crystal and thus preventing the undesired phase transition is to replace the small Li^+ ions (diameter $d \sim 2$ Å) with larger cations, such as quaternary ammonium ions [32]. Duan et al. developed an electrochemical exfoliation method via intercalating large-size tetramethyl ammonium bromide (THAB; $d \sim 20$ Å) molecules into the MoS_2 crystal [35]. Driven by the negative electrochemical potential, THA^+ cations were inserted into the MoS_2 layer, causing a substantial volume expansion. Then, the expanded MoS_2 crystal was immediately sonicated in the organic solvent to disperse thin MoS_2 nanosheets within several minutes. This approach can achieve high-quality MoS_2 nanosheets on a large scale with a relatively uniform thickness that have been applied to constructing large-area thin-film electronic devices, and this method has been extended to prepare other 2D TMDs [20, 36].

To sum up, the LPE method can prepare large-scale 2D materials with monolayer and few-layer, but it still needs to improve the controllability of thickness and morphology. In addition, high-power ultrasonic energy, organic solvents, or chemical intercalators inevitably degenerate the intrinsic structure and properties of 2D materials, which hinders their high-performance optical and electronic applications. Hence, the 2D materials prepared by LPE methods are favorable to some particular fields, such as supercapacitors, lithium-ion batteries, catalysis, printer ink, and composite materials.

1.3 Chemical Vapor Deposition Growth of 2D Materials

Comparing with the above top-down methods, bottom-up synthesis by chemical vapor deposition (CVD) proves to be an effective method to grow large-scale 2D materials with controllable domain size, morphology, thickness, and electronic-grade quality, which shows great potential for fabricating high-performance optical, electronic, and optoelectronic devices for commercial applications. The CVD method used to synthesize 2D materials refers to the gas-vapor growth of material on a substrate within the tube furnace system (Figure 1.3a) [37, 41]. To be specific, two or more precursors are introduced into the high-temperature reaction zone in quartz tube by carrier gas for chemical reaction, nucleation, growth, and final formation of 2D flakes or continuous films on the substrate.

In the past decades, CVD has been widely used in the preparation of various 2D materials, such as elemental graphene and BP, binary h-BN, TMDs (MoS_2, WSe_2, $MoTe_2$, VS_2, TiS_2, $ReSe_2$, etc.), ternary chalcogenides, and their alloys and heterostructures [42–47]. Take the CVD growth of MoS_2 as an example, Van der Zande et al. synthesized monolayer MoS_2 nanosheets with uniform triangle morphology by using CVD growth with sulfur (S) powder and molybdenum trioxide (MoO_3) as precursors [48]. The obtained MoS_2 possesses high-quality, uniform monolayer and large grain size up to 120 μm, which enables comparable electrical and photoelectric properties with those of mechanically exfoliated MoS_2.

During the process of 2D material growth, many factors affect the quality and properties of the product, such as precursor, substrate, growth temperature, carrier gas flow, and additives [49–51]. Thus, great efforts have been made to optimize the CVD growth method to achieve the preparation of large-area, high quality, homogeneous-thickness 2D materials [49]. For instance, to achieve a stable and continuous supply of Mo source, electrochemically oxidized arched Mo foil was utilized as the precursor instead of the traditional powder, and the growth of MoS_2 with controllable thickness was realized (Figure 1.3b) [38]. In addition, researchers have confirmed that the addition of synergistic additives can modulate the growth chemistry of 2D materials, facilitate the growth rate, and adjust the morphology and

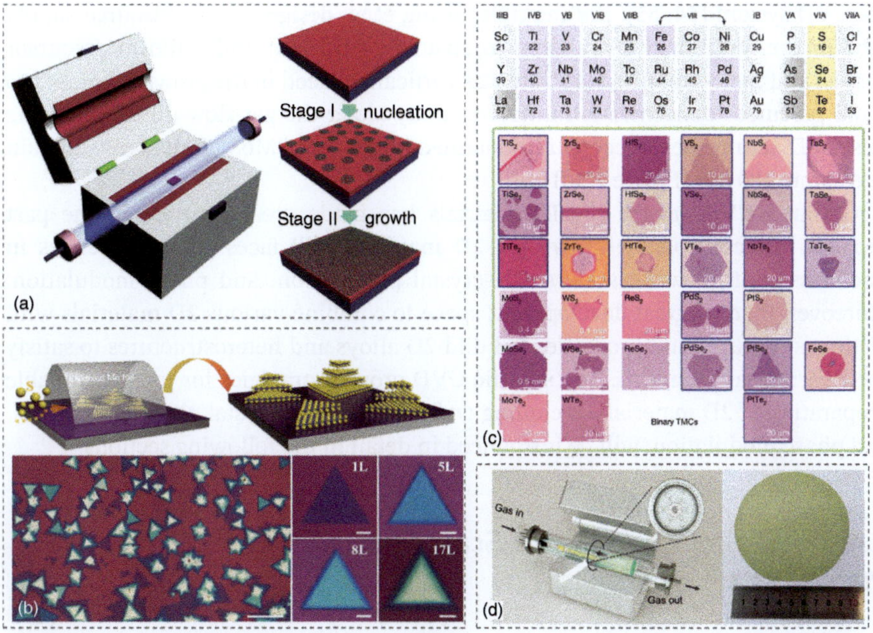

Figure 1.3 CVD growth of 2D materials. (a) Schematic diagram of the CVD setup and the main growth process [37]. (b) Schematics of the CVD growth of MoS_2 using arched oxidized Mo foil as precursor and optical images of the growth products with varied thickness. Source: Reproduced with permission from Jingying Zheng et al. [38]/John Wiley & Sons. (c) The transition metals and chalcogens used for molten salt-assisted CVD growth and optical images of the different atomically thin TMCs. Source: Reproduced with permission from Jiadong Zhou et al. [39]/Springer Nature. (d) Schematic diagram of the multi-source CVD setup and photograph of a monolayer MoS_2 film at a 4 in. wafer scale. Source: Reproduced with permission from Qinqin Wang et al. [40]/American Chemical Society.

structure. For many 2D TMDs, the high melting point of metal precursors leads to a challenge for low-temperature growth through CVD. In view of this, Zhou et al. developed a molten salt-assisted CVD method and demonstrated that salt decreases the melting point of the reactants and increases the overall reaction rate, which is favorable for highly efficiently achieving various 2D TMDs (Figure 1.3c) [39]. Moreover, by introducing fluorine into the growth system, Liu et al. demonstrated that the growth rate of graphene, h-BN, and WS_2 could be greatly accelerated, which is beneficial for growing large domain-size 2D materials [52].

Owing to the inherent difficulty in controlling the concentration of solid precursors during the entire growth time, it remains a challenge to prepare large-area monolayer films with spatial homogeneity and structural continuity. Kang et al. adopted a metal-organic chemical vapor deposition (MOCVD) technique, with $Mo(CO)_6$, $W(CO)_6$, $(C_2H_5)_2S$, and H_2 as independent gas-phase precursors, to grow large-area and uniform 2D TMD films [53]. They demonstrated that maintaining a low partial pressure at the Mo source by controlling the partial pressure is the key to the layer-by-layer growth and the formation of a wafer-scale TMDs film. Furthermore, since the configuration of the CVD instrument may play an important role in 2D material growth, many studies have also focused on the optimization of the CVD system [54, 55]. For example, Wang et al. designed a multisource supply tube in the traditional CVD setup to ensure a sufficient and uniform precursor supply, and the c-sapphire substrate was vertically placed in the growth chamber to avoid precursor concentration inhomogeneity along the gas flow direction. Using this CVD setup, they successfully obtained wafer-scale MoS_2 films with domain sizes up to 180 μm (Figure 1.3d) [40].

Although CVD-growth of 2D materials has made great progress in the past decades, the controllable growth of 2D materials still faces great challenges in thickness control, wafer-scale single-crystal preparation, and phase modulation. Moreover, there is great development space to combine various 2D materials with distinct structures and properties to build 2D alloys and heterostructures to satisfy their various applications. The specific CVD growth strategies for the controllable preparation of 2D materials, including wafer-scale single crystal, thickness control, and phase modulation, will be introduced in detail in the following sections.

1.4 CVD Growth of Wafer-Scale Single Crystal 2D Materials

In most of the present works, amorphous SiO_2/Si was primarily used as the substrate in CVD growth of 2D materials, which results in the random orientation of 2D grains [48]. Thus, CVD-grown 2D films are usually polycrystalline with many grain boundaries, which seriously deteriorates their optical, electrical, mechanical, chemical, stability, etc. To improve these properties, it is highly desired to prepare 2D films with uniform thickness, high crystallinity, defect-free, and limited grain boundaries, especially wafer-scale single crystal 2D films [56]. Substrate-guided epitaxial growth is widely utilized to achieve these 2D films. The idea is based on the effective control of the orientation of 2D materials grown on a substrate, where the grains merge seamlessly with the same (or small misorientation) angle, which is considered to be an essential way to decrease or eliminate the grain boundaries [57].

Most 2D TMDs have a threefold symmetry structure, thus the lattice-induced epitaxial can be used to achieve wafer-scale 2D films. So, the c-plane sapphire, which has the same symmetry and compatible lattice constant with TMDs, has been widely used as a substrate to align the TMDs [40, 58]. However, the epitaxial relationship indicates that the $[11\bar{2}0]$ direction of the c-plane sapphire is perpendicular to the zig-zag edge of TMD domains, which equalizes the two antiparallel domains and prevents the growth of single crystals. To solve the above problem, Wang et al. designed a c-plane sapphire wafer with a major miscut angle toward the A axis (defined as C/A), which produced surface steps along $< 10\bar{1}0 >$ direction for unidirectionally aligning MoS_2, as shown in Figure 1.4a [59]. In principle, a good epitaxial relationship between MoS_2 and c-plane sapphire is formed (Figure 1.4b) when the armchair and zigzag directions of MoS_2 are parallel to $< 11\bar{2}0 >$ and $< 10\bar{1}0 >$ of Al_2O_3, respectively. An OM image of as-grown triangular MoS_2 domains on c-plane sapphire shows an obvious unidirectional alignment. Second harmonic generation (SHG) mapping of two adjacent MoS_2 grains (Figure 1.4d)

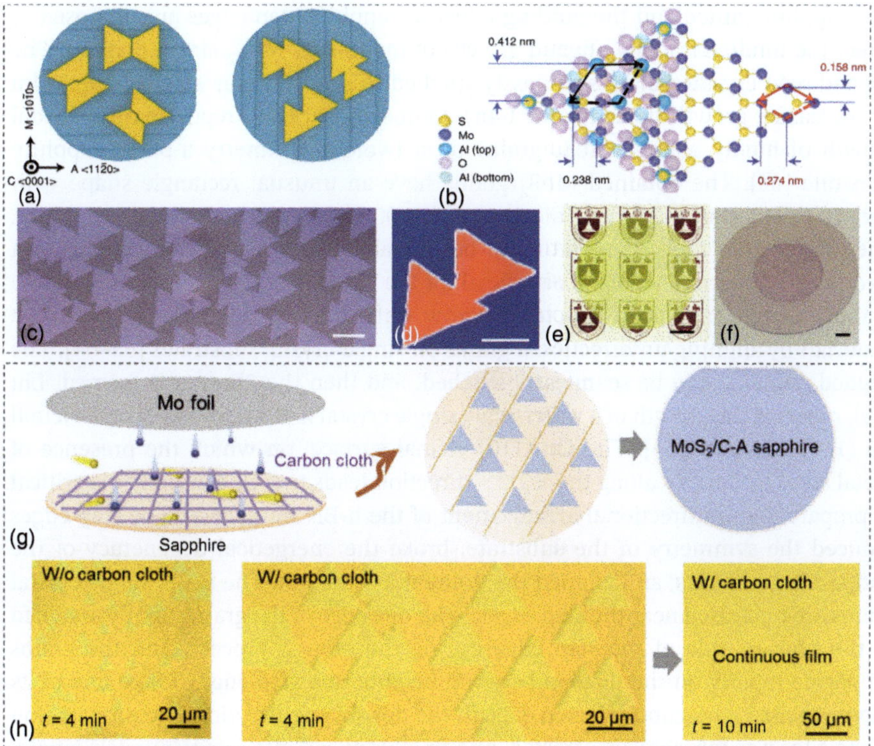

Figure 1.4 CVD growth of wafer-scale single-crystal 2D materials. (a) Schematically show the epitaxial relationship between MoS_2 grain and c-plane sapphire with different step orientations. Source: Taotao Li et al. [59]/Reproduced with permission of Springer Nature. (b) The epitaxial relationship of MoS_2 on c-plane sapphire in atomic scale. Source: Taotao Li et al. [59]/Reproduced with permission of Springer Nature. (c) Unidirectional alignment of MoS_2 domains on a C/A sapphire (0001) substrate. Source: Taotao Li et al. [59]/Reproduced with permission of Springer Nature. (d) Polarized SHG mapping of two merging MoS_2 domains on the C/A sapphire (0001) substrate. Source: Reproduced with permission from Taotao Li et al. [59]/Springer Nature. (e, f) Photographs of 2-inch monolayer MoS_2 single crystal on C/A sapphire and SiO_2/Si substrates. Source: Taotao Li et al. [59]/Reproduced with permission of Springer Nature. (g) Schematic illustration of the carbon-cloth shield over sapphire template-assisted growth of monolayer MoS_2 single crystal. Source: Pengfei Yang [60]/Reproduced with permission from John Wiley & Sons. (h) Epitaxial growth of 2-inch uniform monolayer MoS_2 single crystal on sapphire. Source: Reproduced with permission from Pengfei Yang [60]/John Wiley & Sons.

shows uniform intensity, suggesting the same grain orientations without an obvious grain boundary. By extending the growth time, they obtained wafer-scale (2-inch) monolayer MoS_2 films that can be easily transferred to arbitrary substrates (Figure 1.4e,f). Zhang et al. reported the epitaxial growth of wafer-scale MoS_2 single crystal on sapphire with Mo foil and S powder as precursors [60]. The carbon cloth layer placed between Mo film and sapphire substrate ensures the homogenous and continuous Mo supply (Figure 1.4g), which enables uniform nucleation and growth of monolayer MoS_2 film. In addition, the commensurability between MoS_2 lattice and sapphire lattice and the guiding effect of sapphire step edges are proposed to direct the unidirectionally aligned growth of monolayer MoS_2 single crystals. The aligned seed crystal can be seamlessly stitched to a continuous single crystal film by increasing growth time (Figure 1.4h). Moreover, Xu et al. reported an epitaxial growth of highly aligned MoS_2 grains on a twofold symmetry a-plane sapphire substrate [61]. The obtained MoS_2 grains have an unusual rectangle shape with perfect orientation alignment along the $< 1\bar{1}00 >$ direction of a-plane sapphire. They found that high temperature is beneficial to the MoS_2 seeds rotation to a favorable orientation, which is beneficial for the unidirectional alignment of MoS_2.

Many efforts have been devoted to grow wafer-scale non-centrosymmetric 2D materials by aligning unidirectional grains on single-crystal substrates [62–64]. The aligned domains can be seamlessly stitched, and then the 2D film is formed. Liu et al. reported the growth of a wafer-scale single crystal h-BN monolayer on a vicinal Cu (110) substrate [64]. The Cu (110) vicinal surface, on which the presence of metal steps uniformly along the <211> direction leads to a C_1 symmetry, is critical to prepare the unidirectional arrangement of the h-BN monolayer. The step edges reduced the symmetry of the substrate, broke the energetical degeneracy of the antiparallel domains, and aligned the domain orientation. The h-BN single crystal grains are nucleated near the step edges, with one edge of the grains tightly attached to the upward side of the step edge during the growth process, and the grains propagate rapidly on the plateau between neighboring step edges. Once one of its edges reached a neighboring step edge in the downward direction, the propagation of the edge was temporarily arrested, and finally spliced into a continuously single-crystalline 2D film. Using this method, the unidirectional alignment of h-BN grains can be achieved on a large scale.

Beside the substrate-guided epitaxial growth strategy, a seed-induced epitaxial growth strategy was also developed for controllable fabrication of wafer-scale single crystal 2D TMDs films. For example, Ye et al. reported the synthesis of a large-area 2H-$MoTe_2$ single crystal via solid-to-solid transition and recrystallization process, where a piece of 2H-$MoTe_2$ as a seed crystal was implanted into the 1T′-$MoTe_2$ film to trigger the in-plane 2D-epitaxy growth [65]. The resulting single-crystalline film completely covered a 2.5-cm wafer with excellent uniformity.

1.5 Thickness Control in CVD Growth of 2D Materials

As we all know, the properties of 2D materials are closely related to their layer numbers [9, 38]. For example, compared with the single layer, the bilayer TMDs have good electrostatic control, a smaller band gap, and higher mobility, which will improve the energy-delay product of the transistor [66–69]. Hence, the preparation of high-quality 2D films with controllable layers is another important project in this field. In the traditional CVD growth process, the continuous flow of carried gas in the same direction leads to the random growth of many small triangular grains (the second layer) on the surface of monolayer crystal (the first layer), which makes the controllable growth of uniform multiple TMD layers much more

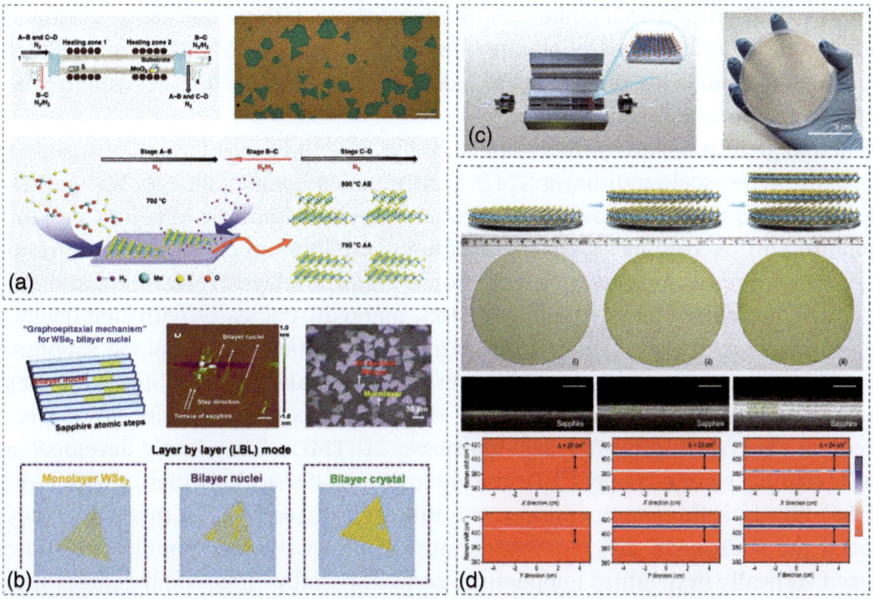

Figure 1.5 Thickness control for CVD growth of 2D materials. (a) Schematic diagram of reverse-flow CVD process for growing bilayer MoS_2 and corresponding OM image of obtained bilayer MoS_2. Source: Xiumei Zhang [71]/Springer Nature/CC BY 4.0. (b) Schematic diagram of the bilayer WSe_2 epitaxy growth process on c-plane sapphire and the corresponding AFM and OM images. Source: Reproduced with permission from Ali Han et al. [72]/Reproduced with permission from Royal Society of Chemistry. (c) Diagram of the CVD growth of wafer-scale (4-inch) $MoSe_2$ in three-temperature-zone tube furnace. Source: Jiawei Li et al. [73]/John Wiley & Sons/CC BY 4.0. (d) Schematic diagram of layer-by-layer epitaxy of multilayer wafer MoS_2 and corresponding photographs, STEM images (cross-section), and Raman spectra line mapping images of the wafers. Source: Qinqin Wang et al. [74]/Oxford University Press/CC BY 4.0.

difficult [70]. Ostrikov et al. developed a reverse-flow CVD growth strategy to prepare the uniform bilayer MoS_2 [71]. The detailed growth produced is shown in Figure 1.5a. Different from the traditional CVD method with constant gas flow direction during the entire growth process, here they introduced a reverse airflow in the variable temperature section (b–c section). This reverse gas flow reduces uncontrolled nucleation and promotes uniform epitaxy of the second monolayer from the active nucleation center on the first monolayer. This approach enables high-quality, uniform bilayer MoS_2 crystals with high yield, controllability, and reliability, and it provides a possible route for the subsequent large-scale growth of 2D materials with controllable layers. Wu et al. successfully grew a controllable three-layer MoS_2 with high mobility and large single crystals on a sodium-lime glass substrate by using a CVD strategy [75]. In addition, Zhang et al. used a similar reverse-flow CVD growth strategy to inhibit the uncontrolled nucleation and thus achieved the highly robust epitaxial growth of various 2D heterogeneous structures and superlattices [51]. Li et al. reported the controlled growth of 2H-stacked bilayer

WSe_2 by CVD growth on a c-plane sapphire substrate with atomic steps, as shown in Figure 1.5b [72]. They demonstrate that the nuclei growth of bilayer WSe_2 slides along the pronounced atomic steps induced by WSe_2 crystals atop, resembling the graphoepitaxy mechanism.

Though great progress has been made in the growth of multilayer TMD grains, achieving wafer-scale multilayer TMD films remains a big challenge. Wang et al. report the step-induced uniform nucleation (>99%) of bilayer MoS_2 on c-plane sapphire [76]. According to DFT calculations, a bilayer nucleation with aligned edges is required before merging to obtain a uniform bilayer TMDs film, and the interfacial formation energy of bilayer MoS_2 markedly decreases with step height. In an experiment, they explored the atomic terrace height on c-plane sapphire to enable an edge-nucleation mechanism and the coalescence of MoS_2 domains into continuous, centimeter-scale films. Moreover, Zhang's group devotes a great effect on the growth of wafer-scale multilayer 2D TMDs. They firstly developed a multi-channel source-supply CVD strategy to grow wafer-scale (4-inch) single-layer $MoSe_2$ film in a three-temperature zone tubular furnace [73]. In this work, they placed the sources in different temperature zones, with the sapphire substrate placed vertically in the third temperature zone. Especially, three small quartz tubes in the growth chamber act as containers for MoO_3, each capable of independently transporting carrier gas. This multi-channel design provides an even and continuous precursor supply, allowing uniform nucleation of $MoSe_2$ across the entire wafer with high nucleation density and the ability to grow wafer-scale monolayers in a short period of time. This work provides an important foundation for the subsequent wafer-scale growth of multilayer TMD films. Zhang et al. report the layer-by-layer epitaxy process growth method for preparing high-quality 4-inch multilayer MoS_2 wafers on sapphire substrate [74]. Firstly, monolayer MoS_2 is prepared by domain-domain coalescence in a multi-channel oxygen-enhanced CVD system. Then, additional epitaxial layers are grown on top of the first layer using the same technique to control the number of layers, resulting in multilayers of MoS_2 with clean, sharp interface atoms. The layer-by-layer epitaxial growth process enables a well-defined stack sequence with precise control over the number of layers, up to six. Furthermore, recent studies have demonstrated the successful synthesis of high-quality wafer-scale multilayer 2D films such as graphene and h-BN by reasonably designing the CVD growth conditions and processes [77–79]. Above progresses on thickness-controllable preparation of 2D materials provides the material foundation for exploring their thickness-dependent properties and device applications. With the development of new CVD technology, it is believed that the thickness-controllable preparation of many other 2D materials will be realized in the near future.

1.6 Phase Control in CVD Growth of 2D Materials

The phase multiplicity of 2D materials is pivotal for exploring their novel physical and chemical properties [80]. For example, most 2D TMDs (MoS_2, WS_2, WSe_2, etc.)

possess a stable 2H phase and a metastable 1T and 1T′ phase, which exhibit semiconducting and metallic properties, respectively [5]. Moreover, researchers found that the semiconducting 2H $MoTe_2$ might be commendable in thermoelectricity [81], and the metallic 1T′ $MoTe_2$ has extremely large magnetoresistance and quantum spin Hall effect [82, 83]. Interestingly, the hexagonal and tetragonal FeTe nanosheets were demonstrated to possess ferromagnetism and antiferromagnetism, respectively [84].

Phase engineering of 2D materials during the CVD growth process is much important to explore their various properties and device applications. One of the important approaches for phase engineering is to realize the phase transformation of existing 2D materials in a CVD system. Ye et al. reported a route for synthesizing wafer-scale single-crystalline 2H $MoTe_2$ by in-plane epitaxial tellurizing, which was triggered by a deliberately implanted single seed crystal [65]. Hu et al. realized a large-scale selective growth of the 1T′/2H/1T′ $MoTe_2$ multiphase structure, with the 1T′ and 2H phases seamlessly stitched [85]. The various phase transformation methods not only provide convenient and effective approaches readily applicable in many applications but also play critical roles in understanding the fundamentals of how crystal phases impact their properties.

Compared with the phase transformation method, the direct synthesis of 2D materials with distinct phase structures is more favorable to achieve a high-purity phase structure. CVD growth has great potential in the phase-controllable synthesis of 2D materials because of its diverse growth parameters (temperature, precursor, carrier gas, composition, etc.) [86]. For example, Jiao et al. reported the phase-selective growth of 1T′ and 2H MoS_2 monolayers and 1T′/2H heterophase bilayers using a potassium (K)-assisted CVD method [87]. This was realized by using K_2MoS_4 as a precursor and tuning the concentration of K in the growth products to invert the stability of the 1T′ and 2H phases. In Figure 1.6a, Xu et al. developed a facile CVD method to synthesize high-quality $Mo_xRe_{1-x}S_2$ alloys with tunable composition and phase structure [88]. The 1T′ phase $Mo_xRe_{1-x}S_2$ alloys were obtained for x in the range of 0–0.25, while the 2H phase $Mo_xRe_{1-x}S_2$ alloys were achieved for x in the range of 0.75–1. Liu et al. selectively synthesized the hexagonal phase and the tetragonal phase FeTe nanosheets on SiO_2/Si by controlling the growth temperature during the CVD process (Figure 1.6b) [84]. The phase-controllable growth of FeTe originates from the formation energy difference between the hexagonal and tetragonal phases, and maintaining a relatively high temperature is essential for obtaining the thermodynamically stable hexagonal phase, while a low temperature is favorable to the tetragonal phase. Most recently, Zhao et al. synthesized both pure β and β′ In_2Se_3 by means of controlling whether to add InSe into the In_2O_3 precursor [89]. Using DFT calculations and in situ TEM experiments, they confirm that the Se deficiency triggers the $\beta \rightarrow \beta'$ phase transition, which effectively explains the seeding effect of InSe additive in CVD precursors for the β′-phase growth. The above results demonstrate the feasibility of CVD synthesis of large-area, highly crystalline 2D materials with controllable phase structures, which is highly desirable for their promising wide applications.

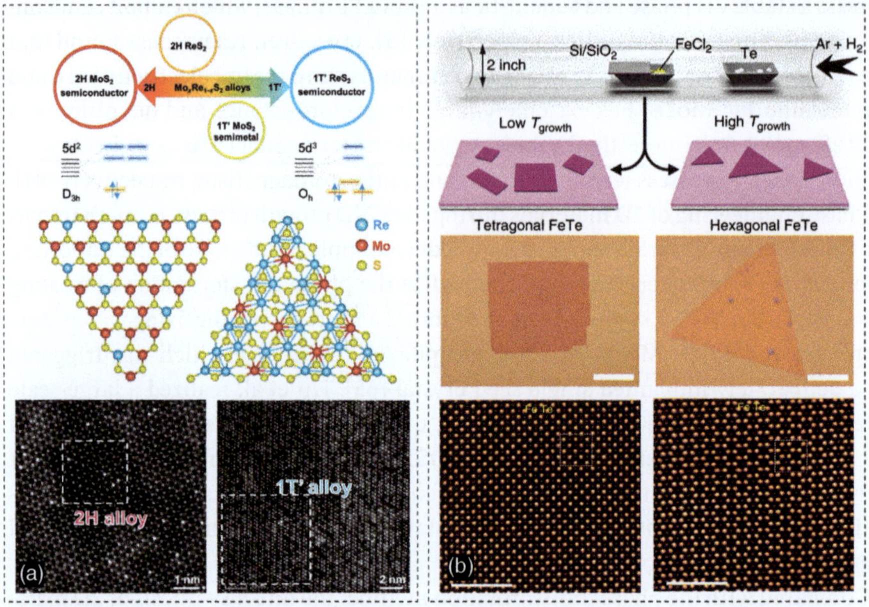

Figure 1.6 Phase-tunable synthesis of 2D materials. (a) Schematic illustration of the structure, phase, and energy band engineering of $Mo_xRe_{1-x}S_2$ alloys, the atomic structure of 2H and 1T′ $Mo_xRe_{1-x}S_2$ alloys, and the ADF-STEM images of 1T′ and 2H $Mo_xRe_{1-x}S_2$ alloys. Source: Reproduced with permission from Qixin Deng et al. [88]/John Wiley & Sons. (b) Schematic view for the temperature-modulated phase selective growth process of FeTe, the OM images (scale bar: 20 μm) and atomic-resolution STEM-ADF images (scale bar: 5 nm) of tetragonal and hexagonal FeTe nanosheets. Source: Lixing Kang et al. [84]/Springer Nature/CC BY 4.0.

1.7 Summary and Prospect

Over the past two decades, the preparation of 2D materials has made tremendous progress, which greatly promotes the fast development of the 2D field. To realize the high-efficiency synthesis of 2D materials with controllable thickness, domain size, crystal quality, and phase structures, researchers have developed a series of preparation approaches, including mechanical exfoliation, LPE, and CVD. The various preparation methods possess their own advantages and disadvantages, and thus the obtained 2D materials exhibit distinct features that fulfill the requirements of their diverse applications. In brief, the mechanical cleavage method can prepare most of the target 2D materials with high crystal quality, but the samples acquired via this approach also possess several problems, such as irregular morphology, uncontrollable thickness, small domain size, and low yield. The corresponding 2D materials are primarily used to explore their fundamental properties. LPE is suitable for large-scale production at a low cost, but precise control of size and layer number with preservation of pristine quality is still highly challenging. The obtained 2D materials exhibit prominent superiorities in some special applications, such as lithium-ion batteries, catalysis, printer ink, and composite materials. Bottom-up synthesis via CVD growth has emerged as a versatile and scalable approach enabling precise control over the thickness, morphology, crystallinity, and phase structure, which provides significant opportunities for exploring their fundamental physics and device applications.

References

1 Akinwande, D., Huyghebaert, C., Wang, C.-H. et al. (2019). Graphene and two-dimensional materials for silicon technology. *Nature* 573 (7775): 507–518.

2 Giri, A., Park, G., and Jeong, U. (2023). Layer-structured anisotropic metal chalcogenides: recent advances in synthesis, modulation, and applications. *Chemical Reviews* 123: 3329–3442.

3 Novoselov, K.S., Mishchenko, A., Carvalho, A., and Castro Neto, A.H. (2016). 2D materials and van der Waals heterostructures. *Science* 353 (6298): aac9439.

4 Polyushkin, D.K., Wachter, S., Mennel, L. et al. (2020). Analogue two-dimensional semiconductor electronics. *Nature Electronics* 3 (8): 486–491.

5 Chhowalla, M., Shin, H.S., Eda, G. et al. (2013). The chemistry of two-dimensional layered transition metal dichalcogenide nanosheets. *Nature Chemistry* 5 (4): 263–275.

6 Novoselov, K.S., Geim, A.K., Morozov, S.V. et al. (2004). Electric field effect in atomically thin carbon films. *Science* 306 (5696): 666–669.

7 Huang, Y., Sutter, E., Shi, N.N. et al. (2015). Reliable exfoliation of large-area high-quality flakes of graphene and other two-dimensional materials. *ACS Nano* 9 (11): 10612–10620.

8 Huang, Y., Pan, Y.-H., Yang, R. et al. (2020). Universal mechanical exfoliation of large-area 2D crystals. *Nature Communications* 11 (1): 2453.

9 Mak, K.F., Lee, C., Hone, J. et al. (2010). Atomically thin MoS_2: a new direct-gap semiconductor. *Physical Review Letters* 105 (13): 136805.

10 Radisavljevic, B., Radenovic, A., Brivio, J. et al. (2011). Single-layer MoS_2 transistors. *Nature Nanotechnology* 6 (3): 147–150.

11 Xin-Yu, H., Xu, H., Hui, C. et al. (2022). New progress and prospects of mechanical exfoliation technology of two-dimensional materials. *Acta Physica Sinica* 71 (10): 108201.

12 Guo, Q.B., Qi, X.-Z., Zhang, L.S. et al. (2023). Ultrathin quantum light source with van der Waals $NbOCl_2$ crystal. *Nature* 613 (7942): 53–59.

13 Deng, Y.J., Yu, Y.J., Song, Y.C. et al. (2018). Gate-tunable room-temperature ferromagnetism in two-dimensional Fe_3GeTe_2. *Nature* 563 (7729): 94–99.

14 Xu, X.L., Yang, S.Q., Wang, H. et al. (2022). Ferromagnetic-antiferromagnetic coexisting ground state and exchange bias effects in $MnBi_4Te_7$ and $MnBi_6Te_{10}$. *Nature Communications* 13 (1): 7646.

15 Wang, M.X., Zhang, J., Wang, Z.P. et al. (2019). Broadband CrOCl saturable absorber with a spectral region extension to 10.6 μm. *Advanced Optical Materials* 8 (2): 1901446.

16 Xi, X.X., Wang, Z.F., Zhao, W.W. et al. (2015). Ising pairing in superconducting $NbSe_2$ atomic layers. *Nature Physics* 12 (2): 139–143.

17 Zhang, B.Y., Xu, K., Yao, Q. et al. (2021). Hexagonal metal oxide monolayers derived from the metal–gas interface. *Nature Materials* 20 (8): 1073–1078.

18 Jiang, K., Ji, J.P., Gong, W.B. et al. (2022). Mechanical cleavage of non-van der Waals structures towards two-dimensional crystals. *Nature Synthesis* 2 (1): 58–66.

19 Pinilla, S., Coelho, J., Li, K. et al. (2022). Two-dimensional material inks. *Nature Reviews Materials* 7 (9): 717–735.

20 Coleman, J.N., Lotya, M., O'Neill, A. et al. (2011). Two-dimensional nanosheets produced by liquid exfoliation of layered materials. *Science* 331 (6017): 568–571.

21 Li, H.L., Jing, L., Liu, W.W. et al. (2018). Scalable production of few-layer boron sheets by liquid-phase exfoliation and their superior supercapacitive performance. *ACS Nano* 12 (2): 1262–1272.

22 Kaur, H. and Coleman, J.N. (2022). Liquid-phase exfoliation of nonlayered non-Van-Der-Waals crystals into nanoplatelets. *Advanced Materials* 34 (35): 2202164.

23 Cullen, P.L., Cox, K.M., Bin Subhan, M.K. et al. (2017). Ionic solutions of two-dimensional materials. *Nature Chemistry* 9 (3): 244–249.

24 Grayfer, E.D., Kozlova, M.N., and Fedorov, V.E. (2017). Colloidal 2D nanosheets of MoS_2 and other transition metal dichalcogenides through liquid-phase exfoliation. *Advances in Colloid and Interface Science* 245: 40–61.

25 Eda, G., Yamaguchi, H., Voiry, D. et al. (2011). Photoluminescence from chemically exfoliated MoS_2. *Nano Letters* 11 (12): 5111–5116.

26 Lam, D., Lebedev, D., Kuo, L. et al. (2022). Liquid-phase exfoliation of magnetically and optoelectronically active ruthenium trichloride nanosheets. *ACS Nano* 16 (7): 11315–11324.

27 Huo, C.X., Yan, Z., Song, X.F., and Zeng, H.B. (2015). 2D materials via liquid exfoliation: a review on fabrication and applications. *Scientific Bulletin* 60 (23): 1994–2008.

28 Ma, S., Li, G., Li, Z. et al. (2022). 2D Magnetic semiconductor Fe(3)GeTe(2) with few and single layers with a greatly enhanced intrinsic exchange bias by liquid-phase exfoliation. *ACS Nano* 16 (11): 19439–19450.

29 Yang, R., Mei, L., Zhang, Q. et al. (2022). High-yield production of mono- or few-layer transition metal dichalcogenide nanosheets by an electrochemical lithium ion intercalation-based exfoliation method. *Nature Protocols* 17 (2): 358–377.

30 Zeng, Z.Y., Yin, Z.Y., Huang, X. et al. (2011). Single-layer semiconducting nanosheets: high-yield preparation and device fabrication. *Angewandte Chemie, International Edition* 123 (47): 11289–11293.

31 Kappera, R., Voiry, D., Yalcin, S.E. et al. (2014). Phase-engineered low-resistance contacts for ultrathin MoS_2 transistors. *Nature Materials* 13 (12): 1128–1134.

32 Li, J., Song, P., Zhao, J.P. et al. (2021). Printable two-dimensional superconducting monolayers. *Nature Materials* 20 (2): 181–187.

33 Li, Y., Duerloo, K.A., Wauson, K., and Reed, E.J. (2016). Structural semiconductor-to-semimetal phase transition in two-dimensional materials induced by electrostatic gating. *Nature Communications* 7: 10671.

34 Wang, Y., Xiao, J., Zhu, H.Y. et al. (2017). Structural phase transition in monolayer $MoTe_2$ driven by electrostatic doping. *Nature* 550 (7677): 487–491.

35 Lin, Z.Y., Liu, Y., Halim, U. et al. (2018). Solution-processable 2D semiconductors for high-performance large-area electronics. *Nature* 562 (7726): 254–258.

36 Ciesielski, A. and Samori, P. (2016). Supramolecular approaches to graphene: from self-assembly to molecule-assisted liquid-phase exfoliation. *Advanced Materials* 28 (29): 1505371.

37 Chen, J.Y., Guo, Y.L., Wen, Y.G. et al. (2013). Two-stage metal-catalyst-free growth of high-quality polycrystalline graphene films on silicon nitride substrates. *Advanced Materials* 25 (7): 992–997.

38 Zheng, J.Y., Yan, X.X., Lu, Z.X. et al. (2017). High-mobility multilayered MoS_2 flakes with low contact resistance grown by chemical vapor deposition. *Advanced Materials* 29 (13): 1604540.

39 Zhou, J.D., Lin, J.H., Huang, X.W. et al. (2018). A library of atomically thin metal chalcogenides. *Nature* 556 (7701): 355–359.

40 Wang, Q., Li, N., Tang, J. et al. (2020). Wafer-scale highly oriented monolayer MoS_2 with large domain sizes. *Nano Letters* 20 (10): 7193–7199.

41 Sun, L.Z., Yuan, G.W., Gao, L.B. et al. (2021). Chemical vapour deposition. *Nature Reviews Methods Primers* 1 (1): 5.

42 Cai, Z.Y., Liu, B.L., Zou, X.L., and Cheng, H.-M. (2018). Chemical vapor deposition growth and applications of two-dimensional materials and their heterostructures. *Chemical Reviews* 118 (13): 6091–6133.

43 Wang, H., Chen, Y., Duchamp, M. et al. (2018). Large-area atomic layers of the charge-density-wave conductor $TiSe_2$. *Advanced Materials* 30 (8).

44 Hafeez, M., Gan, L., Li, H.Q. et al. (2016). Chemical vapor deposition synthesis of ultrathin hexagonal ReSe2 flakes for anisotropic raman property and optoelectronic application. *Advanced Materials* 28 (37): 8296–8301.

45 Wu, J.X., Tan, C.W., Tan, Z.J. et al. (2017). Controlled synthesis of high-mobility atomically thin bismuth oxyselenide crystals. *Nano Letters* 17 (5): 3021–3026.

46 Gao, T., Zhang, Q., Li, L. et al. (2018). 2D ternary chalcogenides. *Advanced Optical Materials* 6 (14): 1800058.

47 Tan, C.L. and Zhang, H. (2015). Epitaxial growth of hetero-nanostructures based on ultrathin two-dimensional nanosheets. *Journal of the American Chemical Society* 137 (38): 12162–12174.

48 van der Zande, A.M., Huang, P.Y., Chenet, D.A. et al. (2013). Grains and grain boundaries in highly crystalline monolayer molybdenum disulphide. *Nature Materials* 12 (6): 554–561.

49 Shi, Y.M., Li, H.N., and Li, L.-J. (2015). Recent advances in controlled synthesis of two-dimensional transition metal dichalcogenides via vapour deposition techniques. *Chemical Society Reviews* 44 (9): 2744–2756.

50 Gao, Y., Hong, Y.L., Yin, L.C. et al. (2017). Ultrafast growth of high-quality monolayer WSe_2 on Au. *Advanced Materials* 29 (29): 1700990.

51 Zhang, Z.W., Chen, P., Duan, X.D. et al. (2017). Robust epitaxial growth of two-dimensional heterostructures, multiheterostructures, and superlattices. *Science* 357 (6353): 788–792.

52 Liu, C., Xu, X.Z., Qiu, L. et al. (2019). Kinetic modulation of graphene growth by fluorine through spatially confined decomposition of metal fluorides. *Nature Chemistry* 11 (8): 730–736.

53 Kang, K., Xie, S.E., Huang, L.J. et al. (2015). High-mobility three-atom-thick semiconducting films with wafer-scale homogeneity. *Nature* 520 (7549): 656–660.

54 Chen, C., Chen, X.D., Wu, C.W. et al. (2022). Air-stable 2D Cr_5Te_8 nanosheets with thickness-tunable ferromagnetism. *Advanced Materials* 34 (2): 2107512.

55 Yan, C.Y., Gan, L., Zhou, X. et al. (2017). Space-confined chemical vapor deposition synthesis of ultrathin HfS2 flakes for optoelectronic application. *Advanced Functional Materials* 27 (39): 1702918.

56 Wang, Q., Shi, R., Zhao, Y.X. et al. (2021). Recent progress on kinetic control of chemical vapor deposition growth of high-quality wafer-scale transition metal dichalcogenides. *Nanoscale Advances* 3 (12): 3430–3440.

57 Dong, J.C., Zhang, L.N., Dai, X.Y., and Ding, F. (2020). The epitaxy of 2D materials growth. *Nature Communications* 11 (1): 5862.

58 Suenaga, K., Ji, H.G., Lin, Y.-C. et al. (2018). Surface-mediated aligned growth of monolayer MoS_2 and in-plane heterostructures with graphene on sapphire. *ACS Nano* 12 (10): 10032–10044.

59 Li, T.T., Guo, W., Ma, L. et al. (2021). Epitaxial growth of wafer-scale molybdenum disulfide semiconductor single crystals on sapphire. *Nature Nanotechnology* 16 (11): 1201–1207.

60 Yang, P.F., Liu, F.C., Li, X. et al. (2023). Highly reproducible epitaxial growth of wafer-scale single-crystal monolayer MoS_2 on sapphire. *Small Methods* 19: 2300165.

61 Ma, Z.P., Wang, S.Y., Deng, Q.X. et al. (2020). Epitaxial growth of rectangle shape MoS_2 with highly aligned orientation on twofold symmetry a-plane sapphire. *Small* 16 (16): 2000596.

62 Bian, R.J., Li, C.C., Liu, Q. et al. (2022). Recent progress in the synthesis of novel two-dimensional van der Waals materials. *National Science Review* 9 (5), nwab164.

63 Chen, T.A., Chuu, C.P., Tseng, C.C. et al. (2020). Wafer-scale single-crystal hexagonal boron nitride monolayers on Cu (111). *Nature* 579 (7798): 219–223.

64 Wang, L., Xu, X.Z., Zhang, L.N. et al. (2019). Epitaxial growth of a 100-square-centimetre single-crystal hexagonal boron nitride monolayer on copper. *Nature* 570 (7759): 91–95.

65 Xu, X.L., Pan, Y., Liu, S. et al. (2021). Seeded 2D epitaxy of large-area single-crystal films of the van der Waals semiconductor 2H $MoTe_2$. *Science* 372 (6538): 195–200.

66 Cheng, R., Jiang, S., Chen, Y. et al. (2014). Few-layer molybdenum disulfide transistors and circuits for high-speed flexible electronics. *Nature Communications* 5: 5143.

67 Li, S.L., Wakabayashi, K., Xu, Y. et al. (2013). Thickness-dependent interfacial coulomb scattering in atomically thin field-effect transistors. *Nano Letters* 13 (8): 3546–3552.

68 Liu, Y.C. and Gu, F.X. (2021). A wafer-scale synthesis of monolayer MoS_2 and their field-effect transistors toward practical applications. *Nanoscale Advances* 3 (8): 2117–2138.

69 Kim, S., Konar, A., Hwang, W.S. et al. (2012). High-mobility and low-power thin-film transistors based on multilayer MoS_2 crystals. *Nature Communications* 3: 1011.

70 Yan, J., Xia, J., Wang, X. et al. (2015). Stacking-dependent interlayer coupling in trilayer MoS_2 with broken inversion symmetry. *Nano Letters* 15 (12): 8155–8161.

71 Zhang, X.M., Nan, H.Y., Xiao, S.Q. et al. (2019). Transition metal dichalcogenides bilayer single crystals by reverse-flow chemical vapor epitaxy. *Nature Communications* 10 (1): 598.

72 Han, A., Aljarb, A., Liu, S. et al. (2019). Growth of 2H stacked WSe_2 bilayers on sapphire. *Nanoscale Horizons* 4 (6): 1434–1442.

73 Li, J.W., Wang, S.P., Li, L. et al. (2022). Chemical vapor deposition of 4 inch wafer-scale monolayer $MoSe_2$. *Small Science* 2 (11): 2200062.

74 Wang, Q.Q., Tang, J., Li, X.M. et al. (2022). Layer-by-layer epitaxy of multi-layer MoS_2 wafers. *National Science Review* 9 (6), nwac077.

75 Li, X.F., Zhang, Z.F., Gao, T.T. et al. (2022). Van der Waals epitaxial trilayer MoS_2 crystals for high-speed electronics. *Advanced Functional Materials* 32 (46): 2208091.

76 Liu, L., Li, T.T., Ma, L. et al. (2022). Uniform nucleation and epitaxy of bilayer molybdenum disulfide on sapphire. *Nature* 605 (7908): 69–75.

77 Ma, W., Chen, M.L., Yin, L.C. et al. (2019). Interlayer epitaxy of wafer-scale high-quality uniform AB-stacked bilayer graphene films on liquid Pt_3Si/solid Pt. *Nature Communications* 10 (1): 2809.

78 Zhang, Z.B., Ding, M.C., Cheng, T. et al. (2022). Continuous epitaxy of single-crystal graphite films by isothermal carbon diffusion through nickel. *Nature Nanotechnology* 17 (12): 1258–1264.

79 Ma, K.Y., Zhang, L., Jin, S. et al. (2022). Epitaxial single-crystal hexagonal boron nitride multilayers on Ni (111). *Nature* 606 (7912): 88–93.

80 Li, W.B., Qian, X.F., and Li, J. (2021). Phase transitions in 2D materials. *Nature Reviews Materials* 6 (9): 829–846.

81 Park, J.C., Yun, S.J., Kim, H. et al. (2015). Phase-engineered synthesis of centimeter-scale 1T′- and 2H-molybdenum ditelluride thin films. *ACS Nano* 9 (6): 6548–6554.

82 Qian, X.f., Liu, J.w., Fu, L., and Li, J. (2014). Quantum spin Hall effect in two-dimensional transition metal dichalcogenides. *Science* 346 (6215): 1344–1347.

83 Keum, D.H., Cho, S., Kim, J.H. et al. (2015). Bandgap opening in few-layered monoclinic $MoTe_2$. *Nature Physics* 11 (6): 482–486.

84 Kang, L.X., Ye, C., Zhao, X.X. et al. (2020). Phase-controllable growth of ultrathin 2D magnetic FeTe crystals. *Nature Communications* 11 (1): 3729.

85 Zhang, S.C., Wu, Y., Gao, F. et al. (2022). Field effect transistor sensors based on in-plane 1T′/2H/1T′ $MoTe_2$ heterophases with superior sensitivity and output signals. *Advanced Functional Materials* 32 (41): 2205299.

86 Tang, L., Tan, J.Y., Nong, H.Y. et al. (2020). Chemical vapor deposition growth of two-dimensional compound materials: controllability, material quality, and growth mechanism. *Accounts of Materials Research* 2 (1): 36–47.

87 Liu, L., Wu, J.X., Wu, L.Y. et al. (2018). Phase-selective synthesis of 1T' MoS_2 monolayers and heterophase bilayers. *Nature Materials* 17 (12): 1108–1114.

88 Deng, Q.X., Li, X.B., Si, H.Y. et al. (2020). Strong band bowing effects and distinctive optoelectronic properties of 2H and 1T′ phase-tunable $MoxRe_1–xS_2$ Alloys. *Advanced Functional Materials* 30 (34): 2003264.

89 Han, W., Zheng, X.D., Yang, K. et al. (2023). Phase-controllable large-area two-dimensional In_2Se_3 and ferroelectric heterophase junction. *Nature Nanotechnology* 18 (1): 55–63.

2

An Introduction to the Nonlinear Optical Properties of 2D Materials

Bolong Wang and Hao-Li Zhang

Lanzhou University, State Key Laboratory of Applied Organic Chemistry (SKLAOC), Key Laboratory of Nonferrous Metal Chemistry and Resources Utilization of Gansu Province, College of Chemistry and Chemical Engineering, Key Laboratory of Special Function Materials and Structure Design, Ministry of Education, No. 222 South Tianshui Road, 730000, Lanzhou, China

2.1 Introduction

Nonlinear optics (NLO) refers to the observation that the induced polarization nonlinearly depends on the external electric field for light–matter interaction. The propagation of light in the medium is a process of interaction. If the medium response is linear, light propagation follows the principle of independent propagation and linear superposition [1–3]. Otherwise, the response is nonlinear and intensity-dependent and is thus considered an NLO [2, 4]. A large range of nonlinear optical applications has prevailed in versatile fields, such as nonlinear two-photon and three-photon processes, super Rayleigh scattering [5], sum frequency generation (SFG) and difference frequency generation (DFG) [6], second harmonic generation (SHG) and third harmonic generation (THG) [7, 8], and electro-optic effect [8].

However, currently, most of the materials used are still conventional nonlinear optical ones, the nonlinearity, fabrication, and integration technologies of which fall short of the ever-increasing needs [2]. Hence, novel materials with large nonlinear optical responses, new functionalities, and high performance are in great demand for future photonics and optoelectronics [1, 9]. Under this context, the emerging of the 2D materials offers unprecedented opportunities for NLO with numerous new physical insights. 2D materials feature atomic thickness, the quantum confinement effect of which endows the materials with intriguing properties absent in their bulk counterparts. For instance, the confinement of electron motion enhances light–matter interaction and hence enlarged nonlinear optical responses. This explains why nanomaterials oftentimes exhibit much stronger nonlinear optical limiting and saturable absorption than the corresponding bulk ones [10, 11]. As a result, 2D materials have been used to explore their potential in a plethora of nonlinear optical applications including optical switches/modulators [12], all-optical photonic device [13], ultrashort pulse lasers [14, 15], and optical limiting [16, 17].

Two-Dimensional Materials for Nonlinear Optics: Fundamentals, Preparation Methods, and Applications, First Edition. Edited by Qiang Wang and Hao-Li Zhang.

In this chapter, we summarize the latest development in the field of 2D materials for NLO. We first briefly describe the fundamental of NLO, mainly including second- and third-order nonlinearity. We then introduce related measurement techniques of NLO, including harmonic generation, Z-scan techniques, pump–probe techniques, and nonlinear optical imaging. Thereafter, the applications of 2D materials in various fields, such as optical limiting, mode locking, optical switch, and nonlinear frequency conversion are numerated in detail. We also provide our insight on the challenges and future development of NLO of 2D materials.

2.2 Nonlinear Optics of 2D Materials

Equation (2.1) depicts the optical response of a material to an applied optical field:-

$$P(t) = \varepsilon_0(\chi^{(1)}E(t) + \chi^{(2)}E^2(t) + \chi^{(3)}E^3(t) + \cdots \chi^{(n)}E^n(t)) \tag{2.1}$$

where $P(t)$ is the induced polarization intensity, $E(t)$ the photoelectric field, ε_0 the vacuum permittivity, and $\chi^{(n)}$ the nth-order susceptibility. $\chi^{(1)}$ is for linear optical effects, but when the optical field is strong enough or the susceptibility is large, the higher-order ($n \geq 2$) terms cannot be neglected any more [2]. As a result, nonlinear optical phenomena such as sum frequency and multiphoton absorption occur. Meanwhile, a tensor depicts the relation between different vectors. For instance, $\chi^{(2)}$ is a third-rank tensor, including 27 ($3 \times 3 \times 3$) elements [2, 26]. Usually the number of tensor of elements is very large, which however can be decreased if the system symmetry is considered, as the Neumann principle demands the sign and the amplitude of the physical properties remain the same under any symmetry operation on the system [26, 27].

Typical second-order nonlinearities include SHG, SFG, DFG, and optical rectification effects (Figure 2.1). Second-order and other even-order nonlinear optical effects can only occur in materials without inversion of symmetry centers:

$$P^{(2)} = \chi^{(2)}E^{(2)} \tag{2.2}$$

The electric field and the polarization convert as $E \rightarrow -E$ and $P \rightarrow -P$ under inversion. Equation (2.2) then changes as follows:

$$-P^{(2)} = \chi^{(2)}(-E)^2 = \chi^{(2)}E^2 \Rightarrow P^{(2)} = \chi^{(2)}E^{(2)} \tag{2.3}$$

Clearly, under inversion the second-order polarization sign changes. Based on Neumann principle, $\chi^{(2)}$ has to be zero, i.e. no second-harmonic generation for a centrosymmetric system.

Alternatively, we compare Eqs. (2.2) and (2.3):

$$P^{(2)} = \chi^{(2)}E^2 = -\chi^{(2)}E^2 \tag{2.4}$$

$\chi^{(2)}$ should be zero to make it valid. Hence, second-harmonic response only occurs for noncentrosymmetric systems [26].

Unlike its second-order nonlinearity counterpart, the third-order nonlinear response of materials ($\chi^{(3)}$ in Eq. (2.1) is the third-order susceptibility) is allowed

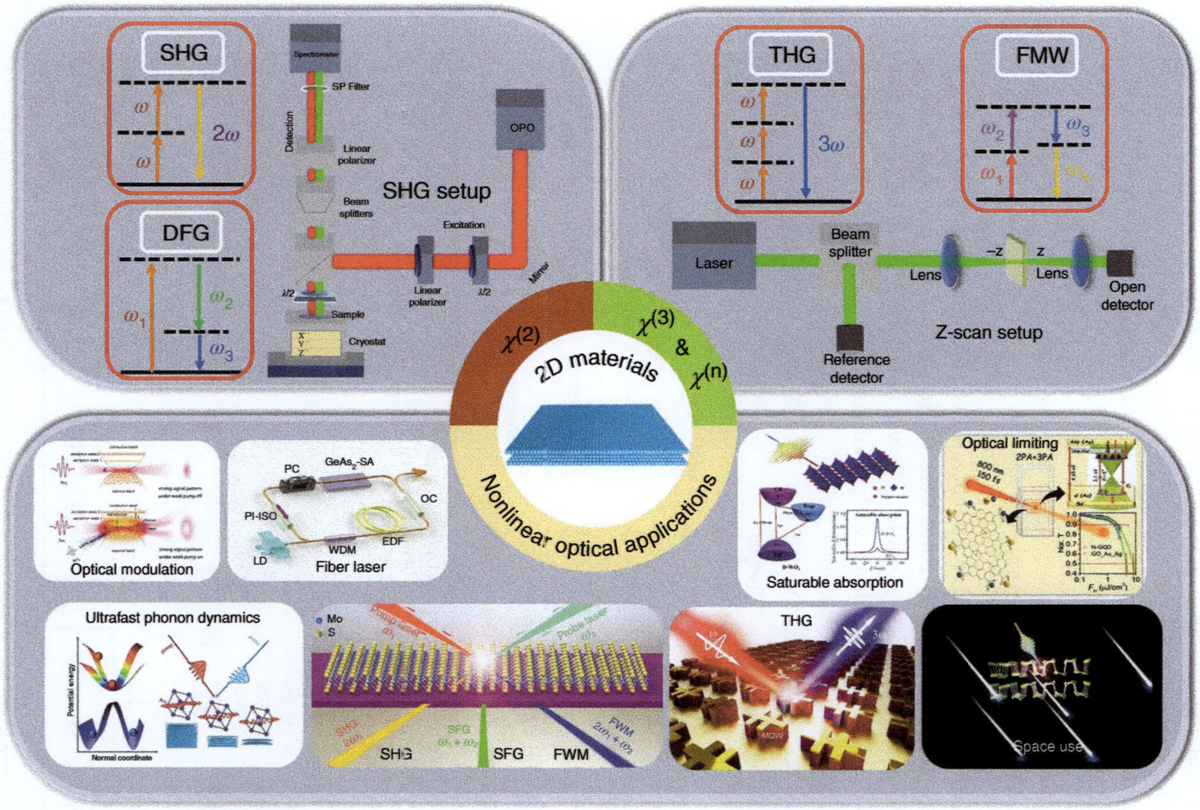

Figure 2.1 Schematic illustration of various nonlinear optical phenomena and applications. (Optical modulation: Adapted with permission Zhang et al. [18]. Copyright © 2020, WILEY-VCH Verlag GmbH & Co. KGaA, Weinheim. Fiber laser: Adapted with permission Liu et al. [19]. Copyright © 2022, WILEY-VCH Verlag GmbH & Co. KGaA, Weinheim. Ultrafast phonon dynamics: Adapted with permission Ma et al. [20]. Copyright © 2022, American Chemical Society. SHG, SFG and FWM: Adapted with permission Li et al. [21]. Copyright © 2016, American Chemical Society. THG: Adapted with permission Yu et al. [22]. Copyright © 2019, WILEY-VCH Verlag GmbH & Co. KGaA, Weinheim. Space use: Adapted with permission Wang et al. [23]. Copyright © 2022, Royal Society of Chemistry. Saturable absorption: Adapted with permission Zhai et al. [24]. Copyright © 2022, WILEY-VCH Verlag GmbH & Co. KGaA, Weinheim. Optical limiting: Adapted with permission Mondal et al. [25]. Copyright © 2022, Elsevier).

regardless of crystal symmetry, which manifests as third-harmonic generation (THG), four-wave mixing (FWM), optical Kerr effect, etc. Even higher order nonlinear effects, with nth-order NLO susceptibility $\chi^{(n)}$ ($n > 5$), demonstrates as multiphoton absorption, emission and high harmonic generation (HHG), etc. The various nonlinear optical phenomena, related parameters and applications are summarized in Figure 2.1. In the following sections, we introduce typical setups for characterizing second-order, third-order and even higher-order nonlinearities of 2D materials.

2.2.1 SHG, THG, and HHG Setups

Figure 2.2a demonstrates an SHG setup that allowed simultaneous multiphoton fluorescence measurement [28]. SHG was detected in the forward direction (using a 450-nm narrow band-pass filter), while TPF was detected in the backward direction (via a 470-nm long-pass filter), with photomultiplier tubes being applied to

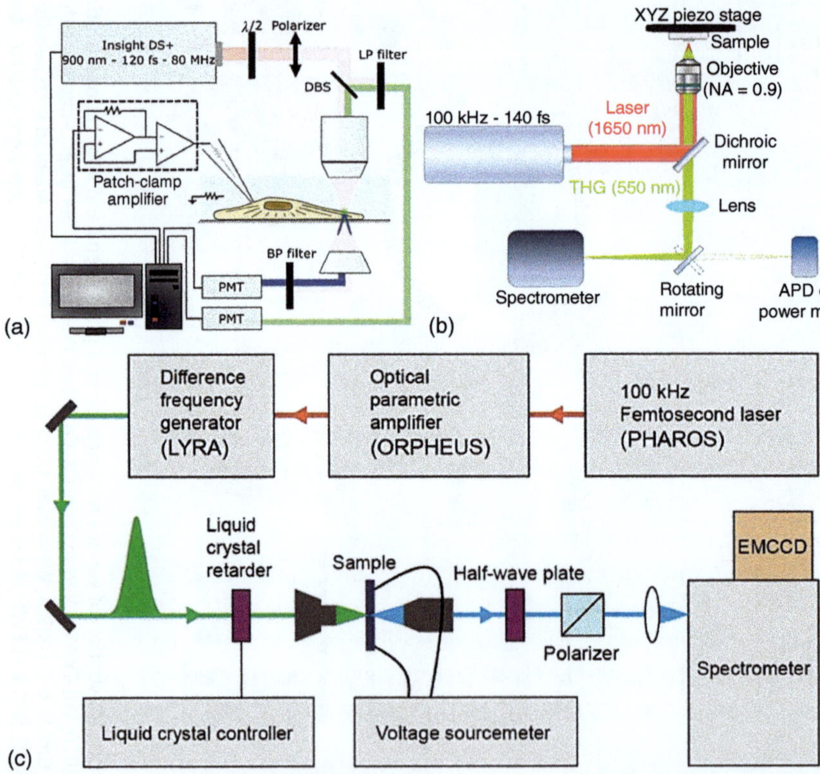

Figure 2.2 A summary of representative SHG, THG and HHG setups. (a) simultaneous imaging of two-photon fluorescence (TPF, reflection, green) and SHG (transmission, blue). Source: de Coene et al. [28]/with permission of John Wiley & Sons. (b) Schematic diagram of the THG microscope. Source: Adapted from Grinblat et al. [29]. Copyright © 2016, American Chemical Society. (c) Schematic diagram of the experimental setup of the HHG. Source: Adapted from Cha et al. [30]. Copyright © 2022, Springer.

amplify the signal-to-noise (S/N) ratio. The lack of inversion symmetry led to strong SHG signals upon pump excitation [1]. Shree et al. observed exciton-mediated SHG in MoS_2 bilayers and trilayers [31]. Zhao et al. reported generation and control of SHG from $LaTiO_3/SrTiO_3$ superlattices [32], where the nonlinear optical susceptibility $\chi^{(2)}$ was tuned by tailoring the number of superlattice interfaces. Spreyer et al. showed the symmetry effects on the SHG process of hybrid metasurfaces of WS_2 monolayer/Au nanorod antennas, where the localized plasmonic resonances demonstrated significant influence on the SHG signals [33]. Strong SHG signals have also been observed for 2D perovskites [34, 35]. Yuan et al. reported a new type of 2D organic/inorganic hybrid perovskite of $(MPEA)_{1.5}PbBr_{3.5}(DMSO)_{0.5}$ [36], where the chiral β-methylphenethylammonium (MPEA) salt induced the noncentrosymmetric assembly with a second-order NLO coefficient of ~0.68 pm V^{-1} at 850 nm. Zhang et al. reported a novel and inversion symmetry broken 2D layered SnP_2S_6 with strong second-order susceptibility $\chi^{(2)}$ of 4.06×10^{-9} m V^{-1} at 810 nm excitation [37], 1–2 orders of magnitude higher than that of conventional 2D materials.

In contrast to SHG, THG can be generated in materials regardless of symmetry [2]. A representative THG setup is shown in Figure 2.2b. With this setup, Grinblat et al. found all-dielectric germanium nanostructures demonstrated efficient THG [29]. A large third order susceptibility of $\chi^{(3)}$ of 4.3×10^{-9} esu was estimated for the germanium nanodisk, implying engineering dielectric materials could optimize nanoscale upconversion processes. Twisted bilayer graphene exhibited unconventional physical properties due to Moiré superlattices. Applying electrically tunable optical nonlinearity, Ha et al. realized a THG enhanced up to 60 times compared to monolayer graphene with a gate-control on bilayer graphene [38]. Natural/getchellite is a new type of ternary layered material, which Tripathi et al. mechanically exfoliated to obtain large-area getchellite thin flakes with large $\chi^{(3)}$ of 2.89×10^{-20} m^2 V^{-2} [39]. Strong anisotropic linear and nonlinear optical responses of the novel 2D materials imply their potential use in photodetectors, optical sensors, etc. Abdelwahab et al. discovered strong THG with a maximum effective $\chi^{(3)}$ of 1.12×10^{-17} m^2 V^{-2} for 2D perovskite nanosheets $(C_4H_9NH_3)_2(CH_3NH_3)Pb_2I_7$ [40], which suggested their potential for all optical processing photonic applications as on-chip components.

Figure 2.2c demonstrates a typical HHG setup [30]. Mid-infrared pulses were obtained with an optical parametric amplifier (OPA) and a difference frequency generator. The HHG spectra were collected through a CCD detector and a grating spectrometer. With this setup, Cha et al. discovered the fifth and seventh harmonics of graphene at 1.38 and 1.93 eV, respectively under excitation of 0.28 eV. Yoshikawa et al. excited a single-layer graphene and obtained up to the 9th harmonic ultraviolet light [41]. In monolayer MoS_2, even higher order harmonics of 18th was generated. Hirori et al. generated HHG through a MIR laser excitation on three hybrid organic-inorganic perovskites ($MAPbX_3$, X being either I, Br, or Cl) [42]. Harmonics up to the 13th order with a photon energy of 4.6 eV were obtained from the $MAPbCl_3$ film. And the HHG signals from $MAPbBr_3$ were over 10 times stronger than those from GaSe control sample. Langer et al. used a femtosecond pulse excited coherent excitons in WSe_2 with direct band gap [43]. A co-propagating

THz pulse modulated the excitonic polarization and high-order sideband generation was observed. The work indicated coherent interband polarization emitted photons at any point in momentum space. Chen et al. used ab initio simulations based on real-time-dependent density-functional theory predicted HHG in van der Waals heterostructures of graphene/hBN [44]. They found the interlayer spacing was one useful degree of freedom to tailor the HHG generation, with the HHG of the bilayer distinguishing from those of the constituent layers. The findings offered new avenues to control ultrafast physical properties of 2D materials.

Note HG is easily developed into powerful nonlinear imaging techniques for 2D material characterization with non-label and noninvasive advantages. Other imaging methods include fluorescence-labeled imaging, such as two-photon absorption TPA, multiphoton absorption (MPA) imaging, and complementary non-label Raman imaging is also frequently used, which together with HG imaging will be introduced in Section 2.2.4.

2.2.2 Four-Wave Mixing

The FWM is based on the third-order nonlinear optical effect described by $\chi^{(3)}$ coefficient. In short, the FWM effect may occur when light with at least two different frequency components propagates in a nonlinear medium.

Figure 2.3a shows the setup for FWM measurement in the microring resonators (MRRs). Two CW tunable lasers acted as the pump and signals respectively, combined and coupled into the MRRs [45]. For an MRR coated with 1 layer of graphene oxide (GO), up to ≈7.6 dB enhancement in the FWM efficiency was obtained, while for a device patterned with 50 layers of GO the enhancement was ≈ 10.3 dB. The results inform strategies that use the high Kerr nonlinearity of 2D GO to enhance FWM efficiency. Sun et al. exploited plasmonic effect to enhance broadband FWM efficiency in monolayer MoS_2 [46]. The microscopic setup in Figure 2.3b shows the combined pump and idler beams focused on the sample via an objective lens. With the facilitation of the strongly confined electric field

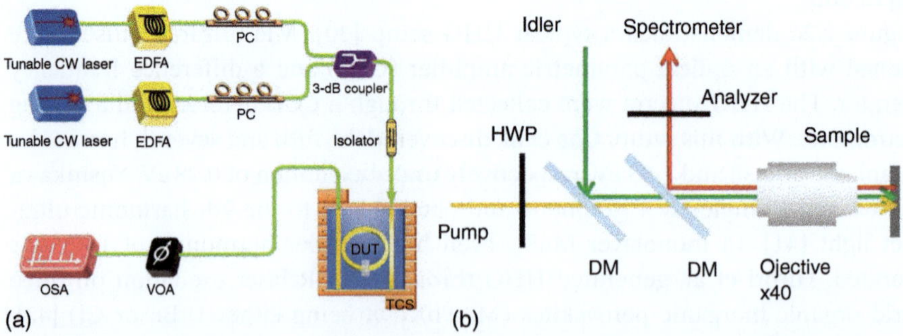

Figure 2.3 (a) Typical experimental setup for FWM measurements in integrated MRRs. Source: Jiayang et al. [45]/with permission of John Wiley & Sons. (b) Schematic of the experimental setup for nonlinear optical measurements. Source: Adapted from Dai et al. [46]. Copyright © 2021, American Chemical Society.

by the plasmonic effect, third-order nonlinearity in the order of 10^{-17} m^2 V^{-2} for hybrid MoS_2-plasmonic structure was obtained, ~10–100 times larger than that of conventional bulk materials such as $LiNbO_3$. Similarly, You et al. used a graphene plasmonic metasurface and observed a FWM process at an ultralow pump power due to large optical near-field enhancement and long lifetime of graphene plasmons in the metasurface [47]. This work offered new insights in creating compact integrated photonic platform with low power consumption for quantum communications.

Ko et al. observed simultaneous FWM and two-photon luminescence (TPL) of 2D perovskites of $(BA)_2(MA)_2Pb_3Br_{10}$ [48]. In the spectra, the green 525 nm emission was attributed to TPL arising from two-photon absorption, while the red 650–785 nm emission stemmed from FWM. Bauer et al. studied exciton-based quantum NLO by observing the second- and third-order optical waves mixing in a transition metal chalcogenide (TMD) monolayer [49]. The experiment was based on a two-color detection scheme, in which the pump light resonated with A exciton while the probe detuned, and the time difference between them was $\Delta t = t_{probe} - t_{pump}$. The result showed that that to observe SFG and four-wave mixing for the monolayer WSe_2 sample, a significant time delay was necessary between the two light pulses.

2.2.3 Z-Scan Techniques

The third-order nonlinear optical phenomena mainly include nonlinear absorption, refraction, and scattering [50]. Saturable absorption (SA), reverse saturable absorption (RSA), two-photon absorption (TPA), multiphoton absorption (MPA), and free carrier absorption (FCA) are all typical nonlinear absorption behaviors. Among them, reverse saturation absorption, two-photon absorption, and free carrier absorption of materials can be harnessed to fabricate optical limiters, while saturable absorption is very useful for obtaining passive *Q*-switching and mode-locking lasers with ultrashort pulse output [51].

Generally, the change of absorption coefficient (α) and refractive index (n) of materials with the varied input energy can be described by the following equations:

$$\alpha = \alpha_0 + \beta I \tag{2.5}$$

$$n = n_0 + \gamma I \tag{2.6}$$

where α is the linear absorption coefficient, n_0 the linear refractive index, β the nonlinear absorption (NLA) coefficient and γ the nonlinear refractive (NLR) index.

Z-scan techniques are powerful tools to measure third-order nonlinear optical parameters [51–53]. As shown in Figure 2.4a, the normalized transmittance of samples is measured as a function of position (Z). The incident beam is split into one reference and the other probe beam passing through the sample. The sample moves around the focus along the Z axis of laser propagation, with the transmitted light being collected. Dependent on whether the aperture before the detector is open or not, the Z-scan setup can be divided into open-aperture and closed-aperture configuration, respectively. In the former nonlinear absorption of the samples is measured

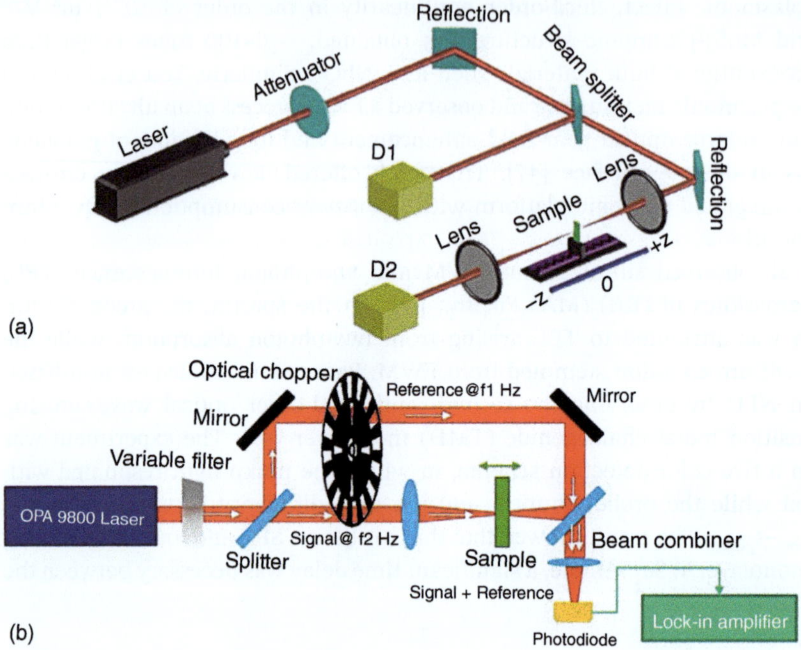

Figure 2.4 (a) The simplified scheme of the ns Z-scan setup. Source: Adapted from Li et al. [51]. Copyright © 2019, WILEY-VCH Verlag GmbH & Co. KGaA, Weinheim. (b) The simplified scheme of the fs Z-scan setup. Source: Adapted from Wang et al. [54]. Copyright © 2016, American Chemical Society.

whereas in the latter both nonlinear absorption and nonlinear refraction signals are recorded. For the open aperture Z-scan curves, if there is one valley at the focus point ($z \to 0$), the sample exhibits reverse saturable absorption; or if there is a peak at the focus, saturable absorption occurs.

The following equations are used to fit the Z-scan curves:

$$\alpha = \beta I + \frac{\alpha_0}{1 + I/I_s} \tag{2.7}$$

where I is laser intensity and I_s the saturable light intensity. I_s defined as the laser intensity at which the linear absorption of materials is reduced to half of the initial value. Note β is related to the reverse saturable absorption or two-photon absorption and dependent on the sample concentration.

The following differential equation implies the input intensity changes with the propagation depth z':

$$\frac{dI}{dz'} = -\alpha(I) \cdot I \tag{2.8}$$

The above equations are combined to extract parameters of β and I_s [51].

A representative fs Z-scan apparatus is shown in Figure 2.4b. Compared with ns Z-scan measurements, the laser source was replaced with a femtosecond laser [51, 54, 55]. The reference and signal beam were modulated by two optical choppers but measured by the same photodiode to reduce noise. A lock-in amplifier in dual-reference mode was exploited to separate the intensities of two beams to obtain high noise-signal ratio.

Wang et al. reported a systematic investigation of the nonlinear optical properties of BP nanosheets and other 2D nanomaterials across the NIR and MIR region [54]. Using open aperture Z-scans measurement, they discovered broad saturable absorption responses of BP nanosheets at multiple wavelengths of 800, 1330, 1420, 1550, and 2100 nm. Menezes et al. exploited Z-scan techniques to measure nonlinear absorption and nonlinear refraction of NbS_2 [56]. The NLA coefficient of 2.1×10^1 cm GW^{-1} and a saturation intensity $I_S = 52$ GW cm^{-2} was obtained. The nonlinear refractive index n_2 changed from negative to positive above the critical intensity of ~22 GW cm^{-2} with the increase of the input intensity, and a value of $(3.0 \pm 0.2) \times 10^{-16}$ cm^2 W^{-1} was derived. Chen et al. reported the reverse saturable absorption of $MAPbBr_3$/polymethyl methacrylate (PMMA) organic glasses by ns Z-scan [57]. The β value increased with the increase of perovskites content, with the largest β value being 3.8 m GW^{-1}.

2.2.4 Nonlinear Optical Imaging

Besides HG imaging, other nonlinear optical imaging techniques have become powerful tools for 2D material characterization, including Raman spectroscopy-related methods such as non-destruction, noninvasive stimulated Raman scattering (SRS) [60]. Cheng et al. studied the SERS of 2D hexagonal boron nitride (h-BN) by using multimode nonlinear optical microscope, including simultaneous SHG, anti-Stokes Raman scattering (CARS), and SRS imaging [61]. In Figure 2.5a, one picosecond laser with 913 nm output was used as the pump, and the other picosecond 1040 nm pulse was the Stokes beam. The overlapped two beams illuminated the samples and the nonlinear optical responses were collected by a microscope. The SHG signal of the 1040 nm beam was directly recorded by a photomultiplier tube. The CARS signal was monitored in the epi direction, while the SRS signal collection was more complicated. The Stokes beam needed to be modulated and then demodulated by a lock-in amplifier. The thickness of the h-BN flakes were determined by AFM to be 2.5–70 nm, which were then characterized by this setup via SRS and CARS modes, with the obtained images shown in Figure 2.5a. He et al. reported the facile synthesis of bimetallic solid-solution TiVC (MXene) and its application in SERS [58]. With the laser confocal Raman spectrometer, TiVC film demonstrated high SERS sensitivity for rhodamine 6G with a detection limit on the femtomolar level. Li et al. set up nonlinear optical microscopies with different modes of CARS, two-photon excitation fluorescence(TPEF) and SHG measurement [62]. As shown in Figure 2.5b, a Ti:sapphire femtosecond laser was applied for the CARS measurement. One beam was the pump and probe and the other produced white light as the Stokes light. The two beams combined and illuminated the sample through a multiphoton laser scanning microscope to provide nonlinear optical images. The optical microscope (OM), CARS, TPEF + SHG and OM + CARS + TPEF + SHG images of the monolayer graphene, respectively are also shown in Figure 2.5b.

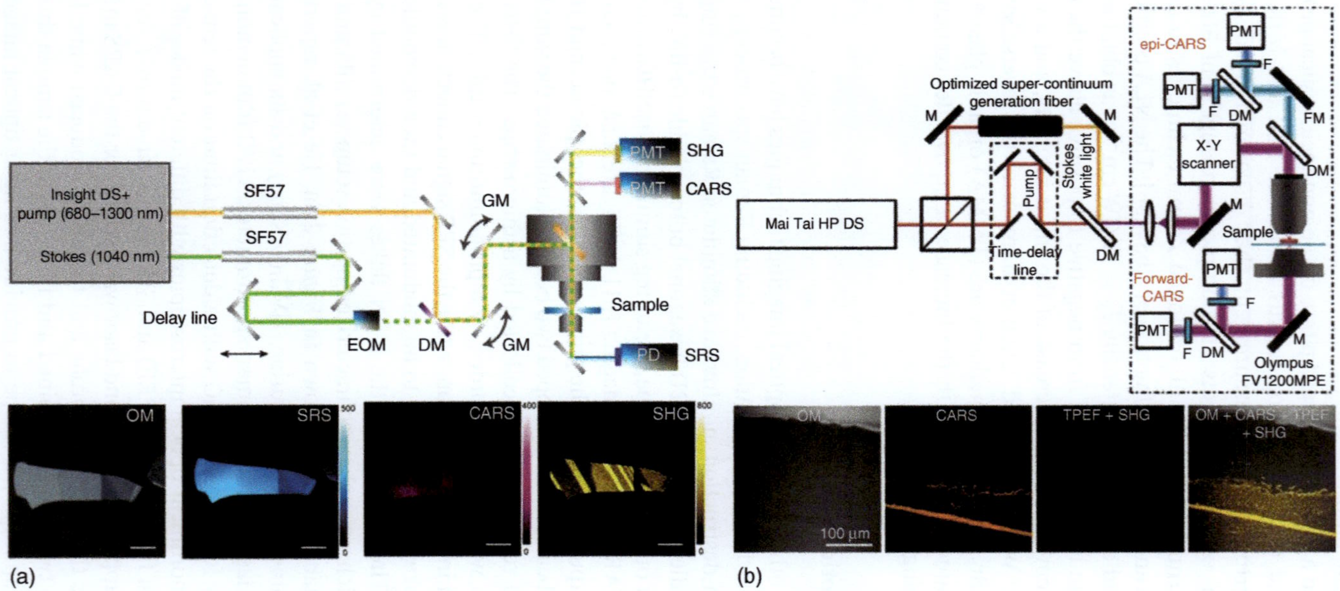

Figure 2.5 (a) The microscope system and Bright field, SRS, CARS, and SHG images of an h-BN thin flake. Scale bar: 20 µm. Source: Adapted from He et al. [58]. © 2019, American Chemical Society. (b) The nonlinear optical microscopy and OM of monolayer graphene, CARS, TPEF + SHG, OM + CARS + TPEF + SHG at Raman mode 1560 cm^{-1}. Source: Adapted with permission Hontani et al. [59]. Copyright © 2018, De Gruyter.

The bright field and CARS images clearly displayed the surface topography of graphene. By contrast, the TPEF + SHG signal is very weak due to the high centrosymmetry of the sample.

Another nonlinear optical imaging technique is based on fluorescent-labelled imaging techniques, including TPEF [59, 62], three-photon microscopy (3PM) [59], and MPA. As early as 1996 [63], Hell et al. had shown the imaging resolution of a three-photon excitation microscope was more potential than that of a similar two-photon excitation one. Fan et al. studied multiphoton luminescence from $MAPbBr_3$ perovskite surfaces [64]. The main light source was a tunable beam generated by pumping one oscillating parametric amplifier (290–2600 nm) with a regenerative amplified femtosecond laser (800 nm). The enhancement for all of the two-, three-, and four-photon processes was observed.

Li et al. reported giant and broadband MPA properties of 2D hybrid perovskite of $(IA)_2(MA)_2Pb_3Br_{10}$ [65], which showed remarkable multiphoton-excited optical nonlinearities covering a wide range of 550–2400 nm. The five-photon absorption cross-section (σ_5) was calculated to be 1.2×10^{-132} $cm^{10}s^4/photon^4$ (at 2400 nm). He et al. prepared high-quality single crystal perovskites of $MAPbBr_3$ and $CsPbBr_3$, which exhibited excellent multiphoton excitation behaviors [66]. The three-photon pumped stimulated emission threshold of $CsPbBr_3$ and $MAPbBr_3$ were 159 mJ cm^{-2} and 200 mJ cm^{-2}, respectively. Lu et al. predicted 2D excitons would enhance three-photon absorption in layered hybrid perovskites $(C_4H_9NH_3)_2PbBr_4$ based on a quantum perturbation theory [67]. The experiment corroborated the calculation and 3PA coefficients for the 2D perovskites were measured as 2–7 cm^3 GW^{-2} at ≈1100 nm. Han et al. demonstrated the three-photon luminescence of $PtSe_2$ atomic layers [68]. Under resonant excitation (880 nm), the luminescence was about sixfold of that under non-resonant conditions (750 nm), as local field was enhanced in the former case. Pettine et al. combined scanning photoemission microscopy with velocity map imaging to measure transverse photoelectron velocity distributions on 10-nm Au film following multiphoton excitation/photoemission [69]. They found the film photoemission proceeded through a three-photon channel from the conduction band at low photon energies, while the two-photon processes dominated at higher photon energies.

The application of MPA is not limited to imaging. Indeed, multiphoton lithography (MPL) is one of the most advanced micro processing technologies. Topolniak et al. reported surface modification via multiphoton lithography [70]. A benzylidene cyclopentanone-based photoinitiator was employed to trigger the formation of poly dopamine. Through tuning laser power and beam scanning velocity, the morphology and thickness of poly dopamine microstructure were conveniently controlled. For instance, increasing the fabrication speed to 200 μm s^{-1} led to thinner structures with 30–150 nm. This work offered a new methodology for precise and controllable surface functionalization.

2.2.5 Pump–Probe Techniques

To explore the underlying mechanism accounting for the excellent opto-electronic properties of 2D materials, ultrafast pump–probe techniques elucidating their excited-state dynamics is one indispensable technique. In fact, photoinduced ultrafast dynamics occurs in many important chemical reactions such as photosynthesis, which utilizes solar energy to convert carbon dioxide and water into oxygen and sugar. To disentangle these complicated processes, various time-resolved spectroscopic techniques, including laser photolysis (ns-ms), fluorescence lifetime measurement and ultrafast femtosecond broadband pump–probe spectroscopy in particular are powerful tools. Our group used time-resolved spectroscopy to investigate the photoinduced excited states of a series of 2D materials, including black phosphorus, bismuthene, and metal halide perovskites. The research shed lights on the photocatalytic, luminescent and nonlinear optical response mechanism of these materials, which have been summarized in our previous reviews [51, 71].

We recently systematically investigated the saturable absorption properties of ultrathin 2D metal oxides of WO_3 and the underlying mechanism by time-resolved spectroscopy [24]. Through defect engineering, different types WO_3 with tailored band gaps were synthesized and exhibited broad and enhanced nonlinear optical responses. fs transient absorption and laser photolysis measurements combined disclosed initial rapid fs-to-ps carrier dynamics. Subsequently, slower exciton bleaching on the μs time scale for the vacancy-rich WO_3 was observed. The results explained the origin of the strong saturable absorption of the 2D metal oxides.

We also unraveled the origin of the nonlinear responses of tellurium from the perspective of excited-state dynamics [17]. Using electrochemical exfoliation method, we fabricated Te nanostructures with various dimensions, including one-dimensional (1D) Te nanowires, quasi-1D Te nanorods, zero-dimensional (0D) Te nanodots, and 2D Te nanosheets. Femtosecond (fs) open-aperture Z-scan measurements found the nonlinear optical responses transited from saturable to reverse saturable absorption with the reduction of the critical dimension of the samples: 1D Te nanowires displayed stronger saturable absorption than that of quasi-1D Te nanorods, while 0D Te nanodots showed mixed saturable and reverse saturable absorption signals. For the 2D Te nanosheets, however, the nonlinear optical responses transformed to strong reverse saturable absorptions under the same experimental conditions. fs transient absorption measurement was applied to reveal the strong dimension-dependent exciton dynamics. Among the samples, 1D Te nanowires possessed the smallest bandgap of 1.12 eV, for which carriers easily filled the CB edges, leading to Pauli blocking and thus saturable absorption. However, with the decrease of critical dimension and increase of bandgaps, Pauli blocking became more difficult. Eventually, for 2D Te nanosheets with relatively largest bandgap, lowest critical dimension and hence considerable amount of trap and edge states, faster exciton relaxation, and stronger excited-state absorption (ESA) led to strong reverse saturable absorption.

We produced size-controllable Sb nanosheets through a lithium ion intercalation method, and samples with different size were separated by centrifugation at varied speeds [72]. ns and fs transient absorption measurements illustrated the carrier dynamics of the samples upon photo excitation: generation of initial excitons on the fs scale, followed by a faster intraband thermalization on several ps timescale, and a slower electron–hole recombination process that could reach the ns time window. With the decrease of the sample size, the interband carrier recombination component became faster. Eventually, when the average size transited from 3 μm to 50 nm, the nonlinear responses of the Sb nanosheets underwent a clear transition from saturable absorption to reverse saturable absorption.

Ultrafast spectroscopy is also a powerful tool to disentangle photocatalytic mechanism. We utilized a "cascade electron transfer" channel to boost the efficiency of CO_2 photocatalytic reduction [73]. For the 0D/2D photocatalyst of $CsPbBr_3$/CuTCPP MOF, femtosecond transient absorption measurement disclosed an ultrafast interfacial electron transfer rate of 1.4 ps from $CsPbBr_3$ to CuTCPP MOF, which was followed by rapid internal electron transfer of 21 ps from CuTCPP ligand to the Cu nodes. This led to efficient charge transfer and separation steps and therefore for high photocatalytic efficacy.

Currently, one developing trend for conventional pump–probe techniques is to expand them to even wider horizons. For instance, expanding the probe to extreme short (X-ray) or extreme long wavelength region (MID IR and THz) or developing pump–probe imaging techniques.

Expansion of probe to extreme short wavelength (X-ray). As early as 2004, Bargheer et al. applied spatiotemporal X-ray diffraction technique (sub-ps and sub-ps resolution) nondestructively explored the reversible structural changes of a gallium arsenide/aluminum gallium arsenide superlattice with subpicometer spatial and subpicosecond [75]. Coherent lattice motions with a 3.5-ps period was observed by femtosecond X-ray pulses within the nanostructure. Wieghold et al. coupled nanoscale characterization and time-resolved wide-angle X-ray scattering to measure photocarriers dynamics of CsFAMA perovskites, where photocarriers distorted the perovskite and nonthermal lattice deformations was observed [76]. A characteristic time of $\tau_{relax} = 700 \pm 50$ ns comparable to the charge carrier lifetime of this material was estimated for the relaxation of the structural expansion.

Expansion of probe to extreme long wavelength (MID IR and THz). The ultrafast spectroscopy in THz regime can be used to study the competition mechanism of electrons and excitons in 2D materials, and the dynamical processes of excitons and free carriers can be identified [77]. Kumar et al. observed the hybridizing of the 2D perovskites $(C_6H_5(CH_2)_2NH_3)_2PbI_4$ with metamaterials on a rigid and a flexible substrate enabled modulation of terahertz fields at 50 GHz modulating speed [77]. Ultrafast relaxation of free carriers in 20 ps due to the quantum confinement of free carriers in a self-assembled quantum well structures that form excitons. Hence, an exciton-based ultrafast response of 2D perovskites offered new avenues for a wide range of scalable dynamic photonic devices. With Figure 2.6, Chen et al. showed the configuration of the scanning near-field optical microscope (s-SNOM) with a AFM tip as a THz near-field probe [74]. With this setup, the THz polaritons in Bi_2Se_3 topological insulators was captured.

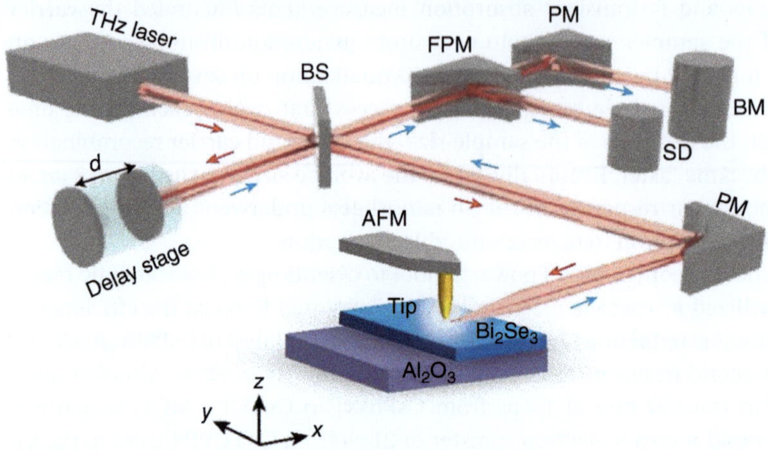

Figure 2.6 THz s-SNOM setup for imaging polaritons in Bi_2Se_3 films. Source: Adapted from Chen et al. [74]. Copyright © 2022, Springer.

Expansion of pump–probe imaging techniques (TAM) by coupling with optical microscopes. Richard's group at the Cavendish Laboratory had done extremely in-depth research on charges, excitons, and phonons of materials, for which ultrafast microscopes played crucial roles. They employed an ultrafast pump–push–probe technique and discovered the quadratic Stark effect for polymer/fullerene blends semiconductors [78]. The group also developed femtosecond transient absorption microscopy (fs-TAM) with a 10-fs temporal resolution and 10-nm spatial precision [79], through which they observed carriers propagated ballistically over 150 nm within 20 fs upon photoexcitation in methylammonium lead iodide perovskite films.

Liu and coworkers applied pump–probe transient reflectivity microscopy to measure carrier mobility of 2D boron arsenide (c-BAs) [80]. Based on the pump–probe configuration, the reflected probe pulse was collected and focused onto a complementary metal-oxide semiconductor (CMOS) detector for imaging. Giant carrier mobility of $1400\,cm^2\,Vs^{-1}$ for electrons and $2100\,cm^2\,Vs^{-1}$ for holes were obtained for c-BAs at room temperature, suggesting the potential applications of the 2D material in high-performance electronics and optoelectronics. Compared with the common Hall effect test, transient reflectivity microscopy provides one important supplement to measure mobility for small size or nonuniform samples. Meanwhile, Chen et al. used one optical transient grating technique and observed thermal conductivity of $1200\,W\,mK^{-1}$ and ambipolar mobility of $1600\,cm^2\,Vs^{-1}$ for c-BAs [81].

Expansion of pump–probe imaging techniques through in combination with electron microscopes, low-energy electron diffraction (LEED), photoemission electron microscopy (PEEM), etc. The spatial resolution of optical spectroscopy is limited by diffraction, which hinders dynamic exploration at the nanoscale. In combination with recent nanoscale imaging techniques such as aberration-corrected and monochromated-scanning transmission electron microscopy, pump–probe techniques allow high spatial and time resolution simultaneously [82]. With atomic indium wires on the (111) surface of silicon as a model system, Ropers et al. used a femtosecond double-pulse excitation scheme to switch the system from the insulating to a metastable metallic state and monitored structural dynamics by ultrafast LEED [83]. Ultrashort electron pulses were generated by focusing the ultraviolet pulses onto the tungsten needle inside the electron gun (e^--gun). The microchannel plate assembly recorded the LEED pattern changes, implying an ultrafast 350-fs and ballistic transition. The work thus offers strategies for coherent control of solids and surfaces.

Schmitt et al. used femtosecond photoemission momentum microscopy to study moiré interlayer excitons [84]. The setup is composed of a time-of-flight momentum microscope and a femtosecond laser as the light source, through which the 100-μm^2 region of the WSe_2/MoS_2 heterobilayer with a twist angle of $9.8 \pm 0.8°$ could be identified and the occupied band structure was selectively probed. This technique allows direct spatiotemporal observation of correlated moiré and exciton physics that is crucial for future harness of material quantum phases.

Other vigorous development in ultrafast low-frequency lasers, X-ray free electron lasers, electron pulses for ultrafast imaging of molecular dynamics [85], and time-resolved measurements using attosecond soft X-ray pulses [86], all expect to open up new vistas in investigating the NLO of 2D materials.

2.3 Application of 2D Nonlinear Materials

2.3.1 Optical Limiting

Optical limiting (OL) materials can protect precision optical elements and human eyes from laser damage and also have important applications in the defense field [87, 88]. Compared with traditional OL effects by changing the linear transmittance, the regulation method based on the third-order optical nonlinearity usually stems from further absorption of photons in the excited states, such as ESA, 2PA, MPA, and FCA [2, 71, 72].

Our group is among the earlier research groups and has been keeping investigating optical limiting via 2D materials. We optimized an intercalating exfoliation method, where different rotational speeds corresponding to gravity acceleration were used to prepare Sb nanowires with controllable size [72]. All nanosheets obtained at centrifugal speed from 1000 to 12 000 rpm ranges showed RSA phenomena, and the third-order nonlinear absorption coefficient β changed from -2.4×10^{-10} to $-0.43 \times 10^{-10}\,\mathrm{m\,W^{-1}}$. At the same time, we find that VIA group element-based 2D materials have strong nonlinear optical responses [17]. A series of Te nanostructures (nanosheets, nanowires, nanorods and nanodots) were prepared by electrochemical exfoliation/liquid phase exfoliation methods, and fs Z-scan measurements revealed their dimension-dependent nonlinear optical properties. Chromium (Cr) is the hardest metal in nature [89]. Under picosecond laser excitation, Cr nanosheets exhibited OL behavior ranging from UV to NIR with a typical β value of $2.58 \pm 0.04\,\mathrm{cm\,GW^{-1}}$ at 1064 nm (Figure 2.7a). 2D tungsten (W) nanosheets

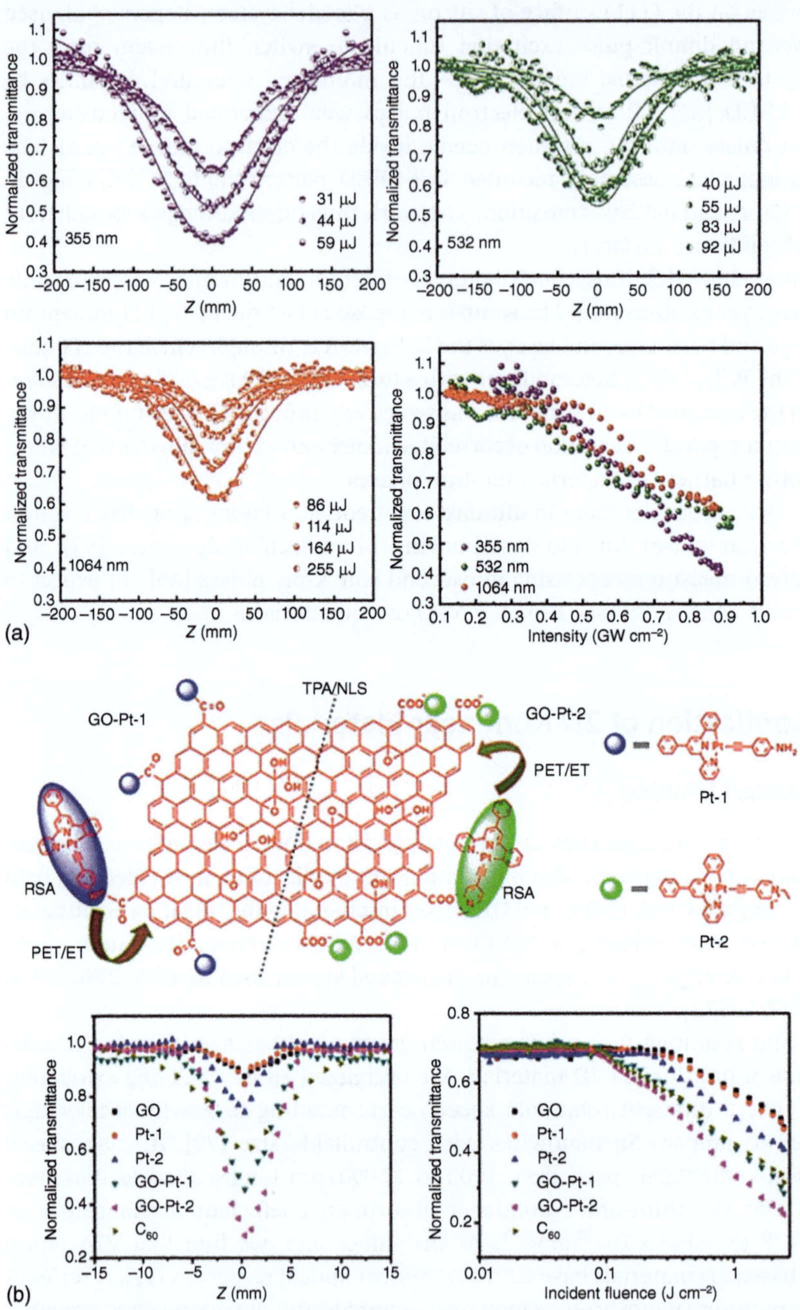

Figure 2.7 (a) Z-scan results of Cr at 355, 532, and 1064 nm and relationships between the normalized transmittance and the excitation energy density. Source: Yuan et al. [89]/with permission of Royal Society of Chemistry. (b) Schematic representation of the Structures for GO-Pt-1 and GO-Pt-2; Open-aperture Z-scan curves and optical limiting performance of GO, Pt-1, Pt-2, GO-Pt-1, GO-Pt-2, and C_{60}. Source: Liu et al. [90]/with permission of Royal Society of Chemistry.

exhibited a large two-photon absorption cross-section of $4.29 \pm 0.05 \times 10^{-45}$ cm^4 s at 355 nm, leading to their strong optical limiting in the ultraviolet band [91]. The saturation intensity I_s at 355 nm was about 0.22 ± 0.01 GW cm^{-2}, and the corresponding value of β was about 2.50 ± 0.03 cm GW^{-1}. Likewise, 2D germanium (Ge) nanosheets showed excellent RSA properties due to TPA in the infrared regime ($I_s = 0.78 \pm 0.03$ GW cm^{-2} at 1064 nm) [92].

Dissanayake et al. reviewed the OL characteristics of covalently functionalized GO and reduced graphene oxide (rGO) [93], expounding the relationship between NLR, nonlinear scattering (NLS) and OL. Liu et al. reported that two GO–platinum complex (GO-Pt) hybrid materials, which demonstrated significantly enhanced optical limiting behavior at 532 nm [90] (Figure 2.7b). The $\chi^{(3)}$ of GO-Pt-1 and GO-Pt-2 were $22.1 \pm 0.14 \times 10^{-12}$ and $26.6 \pm 0.08 \times 10^{-12}$ esu, respectively, about an order of magnitude higher than that of pristine GO ($3.13 \pm 0.07 \times 10^{-12}$ esu).

Chalcogenide materials are also widely used for optical limiting. Dong et al. discovered broadband nonlinear optical behaviors of TMD nanosheet dispersions caused by Mie scattering [94]. Among the four TMDs, the optical limiting responses follow the order of $MoSe_2 > WSe_2 > MoS_2 \sim WS_2$. Varma et al. discovered the optical limiting responses of 2D TiS_2 and graphene by photoacoustic Z-scan measurements [95]. At 0.713 GW cm^{-2}, the optical limiting effect of TiS_2 was enhanced to 77%, which was attributed to the TPA instead of scattering effects. Liu et al. also reported the excellent OL behavior of TiS_2 [96]. When the TiS_2 nanosheets was incorporated into PMMA, the TiS_2-PMMA glass with an extremely low content of 0.066 wt‰ TiS_2 displayed a low onset threshold (defined as the input energy that the normalized transmittance starting to drop) of 0.03 J cm^{-2} and a low optical limiting threshold (defined as the input energy that the normalized transmittance dropping to 50%) of 0.57 J cm^{-2}.

Kushavah et al. constructed 2D CdSe structures on MoS_2 nanosheets using the latter as a template [97]. The fs Z-scan measurement disclosed the optical limiting responses of the CdSe/MoS_2 film with a limiting threshold of ~1.9 mJ cm^{-2}. Abith and Sabari Girisun proposed a facile hydrothermal method to prepare MoS_2 and rGO nanocomposite [98]. The strong ESA enhanced nonlinear absorption of the composite implied its potential use for laser protection against ns green lasers. For instance, a large 2PA coefficient of 9.42×10^{-10} m W^{-1} was deduced for MoS_2-(30 wt%)/rGO, which led to a low onset limiting threshold of 3.15×10^{12} W m^{-2} for the composite. Zu et al. reported that a WSe_2/rGO composite demonstrated significantly enhanced OL responses compared with that of WSe_2, rGO and WSe_2/GO composite [99]. A large nonlinear absorption coefficient of 3.5×10^5 cm GW^{-1} and a low threshold value (0.25 J cm^{-2}) with the input energy of 27 μJ at 532 nm was estimated for the WSe_2/rGO/PVA film.

Liu et al. reported a novel F_{12}PcZn-BP nanohybrid consisting of covalently linked PcZn and BP nanosheets and was embedded into a PMMA matrix [100]. The F_{12}PcZn-BP/PMMA film showed excellent OL properties with a large β of 177.10 cm GW^{-1} and a small limiting threshold of 0.72 GW cm^{-2}. Shen et al. reviewed the nonlinear optical phenomena and mechanism of lead halide perovskites and summarized their applications in passive Q-switched lasers, upconversion lasers, and optical limiting [101]. Liu et al. investigated nonlinear optical properties of 2D perovskite of phenylethylamine lead iodide of $(PEA)_2PbI_4$ by Z-scan measurement, which exhibited strong optical limiting performance due to a giant TPA with a coefficient of 211.5 cm MW^{-1} under an 800-nm fs pulse excitation [102]. Similarly, Lim et al. synthesized a quasi-2D hybrid perovskite composed of aggregation-induced enhanced emission luminophores of organic cation *(Z)*-2-([1,1′-biphenyl]-4-yl)-3-(4-(3-aminopropoxy)phenyl)acrylonitrile (BPCSA) and multi-layered 2D perovskite of $MA_{n-1}Pb_nI_{3n+1}$ alternatively [103]. The hybrid exhibited strong optical limiting effects with a large TPA coefficient β of 11.25 ± 2.0 cm GW^{-1}, and a nonlinear Kerr coefficient n_2 of $1.2 \pm 0.3 \times 10^{-13}$ $cm^2\ W^{-1}$ with a positive (self-focusing) sign.

2.3.2 Q-Switched and Mode-Locked Lasers

The SA of 2D materials is closely related to the practical use in Q-switching or mode-locking lasers [105–108]. TMDs are among the earliest reported use as SAs. Few-layer/multilayer WSe_2 prepared by the chemical vapor deposition (CVD) method was applied as SAs, and the modulation depth of which could be conveniently optimized by controlling layer numbers [104]. With the increase of the modulation depth of the SA, the pulse became narrower. The maximum optical damage threshold of the 1.5-nm-thick, 5.7-nm-thick, and 11-nm-thick WSe_2 Sas were 4.01, 4.24, and 4.49 mJ cm^{-2}, respectively. Hence, a stable mode-locked all-fiber laser was fabricated, with a wavelength centered at 1562 nm and a transform-limited pulse of 182 fs. Chen et al. reported a WS_2-MoS_2-WS_2 heterostructure, which exhibited a small saturable intensity (6.23 MW cm^{-2}) and a large modulation depth (~16.99%), and was utilized as SAs in an erbium-doped fiber laser [109]. Stable output at 1.55 μm with a short duration of 296 fs was achieved, with a high average power of 25 mW.

Ma et al. created a high-quality monolayer graphene via CVD as a SA to mode-lock a Yb: $CaYAlO_4$ fiber laser, realizing a stable 1068-nm output with a breakthrough ultrashort 30 fs pulse duration [110]. Lee et al. used one all-fiber graphene device as a gate-variable SA in a fiber laser cavity to fine-tune the output features by varying the gate voltage. The structure of the electrically controllable all-fiber graphene device is shown in Figure 2.8a [108]. At the applied V_G of −1.05 V, a mode-locking output of 1609 nm with a short pulse duration of 423 fs was obtained, with an average output power of 3.1 mW. An electrically controlled mode-locked all-fiber laser was created, providing new insights on the construction of compact micro/nano optoelectronic and nonlinear photonic devices.

The 2D materials prepared by liquid phase stripping method are frequently used for investigating nonlinear optical effects [17, 19]. Liu et al. fabricated $GeAs_2$ nanosheets via liquid-phase exfoliation methods as SAs in mode-locked fiber lasers, (Figure 2.8b) [19]. The device mainly comprises laser diode (LD), wavelength division multiplexer (WDM), erbium-doped fiber (EDF), output coupler (OC), $GeAs_2$-SA, polarization controller (PC), and polarization independent isolator (PI-ISO), respectively. The laser showed a stable output at 1560 nm with a short pulse duration of 371 fs and a high signal-to-noise (80 dB).

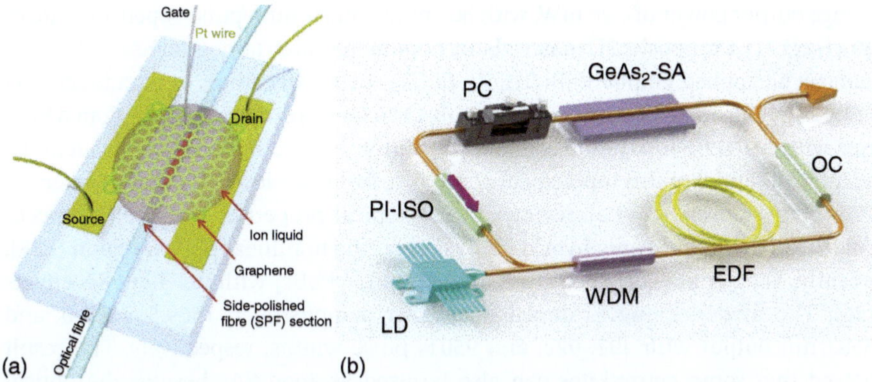

Figure 2.8 (a) Schematic diagram of a gate-variable all-fiber graphene device as SA. Source: Adapted from Liu et al. [104]. Copyright © 2015, Springer. (b) The $GeAs_2$ mode-locking erbium-doped fiber laser. Source: Reproduced with permission from Liu et al. [19]. © 2022, John Wiley & Sons.

Ma et al. utilized a novel MXene of V_2CT_x for mode-locking fiber lasers [111]. A 72-fs pulse duration pulse duration was easily achieved from this hybrid mode-locked fiber laser system. The V_2CTx showed a high modulation depth of 23.7% as SA, based on which an ultrastable with a 1064-nm output and a low threshold was obtained. The Yb-doped fiber laser could work 24 hours continuously. The study of 2D materials for solid-state lasers have also attracted much attention [105]. A MXene of molybdenum carbide (Mo_2C) thin films displaying large nonlinear saturable absorption at 1064 and 1342 nm were prepared via a magnetron sputtering method [112]. When they were used in Nd:YVO_4 solid-state lasers as SAs, stable and short Q-switched pulses at 1.34 μm were achieved. The applications of Mo_2C in other solid-state lasers had also been reported [113, 114]. Similarly, also a member of the MXene family, Nb_2C was exploited as a broadband SA for the solid-state lasers [115]. Recently, Li et al. tailored the nonlinear optical absorption of carbide MXene ($Ti_3C_2T_x$) by modulating its surface terminations via electrochemical methods [116]. Upon a 515-nm laser excitation, $Ti_3C_2T_x$ (=O rich) exhibited larger SA with a β value of -1020 ± 136.2 cm GW^{-1}. By contrast, $Ti_3C_2T_x$ (pristine) and $Ti_3C_2T_x$ (–OH showed RSA behaviors, offering a convenient method to tune the optical nonlinearity of MXenes.

Photonic crystal fiber have also been employed as platform of SA [105, 117, 118]. Yan et al. reported topological insulator (TI: Bi_2Te_3) nanosheets dispersed in photonic crystal fiber could operate as an effective SA with the merits of low-insertion loss (~0.42 dB), long interaction length (>10 cm), and high-power tolerance [119]. This SA device exhibited a saturable intensity of 14.9 MW cm^{-2}, modulation depth of 19.1%, and nonsaturable loss of 25% at 1060 nm. Ma et al. used a GO-PVA (polyvinyl alcohol) absorber for mode-locking Yb-doped fiber laser [120]. As no surfactant was in the GO-PVA solution, non-saturable losses and strong scatterings were significantly reduced in the SA.

2D topological materials are also suitable for SAs. Bi_2Se_3 was used for passive Q-switching of a diode-pumped Pr: $LiYF_4$ laser at 604 nm, achieving a maximum average output power of ~26 mW, with 802 ns at a maximum pulse repetition rate of 130 kHz [121]. Composite 2D materials are popular research targets for SAs [105, 122]. Zhang et al. applied graphene-Bi_2Te_3 (G-Bi_2Te_3) heterostructures as broadband SAs [123, 124]. Both an Er-doped fiber laser with a 1565.6-nm output and a Yb-doped fiber laser with a 1049.1-nm output were successfully realized by G-Bi_2Te_3, evidencing the huge potential of this topological composite for mode-locking ultrafast lasers.

For perovskites, Han et al. reviewed some typical properties and applications of some 2D organic–inorganic hybrid perovskite in the nonlinear optical region [125]. Recently, Yang et al. fabricated $(PEA)_2(CsPbBr_3)_{n-1}PbBr_4$ with excellent SA effects [126]. The mode-locking Er-doped fiber laser demonstrated 1569.5, 1562.0, and 1558.8 nm output with 492, 922, and 950 fs pulse widths, respectively. The result showed that some perovskites can also be used as good SAs besides the optical limiting applications of the perovskite family discussed in Section 2.3.1.

2.3.3 Optical Switch and Modulation

Optical switch and modulation, i.e. manipulating amplitude, frequency, temporal phase or polarization of light in free space or optical waveguide, is crucial for varied applications including sensing, communication, and signal processing. 2D materials excel in optical modulation as tunable waveguides [127, 128], electro-optic modulators [129–131], and switches [132–134], opening up new avenues for manipulating and guiding light and meeting the integration needs of new generation optical networks. For all-optical control, the light excitation induces the change of the carrier densities and distribution in 2D materials, which thus leads to the real and imaginary refractive index change. Third order nonlinear responses like SA, RSA and Kerr effects, and thermo-optic effect can all be used for optical modulation [133, 135–137].

Optical switches using bismuthene based on cross phase modulation (XPM) were reported by Lu et al. [138]. Wu et al used 2D cesium lead bromine iodine ($CsPbBr_xI_{3-x}$) nanosheets (NSs) for all-optical switching [139]. Based on the spatial self-phase modulation effect, the "on" and "off" states were switched by controlling the phase change of the signal beam at 457 nm with the 532 nm light.

Through raising the optical nonlinear absorption of graphene, Ono et al. fabricated one all-optical switching with an ultrafast switching time of 260 fs [140]. The switching energy of the device is as small as 35 fJ and could be conveniently embedded into silicon photonic integrated circuits. Qiu et al. also performed all-optical control on graphene-on-Si_3N_4 integrated platform [141]. The device in a dimension of $60 \times 60\,\mu m$ realized a switching response time constant of 253 ns and a switching energy of ~50 nJ, shedding light on a large-scale integrated all-optical signal processor.

Xu et al. used ultrathin 2D Ga_2S_3 to fabricate an all-optical on-chip switch based Mach-Zehnder interferometer (MZI) [142]. The 532-nm excitation induced excessive photocarriers in Ga_2S_3, leading to the refractive index change. Thereafter, a phase variation between MZI arms at the 1550 nm caused optical switching. The device exhibited a small optical power density of $0.12\,W\,cm^{-2}$, a fast response and recovery time of 26.3 and 43.5 μs, respectively. Wang et al. synthesized ultrathin SnSe nanosheets by a Li-intercalation-assisted liquid exfoliation method [143]. The 2D SnSe was used for optical switch, exhibiting an extinction ratio of 20 dB and an operation range over 6 nm. Jiang et al. also showed the potential of SnSe for the applications in all-optical switching [144]. SnS belong to the same IV-VI compound-based 2D nanomaterials as SnSe and are also excellent optical switching materials. The nonlinear refractive index and third-order nonlinear susceptibility of SnS were measured to be $\sim 10^{-5}$ $cm^2\ W^{-1}$ and $\sim 10^{-10}$ esu, respectively [145].

As early as 2014 [146], graphene had been used as ultrafast all-optical modulators. A modulation depth of 38% and a response time of ~2.2 ps were achieved, potentially used for ultrafast optical signal processing. Recently, the graphdiyne/graphene (GDY/G) was very effective for laser modulation, not only as SAs but also for its spatial self-phase modulation properties [147]. Due to synergistic effect, GDY/G possessed a large nonlinear refractive index in the order of 10^{-5} $cm^2\ W^{-1}$. Wang et al. developed all-optical modulator based on phosphorene with the potential of being integrated into optical fiber systems [148]. The modulator exhibited a conversion efficiency of $0.029\,\pi\ mW^{-1}$, with a maximal phase shift of 8π at a control light of 290 mW.

Grinblat et al. used 2D perovskite of $(C_4H_9NH_3)_2(CH_3NH_3)_3Pb_4I_{13}$ for all-optical modulation based on nonlinear optical Kerr effect and TPA [149]. Using magneto-optical Kerr microscopy, Powalla et al. investigated gate-tunable local spin polarization in graphene/WTe_2 heterostructures [150], indicative of the potential of 2D heterostructure for spintronics. Fan et al. studied the interplay between magneto-optical properties and ferroelectric polarization in a perovskite/metal-organic framework structure [151]. They found external electric field coupled to the ferroelectric polarization could be used to control the Kerr rotation.

2.3.4 Other Nonlinear Optical Phenomena

A plethora of new physics related to 2D materials has also appeared, such as ultrafast phonon dynamics, novel nonlinear quantum optics, and metamaterials.

Ultrafast phonon dynamics. Phonons are collective atomic motions in solid materials and are regarded as quasi-particles. Phonon dynamics provides valuable information on electronic and thermal properties of materials. In particular, for 2D materials, phonon dynamics offers information on interlayer interactions, layer thickness, stacking order, etc. For example, phonon dynamics has been used to explore interlayer phonons mediated conductance and thermal transport in twisted bilayer graphene [153], spin phonon coupling [154], and optical manipulation of heterostructure devices [155].

Yang and coworkers reported coherent interlayer phonons in 2D magnetic topological insulator of $MnBi_2Te_4$ [152]. With time-resolved pump–probe reflectivity spectroscopy, they observed coherent phonon oscillations as a function of layer number. Figure 2.9a shows the phonon decay rate increased with decreasing thickness for the sample <10 layers. However, for thicker samples >10 layers, the phonon decay rate was independent of the thickness.

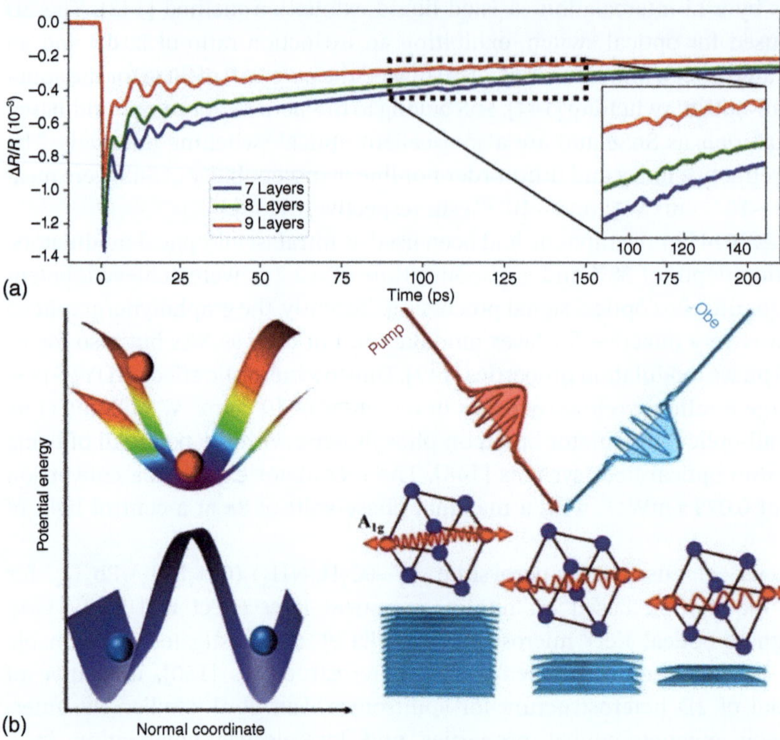

Figure 2.9 (a) Coherent phonon dynamics of 2D magnetic topological insulator of $MnBi_2Te_4$. Source: Michael Bartram et al. [152]/Springer Nature / CC BY 4.0. (b) The potential energy surface evolution for Bi upon fs laser pulse excitation. Source: Ma et al. [20]/with permission of American Chemical Society.

Using transient reflectance spectroscopy, our group observed size-dependent phonon coherence for bismuthene (Figure 2.9b) [20]. Under the same energy excitation, the oscillation frequency gradually decreased and decoherence time gradually shortened from bulk Bi to multilayer Bi and few-layer Bi. The phenomenon was attributed to the phonon confinement and eventual bond softening of 2D materials due to quantum size effect in comparison with bulk materials. The work thus shed light on utilizing coherent phonons in 2D materials for quantum technologies.

NLO of 2D metamaterials and plasmonics. The peculiar properties of metamaterials mainly come from their special geometric structure and size parameters. Some limitations of the nonlinear optical effects of natural materials can be broken through the reasonable design of the geometric parameters and spatial configuration of metamaterials. Hence, metamaterials provide new routes to enhance nonlinear optical interactions [156] and metasurfaces are the ultrathin version of metamaterials.

Abir et al. demonstrated strong collective resonances from coupling the metasurface to an optical waveguide [157], which consisted of a thin TiO_2 film sputtered on a fused silica substrate and a metasurface of gold split-ring resonators on the top. An enhancement factor of 8 was obtained for the transmitted second harmonic. Raju et al. attempted to construct a plasmonic structure to induce SHG of an $LiNbO_3$ nanolaminate via inverse design [158]. Alam et al. developed RLC model for design of nonlinear plasmonic metasurfaces for SHG based on gold elongated nanoprisms [159]. The detected SHG intensities exhibited β values within $\sim 10^{-30}$ m^4 V^{-1}. Sarma et al. fabricated an all-dielectric polaritonic metasurface made of $In_{0.53}Ga_{0.47}As/A_{l0.52}In_{0.48}As$ [160]. A maximum SHG power conversion factor of 0.5 mW W^{-2} and a power conversion efficiency of 0.015% were obtained.

Quantum NLO. The development of quantum NLO enables the generation of optical nonlinearities at individual photon levels. 2D materials with strong photon-photon interactions promote a large range of promising applications from optical technology to quantum information processing, such as single-photon switches and transistors and quantum control of light [161]. Wilson et al. reviewed the relationship between stacked 2D semiconductors (TMDs) and excitons and emergent photon quantum phenomena [162]. 2D materials-based van der Waals (vdW) structures demonstrate unique advantages in the quantum NLO field, such as high sample quality immune from chemical and structural defects that are oftentimes present in conventional epitaxial heterostructures due to atomic diffusion and lattice mismatch. Indeed, many novel quantum phenomena based on vdW structures have been reported, including moiré excitons, many-body interactions and correlated states in superlattices, and quantum light source.

2.4 Prospect

We have introduced the fundamental NLO revolving 2D materials, summarized representative experimental setups for nonlinearity characterization, and reviewed recent developments in applications. The following are the future trends for the research field involving nonlinear optical applications of 2D materials.

2.4.1 Precise Fabrication and Functionalization of 2D Materials

High-quality 2D materials with high stability, uniform physical, and chemical properties in large sizes are the foundation for their practical use, and hence development of more novel 2D materials and enriching their functions is the persistent goal. A promising trend is to build vdW heterojunctions or organic/hybrid structures to realize more function. For instance, assisted by the smooth surface, strong interface coupling of 2D materials, thin vdW heterojunctions can be flexibly designed with a large tolerance on lattice mismatch. Utilizing methods such as CVD, physical vapor deposition (PVD), molecular beam epitaxy (MBE), metal organic chemical vapor deposition (MOCVD), and vdW heterojunctions with tunable nonlinear optical properties can be conveniently constructed. On the other hand, recently emerging artificial intelligence (AI)-assisted functionalization may also facilitate the rational design of 2D materials composites.

2.4.2 High Resolution Spatiotemporal Characterizing Techniques

The photoinduced excited-state dynamics of 2D materials are closely associated with their nonlinear optical behaviors, the deep understanding of which offers general guidance for design materials with better performance in optical limiting, mode-locking, harmonic generation, etc. Femtosecond transient spectroscopy has joined other time-resolved techniques including streak cameras, transient photocurrent spectroscopy, time-resolved terahertz spectroscopy, time-resolved optical Kerr effect technique, ultrafast 2D electronic spectroscopy, etc., in the toolkits for exploring the ultrafast dynamics of 2D materials. Additionally, a new trend of combining time and space resolution, i.e. spatiotemporal technology provides high spatial resolution beyond time-resolved information. For instance, coupling pump–probe techniques with transmission electronic microscopes offers opportunities for direct visualization of 2D materials with atomic spatial resolution (angstrom) and femtosecond time resolution as well. Multidimensional, in situ, high resolution spatiotemporal techniques are expected to play more crucial roles in characterizing 2D nonlinear optical materials.

2.4.3 New Physics of 2D Materials

With the ever-expanding of novel 2D materials, a plethora of related new physics has also emerged. For instance, construction of layered heterostructures through stacking 2D materials at a defined twist angle leads to exotic twist-dependent nonlinear optical properties, such as moiré excitons related to quantum spin. Fine-tuning the twist angle thus offers new avenues to build integrated nonlinear optical devices. Further study on topics like moiré superlattices based on 2D materials, nonlinear twistronics is expected to give an additional knob to engineer the nonlinear optical properties of layered materials. Another example is quantum NLO based 2D materials. Many TMD materials can undergo multiple nonlinear processes like SHG, SFG, and FWM simultaneously under intense excitation, which offers unprecedented opportunities to explore exciton-based NLO in these monolayers.

2.4.4 Ever-Lasting Expansion of Applications

The demand for high quality 2D nonlinear optical materials and devices has kept expanding to even broader field. Exceling in the myriad applications such as high-speed modulators, ultrafast lasers, ultrasensitive photodetectors, and 2D materials demonstrate unprecedented potential for fabricating integrated all-optical photonic chips. Moreover, the emerging usage of 2D materials include single photon emitter and detectors, artificial intelligence devices, phonon lasers, and even space use. On the other hand, large-scale production is also essential for 2D nonlinear optical materials to realize practical use.

Acknowledgment

This work is supported by the National Natural Science Foundation of China (NSFC.22221001, 22073038, 92256202, U22A20399) and Supercomputing Center of Lanzhou University.

References

1 Yin, P., Jiang, X., Huang, R. et al. (2021). 2D materials for nonlinear photonics and electro-optical applications. *Advanced Materials Interfaces* 8: 2100367.
2 Autere, A., Jussila, H., Dai, Y. et al. (2018). Nonlinear optics with 2D layered materials. *Advanced Materials* 30: 1705963.
3 Krasnok, A., Tymchenko, M., and Alù, A. (2018). Nonlinear metasurfaces: a paradigm shift in nonlinear optics. *Materials Today* 21: 8–21.
4 Wang, W., Zhang, W., Chu, S.T. et al. (2017). Repetition rate multiplication pulsed laser source based on a microring resonator. *ACS Photonics* 4: 1677–1683.
5 Inouye, S., Chikkatur, A.P., Stamper-Kurn, D.M. et al. (1999). Superradiant Rayleigh scattering from a Bose-Einstein condensate. *Science* 285: 571–574.
6 Yao, K., Yanev, E., Chuang, H.-J. et al. (2019). Continuous wave sum frequency generation and imaging of monolayer and heterobilayer two-dimensional semiconductors. *ACS Nano* 14: 708–714.
7 Dasgupta, A., Gao, J., and Yang, X. (2020). Anisotropic third-harmonic generation in layered germanium selenide. *Laser & Photonics Reviews* 14: 1900416.
8 Wang, H. and Qian, X. (2017). Giant optical second harmonic generation in two-dimensional multiferroics. *Nano Letters* 17: 5027–5034.
9 Ma, Q., Ren, G., Mitchell, A., and Ou, J.Z. (2020). Recent advances on hybrid integration of 2D materials on integrated optics platforms. *Nano* 9: 2191–2214.
10 Liu, X., Guo, Q., and Qiu, J. (2017). Emerging low-dimensional materials for nonlinear optics and ultrafast photonics. *Advanced Materials* 29: 1605886.
11 Dini, D., Calvete, M.J.F., and Hanack, M. (2016). Nonlinear optical materials for the smart filtering of optical radiation. *Chemical Reviews* 116: 13043–13233.
12 Leiming, W., Jiang, X., Zhao, J. et al. (2018). MXene-based nonlinear optical information converter for all-optical modulator and switcher. *Laser & Photonics Reviews* 12: 1800215.
13 Shan, Y., Li, Z., Ruan, B. et al. (2019). Two-dimensional Bi_2S_3-based all-optical photonic devices with strong nonlinearity due to spatial self-phase modulation. *Nano* 8: 2225–2234.
14 Shu, Y., Guo, J., Fan, T. et al. (2020). Two-dimensional black arsenic phosphorus for ultrafast photonics in near-and mid-infrared regimes. *ACS Applied Materials & Interfaces* 12: 46509–46518.

15 Zhang, M., Wu, Q., Zhang, F. et al. (2019). Black-phosphorous-based pulsed lasers: 2D black phosphorus saturable absorbers for ultrafast photonics. *Advanced Optical Materials* 7: 1970001.

16 Biswas, S., Tiwary, C.S., Vinod, S. et al. (2017). Nonlinear optical properties and temperature dependent photoluminescence in hBN-GO heterostructure 2D material. *The Journal of Physical Chemistry C* 121: 8060–8069.

17 Xiao, Q., Ma, B., Fei, X. et al. (2021). Unveiling the dimension-dependence of femtosecond nonlinear optical properties of tellurium nanostructures. *Nanoscale Horizons* 6: 918–927.

18 Zhang, C., Wu, H., Xu, Y. et al. (2020). Plasmon-assisted broadband all-optical control of highly intense femtosecond laser by weak continuous-wave laser. *Advanced Optical Materials* 8: 2000560.

19 Liu, S., Li, G., Zhu, F. et al. (2022). $GeAs_2$ saturable absorber for ultrafast and ultranarrow photonic applications. *Advanced Functional Materials* 32: 2112252.

20 Ma, B., Liu, R.-T., Zhang, X.-D. et al. (2022). Ultrafast generation of coherent phonons in two-dimensional bismuthene. *The Journal of Physical Chemistry Letters* 13: 3072–3078.

21 Li, D., Xiong, W., Jiang, L. et al. (2016). Multimodal nonlinear optical imaging of MoS_2 and MoS_2-based van der Waals heterostructures. *ACS Nano* 10: 3766–3775.

22 Yu, J., Park, S., Hwang, I. et al. (2019). Third-harmonic generation from plasmonic metasurfaces coupled to intersubband transitions. *Advanced Optical Materials* 7: 1801510.

23 Wang, S.-Y., Zhen, X.J., Feng, Z. et al. (2022). Third-order optical nonlinearities of two-dimensional SnS under irradiation: implications for space use. *Journal of Materials Chemistry C* 10: 18025–18032.

24 Zhai, X.-P., Gao, L.-F., Zhang, H. et al. (2022). Defect engineering of ultrathin WO_3 nanosheets: implications for nonlinear optoelectronic devices. *ACS Applied Nano Materials* 5: 1169–1177.

25 Mondal, K., Biswas, S., Pramanik, A. et al. (2022). Giant femtosecond nonlinear optical response in bi-metallic GO nanocomposites for photonic applications. *Applied Surface Science* 578: 151966.

26 Sioncke, S., Verbiest, T., and Persoons, A. (2003). Second-order nonlinear optical properties of chiral materials. *Materials Science and Engineering R: Reports* 42: 115–155.

27 Nye, J.F. (1985). *Physical Properties of Crystals: Their Representation by Tensors and Matrices*. Oxford University Press.

28 de Coene, Y., Jooken, S., Deschaume, O. et al. (2022). Label-free imaging of membrane potentials by intramembrane field modulation, assessed by second harmonic generation microscopy. *Small* 18: 2200205.

29 Grinblat, G., Li, Y., Nielsen, M.P. et al. (2016). Enhanced third harmonic generation in single germanium nanodisks excited at the anapole mode. *Nano Letters* 16: 4635–4640.

30 Cha, S., Kim, M., Kim, Y. et al. (2022). Gate-tunable quantum pathways of high harmonic generation in graphene. *Nature Communications* 13: 1–10.

31 Shree, S., Lagarde, D., Lombez, L. et al. (2021). Interlayer exciton mediated second harmonic generation in bilayer MoS_2. *Nature Communications* 12: 1–9.

32 Zhao, H., An, Q., Ye, X. et al. (2021). Second harmonic generation in AB-type $LaTiO_3/SrTiO_3$ superlattices. *Nano Energy* 82: 105752.

33 Spreyer, F., Ruppert, C., Georgi, P., and Zentgraf, T. (2021). Influence of plasmon resonances and symmetry effects on second harmonic generation in WS_2-plasmonic hybrid metasurfaces. *ACS Nano* 15: 16719–16728.

34 Gan, Z., Cheng, Y., Chen, W. et al. (2021). Photophysics of 2D organic–inorganic hybrid lead halide perovskites: progress, debates, and challenges. *Advanced Science* 8: 2001843.

35 Song, T., Ma, Q.-X., Wang, Q., and Zhang, H.-L. (2022). Design of two-dimensional halide perovskite composites for optoelectronic applications and beyond. *Materials Advances* 3: 756–778.

36 Yuan, C., Li, X., Semin, S. et al. (2018). Chiral lead halide perovskite nanowires for second-order nonlinear optics. *Nano Letters* 18: 5411–5417.

37 Zhang, Y., Wang, F., Feng, X. et al. (2022). Inversion symmetry broken 2D SnP_2S_6 with strong nonlinear optical response. *Nano Research* 15: 2391–2398.

38 Ha, S., Park, N.H., Kim, H. et al. (2021). Enhanced third-harmonic generation by manipulating the twist angle of bilayer graphene. *Light: Science & Applications* 10: 1–10.

39 Ravi, P., Tripathi, N., Yang, X., and Gao, J. (2021). Van der Waals layered mineral getchellite with anisotropic linear and nonlinear optical responses. *Laser & Photonics Reviews* 15: 2100182.

40 Abdelwahab, I., Grinblat, G., Leng, K. et al. (2018). Highly enhanced third-harmonic generation in 2D perovskites at excitonic resonances. *ACS Nano* 12: 644–650.

41 Yoshikawa, N., Tamaya, T., and Tanaka, K. (2017). High-harmonic generation in graphene enhanced by elliptically polarized light excitation. *Science* 356: 736–738.

42 Hirori, H., Xia, P., Shinohara, Y. et al. (2019). High-order harmonic generation from hybrid organic–inorganic perovskite thin films. *APL Materials* 7: 041107.

43 Langer, F., Hohenleutner, M., Schmid, C.P. et al. (2016). Lightwave-driven quasiparticle collisions on a subcycle timescale. *Nature* 533: 225–229.

44 Chen, Z.-Y. and Qin, R. (2020). High harmonic generation in graphene–boron nitride heterostructures. *Journal of Materials Chemistry C* 8: 12085–12091.

45 Jiayang, W., Yang, Y., Yang, Q. et al. (2020). 2D layered graphene oxide films integrated with micro-ring resonators for enhanced nonlinear optics. *Small* 16: 1906563.

46 Dai, Y., Wang, Y., Das, S. et al. (2021). Broadband plasmon-enhanced four-wave mixing in monolayer MoS_2. *Nano Letters* 21: 6321–6327.

47 You, J.W., Lan, Z., and Panoiu, N.C. (2020). Four-wave mixing of topological edge plasmons in graphene metasurfaces. *Science Advances* 6: eaaz 3910.

48 Ko, B.A., Berry, K., Qin, Z. et al. (2021). Resonant degenerate four-wave mixing at the defect energy levels of 2D organic–inorganic hybrid perovskite crystals. *ACS Applied Materials & Interfaces* 13: 57075–57083.

49 Bauer, J., Chen, L., Wilhelm, P. et al. (2022). Excitonic resonances control the temporal dynamics of nonlinear optical wave mixing in monolayer semiconductors. *Nature Photonics* 16: 777–783.

50 Wang, S., Yu, H., Zhang, H. et al. (2014). Broadband few-layer MoS_2 saturable absorbers. *Advanced Materials* 26: 3538–3544.

51 Li, N., Wang, Q., and Zhang, H.-L. (2020). 2D Materials in light: excited-state dynamics and applications. *The Chemical Record* 20: 413–428.

52 Zhao, M., Chang, M.-J., Wang, Q. et al. (2015). Unexpected optical limiting properties from MoS_2 nanosheets modified by a semiconductive polymer. *Chemical Communications* 51: 12262–12265.

53 Zhao, M., Peng, R., Zheng, Q. et al. (2015). Broadband optical limiting response of a graphene–PbS nanohybrid. *Nanoscale* 7: 9268–9274.

54 Wang, K., Szydłowska, B.M., Wang, G. et al. (2016). Ultrafast nonlinear excitation dynamics of black phosphorus nanosheets from visible to mid-infrared. *ACS Nano* 10: 6923–6932.

55 Zheng, X., Chen, R., Shi, G. et al. (2015). Characterization of nonlinear properties of black phosphorus nanoplatelets with femtosecond pulsed Z-scan measurements. *Optics Letters* 40: 3480–3483.

56 Maldonado, M., da Silva Neto, M.L., Vianna, P.G. et al. (2020). Femtosecond nonlinear optical properties of 2D metallic NbS_2 in the near infrared. *The Journal of Physical Chemistry C* 124: 15425–15433.

57 Chen, B., Zhang, Q., Shen, X. et al. (2022). Nonlinear optical and optical limiting properties of $MAPbBr_3$/Polymethyl methacrylate organic glasses. *The Journal of Physical Chemistry C* 126: 5743–5750.

58 He, Z., Rong, T., Li, Y. et al. (2022). Two-dimensional TiVC solid-solution MXene as surface-enhanced Raman scattering substrate. *ACS Nano* 16: 4072–4083.

59 Hontani, Y., Xia, F., and Xu, C. (2021). Multicolor three-photon fluorescence imaging with single-wavelength excitation deep in mouse brain. *Science Advances* 7: eabf3531.

60 Cheng, Q., Miao, Y., Wild, J. et al. (2021). Emerging applications of stimulated Raman scattering microscopy in materials science. *Matter* 4: 1460–1483.

61 Ling, J., Miao, X., Sun, Y. et al. (2019). Vibrational imaging and quantification of two-dimensional hexagonal boron nitride with stimulated Raman scattering. *ACS Nano* 13: 14033–14040.

62 Li, R., Zhang, Y., Xu, X. et al. (2018). Optical characterizations of two-dimensional materials using nonlinear optical microscopies of CARS. *TPEF, and SHG, Nanophotonics* 7: 873–881.

63 Hell, S.W., Bahlmann, K., Schrader, M. et al. (1996). Three-photon excitation in fluorescence microscopy. *Journal of Biomedical Optics* 1: 71–74.

64 Fan, Y., Tonkaev, P., Wang, Y. et al. (2021). Enhanced multiphoton processes in perovskite metasurfaces. *Nano Letters* 21: 7191–7197.

65 Li, M., Xu, Y., Han, S. et al. (2020). Giant and broadband multiphoton absorption nonlinearities of a 2D organometallic perovskite ferroelectric. *Advanced Materials* 32: 2002972.

66 He, H., Ma, E., Chen, X. et al. (2019). Single crystal perovskite microplate for high-order multiphoton excitation. *Small Methods* 3: 1900396.

67 Lu, S., Zhou, F., Zhang, Q. et al. (2019). Layered hybrid perovskites for highly efficient three-photon absorbers: theory and experimental observation. *Advanced Science* 6: 1801626.

68 Han, J., Wang, Y., He, J. et al. (2021). Fabry–Perot cavity enhanced three-photon luminescence of atomically thin platinum diselenide. *Nanoscale* 13: 9031–9038.

69 Pettine, J., Grubisic, A., and Nesbitt, D.J. (2018). Angle-and momentum-resolved photoelectron velocity map imaging studies of thin Au film and single supported Au nanoshells. *The Journal of Physical Chemistry C* 122: 3970–3984.

70 Topolniak, I., Elert, A.M., Knigge, X. et al. (2022). High-precision micropatterning of polydopamine by multiphoton lithography. *Advanced Materials* 34: 2109509.

71 Zhai, X.-P., Ma, B., Wang, Q., and Zhang, H.-L. (2020). 2D materials towards ultrafast photonic applications. *Physical Chemistry Chemical Physics* 22: 22140–22156.

72 Zhang, L., Fahad, S., Wu, H.-R. et al. (2020). Tunable nonlinear optical responses and carrier dynamics of two-dimensional antimonene nanosheets. *Nanoscale Horizons* 5: 1420–1429.

73 Li, N., Zhai, X.-P., Yan, W.-K. et al. (2021). Boosting cascade electron transfer for highly efficient CO_2 photoreduction. *Solar RRL* 5: 2100558.

74 Chen, S., Bylinkin, A., Wang, Z. et al. (2022). Real-space nanoimaging of THz polaritons in the topological insulator Bi_2Se_3. *Nature Communications* 13: 1–9.

75 Bargheer, M., Zhavoronkov, N., Gritsai, Y. et al. (2004). Coherent atomic motions in a nanostructure studied by femtosecond X-ray diffraction. *Science* 306: 1771–1773.

76 Wieghold, S., Cope, E.M., Moller, G. et al. (2022). Stressing halide perovskites with light and electric fields. *ACS Energy Letters* 7: 2211–2218.

77 Kumar, A., Solanki, A., Manjappa, M. et al. (2020). Excitons in 2D perovskites for ultrafast terahertz photonic devices. *Science Advances* 6: eaax8821.

78 Jakowetz, A.C., Böhm, M.L., Sadhanala, A. et al. (2017). Visualizing excitations at buried heterojunctions in organic semiconductor blends. *Nature Materials* 16: 551–557.

79 Sung, J., Schnedermann, C., Ni, L. et al. (2020). Long-range ballistic propagation of carriers in methylammonium lead iodide perovskite thin films. *Nature Physics* 16: 171–176.

80 Yue, S., Tian, F., Sui, X. et al. (2022). High ambipolar mobility in cubic boron arsenide revealed by transient reflectivity microscopy. *Science* 377: 433–436.

81 Shin, J., Gamage, G.A., Ding, Z. et al. (2022). High ambipolar mobility in cubic boron arsenide. *Science* 377: 437–440.

82 Kumar, V. and Camden, J.P. (2022). Imaging vibrational excitations in the electron microscope. *The Journal of Physical Chemistry C* 126: 16919–16927.

83 Horstmann, J.G., Böckmann, H., Wit, B. et al. (2020). Coherent control of a surface structural phase transition. *Nature* 583: 232–236.

84 Schmitt, D., Bange, J.P., Bennecke, W. et al. (2022). Formation of moiré interlayer excitons in space and time. *Nature* 608: 499–503.

85 Li, S., Driver, T., Rosenberger, P. et al. (2022). Attosecond coherent electron motion in Auger-Meitner decay. *Science* 375: 285–290.

86 Zhang, M., Guo, Z., Mi, X. et al. (2022). Ultrafast imaging of molecular dynamics using ultrafast low-frequency lasers, X-ray free electron lasers, and electron pulses. *The Journal of Physical Chemistry Letters* 13: 1668–1680.

87 Zhang, F., Wang, M., Wang, Z. et al. (2018). Excellent nonlinear absorption properties of β-antimonene nanosheets. *Journal of Materials Chemistry C* 6: 2848–2853.

88 Li, Y., Zheng, C., Wang, S.-T. et al. (2022). Record aluminum molecular rings for optical limiting and nonlinear optics. *Angewandte Chemie International Edition* 61: e202116563.

89 Yuan, J., Xin, Y., Li, Z. et al. (2021). Excellent ultraviolet optical limiting properties of 2D chromium nanosheets. *Journal of Materials Chemistry C* 9: 13432–13438.

90 Liu, R., Hu, J., Zhu, S. et al. (2017). Synergistically enhanced optical limiting property of graphene oxide hybrid materials functionalized with Pt complexes. *ACS Applied Materials & Interfaces* 9: 33029–33040.

91 Yuan, J., Xin, Y., Wang, X. et al. (2022). 2D tungsten nanosheets: ascendant nonlinear absorption properties in the ultraviolet band. *Journal of Materials Chemistry C* 10: 6682–6686.

92 Yuan, J., Wang, X., Liu, Y. et al. (2022). Excellent nonlinear absorption properties of 2D germanium nanosheets in the infrared band. *Optical Materials* 125: 112115.

93 Aradhana, D.M., Dissanayake, S., Cifuentes, M.P., and Humphrey, M.G. (2018). Optical limiting properties of (reduced) graphene oxide covalently functionalized by coordination complexes. *Coordination Chemistry Reviews* 375: 489–513.

94 Dong, N., Li, Y., Feng, Y. et al. (2015). Optical limiting and theoretical modelling of layered transition metal dichalcogenide nanosheets. *Scientific Reports* 5: 1–10.

95 Varma, S.J., Kumar, J., Liu, Y. et al. (2017). 2D TiS_2 layers: a superior nonlinear optical limiting material. *Advanced Optical Materials* 5: 1700713.

96 Liu, Y., Li, X., Wang, E. et al. (2021). Exceptional size-dependent property of TiS_2 nanosheets for optical limiting. *Applied Surface Science* 541: 148371.

97 Kushavah, D., Mushtaq, A., Ghosh, S., and Pal, S.K. (2021). Ultrafast and nonlinear optical properties of two-dimensional CdSe nanostructures prepared using MoS_2 nanosheets as template. *Physica E: Low-dimensional Systems and Nanostructures* 130: 114682.

98 Abith, M. and Sabari Girisun, T.C. (2021). Excited state absorption induced optical limiting action of MoS_2-rGO nanocomposites. *Journal of Molecular Liquids* 341: 117337.

99 Zu, Y., He, C., Liu, D. et al. (2021). Nonlinear optical properties of poly (vinyl alcohol) thin films doped with in-situ WSe_2/rGO composite. *Optics & Laser Technology* 142: 107198.

100 Liu, Z., Gan, F., Dong, N. et al. (2019). Fabrication and nonlinear optical characterization of fluorinated zinc phthalocyanine covalently modified black phosphorus/PMMA films using the nanosecond Z-scan technique. *Journal of Materials Chemistry C* 7: 10789–10794.

101 Shen, W., Chen, J., Wu, J. et al. (2020). Nonlinear optics in lead halide perovskites: mechanisms and applications. *ACS Photonics* 8: 113–124.

102 Liu, W., Xing, J., Zhao, J. et al. (2017). Giant two-photon absorption and its saturation in 2D organic–inorganic perovskite. *Advanced Optical Materials* 5: 1601045.

103 Lim, C.-K., Maldonado, M., Zalesny, R. et al. (2020). Interlayer-sensitized linear and nonlinear photoluminescence of quasi-2D hybrid perovskites using aggregation-induced enhanced emission active organic cation layers. *Advanced Functional Materials* 30: 1909375.

104 Liu, W., Liu, M., Yin, J. et al. (2018). Tungsten diselenide for all-fiber lasers with the chemical vapor deposition method. *Nanoscale* 10: 7971–7977.

105 Wang, J., Wang, X., Lei, J. et al. (2020). Recent advances in mode-locked fiber lasers based on two-dimensional materials. *Nano* 9: 2315–2340.

106 Wang, J., Chen, L., Dou, C. et al. (2018). $Mo_{0.5}W_{0.5}S_2$ for Q-switched pulse generation in ytterbium-doped fiber laser. *Nanotechnology* 29: 224002.

107 Jiang, T., Yin, K., Wang, C. et al. (2020). Ultrafast fiber lasers mode-locked by two-dimensional materials: review and prospect. *Photonics Research* 8: 78–90.

108 Lee, E.J., Choi, S.Y., Jeong, H. et al. (2015). Active control of all-fibre graphene devices with electrical gating. *Nature Communications* 6: 1–6.

109 Chen, H., Yin, J., Yang, J. et al. (2017). Transition-metal dichalcogenides heterostructure saturable absorbers for ultrafast photonics. *Optics Letters* 42: 4279–4282.

110 Ma, J., Huang, H., Ning, K. et al. (2016). Generation of 30 fs pulses from a diode-pumped graphene mode-locked Yb: $CaYAlO_4$ laser. *Optics Letters* 41: 890–893.

111 Ma, C., Huang, W., Wang, Y. et al. (2020). MXene saturable absorber enabled hybrid mode-locking technology: a new routine of advancing femtosecond fiber lasers performance. *Nano* 9: 2451–2458.

112 Wang, J., Liu, S., Wang, Y. et al. (2020). Magnetron-sputtering deposited molybdenum carbide MXene thin films as a saturable absorber for passively Q-switched lasers. *Journal of Materials Chemistry C* 8: 1608–1613.

113 Yang, Z., Gao, L., Chen, H. et al. (2021). Broadband few-layer niobium carbide MXene as saturable absorber for solid-state lasers. *Optics & Laser Technology* 142: 107199.

114 Liu, S., Cui, N., Liu, S. et al. (2022). Nonlinear optical properties and passively Q-switched laser application of a layered molybdenum carbide at 639 nm. *Optics Letters* 47: 1830–1833.

115 Wang, J., Wang, Y., Liu, S. et al. (2020). Nonlinear optical response of reflective MXene molybdenum carbide films as saturable absorbers. *Nanomaterials* 10: 2391.

116 Li, H., Chen, S., Boukhvalov, D.W. et al. (2022). Switching the nonlinear optical absorption of titanium carbide MXene by modulation of the surface terminations. *ACS Nano* 16: 494–404.

117 Luo, Z.-C., Liu, M., Liu, H. et al. (2013). 2 GHz passively harmonic mode-locked fiber laser by a microfiber-based topological insulator saturable absorber. *Optics Letters* 38: 5212–5215.

118 Liu, Z.-B., He, X., and Wang, D.N. (2011). Passively mode-locked fiber laser based on a hollow-core photonic crystal fiber filled with few-layered graphene oxide solution. *Optics Letters* 36: 3024–3026.

119 Yan, P., Lin, R., Chen, H. et al. (2014). Topological insulator solution filled in photonic crystal fiber for passive mode-locked fiber laser. *IEEE Photonics Technology Letters* 27: 264–267.

120 Ma, M., Wen, W., Zhang, Y. et al. (2019). Few-layer $ReS_{2(1-x)}Se_{2x}$ nanoflakes for noise-like pulse generation in a mode-locked ytterbium-doped fiber laser. *Journal of Materials Chemistry C* 7: 6900–6904.

121 Cheng, Y., Peng, J., Bin, X. et al. (2017). Passive Q-switching of Pr: $LiYF_4$ orange laser at 604 nm using topological insulators Bi_2Se_3 as saturable absorber. *Optics & Laser Technology* 88: 275–279.

122 Ma, C., Wang, C., Gao, B. et al. (2019). Recent progress in ultrafast lasers based on 2D materials as a saturable absorber. *Applied Physics Reviews* 6: 041304.

123 Wang, Z., Mu, H., Zhao, C. et al. (2016). Harmonic mode-locking and wavelength-tunable Q-switching operation in the graphene–Bi_2Te_3 heterostructure saturable absorber-based fiber laser. *Optical Engineering* 55: 081314.

124 Wang, Z., Mu, H., Yuan, J. et al. (2016). Graphene-Bi_2Te_3 heterostructure as broadband saturable absorber for ultra-short pulse generation in Er-doped and Yb-doped fiber lasers. *IEEE Journal of Selected Topics in Quantum Electronics* 23: 195–199.

125 Han, X., Zheng, Y., Chai, S. et al. (2020). 2D organic-inorganic hybrid perovskite materials for nonlinear optics. *Nano* 9: 1787–1810.

126 Yang, S., Li, J.-Z., Li, L. et al. (2022). Mode-locking operation of Er-doped fiber laser with $(PEA)_2(CsPbBr3)_{n\text{-}1}PbBr_4$ perovskite saturable absorber. *Journal of Materials Chemistry C* 10: 7504–7510.

127 Wu, L., Wei, S., Zhang, Y. et al. (2022). Tunable nonlinearity in 2D graphdiyne oxide for high-performance all-optical modulation, advanced. *Optical Materials* 2102537.

128 Zhang, T., Wang, C., Chen, H. et al. (2019). Controlled multichannel surface plasmon polaritons transmission on atomic smooth silver triangular waveguide. *Advanced Optical Materials* 7: 1900930.

129 Powell, K., Li, L., Shams-Ansari, A. et al. (2022). Integrated silicon carbide electro-optic modulator. *Nature Communications* 13: 1–7.

130 Vella, D., Barbosa, M.B., Trevisanutto, P.E. et al. (2022). In-plane field-driven excitonic electro-optic modulation in monolayer semiconductor. *Advanced Optical Materials* 10: 2102132.

131 Kravets, V.G., Wu, F., Auton, G.H. et al. (2019). Measurements of electrically tunable refractive index of MoS_2 monolayer and its usage in optical modulators. *npj 2D Materials and Applications* 3: 1–10.

132 Gagel, P., Harder, T.H., Betzold, S. et al. (2022). Electro-optical switching of a topological polariton laser. *ACS Photonics* 9: 405–412.

133 Khan, A.R., Zhang, L., Ishfaq, K. et al. (2022). Optical harmonic generation in 2D materials. *Advanced Functional Materials* 32: 2105259.

134 Zhou, Y., Huang, Y., Xinlong, X. et al. (2020). Nonlinear optical properties of halide perovskites and their applications. *Applied Physics Reviews* 7: 041313.

135 Shaltout, A.M., Shalaev, V.M., and Brongersma, M.L. (2019). Spatiotemporal light control with active metasurfaces. *Science* 364: eaat3100.

136 Shieh, W. and Djordjevic, I.B. (2009). *OFDM for Optical Communications.* Academic Press.

137 Chen, H., Wang, C., Ouyang, H. et al. (2020). All-optical modulation with 2D layered materials: status and prospects. *Nano* 9: 2107–2124.

138 Lu, L., Wang, W., Wu, L. et al. (2017). All-optical switching of two continuous waves in few layer bismuthene based on spatial cross-phase modulation. *ACS Photonics* 4: 2852–2861.

139 Wu, L., Chen, K., Huang, W. et al. (2018). Perovskite $CsPbX_3$: a promising nonlinear optical material and its applications for ambient all-optical switching with enhanced stability. *Advanced Optical Materials* 6: 1800400.

140 Ono, M., Hata, M., Tsunekawa, M. et al. (2020). Ultrafast and energy-efficient all-optical switching with graphene-loaded deep-subwavelength plasmonic waveguides. *Nature Photonics* 14: 37–43.

141 Qiu, C., Yang, Y., Li, C. et al. (2017). All-optical control of light on a graphene-on-silicon nitride chip using thermo-optic effect. *Scientific Reports* 7: 1–7.

142 Kai, X., Zhang, B.Y., Hu, Y. et al. (2021). A high-performance visible-light-driven all-optical switch enabled by ultra-thin gallium sulfide. *Journal of Materials Chemistry C* 9: 3115–3121.

143 Wang, C., Li, F., Xu, Y. et al. (2021). Tin selenide: a promising black-phosphorus-analogue nonlinear optical material and its application as all-optical switcher and all-optical logic gate. *Materials Today Physics* 21: 100500.

144 Zhang, C., Ouyang, H., Miao, R. et al. (2019). Anisotropic nonlinear optical properties of a SnSe flake and a novel perspective for the application of all-optical switching. *Advanced Optical Materials* 7: 1900631.

145 Wu, L., Xie, Z., Lu, L. et al. (2018). Few-layer tin sulfide: a promising black-phosphorus-analogue 2D material with exceptionally large nonlinear optical response, high stability, and applications in all-optical switching and wavelength conversion. *Advanced Optical Materials* 6: 1700985.

146 Li, W., Chen, B., Meng, C. et al. (2014). Ultrafast all-optical graphene modulator. *Nano Letters* 14: 955–959.

147 Zhang, Q., Bai, Q., Cai, E. et al. (2022). Nonlinear optical properties of graphdiyne/graphene van der Waals heterostructure for laser modulations. *Results in Physics* 105654.

148 Wang, Y., Zhang, F., Tang, X. et al. (2018). All-optical phosphorene phase modulator with enhanced stability under ambient conditions. *Laser & Photonics Reviews* 12: 1800016.

149 Grinblat, G., Abdelwahab, I., Nielsen, M.P. et al. (2019). Ultrafast all-optical modulation in 2D hybrid perovskites. *ACS Nano* 13: 9504–9510.

150 Powalla, L., Kiemle, J., König, E.J. et al. (2022). Berry curvature-induced local spin polarisation in gated graphene/WTe_2 heterostructures. *Nature Communications* 13: 1–8.

151 Fan, F.-R., Wu, H., Nabok, D. et al. (2017). Electric-magneto-optical Kerr effect in a hybrid organic–inorganic perovskite. *Journal of the American Chemical Society* 139: 12883–12886.

152 Michael Bartram, F., Leng, Y.-C., Wang, Y. et al. (2022). Ultrafast coherent interlayer phonon dynamics in atomically thin layers of $MnBi_2Te_4$. *npj Quantum Materials* 7: 1–8.

153 Mahapatra, P.S., Sarkar, K., Krishnamurthy, H.R. et al. (2017). Seebeck coefficient of a single van der Waals junction in twisted bilayer graphene. *Nano Letters* 17: 6822–6827.

154 Choe, J., Lujan, D., Rodriguez-Vega, M. et al. (2021). Electron–phonon and spin–lattice coupling in atomically thin layers of $MnBi_2Te_4$. *Nano Letters* 21: 6139–6145.

155 Rodriguez-Vega, M., Lin, Z.-X., Leonardo, A. et al. (2022). Light-driven topological and magnetic phase transitions in thin layer antiferromagnets. *The Journal of Physical Chemistry Letters* 13: 4152–4158.

156 Litchinitser, N.M. (2018). Nonlinear optics in metamaterials. In: *Advances in Physics: X*, vol. 3, 1367628. Taylor & Francis.

157 Abir, T., Tal, M., and Ellenbogen, T. (2022). Second-harmonic enhancement from a nonlinear plasmonic metasurface coupled to an optical waveguide. *Nano Letters* 22: 2712–2717.

158 Raju, L., Lee, K.-T., Liu, Z. et al. (2022). Maximized frequency doubling through the inverse design of nonlinear metamaterials. *ACS Nano* 16: 3926–3933.

159 Saad Bin-Alam, M., Baxter, J., Awan, K.M. et al. (2020). Hyperpolarizability of plasmonic meta-atoms in metasurfaces. *Nano Letters* 21: 51–59.

160 Sarma, R., Xu, J., de Ceglia, D. et al. (2022). An all-dielectric polaritonic metasurface with a giant nonlinear optical response. *Nano Letters* 22: 896–903.

161 Chang, D.E., Vuletić, V., and Lukin, M.D. (2014). Quantum nonlinear optics-photon by photon. *Nature Photonics* 8: 685–694.

162 Wilson, N.P., Yao, W., Shan, J., and Xu, X. (2021). Excitons and emergent quantum phenomena in stacked 2D semiconductors. *Nature* 599: 383–392.

3

Modulation and Enhancement of Optical Nonlinearity in 2D Materials

Xinglin Wen[1] *and Qihua Xiong*[2]

[1] *Huazhong University of Science and Technology, Wuhan National Laboratory for Optoelectronics, School of Optical and Electronic Information, Luoyu Road 1037, Wuhan, 430074, China*
[2] *Tsinghua University, State Key Laboratory of Low-Dimensional Quantum Physics, Department of Physics, Shuangqing Road 30, Beijing, 100084, People's Republic of China*

3.1 Introduction

Nonlinear optics (NLO) is of great importance both in practical applications and fundamental research. The invention of laser offers the opportunities to study the light–matter interaction under intense electromagnetic field illumination. The first NLO was discovered by Franken et al at 1961, in which the second harmonic generation (SHG) was demonstrated in crystalline quartz under 694.3 nm laser excitation [1]. Subsequently, third-order NLO like third-harmonic generation (THG), four-wave mixing (FWM) and two-photon absorption (TPA) were discovered. Recently, high-order NLO is becoming essential in attosecond pulse laser generation. As an important branch of optics, NLO has broad applications in light generation, frequency conversion, optical modulators, materials characterization, high resolution imaging, and so forth. NLO definitely plays an important role in the development of physics, chemistry, and engineering science.

Apart from the intense light source, a medium with considerable nonlinear susceptibility is necessary to produce decent nonlinear optical response. Currently, NLO bulk crystals like potassium dihydrogen phosphate (KH_2PO_4, KDP), ammonium dihydrogen phosphate ($NH_4H_2PO_4$, ADP), and beta barium borate (BaB_2O_4, BBO) are mainly used to generate intense nonlinear optical response. Nevertheless, today's state of technology requires the exploration of the minimization of NLO medium for their application in on-chip integration. The on-chip nonlinear optical devices demand that the medium has strong NLO response even the dimensionality is scaled to few nanometers, which is contradictory to the fact that the nonlinear effects will decline dramatically if the thickness of the NLO crystals is reduced. Therefore, it is desirable to explore the novel nanomaterials with high nonlinear susceptibility.

The artificial plasmonic and metamaterial present strong optical nonlinearity when the size is down to hundreds nanometers due to that the electromagnetic field can be confined in a small volume when the structure is in resonance

Two-Dimensional Materials for Nonlinear Optics: Fundamentals, Preparation Methods, and Applications, First Edition. Edited by Qiang Wang and Hao-Li Zhang.

with the incidence light. These "hot-spots" can enhance the optical nonlinearity significantly considering $P_{NL}^{(n)} \propto E^n$, where $P_{NL}^{(n)}$ is the n-th order nonlinear polarization. Although plasmonic structure and metamaterials can provide strong nonlinear response in tiny size, their structures are complex and incompatible with the complementary-metal–oxide–semiconductor (CMOS) technology. In addition, the large loss of the metals constrains them from practical applications. Also, their working wavelength is highly dependent on the size of the units of plasmonic or metamaterials structures. Consequently, they can only function within a small spectrum range once the device fabrication is finished.

The emergence of two-dimensional (2D) materials offers the opportunities to realize an efficient nonlinear optical response in a chip platform. In 2004, successfully exfoliated single layer graphene from the graphite, which triggered the research of 2D materials [2]. In addition to the flexibility, environmental stability, and excellent physical and electrical properties, graphene also has a broad application in optics. Although graphene cannot behave like a semiconductor because of the gapless band structure, graphene exhibits exceptional optical nonlinearity in the near to middle infrared regime like THG, FWM, saturable absorber (SA), reverse saturable absorption (RSA) and optical limiting [3–6]. After graphene, some other semiconducting 2D materials such as transition metal dichalcogenides (TMDs) and black phosphorus (BP), which also exhibit excellent nonlinear optical properties [7–9]. The nonlinear susceptibility and working wavelength regime of those 2D materials can be tailored by changing the layer numbers, stacking order, elements substitution and excitation polarization, offering more flexibility to change their NLO as desired. Overall, the susceptibility of 2D materials are orders of magnitude larger than the commercial nonlinear crystals. In addition, phase-mismatch problem is absent in 2D materials because of the atomic thickness, which is advantageous over the bulk materials. More importantly, the adjacent layers of 2D materials interact with each other by van der Waals (vdWs) force, thus the possible integration with other materials like silicon without lattice mismatch. In short, the strong nonlinear susceptibility and heterogeneous integration to other materials make 2D materials promising in on-chip nonlinear optical devices.

Although 2D materials can provide decent nonlinear susceptibility, the short light–matter interaction length in 2D materials imposes a limitation on efficiency of nonlinear optical processes. Effective strategies must be developed to modulate and enhance NLO of 2D materials. In this chapter, the strategies of NLO manipulation will be presented and discussed. Firstly, NLO of 2D materials is highly dependent on the symmetry of the materials. Consequently, NLO can be manipulated by engineering the structure such as layer numbers, stacking order, twisted angle, and edge states. Secondly, NLO can be enhanced dramatically by using optical cavity to confine the electromagnetic field. In addition, the excitonic effects and valley degree of freedom play an important role in the NLO manipulation of TMDs. Thirdly, NLO of 2D materials also can be manipulated by applying an external electric field, which provides an effective approach to realize electric-controlled optical nonlinearity. Finally, the nonlinearity of the ferromagnetic 2D materials can be manipulated with an external magnetic field. In summary, NLO of 2D materials can be manipulated by symmetry, optical cavity, and electric and magnetic fields, which provides a great flexibility to construct the on-chip nonlinear devices based on 2D materials.

3.2 Nonlinear Optics in 2D Materials

3.2.1 Basics of Nonlinear Optics

All the light–matter interactions can be described by the light induced polarization in the medium as following equation:

$$P = \varepsilon_0(\chi^{(1)}E + \chi^{(2)}E^2 + \chi^{(3)}E^3 + \cdots) = P_{\mathrm{L}} + P_{\mathrm{NL}} \tag{3.1}$$

where ε_0 is the permittivity of vacuum, $\chi^{(n)}$ is the n-th order susceptibility, P_{L} is the linear light-induced-polarization corresponding to the first term, while P_{NL} is the nonlinear light-induced-polarization corresponding the summary of the rest nonlinear terms. Normally, $\chi^{(n)}(n \geq 2)$ is much smaller than the linear susceptibility $\chi^{(1)}$. Consequently, NLO usually cannot occur under weak light excitation. In addition, $\chi^{(n)}$ decreases with increased n value, which makes it difficult to observe a high-order nonlinear optical effect. Although high-order harmonics (HHG) can be observed under some specific conditions, second- and third-order nonlinearity are primarily dominant in 2D materials corresponding to $\chi^{(2)}$ and $\chi^{(3)}$ respectively.

The second-order nonlinearity includes sum-frequency generation (SFG), difference-frequency generation (DFG), and SHG. Third-order nonlinearity consists of FWM, THG, TPA, self-phase modulation, self-focusing, and optical Kerr effects. Equation (3.1) indicates that $\chi^{(2)}$ is non-vanished only for crystal without spatial-inversion symmetry, which can be also applied to other even-order nonlinearity $\chi^{(2n)}$. The odd-order nonlinearity can present regardless the symmetry. The symmetry dependent second-order nonlinearity leads to the opportunities to modulate NLO by engineering their structures. For the bulk or even-layer TMDs materials, the inversion symmetry is preserved and thus the SHG is vanished. Meanwhile, SHG will occur if the TMDs materials are exfoliated to odd-layers. Specifically, the second-order nonlinearity $\chi^{(2)}$ belongs to $D_{3\mathrm{h}}$ space group for the monolayer TMDs and it is a third-order tensor with 27 Cartesian components. For the monolayer case, $\chi^{(2)}$ only has one independent component and the analysis can be simplified with following formula [9]:

$$\chi^{(2)} = \chi^{(2)}_{xxx} = -\chi^{(2)}_{yyx} = -\chi^{(2)}_{xyy} = -\chi^{(2)}_{yxy} \tag{3.2}$$

Nonlinear processes in 2D materials can be classified into parametric and nonparametric NLO. Parametric NLO is a pure scattering process in which only the virtual states are involved and the absorption is absent. In contrast, the nonparametric process is usually accompanied by absorption and charge transfer because the real energy states are involved, making the final states are different from the initial states. In addition, a complex nonlinear susceptibility $\chi = \chi_1 + i\chi_2$ is necessary to describe the nonparametric process. However, the parametric process can be fully described by a real nonlinear susceptibility. All the frequency conversion processes such as

SHG, THG, FWM, SFG, and DFG belong to the parametric NLO. The nonparametric processes in 2D materials mainly include multi-photon absorption, SA, and RSA.

3.2.2 Parametric Nonlinear Optics

SHG is a typical parametric NLO and widely studied in TMDs. SHG is nonzero only when the inversion symmetry is broken. SHG of bulk materials and even-layer flakes is vanished. There is no SHG for single layer graphene because the lattice of single layer graphene belongs to D_{6h} point group, which is different from the monolayer TMDs. SHG has been investigated in various TMDs such as MoS_2, WS_2, $MoSe_2$, WSe_2, and $MoTe_2$ [10–13]. The second-order susceptibility $\chi^{(2)}$ in these studies are on the order of magnitude of nm/V. Therefore, the second-order susceptibility of the 2D layers is much higher than that of commercial nonlinear crystals such as BBO with $\chi^{(2)}$ on the order of pm/V. An important issue regarding the second-order susceptibility is that the measured $\chi^{(2)}$ deviates across studies in the range of 10–10^5 pm/V. Several factors are responsible for the deviated $\chi^{(2)}$. The most import reason is that susceptibility is related to the polarization in unite volume and it can only be well defined in bulk materials. Consequently, the $\chi^{(2)}$ will vary with each other for differed methods used to estimate $\chi^{(2)}$ of the atomically thin materials. The second factor is that SHG is dependent on temperature, in which the temperature will affect the distance between the adjacent layers and thus modify the SHG intensity. Thirdly, the sample quality and synthesis method also have significant influence on SHG. For instance, the SHG of exfoliated monolayer MoS_2 is 20 times large than the CVD grown monolayer MoS_2 [14]. Last but not the least, the substrate will dramatically affect the measured SHG intensity because of the constructive or destructive interference effects.

SHG has another important feature that the polarization depends on the lattice symmetry. Sixfold pattern is resolved as the sample is rotated from 0 to 360°. In addition, SHG with polarization parallel to the polarization of incidence is different from the perpendicular one, which can be described by $I_{SHG} = I_0 \cos^2(3\theta + \varphi_0)$ and $I_{SHG} = I_0 \sin^2(3\theta + \varphi_0)$ respectively. The orientation of the structure can be precisely determined by resolving φ_0 via SHG measurement. This is very useful in the construction of moiré superlattice, which requires to control the twisted angle of the adjacent layers precisely. Similar to TMDs layers, the lattice orientation of hexagonal boron nitride (h-BN) can also be determined by SHG.

In contrast to second-order nonlinearity, third-order nonlinear process can occur in any 2D materials regardless of the layer symmetry. FWM is a typical third-order nonlinear process which combine three photons to generate one photon. THG is a special case of FWM which converts three photons with energy of $\hbar\omega$ to one photon with energy of $3\hbar\omega$. THG and FWM are widely investigated in TMDs, graphene, h-BN and BP [5, 15–17]. As same as $\chi^{(2)}$, the reported third-order susceptibility $\chi^{(3)}$ of 2D materials also vary significantly depending on measurement condition and sample quality. For TMDs, $\chi^{(3)}$ can vary by two orders of magnitude ($\sim 10^{-19}$ m^2V^{-2}). One important factor contributing to the varied $\chi^{(3)}$ is the excitonic effect enhanced THG. The measured $\chi^{(3)}$ of graphene also show discrepancies from 1×10^{-19} to

4×10^{-15} m^{2}V^{-2} in different process like THG, FWM and self-phase modulation [18]. The variation of $\chi^{(3)}$ in graphene is primarily ascribed to the frequency-dependent resonance effects. $\chi^{(3)}$ of graphene is several orders of magnitude higher than monolayer TMDs, which is due to that the Dirac point structure of graphene make it easy to form resonance condition for interband electron transition.

Different from SHG, polarization of THG in monolayer TMDs is independent on the crystal structure [19]. However, THG is sensitive to the grains and boundaries, which is useful to recognize the grains, boundaries, and defects of 2D flakes via THG mapping [20]. The THG mapping provides a higher resolution in grain recognition than the Raman, PL, or SHG mapping.

Although SHG and THG can be routinely realized in TMDs or graphene, their atomic thickness limits the light–mater interaction length. Recently, 2D perovskites have the potential for SHG and THG radiation. Chiral molecules can be incorporated into the 2D inorganic framework to break the inversion symmetry to induce SHG. SHG in nanowires demonstrates SHG anisotropy up to 96%, and the second-order NLO coefficient is 0.68 pm V^{-1} [21]. Furthermore, THG of 2D perovskites are highly dependent on the incidence wavelength and the thickness of the inorganic layer. The third-order susceptibility $\chi^{(3)}$ of 2D perovskites is on the order of 10^{-17} m^2 V^{-2} [22], which is many times higher than 3D perovskites because of the quantum confinement effects in 2D counterparts.

Although HHG is commonly reported in bulk materials or gases, 2D materials hold the potential in HHG because of their distinctive electronic properties and strong many-body interactions, demonstrating HHG in both graphene and monolayer MoS_2 [23, 24]. 13-th order harmonics were reported in MoS_2 under the pump power intensity of several TW/cm^2 [24]. In contrast to the perturbative process that intensity of n-th order harmonics is proportional to the incidence power with $I \propto P^n$, the HHG in monolayer MoS_2 is a nonperturbative process with $I \propto P^{3.3}$ for all the HHG process. In addition, the polarization of the HHG radiation is dependent on the crystal symmetry, similar to the SHG. The realization of HHG in 2D materials make it feasible to construct the ultraviolet and attosecond light source in chip scale.

3.2.3 Nonparametric Nonlinear Optics

The nonparametric processes in 2D materials are mainly ascribed to the third-order nonlinear such as two-photon absorption, saturable absorption and reversed saturable absorption. Nonparametric process in meaningful in up-conversion luminescence, Q-switch, and all-optical light modulation.

TPA refers to the absorption of two photons to induce one electron transition. TPA is widely demonstrated in TMDs materials that normally have the TPA coefficient at the order of 10^3–10^4 GW cm^{-2}, which is 10^4 times larger than the standard bulk semiconductor like GaAs [25]. The large TPA coefficient is attributed to the quantum confinement effect and strong excitonic effect of TMDs. TPA can enable the detection of infrared light in which 2D materials can absorb two photons with low energy to generate the free carriers, which provide an effective approach to detect infrared light at room temperature. In addition, the selection rule of TPA is different from the single photon absorption. Consequently, TPA can be used to detect the dark states that cannot be accessible by one photon absorption to resolve the fine band structures of the materials [26].

Graphene exhibit unique TPA properties in which the TPA coefficient is highly dependent on the layer numbers and stacking order [27]. The TPA coefficient of bilayer graphene is much larger than the single layer in the near infrared regime. The TPA coefficient of bilayer graphene is proportional to ω^{-3}, while it is ω^{-4} for single layer graphene [27]. Moreover, the TPA coefficient of AB-stacked graphene is two orders of magnitude larger than other bilayer graphene [27]. Recently, TPA was also demonstrated in 2D perovskites [28]. Apart from TPA, multi-photon absorption such as three-photon absorption and four-photon absorption also can be realized in 2D materials, which make it possible to detect infrared light with a longer wavelength.

SA refers to the process that the absorption is reduced when the intensity of the incidence increase, which is attributed to the imaginary part of $\chi^{(3)}$. Under the illumination with high intensity, the generation rate is larger than the decay rate, and electrons are accumulated in the conduction band. Consequently, the electron accumulation prevents the absorption because of the Pauli blocking effect, leading to the SA. SA can be demonstrated in most of the 2D materials with various saturable intensity. SA-enabled *Q*-switch and mode-locking is frequently used to generation pulse light. For instance, graphene-based mode-locking laser can generate pulse with duration down to 800 ps [29]. However, the pulse duration of TMDs-based mode locking laser is limited by the lifetime of carriers.

Compared to TMDs, graphene provides better SA performance. Firstly, SA can be achieved in a wide wavelength range from UV to mid infrared because graphene has no energy bandgap. Secondly, carriers in graphene will undergo interband transition via electron–electron and electron-phonon scattering with the lifetime of hundreds of femtoseconds, which can provide much higher processing speed compared to semiconductors. In short, the broad frequency band and high-speed processing make graphene promising in ultrafast pulse generation and ultrafast optical modulation. A laser pulse under 200 fs could be realized in a graphene-based SA device [30]. Moreover, a graphene-based ultrafast all-optical switch was achieved in a graphene with the switching time of 260 fs [31].

RSA is an opposite process of SA in which the transmittance is reduced under high incidence intensity. As a high intensity will pump the excited electrons to higher energy states, the reduced electron concentration in excited states leads to a stronger absorption. RSA is a typical photo-induced absorption process and usually observed in graphene [32].

3.3 Nonlinearity Modulation in 2D Materials

3.3.1 Structural Engineering

3.3.1.1 Layer Number

NLO of 2D materials can be engineered by changing the layer numbers to change the crystal symmetry or light–matter interaction length. SHG only occurs in odd-layer

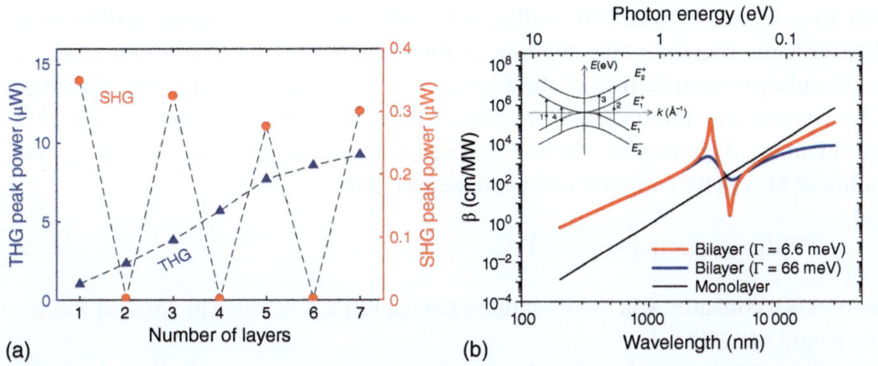

Figure 3.1 (a) SHG and THG of MoS_2 with different layers, (b) TPA spectra of monolayer and bilayer graphene. Source: (a) reproduced with permission [15]. Copyright © 2017, Springer Nature; (b) Reproduced with permission from Yang et al. [27]/American Chemical Society.

TMDs that breaks inversion symmetry. The 1-, 3-, and 5-layer TMDs almost have the same SHG intensity. Although THG of TMDs is not dependent on the symmetry, the light–matter interaction length is enlarged for thick materials. As a result, THG intensity gradually increases with thick TMDs (Figure 3.1a). NLO is complicated in graphene with different layers. For instance, TPA of bilayer graphene is significantly larger than it of single layer graphene, which is caused by the different electronic structure. Monolayer graphene has the Dirac-cone band structure with two bands, and thus absorption of two photons creates an electron in the conduction band and a hole in the valence band (Figure 3.1b). By adding all possible intermediate states in the two bands together, the TPA coefficient is expressed as [27]

$$\beta_{1L} = \frac{4\pi^2}{\varepsilon_\omega \omega^4 \hbar^3}\left(\frac{v_F e^2}{c}\right)^2 \tag{3.3}$$

However, there are two more bands for bilayer graphene. TPA can occur in four possible interband transition, namely, from the lower valence (E_2^-) band to two conduction bands $(E_1^+$ and $E_2^+)$ and from higher valence band (E_1^-) to two conduction bands $(E_1^+$ and $E_2^+)$. In the case of $\hbar\omega > \gamma_1$, the TPA coefficient of bilayer graphene is expressed as [27]

$$\beta_{2L} = 8\beta_{1L}\frac{\hbar\omega}{\gamma_1}\left[16\hbar^2\omega^2\left(\frac{1}{\hbar\omega}+\frac{1}{\gamma_1+\hbar\omega}\right)^2+(\hbar\omega-\gamma_1)^2\left(\frac{2}{\hbar\omega}+\frac{1}{\hbar\omega+\gamma_1}+\frac{1}{\hbar\omega-\gamma_1}\right)^2\right.$$
$$\left.+16(\hbar\omega-\gamma_1)^2\left(\frac{1}{\hbar\omega}+\frac{1}{\hbar\omega-\gamma_1}\right)^2+2(\hbar\omega-\gamma_1)^2\left(\frac{2}{\hbar\omega}+\frac{1}{\hbar\omega+\gamma_1}+\frac{1}{\hbar\omega-\gamma_1}\right)^2\right] \tag{3.4}$$

where γ_1 is the e hopping energy between two atoms. It is the additional interband transitions contribute to the larger TPA in bilayer graphene. In addition, TPA of bilayer graphene has the resonant feature at a specific wavelength, which is significantly different form the single layer.

For thick 2D flakes, their nonlinearity also can be engineered by changing thickness. THG intensity of SiP flakes increases firstly when the thickness increases from 10 to 33 nm. While, THG will reduce if the thickness increases further from 40 to 100 nm. The dramatic increase in the thickness of 10–33 nm is a result of the cumulative contribution of the increased SiP layer number. For the thickness of 40–100 nm, the dominant absorption effects reduce the THG significantly [33]. The thickness dependence THG of 2D materials can be obtained by solving the nonlinear Maxwell's equation and expressed as [33]

$$P^{3\omega}(d) \approx Ad^2 \exp\left(-\frac{4\pi kd}{\lambda}\right) \tag{3.5}$$

where A is a constant, k is the imaginary part of refractive index at 3ω, and λ is the wavelength of THG.

Recently, a layer number-dependent NLO was demonstrated in 2D $PdSe_2$ [34]. The bandgap of $PdSe_2$ decreases from 0.8 to 0 eV if the layer number increases from 15 to 50. The band narrowing and semiconductor–semimetal phase transition modify the nonlinear optical characteristics and carrier dynamics significantly. The spectral range of SA can be extended to near infrared by increasing the layer number.

3.3.1.2 Twisted Angle

Although 2H-stacked TMDs bilayer presents inversion symmetry, the symmetry can be broken by staking two monolayer TMDs with a specific angle. In other words, the NLO of TMDs can be modulated and enhanced by constructing a homo- or heterostructural bilayer with a specific angle. SHG of artificially homo-stacked MoS_2/MoS_2 bilayer is highly dependent on the twisted angle (Figure 3.2a), which is also applicable to the WSe_2/MoS_2 and WSe_2/WS_2 hetero-stacked bilayers and insensitive to the constituent layered materials [35]. The twisted angle-dependent SHG is attributed to the interference of the second-harmonic fields between the constituent layers and their phase difference depends on the twisted angle. SHG of TMDs bilayers with arbitrary twisted angle can be fully described by the mode proposed by Hsu et al. [35] Considering the polarization of the incident electric field $\vec{E}(\omega)$ has an angle of $\varphi_1(\varphi_2)$ with respect to armchair direction of the constituent layers, the second-harmonic fields $\vec{E}_1(2\omega)[\vec{E}_2(2\omega)]$ from the constituent layers then have an angle of $3\varphi_1(3\varphi_2)$ with respect with then incident electric field (Figure 3.2b). Consequently, the second-harmonic field in the stacking region is the coherent superposition of the second-harmonic fields of the constituent layers: $\vec{E}_{2L}(2\omega) = \vec{E}_1(2\omega) + \vec{E}_2(2\omega)$. Total SHG intensity $\left|\vec{E}_{2L}(2\omega)\right|^2$ of the stacked bilayer can be expressed as

$$I_{2L}(\theta) = I_1 + I_2 + 2\sqrt{I_1 I_2}\cos 3\theta \tag{3.6}$$

where I_1 and I_2 are SHG intensity of the two constituent monolayers, $\theta = \varphi_1 - \varphi_2$ is the twisted angle between two individual monolayers. The twisted angle modulated SHG provides an efficient approach to detect stacking orientation, crystal polarity, and domain boundary of vdWs heterostructure with broken inversion symmetry. The coherent superposition is not applicable to heterostructures with interlayer coupling where charge transfer may occur. The interlayer coupling will modify band structure and thus affect the SHG intensity via excitonic resonance effect. For

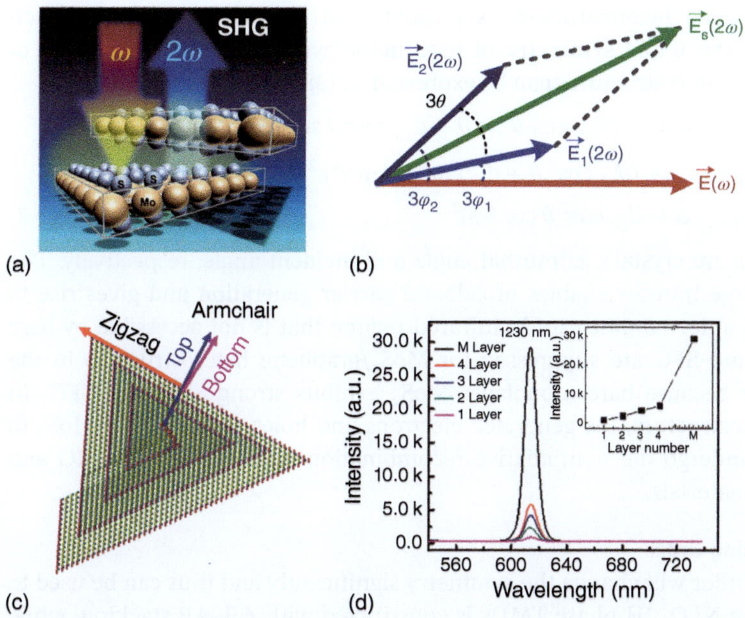

Figure 3.2 (a) SHG of bilayer with a twisted angle, (b) interference of the second-harmonic fields, (c) spiral WS_2, and (d) layer number dependent SHG of spiral WS_2. Source: (a) and (b) Reproduced with permission from Zhao et al. [35]/American Chemical Society; (c) and (d) Reproduced with permission from Fan et al. [36]/American Chemical Society.

instance, SHG is systematically investigated in strongly coupled, weekly coupled, and decoupled $MoS_2/MoS_{2(1-x)}Se_{2x}$ [37]. The superposition of second-harmonic fields are only applicable to the decoupled heterostructures. For heterostructures with interlayer coupling, the band offset causes a significant shift of the exciton wavelength and the efficient SHG wavelength window change accordingly.

Furthermore, significant enhancement of SHG and THG can be achieved in a spiral TMDs with a twisted angle of 5° between adjacent layers (Figure 3.2c) [36]. SHG of the spiral WS_2 intensity quadratically increases with layer numbers (Figure 3.2d), which is in contrast to the vanished SHG of thick 2H-staked WS_2 layers. The anomalous SHG performance comes from the broken centrosymmetry of the twisted screw structures. With increased layer numbers, the longer light–matter interaction leads to a stronger SHG signal.

Second-order nonlinearity of TMDs/graphene heterostructure is insensitive to the twisted angle because graphene is a centrosymmetric material. However, the stacking with graphene will change the NLO of TMDs dramatically [38]. First-principle calculations indicate that $\chi^{(2)}$ is can be enhanced in near infrared regime in TMDS/graphene heterostructure. The emergence of second-order nonlinearity in near infrared regime is primarily due to interlayer charge transfer between TMDs and graphene. The redistribution of charge at the interface modifies the electronic band structure as well as the generation of carriers, leading to the change of nonlinear coefficient. In addition, SHG polarization dependence of TMDs/graphene heterostructure is present with threefold symmetry, which is different to the 6-fold symmetry of bare monolayer TMDs. SHG intensity of graphene/WS_2 heterostructures can be expressed as [38]

$$I^{//}_{\text{graphene/WS}_2} \propto \left(-\left(2d_{15}\cos\theta\sin\theta - d_{22}\cos^2\theta\sin 3\phi\cos\theta\right) + (d_{31}\cos^2\theta + d_{33}\sin^2\theta)\sin\theta\right)^2$$
$$I^{\perp}_{\text{graphene/WS}_2} \propto (-d_{22}\cos^2\theta\cos 3\phi)^2 \tag{3.7}$$

where θ and ϕ are crystal's azimuthal angle and incident angle, respectively. The interlayer charge transfer enables broadband carrier generation and gives rise to strong second-order nonlinearity in infrared regime that is not accessible by bare TMDs. SHG and SFG are suppressed for MoS_2/graphene heterostructure in the visible regime because bare monolayer MoS_2 exhibits strong SHG and SFG. In terms of heterostructure, the generated electrons and holes transfer from MoS_2 to graphene to undergo the nonradiative recombination and thus reduce SHG and SFG of MoS_2 materials.

3.3.1.3 Stacking Order

The stacking order will change the symmetry significantly and thus can be used to manipulate the NLO. 2H phase TMDs is constructed with AB–AB stacking, while the 3R phase is ABC–ABC stacking. Unlike the vanished SHG in even-layer 2H TMDs, SHG is present in 3R phase TMDs regardless layer numbers [39]. This is because inversion symmetry is always broken for 3R stacking TMDs. In 3R phase MoS_2 materials, SHG continuously increase with increased layer numbers, which is contrast to 2H phase MoS_2. The estimated second-order susceptibility of MoS_2 $\chi^{(2)}_{\text{MoS}_2}$ are 4.05×10^{-10}, 1.18×10^{-12}, and 3.0×10^{-11} mV^{-1} for monolayer, 2H bulk, and 3R bulk MoS_2 under excitation wavelength at 1200 nm, respectively [40].

Furthermore, the calculation indicates that the different stacking order can change the in-plane and out-of-plane charge density symmetry. As a result, the stacking order will change the dispersion of second-order nonlinear coefficients in a TMDs homobilayer. The AA stacking of homobilayer belongs to D_{3h} space, while the AA′_Mo-60 stacking belongs to C_{3v} space group. The different stacking order will render deviated SHG polarization as [41]

$$I^{//}_{AA} \propto d_{22}^2E_0^4\cos[\theta]^2\left(\cos[\theta]\cos[3\phi]\sin[2\alpha]+ (\cos[\alpha]^2[\theta]^2 - \sin[\alpha]^2)\sin[3\phi]\right)^2$$
$$I^{\perp}_{AA} \propto d_{22}^2E_0^4\left(\cos[\alpha]^2\cos[\theta]^2\cos[3\phi] + \cos[3\phi]\sin[\alpha]^2+ +\cos[\theta]\sin[2\alpha]\sin[3\phi]\right)^2 \tag{3.8}$$

$$I^{//}_{AA'\_Mo-60} \propto E_0^4\left(d_{32}\sin[\alpha]^2\sin[\theta] + d_{33}\cos[\alpha]^2\sin[\theta]^3 + \cos[\theta]^2(d_{22}\cos[3\phi]\sin[2\alpha] + (-2d_{24} + \mathrm{d}_{32})\cos[\alpha]^2\sin[\theta]) + d_{22}\cos[\alpha]^2\cos[\theta]^3\sin[3\phi] - d_{22}\cos[\theta]\sin[\alpha]^2\right)$$
$$I^{\perp}_{AA'\_Mo-60} \propto E_0^4\left(-d_{22}\cos[\alpha]^2\cos[\theta]^2\cos[3\phi] + d_{22}\cos[3\phi]\sin[\alpha]^2 + \sin[2\alpha](d_{24}\sin[\theta] + \mathrm{d}_{22})\cos[\theta]\sin[3\phi])\right)^2 \tag{3.9}$$

where θ, α, ϕ are incident pump angle, polarization angle and azimuthal angle, respectively.

SHG is vanished in single layer graphene because it is centrosymmetric. However, SHG can arise in Bernal ABA–stacked trilayer, which is non-centrosymmetric, whereas SHG vanishes in a rhombohedral ABC-stacked centrosymmetric trilayer. $\chi^{(2)}$ of ABA-stacked trilayer graphene is around 0.9×10^{-10} m/V under 1300 nm excitation, which is comparable to the $\chi^{(2)}$ of monolayer TMDs under excitonic resonance condition [42]. Apart from the broken centrosymmetry, the large $\chi^{(2)}$ of ABA-stacked graphene trilayer also benefits from the nonlinear optical transitions, which is in resonance with the semimetal band structure. In addition, the stronger second-order nonlinearity can be partly attributed to the strong piezoelectric effect in few layers graphene samples. Overall, stacking order engineering is indeed an effective strategy to manipulate and enhance the optical nonlinearity.

3.3.1.4 Edge States

The NLO of materials are highly sensitive to the electronic structure change. For centrosymmetric 3D bulk materials, breaking the translational symmetry at 2D surface leads to the dipolar contribution to surface optical nonlinearity, which is referred as surface-enhanced NLO and servers as an effective approach to probe the surface properties of various materials [43, 44]. Similarly, breaking the translational symmetry at edges of 2D TMDs yields 1D atomic edges with localized electronic structure different from the 2D lattice, which results in strong resonant nonlinear optical susceptibility at the edge. Indeed, the observation of resonant enhancement of SHG of MoS_2 monolayer at frequencies about the half of the bandgap is attributed to the edge states. The density functional theory (DFT) also predicts that there exist localized edge states within the bandgap of 2D crystal [45]. Yin et al. observed that monolayer MoS_2 exhibits a much stronger SHG under 1300 nm laser excitation compared with 1280 nm excitation (A exciton resonance), which is the result of the deviation of electronic structure between 2D crystal and the sharp edge [46]. Furthermore, they calculated $\chi^{(2)}$ of monolayer MoS_2 and nanoribbon with Mo-zigzag edges with DFT. The monolayer shows no features below the band gap (~1.9 eV). Meanwhile, the zigzag nanoribbon shows a two-photon resonance at ~0.8 eV corresponding to the sub-band transitions from valence bands to localized edge states of Mo-zigzag edges.

Moreover, S-zigzag edge and S–Mo Klein edge terminations produce different SHG response [47]. The fundamental wavelength dependent SHG indicates that S-zigzag edge show highest SHG under the 1375 nm excitation, while S–Mo Klein edge has a resonance peak at excitation wavelength of 1300 nm. DFT calculation explains that the S-zigzag edges has a broad peak at 0.8–0.9 eV for the calculated density of states. In the contrast, the S–Mo Klein edge show peaks at 1.2–1.3 eV in the calculated density of states. The local electronic states can be engineered through change of a few atoms at 1D edge. Consequently, NLO of 2D materials can be greatly manipulated via edge states. The edge sensitive SHG provides an effective and convenient approach to detect and probe discontinuities and boundaries of 2D materials in an all-optical manner.

Normally, NLO of 2D materials can be engineered via defects states similar to the edge states. For instance, TPA of monolayer MoS_2 and WS_2 can be enhanced by the reducing detects via chemical treatment [48, 49]. TPA and two-photon luminescence (TPL) are significantly enhanced after the repair of sulfur vacancies of monolayer MoS_2 and WS_2, which is attributed to reduced nonradiative relaxation path after repair of the sulfur vacancies. SA can also be enhanced by reducing the detects density of the TMDs film [50]. In addition, $\chi^{(2)}$ of TMDs can be engineered by modifying the stoichiometry by changing the concentration of Se in $Mo_xSe_yS_z$ alloys [51].

3.3.2 Optical Modulation

3.3.2.1 Excitonic Effects Enhanced Nonlinearity

Initially, harmonics generation is described by electric-dipole approximation with band theory or first-principle calculations, which excludes the exciton contribution. Nevertheless, experiments reveal that harmonic generation output depends on the excitation wavelength and highest enhancement can be observed when the harmonic wavelength is in resonance with exciton. Excitonic effects have to be considered in the nonlinear interaction of coherent light with semiconductors beyond electric dipole approximation. Excitonic effects enhanced optical nonlinearity are wildly observed in conventional semiconductors such as ZnTe, ZnSe, ZnS, ZnO, and GaAs [52–55]. Experimentally, resonant SHG is observed for 1*s* (A,B), 2*s* (A,B), 2*p* (A,B), and 1*s* (C) excitons in the photon energy range of 3.2–3.5 eV [54]. Theory of excitonic effects modulated NLO is firstly developed in a wide bandgap semiconductor ZnO. SHG can originate from the electric and magnetic field-induced perturbations of excitons due to spin, Stark effect as well as Zeeman effects. This theory is also applicable for other semiconductors to explore the excitonic effects on optical nonlinearity.

For a resonant SHG process with exciton energy $E_{exc} = 2\hbar\omega$ that involves the ground state $|g\rangle$, immediate state $|i\rangle$, and the final exciton state $|e\rangle$, the second-order nonlinear susceptibility can be written as [54]

$$\chi_{ijl}(E_{exc}, \mathbf{k}_{exc}, \mathbf{B}, \mathbf{E}) \propto \sum_{v} \frac{\left\langle g \mid \hat{V}_i^{2\omega} \mid e \right\rangle \left\langle e \mid \hat{V}_j^{2\omega} \mid i \right\rangle \left\langle i \mid \hat{V}_l^{2\omega} \mid g \right\rangle}{(E_{exc} - 2\hbar\omega - i\Gamma_{exc})(E_i - \hbar\omega)} \approx \frac{i}{\Gamma_{exc}} \left\langle g \mid \hat{V}_i^{2\omega} \mid e \right\rangle M_{exc,g}^{2ph} \tag{3.10}$$

Here Γ_{exc} is exciton damping constant. $M_{exc,g}^{2ph}$ describes the summarization of all the possible immediate states satisfying the selection rules for two-photon transition from $|g\rangle$ to $|e\rangle$. **B** and **E** indicate that nonlinear susceptibility can be modified by external magnetic and electric fields. From Eq. (3.10), we can see a nonlinear susceptibility becomes much larger when the SHG output is approaching the exciton energy $2\hbar\omega$, leading to excitonic effects enhanced NLO.

To go beyond the electric-dipole approximation, exciton can be expended as the combination of electric dipole (ED), electric quadrupole (EQ), magnetic dipole (MD), and so on [56]

$$\hat{V}_i^{\omega} = \hat{V}_i^{ED}(\omega) + \hat{V}_i^{EQ}(\omega) + \hat{V}_i^{MD}(\omega) + H.c. \tag{3.11}$$

where $\mathbf{e}_\perp$ represents the in-plane polarization vector of incidence light and $q_z = q = \omega/c$.

$$\hat{V}_\perp^{\mathrm{ED}}(\omega) = -i\frac{\varepsilon_0}{2m_0c}\mathrm{e}^{i\omega t}\mathbf{e}_\perp \cdot \hat{\mathbf{p}}_\perp$$
$$\hat{V}_\perp^{\mathrm{EQ}}(\omega) = -\frac{q_z\varepsilon_0}{2m_0c}\mathrm{e}^{i\omega t}[(\mathbf{e}_z \cdot \hat{\mathbf{r}})(\mathbf{e}_\perp \cdot \hat{\mathbf{p}}_\perp) + (\mathbf{e}_\perp \cdot \hat{\mathbf{r}}_\perp)(\mathbf{e}_z \cdot \hat{\mathbf{p}}_z)]$$
$$\hat{V}_\perp^{\mathrm{MD}}(\omega) = -\frac{q_z\varepsilon_0}{2m_0c}\mathrm{e}^{i\omega t}\mathbf{e}_z \cdot [\mathbf{e}_\perp \times (\hat{L} + 2\hat{S})] \tag{3.12}$$

Equation (3.10) indicates that NLO can be tuned by applying external magnetic and electric fields because magnetic and electric fields can modify the excitonic states via Zeeman and Stark effects, respectively. Equation (3.10) also indicates the susceptibility is inversely proportional to the exciton damping constant Γ_{exc}. The exciton damping constant Γ_{exc} depends on temperature, which is attributed to the inhomogeneous and homogeneous broadening of exciton. Thus, temperature change can manipulate optical nonlinearity in addition to external magnetic and electric fields. For temperature exceeding 20 K, exciton linewidth is primarily determined by homogeneous broadening originating from acoustic phonon scattering. Considering exciton linewidth broaden linearly with temperature, SHG intensity depends inversely quadratically on temperature with $I^{2\omega} \propto \Gamma_{\mathrm{exc}}^{-2} \propto T^{-2}$. However, non-resonant SHG is independent on temperature when excitonic effects are absent.

Apart from the general description, the excitonic effects enhanced SHG is specifically investigated by Troll et al. [57] The sheet susceptibility tensor of monolayer MoS_2 under excitation frequency ω can be expressed as [57]

$$\chi_{abc}^{(2)}(\omega) = -\frac{ie^3}{\varepsilon_0\hbar^2 m_{\mathrm{e}}^3\Omega}\sum_{i,j}\frac{1}{\omega_i\omega_{ij}\omega_j}\left\{\frac{P_{0ij}}{[\omega_i - 2\omega][\omega_j - 2\omega]} - \frac{P_{ji0}}{[\omega_i - \omega][\omega_j + \omega]} + \frac{P_{j0i}}{[\omega_i + \omega][\omega_j + 2\omega]}\right\} \tag{3.13}$$

where $P_{mnk} = P_{mn}^a\left(P_{nk}^b P_{km}^c + P_{nk}^c P_{km}^b\right)/2, P_{mn}^a$ is the Cartesian a component of momentum matrix element between exciton states m and n, 0 denotes the ground states. $\omega_{ij} = \omega_i - \omega_j$ is the energy difference between two exciton states. Ω is the crystal area. By considering $\chi^{(2)} = \chi_{xxx}^{(2)} = -\chi_{yyx}^{(2)} = -\chi_{xyy}^{(2)} = -\chi_{yxy}^{(2)}$, the first term of Eq. (3.13) can be rewritten as $\chi_A^{(2)}$

$$\begin{aligned}\chi_{\mathrm{A}}^{(2)}(\omega) &= \frac{e^3}{\varepsilon_0\hbar^2 m_{\mathrm{e}}^3\Omega}\sum_{i,j}\left\langle P \mid \frac{|i\rangle\langle i|}{[\omega_i - 2\omega]\omega_i}\hat{X}\frac{|j\rangle\langle j|}{[\omega_i - \omega]\omega_j} \mid P\right\rangle \\ &= \frac{e^3}{2\varepsilon_0\omega^2 m_{\mathrm{e}}^3\Omega}\langle P \mid [G(2\omega) - G(0)] \times \hat{X}[G(\omega) - G(0)] \mid P\rangle \\ &= \frac{e^3}{2\varepsilon_0\omega^2 m_{\mathrm{e}}^3\Omega}\sum_{i,j}[G_{0i}(2\omega) - G_{0i}(0)]X_{ij} \times G_{j0}(\omega) - G_{j0}(0)]\end{aligned} \tag{3.14}$$

Here, $G(\omega)$ is the Green's function expressed as $G(\omega) = \sum_i |i\rangle\langle i|[\hbar\omega - E_i]$, $G_{0i}(\omega) = \langle q_0 \mid G(\omega) \mid q_i\rangle$ is the off-diagonal resolvent matrix element in the Lanczos basis, and $X_{ij} = \langle q_i \mid \hat{X} \mid q_j \rangle$ is x component of position matrix element in the Lanczos basis. By adding all the three terms, Eq. (3.13) can be rewritten to Green's function

$$\chi^{(2)}(\omega) = \frac{e^3}{\varepsilon_0 \hbar^2 m_e^3 \Omega} \{F_{0i}(2\omega) X_{ij} F_{j0}(\omega) + 2 [F_{0i}(-\omega) X_{ij} F_{j0}(\omega) - F_{0i}(-\omega) X_{ij} F_{j0}(-\omega) \tag{3.15}$$

where $F_{ij}(\omega) = G_{ij}(\omega) - G_{ij}(0)$. In the one-electron limit and within a complete basis, the SHG response reduces to

$$\chi^{(2)''}_{abc}(\omega) = \frac{e^3}{2\pi\varepsilon_0 \hbar^2 m_e^3 \omega^3} \sum_{c,v,l} \int \left[\frac{P_{vcl}}{\omega - \omega_{lv}} \delta(2\omega - \omega_{cv}) + \left(\frac{P_{vlc}}{\omega + \omega_{cl}} + \frac{P_{clv}}{\omega + \omega_{lv}} \right) \delta(\omega - \omega_{cv}) \right] \mathrm{d}^2 k \tag{3.16}$$

where $P_{ijl} = P^a_{ij} \left(P^b_{jl} P^c_{li} + P^c_{jl} P^b_{li} \right) /2$ and band index l runs over all bands with the restriction $l \neq (c, v)$. The first term of Eq. (3.16) corresponds to the process in which exciton frequency is resonant with SHG photon frequency, leading to enhanced SHG in monolayer MoS_2.

Excitonic effects enhanced NLO are widely reported both in experiment and calculation for various TMDs. SHG enhancement up to 3 orders of magnitude is recorded in monolayer WSe_2 when SHG output is in resonance with 1*s* state of A-exciton, which involves two electric dipole transitions and one magnetic dipole transition [56]. The electric quadrupole contribution vanishes in monolayer WSe_2 due to symmetry. In addition, resonant SHG peaks also can be observed for 2*s*/2*p*, 3*s*/3*p* states of A-exciton and 1*s*, 2*s*/2p states of B-exciton of monolayer WSe_2. Importantly, some exitonic states that are partly forbidden for one- or two-photon absorption can be probed by resonant SHG processes. Excitonic resonance of SHG are observed simultaneously for A-, B-, and C-exciton in monolayer MoS_2, and the C-exciton resonance leads to the highest enhancement [58]. Furthermore, both SHG and FWM exhibit excitonic resonance in few-layer MoS_2 [59]. Excitonic effects can only enhanced the NLO in several specific wavelength, which constrains the application in broad range SHG. To address this problem, elements doping can be utilized to tune the exciton energy in a broad range and thus broaden the wavelength range of excitonic effects enhanced SHG. For instance, $MoS_{2(1-x)}Se_{2x}$ alloy can be prepared via chemical vapor deposition and resonant SHG wavelength can be tuned from 600 to 800 nm if x increases from 0 to 1 [60].

Besides TMDs, other materials should be explored with broad exciton wavelength range and tunable optical nonlinearity. Recently, NLO of 2D perovskite has attracted numerous attention. Various nonlinear optical processes have been demonstrated in 2D perovskites such as SHG, THG, TPA, and SA. 2D perovskite has a unique quantum well structure with alternating inorganic and organic layers. The strong quantum confinement and dielectric confinement effects lead to strong excitonic effect at room temperature for 2D perovskites. In addition, the exciton wavelength and exciton binding energy can be tuned by modifying organic species, inorganic species and thickness of inorganic. Furthermore, the structural symmetry of 2D perovskites can be manipulated by using different organic cations such as chiral molecules. In short, the strong and highly tunable excitonic effect renders 2D perovskites promising in the applications of excitonic effects enhanced NLO. The excitonic effects enhanced THG is firstly demonstrated in 2D perovskite $(C_{10}H_{21}NH_3)_2PbI_4$, in which three-photon resonance is present at the fundamental wavelength of 1530 nm [61]. Recently, significant exciton effect enhanced THG is reported in 2D perovskites $(BA)_2(MA)_{n-1}Pb_nI_{3n+1}$ (n = 1–3) nanoflakes with tunable inorganic layer thickness. The maximum conversion efficiency is 0.006% for $n = 2$ perovskite and the effective third-order susceptibility $\chi^{(3)}$ is estimated as 1.12×10^{-17} m^{2}V^{-2} [62]. Excitonic effects enhanced TPA process is also demonstrated in 2D perovskites $(C_4H_9NH_3)_2(CH_3NH_3)_{n-1}Pb_nI_{3n+1}$ ($n = 1$–4) experimentally and theoretically [63]. The $n = 4$ perovskite exhibits largest TPA coefficient of 2.5 cm MW^{-1} due to stronger resonance effect under 1030 nm fundamental excitation.

3.3.2.2 Valley-Dependent Nonlinearity

Monolayer TMDs exhibit unique valley freedom due to strong spin–orbit coupling, broken inversion symmetry and presence of time reversal symmetry. There exists two degenerate but inequal valleys at the corner of Brillouin edge, namely, +K and −K valley. The freedom of valley degree refers that valley and spin of electrons are locked. Consequently, +K and −K valley specifically absorb right-hand (σ^+) and left-hand (σ^-) circularly polarized photons, respectively. The valley freedom originates from the non-vanished Berry curvature and opposite magnetic moment in different valleys. Berry curvature in n-th energy band can be written as [64]

$$\boldsymbol{\Omega}_n(\mathbf{k}) = i\frac{\hbar^2}{m^2}\sum_{i\neq n}\frac{\mathbf{P}_{n,i}(\mathbf{k})\times\mathbf{P}_{i,n}(\mathbf{k})}{\left[E_n^0(\mathbf{k})-E_i^0(\mathbf{k})\right]^2} \tag{3.17}$$

where $E_n^0(\mathbf{k})$ is the energy of n-th band, $\mathbf{P}_{n,i}(\mathbf{k})$ is the matrix element of velocity operator. $\boldsymbol{\Omega}_n(\mathbf{k}) = -\boldsymbol{\Omega}_n(-\mathbf{k})$ should be satisfied for time reversal symmetry, while $\boldsymbol{\Omega}_n(\mathbf{k}) = \boldsymbol{\Omega}_n(-\mathbf{k})$ is not satisfied for the broken inversion symmetry, which leads to the valley contrasting phenomena. Due to the presence of orbit magnetic moment, the energy of n-th band is modified to

$$E_n(\mathbf{k}) = E_n^0(\mathbf{k}) - \mathbf{m}_n(\mathbf{k})\cdot\mathbf{B} \tag{3.18}$$

where $\mathbf{m}$ is the orbital magnetic moment expressed by

$$m_n(\mathbf{k}) = i\frac{\hbar^2}{2m^2}\sum_{i\neq n}\frac{\mathbf{P}_{n,i}(\mathbf{k})\times\mathbf{P}_{i,n}(\mathbf{k})}{E_n^0(\mathbf{k})-E_i^0(\mathbf{k})} \tag{3.19}$$

For monolayer TMDs lacking inversion symmetry, Berry curvature and orbit magnetic moment are opposite between +K and −K valley. The C_3 symmetry makes $\mathbf{k}\cdot\mathbf{p}$ Hamiltonian at +K and −K valleys has following form [65]

$$\hat{H} = at(\tau_z k_x\sigma_x + k_y\sigma_y) + \frac{\Delta}{2}\sigma_z \tag{3.20}$$

where σ is the Pauli matrix element, a is the lattice spacing, t is the nearest-neighbor hopping integral, Δ is the bandgap, and $\tau_z = \pm1$ is the valley index for ±K valley. Berry curvature at conduction band is given by

$$\boldsymbol{\Omega}_c(k) = -\hat{z}\frac{2a^2t^2\Delta}{(4a^2t^2k^2+\Delta^2)^{3/2}}\tau_z \tag{3.21}$$

Thus, Berry curvature has opposite signs at two different ±K valley. Berry curvature at the valence band has same value but opposite sign with conduction band. The orbit magnetic moments are identical in conduction and valence bands:

$$\mathbf{m}(k) = -\hat{z}\frac{2a^2t^2\Delta}{4a^2t^2k^2 + \Delta^2}\frac{e}{2\hbar}\tau_z \tag{3.22}$$

The opposite orbit magnetic moment and Berry curvature lead to the circularly polarized light selection rule. The relation between the circular dichroism $\eta(\mathbf{k})$, orbit magnetic moment, and Berry curvature is given by

$$\eta(\mathbf{k}) = \frac{\mathbf{m}(\mathbf{k})\cdot\hat{\mathbf{z}}}{\mu_B^*(\mathbf{k})} = \frac{\Omega(k)\cdot\hat{\mathbf{z}}}{\mu_B^*(k)}\frac{e}{2\hbar}\Delta(k) \tag{3.23}$$

where $\mu_B^* = e\hbar/2m^*$, $\Delta(k) = (4a^2t^2k^2 + \Delta^2)^{1/2}$ is bandgap at $\boldsymbol{k}$ point. At the energy minima of +K and −K valley, having $\eta(\mathbf{k}) = -\tau_z$. This indicates that the transition at +K can only couple to σ^+ light, while transition at −K can only couple to σ^- light, which is the circularly polarized light selection rule at ±K valley.

Similar to the one-photon transition, two-photon transition such as TPA and SHG should also follow the valley index dependent selection rule [66]. Two-photon transitions between valence band and conduction band are described by the matrix elements $\langle\psi_c \mid \hat{P}_\pm \mid \psi_j\rangle\langle\psi_j \mid \hat{P}_\pm \mid \psi_v\rangle$, where $\hat{P}$ is moment operator with $\hat{P}_\pm = \hat{P}_x + i\hat{P}_y$, $\langle\psi_c \mid \hat{P}_\pm \mid \psi_j\rangle\langle\psi_j \mid \hat{P}_\pm \mid \psi_v\rangle$ are Bloch states at ±K valley for conduction band minima, valence band maximum, and virtual intermediate state. Because monolayer TMDs has C_3 symmetry, the Bloch states transfers as $\hat{C}_3|\psi_j\rangle = e^{-i\frac{2\pi}{3}l_\alpha}|\psi_j\rangle$ under a $2\pi/3$ rotation along z-axis, where $\alpha = c, v, j$ and l_α is an integer. For rotation about the transition metal atoms, (W or Mo), l_α is equal to the magnetic quantum number of the local atomic orbits. By using $\hat{C}_3\hat{P}_\pm\hat{C}_3^{-1} = e^{i\frac{2\pi}{3}}\hat{P}_\pm$, two-photon transition matrix elements can be expressed as [66]

$$\begin{aligned}\langle\psi_c \mid \hat{P}_\pm \mid \psi_j\rangle\langle\psi_j \mid \hat{P}_\pm \mid \psi_v\rangle &\equiv \left\langle\psi_c \mid \hat{C}_3^{-1}\hat{C}_3\hat{P}_\pm\hat{C}_3^{-1}\hat{C}_3 \mid \psi_j\right\rangle\left\langle\psi_j \mid \hat{C}_3^{-1}\hat{C}_3\hat{P}_\pm\hat{C}_3^{-1}\hat{C}_3 \mid \psi_v\right\rangle \\ &= \left\langle\hat{C}_3\psi_c \mid \hat{C}_3\hat{P}_\pm\hat{C}_3^{-1} \mid \hat{C}_3\psi_j\right\rangle\left\langle\hat{C}_3\psi_j \mid \hat{C}_3\hat{P}_\pm\hat{C}_3^{-1} \mid \hat{C}_3\psi_v\right\rangle \\ &= e^{i\frac{2\pi}{3}(m_c - m_j \mp 1)}\langle\psi_c \mid \hat{P}_\pm \mid \psi_j\rangle e^{i\frac{2\pi}{3}(m_j - m_v \mp 1)}\langle\psi_j \mid \hat{P}_\pm \mid \psi_v\rangle\end{aligned} \tag{3.24}$$

For two-photon transition with circularly polarized photon σ^+ or σ^-, $e^{i\frac{2\pi}{3}(m_c - m_v \mp 2)} = 0$, namely, $m_c - m_v \mp 2 = 3n$ (n is an integer). At +K (−K) valley, $m_c = 0$ is always equal to 0 and $m_v = +2(-2)$, which allows the transition of two $\sigma^-(\sigma^+)$ photons. For SHG process at +K (−K), absorption of two σ^- (σ^+) photons results in harmonic output of one σ^+ (σ^-) photon with twice energy, which is the valley index dependent SHG of monolayer TMDs (Figure 3.3a).

The valley-dependent SHG is demonstrated in monolayer WSe_2 and WS_2, exhibiting unity valley polarized SHG [66, 67]. The unity valley-polarized SHG can be maintained in AA-staked bilayer, AAA-staked trilayer, and spiral TMDs [68]. This is of great importance in the application of nonlinear optical valley switch. In addition, the valley polarized TPL is demonstrated in monolayer WS_2, in which the 1s and 2p excitons have different valley selection rule [26]. Emission of 1s exciton has opposite helicity with the incidence, while 2p exciton emission always has the same helicity with the incidence (Figure 3.3b). The valley-dependent TPL is useful in determining the fine structure of monolayer TMDs. Recently, valley-manipulated FWM and SFG

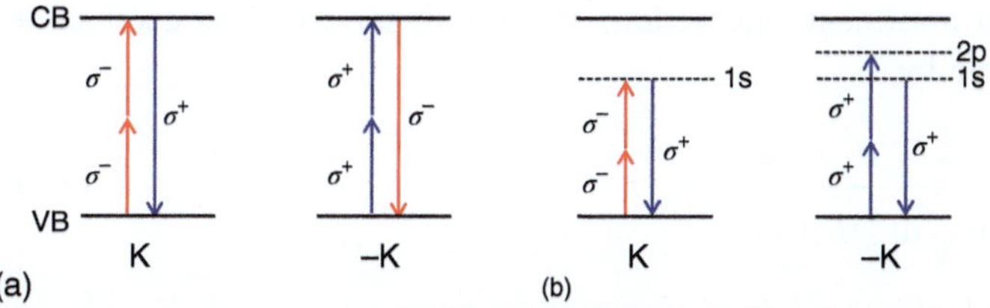

Figure 3.3 (a) Valley index dependent SHG and (b) valley index dependent TPA.

are demonstrated in monolayer MoS_2 [69]. The chiral selection rules of NLO can be utilized to construct ultrafast all-optical chirality logic gates (XNOR, NOR, AND, XOR, OR, and NAND).

3.3.2.3 Optical Cavity-Enhanced Nonlinearity

Although 2D materials exhibit a larger nonlinear susceptibility than conventional bulk nonlinear crystals, the atomic thin feature renders ultrashort light–matter interaction. In order to increase the efficiency of the nonlinear process, an optical cavity is normally used to enhance the NLO of 2D materials. Electromagnetic filed can be confined within a small volume in a resonant optical cavity, which is used to magnify the incidence light or the harmonic output. Apart from the confined light in an optical cavity, the 2D material-waveguide coupled structure can be used to enlarge the light–matter interaction length to enhance the efficiency of nonlinear process of 2D materials.

For 2D materials in a resonant optical cavity, SHG of 2D materials can be expressed as [70]

$$I_{SHG} \propto f^4(\omega)f^2(2\omega)I_{in} \tag{3.25}$$

where I_{in} is intensity of incidence light, $f(\omega)$ and $f(2\omega)$ are the electromagnetic field enhancement factor at fundamental incidence wavelength and SHG wavelength respectively. From Eq. (3.25), we can see SHG of materials coupled to a cavity is fourth power of the bare 2D materials. A resonant optical cavity can be designed to be in resonance with the incidence and SHG wavelength simultaneously to optimize the SHG enhancement, namely, achieving the double resonance enhancement. For a double resonance enhanced SHG, the temporal coupled mode theory [71–73] can be used to describe the SHG under a pair of resonant modes as derived in reference [74]

$$\begin{aligned} \frac{dA}{dt} &= (i\omega_1 - \gamma_1)\cdot A - g_1 \cdot F - i\omega_1\beta_1 A^* B \\ \frac{dB}{dt} &= (i\omega_2 - \gamma_2)\cdot B - i\omega_2\beta_2 A^2 \end{aligned} \tag{3.26}$$

where A and B are amplitude of fundament wavelength (FW) and SHG wavelength (SHW) cavity mode, respectively; F is the amplitude of incidence FW light; ω and γ denote the angular frequency and dissipation rate of cavity mode; and g is the coupling rate between cavity mode and free space and $|g_1|^2 = 2\gamma_1^{rad}$. The last terms in the Eq. (3.26) describe the energy transfer rates between FW and SHW cavity modes due to optical nonlinear interaction. The nonlinear coupling coefficients $\beta_1 = 2\beta_2^*$ and β_2 is given by

$$\beta_2 = \frac{1}{4}\frac{\int_{\text{TMD}} \varepsilon_0 \chi_{xyy}^{(2)} E_{2x}^* E_{1y} E_{1y} \mathrm{d}r}{\int \frac{\varepsilon}{2}|E_1|^2 \sqrt{\mathrm{d}r \frac{\varepsilon}{2}|E_2|^2 \mathrm{d}r}} \tag{3.27}$$

Considering the undepleted pump approximation, having

$$B = \frac{\omega_2 \beta_2}{2\omega - \omega_2 - i\gamma_2} \cdot \frac{g_1^2 F^2}{(\omega - \omega_1 - i\gamma_1)^2} \tag{3.28}$$

Then, the power of SHG output can be derived as

$$P_{\text{SHG}} = 2\gamma_2^{\text{rad}} |B|^2 = \frac{8\gamma_2^{\text{rad}} {\omega_2}^2 |\beta_2|^2}{|2\omega - \omega_2 - i\gamma_2|^2} \cdot \frac{\left(\gamma_1^{\text{rad}}\right)^2 F^2}{|\omega - \omega_1 - i\gamma_1|^4} \tag{3.29}$$

Finally, the enhancement factor of SHG is

$$\begin{aligned} EF_{\text{SHG}} &= \frac{P_{\text{SHG}}}{P_{\text{FW}}} = \frac{P_{\text{SHG}}}{|F|^2} = \frac{8\gamma_2^{\text{rad}} {\omega_2}^2 |\beta_2|^2}{|2\omega - \omega_2 - i\gamma_2|^2} \cdot \frac{\left(\gamma_1^{\text{rad}}\right)^2}{|\omega - \omega_1 - i\gamma_1|^4} \\ &= \frac{64 \left|\chi_{xyy}^{(2)}\right|^2 Q_2 {Q_1}^2 |F|^2}{\varepsilon_0 \lambda^3 \omega} \cdot \frac{\gamma_2^{\text{rad}}}{\gamma_2} \cdot \left(\frac{\gamma_1^{\text{rad}}}{\gamma_1}\right)^2 \cdot Z_{2\omega,\omega}^{\text{TMD}} \end{aligned} \tag{3.30}$$

where Q is the quality factor of cavity modes and $Z_{2\omega,\omega}^{TMD}$ is the overlapping coefficient of paired resonant modes within the 2D materials regime given by

$$Z_{2\omega,\omega}^{\text{TMD}} = \left| \sqrt{{\lambda_\omega}^3} \frac{\int_{\text{TMD}} E_{2x}^* E_{1y} E_{1y} \mathrm{d}r}{\int \varepsilon_1 |E_1|^2 \sqrt{\mathrm{d}r \varepsilon_2 |E_2|^2 \mathrm{d}r}} \right|^2 \tag{3.31}$$

Equation (3.30) indicates that SHG enhancement is proportional to the quality factor of FW and SHW modes (Q_1, Q_2), the ratio of radiative rate to the total dissipation rate $(\gamma_1^{\text{rad}}/\gamma_1, \gamma_2^{\text{rad}}/\gamma_2)$ and overlapping coefficient between FW and SHW modes $\left(Z_{2\omega,\omega}^{\text{TMD}}\right)$.

Apart from the optical cavity, 2D materials also can be integrated with waveguides or optical fibers to enlarge the nonlinear coefficient. The conversion efficiency of 2D materials enhanced by the waveguide can be given by [75]

$$\eta = \frac{P_{\text{SHG}}}{{P_{\text{FW}}}^2} = \xi_{\text{NL}}^2 L^2 \frac{\sin^2(\Delta\beta L/2)}{(\Delta\beta L/2)^2} \tag{3.32}$$

were L is the light-mater interaction length, ξ_{NL} is the nonlinear overlap factor, and $\Delta\beta = 2\beta^{\text{FW}} - \beta^{\text{SHG}}$ is the mismatch of wave vector between FW and SHG. The nonlinear overlap factor ξ_{NL} is define by

$$\xi_{\text{NL}} = \chi_{\text{TMD}} \sin(3\theta) \left(\frac{8\pi^2}{\varepsilon_0 c \lambda_{\text{FW}}^2 n_{\text{SH}} n_{\text{FW}}^2}\right)^{1/2} \frac{\left(E_{\text{SH}}^* E_{\text{FW}}^2\right)\big|_{z=z(\text{TMD})}}{\left(\int_L |E_{\text{FF}}|^2 \,\mathrm{d}z\right) \left(\int_L |E_{\text{SH}}|^2 \,\mathrm{d}z\right)^{1/2}} \tag{3.33}$$

where n_{FW} and n_{SH} are the effective refractive index of FW and SHG modes, z(TMD) means the position of TMD materials, and θ is the angle between guided-mode wave vector and armchair direction of TMDs crystal. From Eq. (3.32), we can see that nonlinear conversion efficiency can be enhanced dramatically by increase light–matter interaction length as demonstrated in $MoSe_2$/Si waveguide and the MoS_2/optical fiber integrated system [75, 76].

Various optical cavities are used to be coupled with 2D materials to enhance their optical nonlinearity, such as metal film, plasmonic array, metamaterials metallic gratings, photonic crystals, Fabry–Pérot (F–P) cavity, distributed Bragg reflector (DBR) cavity, and so on. For instance, 300-fold SHG enhancement is achieved in nanoparticle on mirror (NPoM) system when the surface plasmon resonance of NPoM matches SHG wavelength [77]. This large SHG enhancement is ascribed to interaction between nonlinear dipole of monolayer WS_2 and quadrupole of Ag nanocavity. The single cavity-enhanced SHG enables a large SHG output in a localized position on a nanometer scale. The single nanoparticle also can strongly interact with the monolayer WSe_2 to form a exciton–polariton and thus enhance SHG [78]. A 200-fold SHG enhancement is also realized in a monolayer WSe_2 placed on silicon on insulator (SOI) photonic crystal with quality factor of 2000–3000 [79]. A maximum SHG enhancement of 1600 times is achieved by tuning the incidence wavelength to be resonant with a F–P microcavity mode [80]. This giant SHG enhancement is attributed to the combination of electromagnetic field enhancement in F–P cavity and excitonic resonance effects. The suspended monolayer WS_2 exhibit much stronger exciton oscillation strength, leading to a stronger exciton-enhanced SHG effect. In addition, directional SHG emission with divergence angle of 5° is realized in the presence of the F–P cavity. Exceptionally strong THG of graphene can be realized in the graphene–insulator–metal heterostructures with the enhancement factor of 10^3 compared with bare graphene [81]. The graphene sheet is placed on a gold grating with an insulating layer (Al_2O_3 or h-BN) as spacer. The giant THG enhancement originates from the electromagnetic field confined in the narrow slits between two gold nanoribbons. A maximum THG conversion efficiency of $2\times10^{-7}\%$ at 1.9 mW excitation can be obtained.

3.3.2.4 All-Optical Modulation

NLO of 2D materials can be manipulated in an all-optical manner, in which nonlinearity can be actively controlled by an ultrafast pump laser. The all-optical manipulation can give rise to the ultrafast modulation with high modulation depth. The all-optical modulation is achieved by using two spatially overlap light beam with time delay $\Delta\tau$. Both SHG and THG can be manipulated with an ultrafast pulse laser and the mechanism should be discussed separately.

For SHG of monolayer TMDs, there is only one non-vanishing second-order nonlinear optical susceptibility of $\chi^{(2)} \equiv \chi^{(2)}_{xxx} = -\chi^{(2)}_{yyx} = -\chi^{(2)}_{xyy} = -\chi^{(2)}_{yxy}$ because of the D_{3h} symmetry. For the hexagonal lattice of monolayer TMDs, we can set the Cartesian coordinates x and y are along the armchair (AC) and zigzag (ZZ) directions of the lattice, respectively. In this case, SH intensity of any fundamental incidence along

AC and ZZ direction can be derived as [15]

$$I_x^{2\omega} \propto \left|E_x^2 - E_y^2\right|^2,$$
$$I_y^{2\omega} \propto \left|2E_xE_y\right|^2 \tag{3.34}$$

where E_x and E_y are the fundamental incident fields along AC (x) and ZZ (y) direction respectively. From Eq. (3.34) we can see only SH signal along AC direction will be generated if two fundamental incident fields E_x and E_y are not overlapping in time. In contrast, SH signal along ZZ direction will be observed at zero delay. Thus, 90° polarization switch can be achieved by controlling the delay time between two incident light beams. For instance, all-optical SHG modulation is realized in monolayer MoS_2 with a modulation depth of 100% and the speed is limited by the fundamental pulse duration [82].

For all-optical modulated THG in monolayer TMDs, the scenario is more complicated and the analysis can be found in reference [83]. Considering D_{3h} crystal symmetry of monolayer TMDs, third-order susceptibility $\chi^{(3)}$ can be expressed as [83]

$$\chi^{(3)} = \begin{bmatrix} \chi_{xxxx} & 0 & 0 & 0 & 0 & \chi_{xxzz} & 0 & \chi_{xxxx}/3 & 0 & 0 \\ 0 & \chi_{xxxx} & 0 & \chi_{xxzz} & 0 & 0 & 0 & 0 & \chi_{xxxx}/3 & 0 \\ 0 & 0 & \chi_{zzzz} & 0 & \chi_{zyyz} & 0 & \chi_{zyyz} & 0 & 0 & 0 \end{bmatrix} \tag{3.35}$$

Then, the third-order nonlinear polarization can be written as

$$\begin{bmatrix} P_x(3\omega) \\ P_y(3\omega) \\ P_z(3\omega) \end{bmatrix} = \varepsilon_0 \chi^{(3)} \begin{bmatrix} E_x^3 \\ E_y^3 \\ E_z^3 \\ 3E_yE_zE_z \\ 3E_yE_yE_z \\ 3E_xE_zE_z \\ 3E_xE_xE_z \\ 3E_xE_yE_y \\ 3E_xE_xE_y \\ 3E_xE_yE_z \end{bmatrix} \tag{3.36}$$

where E_x, E_y, and E_z are the incident electric field along x, y and z directions. Assuming the pump beam is in-plane with $E_z = 0$, then

$$\begin{bmatrix} P_x(3\omega) \\ P_y(3\omega) \end{bmatrix} = \varepsilon_0 \chi_{xxxx} \begin{bmatrix} E_x^3 + E_xE_yE_y \\ E_y^3 + E_xE_xE_y \end{bmatrix} \tag{3.37}$$

If the two incident beams are both circularly polarized light (σ^- and σ^+) with power ratio of m, we have $\mathbf{E}_{in} = \mathbf{E}_{\sigma^+} + m\mathbf{E}_{\sigma^-} = (1+m)\mathbf{E}_x + i(1-m)\mathbf{E}_y$. Eq. (3.37) can be derived to

$$\begin{bmatrix} P_x(3\omega) \\ P_y(3\omega) \end{bmatrix} \propto \varepsilon_0 \chi_{xxxx} m \begin{bmatrix} (1+m) \\ i(1-m) \end{bmatrix} \tag{3.38}$$

Thus, the THG intensity $I_{3\omega} \propto |P(3\omega)|^2 \propto \varepsilon_0^2 \chi_{xxxx}^2 m^2(1+m^2)$, the ellipticity of the THG response is $\sigma_{\mathrm{THG}} = \tan\left(\frac{1}{2}\sin^{-1}\frac{1-m^2}{1-m^2}\right)$.

If the two incident beams are linearly polarized and circularly polarized with the power of linearly polarized fixed, we have $\mathbf{E}_{in} = \mathbf{E}_x + m\mathbf{E}_{\sigma^+} = (1+m)\mathbf{E}_x + im\mathbf{E}_y$. Then

$$\begin{bmatrix} P_x(3\omega) \\ P_y(3\omega) \end{bmatrix} \propto \varepsilon_0 \chi_{xxxx}(1+2m) \begin{bmatrix} 1+m \\ im \end{bmatrix} \quad (3.39)$$

In this case, $I_{3\omega} \propto |P(3\omega)|^2 \propto \varepsilon_0^2 \chi_{xxxx}^2 (1+2m)^2(1+2m+2m^2)$, the ellipticity of the THG response is $\sigma_{\text{THG}} = \tan\left\{\frac{1}{2}\sin^{-1}\left[\frac{2m(1+m)}{1+2m+2m^2}\right]\right\}$.

If the two incident beams are linearly polarized and circularly polarized with the power of circularly polarized fixed, we have $\mathbf{E}_{in} = m\mathbf{E}_x + \mathbf{E}_{\sigma^+} = (1+m)\mathbf{E}_x + i\mathbf{E}_y$. Then,

$$\begin{bmatrix} P_x(3\omega) \\ P_y(3\omega) \end{bmatrix} \propto \varepsilon_0 \chi_{xxxx}(2m+m^2) \begin{bmatrix} 1+m \\ i \end{bmatrix} \quad (3.40)$$

In this case, $I_{3\omega} \propto |P(3\omega)|^2 \propto \varepsilon_0^2 \chi_{xxxx}^2 (2m+m^2)^2(2+2m+m^2)$, the ellipticity of the THG response is $\sigma_{\text{THG}} = \tan\left\{\frac{1}{2}\sin^{-1}\left[\frac{2(1+m)}{2+2m+m^2}\right]\right\}$.

From these equations, we can see both the amplitude and polarization of the THG signal can be modulated by controlling the polarization and power of the two incident beams. For instance, coherent modulation of THG is demonstrated in monolayer MoS_2 with chiral light [69]. The modulation depth is close to 100% and the speed is around 130 fs.

In addition to D_{3h} symmetry enabled polarization and amplitude modulation of nonlinear signal, all-optical modulation of NLO in TMDs also can be realized by plasmonic hot carrier injection [84]. Plasmonic hot carrier can be generated via Landan damping with time scale of hundreds of femtoseconds. The plasmonic hot carrier can be injected from metallic structure to bilayer TMDs to break inversion symmetry and thus induce SHG of bilayer TMDs. The amount and generation rate of plasmonic hot carriers can be controlled by the wavelength and power of pump beam, which enables all-optical controlled SHG in an ultrafast manner. Apart from the parametric process, all-optical manipulated TPA is also reported in graphene with a switching energy of 35 fJ and switching speed of 260 fs [31].

3.3.3 Electrical Modulation

Although all-optical controlled NLO provides high speed and large modulation depth, the electric controlled NLO is more convenient for practical applications. NLO of both TMDs and graphene can be tuned electrically, but the mechanism is different from each other. Electrically controlled SHG of TMDs is attributed to the electric-field-induced second harmonics (EFISH) or the excitonic effects modulated SHG, while the THG tuning of graphene is due to the doping effect.

In 1962, Terhune et al. discovered that SHG could be observed in the calcite crystal with inversion symmetry under the applying of a DC electric field, which is termed as EFISH effect [85]. EFISH originates from the broken inversion symmetry at the surface where a charge spacer is built in the presence of an external electric fields. EFISH is also applicable to the centrosymmetric semiconductors such as silicon and germanium. EFISH is a third-order nonlinear optical process determined by $\chi^{(3)}$ rather than the second-order susceptibility which induces conventional SHG. Intensity of EFISH can be given by [86]

$$\begin{aligned} I_{\mathrm{EFISH}}(2\omega) &\propto \left[\chi^{(3)}E_{\mathrm{DC}} + \chi^{(2)}E_{\omega}^{2}\right]^{2} \\ &= (\chi^{(3)}E_{\mathrm{DC}})^{2} + 2\chi^{(3)}\chi^{(2)}E_{\mathrm{DC}}E_{\omega}^{2} + \left(\chi^{(2)}E_{\omega}^{2}\right)^{2} \end{aligned} \tag{3.41}$$

where E_{DC} and E_{ω} are applied external DC electric field and electric field of fundamental incidence, respectively. The third term of Eq. (3.41) is the conventional SHG, while the first term is representative of the EFISH effect. The second term describes the interaction between $\chi^{(2)}$ and $\chi^{(3)}$. The first term indicates intensity of EFISH quadratically depends on the external electric field for centrosymmetric materials with $\chi^{(2)} = 0$. However, the SHG intensity of materials without inversion symmetry depends on the electric field quasi-linearly or quadratically.

Similar to calcite or Si, the SHG of bilayer TMDs can be tuned by applying an external DC electric field. Electric-field switchable SHG of 2H stacked bilayer MoS_2 is realized under a strong external electric field ($|F| = \pm 2.6\ \mathrm{MV \cdot cm^{-1}}$) [87]. EFISH in bilayer MoS_2 also depend on the interlayer coupling in the bilayer. Moreover, electrically tuned SHG of atomically thin ReS_2 flakes is demonstrated, thanks to the distorted 1T crystal structure and interlayer charge transfer [88]. SHG can be tuned from zero to a maximized intensity for centrosymmetric ReS_2 flake and SHG can be enhanced for ReS_2 flake with intrinsic SHG. The first-principle calculations suggest the EFISH of ReS_2 flakes is attributed to the modification of in-plane second-order hyperpolarizability arising from the redistributed interlayer charge transfer. Furthermore, SHG is also induced in bilayer WSe_2 by applying a back gate field, which is ascribed to the accumulated submonolayer screening charge in the tungsten plane [89]. This charge-induced SHG is fundamentally different from the field-induced SHG.

Apart from the field- or charge-induced SHG, SHG of TMDs also can be tuned by tuning the exciton via an external electric field. Because electric field can change the exciton species (neutral or charged) or shift the exciton wavelength via electron doping, SHG can be tuned due to the excitonic effects. SHG of monolayer WSe_2 can be tuned over an order of magnitude at A-exciton resonance by changing the bias [66]. This tuned SHG is attributed to the modulation of exciton strength at exciton resonance, which can serve as a new strategy to achieve electrical modulation in addition to the EFISH.

THG of graphene is highly dependent on the chemical potential because one-photon or multiphoton resonant transitions will be modified under different chemical potential. The one-photon or multiphoton resonant transitions can be switched on or off via external electric field to enhance THG and reduce the linear absorption loss as well. For instance, a maximum THG tuning up to ~30 times is achieved with a chemical potential change of 0.74 eV under 1566 nm excitation [18]. Moreover, THG efficiency of single layer graphene can be enhanced by almost two orders of magnitude by controlling the Fermi energy via an electric field [90].

The microscopic mechanism of how chemical potential affects THG and FWM of graphene is given by Wu et al. $\chi^{(3)}$ dependence on chemical potential of graphene in the THG process ($\omega = \omega_0 + \omega_0 + \omega_0$) is expressed as [18]

$$\begin{aligned}\chi^{(3)}_{xxxx} &= 3\chi^{(3)}_{xxyy} = 3\chi^{(3)}_{xyxy} = 3\chi^{(3)}_{xyyx}\\ &= -\frac{(\hbar v_F e^2)^2}{576\pi\varepsilon_0 d_{gr}(\hbar\omega_0)^5}\left[17G\left(\frac{\hbar\omega_0}{2|\mu|}\right) - 64G\left(\frac{2\hbar\omega_0}{2|\mu|}\right) + 45G\left(\frac{3\hbar\omega_0}{2|\mu|}\right)\right]\end{aligned} \tag{3.42}$$

where μ is the chemical potential, v_F is the Fermi velocity of graphene, and $G(x) = \ln|(1+x)/(1-x)| + i\pi H(|x|-1)$ with H denoting the Heaviside function. The three terms in Eq. (3.42) describe the one-, two-, and three-photon resonant transitions as $2|\mu|$ is larger than $\hbar\omega_0$, $2\hbar\omega_0$, $3\hbar\omega_0$, respectively. The chemical potential dependence of THG in graphene can be explained as following scenarios. If $2|\mu| < \hbar\omega_0$, all three terms are nonzero but they cancel with each other, leading to small $\chi^{(3)}_{xxxx}$. If $\hbar\omega_0 < 2|\mu| < 2\hbar\omega_0$, one-photon resonant transition vanishes. Imperfect cancellation between $G(2\hbar\omega_0/2|\mu|)$ and $G(3\hbar\omega_0/2|\mu|)$ leads to a significant $\chi^{(3)}_{xxxx}$. If $2\hbar\omega_0 < 2|\mu| < 3\hbar\omega_0$, $\chi^{(3)}_{xxxx}$ increases further because only $G(3\hbar\omega_0/2|\mu|)$ exists. If $3\hbar\omega_0 < 2|\mu|$, all three terms vanish, leading to small $\chi^{(3)}_{xxxx}$.

$\chi^{(3)}$ of graphene in sum-frequency mixing (SFM) process ($\omega_t = \omega_1 + 2\omega_2$) can be expressed as [18]

$$\begin{aligned}\chi^{(3)}_{xxxx}(\omega_1 + 2\omega_2) = &-\frac{(\hbar v_F e^2)^2}{4\pi\hbar^5\varepsilon_0 d_{gr}(\omega_1 + 2\omega_2)}\left[\frac{\omega_1\left(2\omega_1^2 + 7\omega_1\omega_2 + 8\omega_2^2\right)}{12\omega_2^2(\omega_1+\omega_2)^2(\omega_1+2\omega_2)}G\left(\frac{\hbar\omega_1}{2|\mu|}\right)\right.\\ &+\frac{10\omega_1^2 + 20\omega_1\omega_2 + 4\omega_2^2}{12\omega_1^2\omega_2(\omega_1+\omega_2)^2(\omega_1+2\omega_2)}G\left(\frac{\hbar\omega_2}{2|\mu|}\right)\\ &-\frac{(\omega_1+\omega_2)^3}{3\omega_1^2\omega_2^4(\omega_1+2\omega_2)}G\left(\frac{\hbar(\omega_1+\omega_2)}{2|\mu|}\right)\\ &\left.-\frac{4\omega_2{}^3}{3\omega_1^2\omega_2^4(\omega_1+2\omega_2)}G\left(\frac{\hbar(2\omega_2)}{2|\mu|}\right)\right]\\ &+\frac{(\omega_1+2\omega_2)^2\left(2\omega_1^2 + \omega_1\omega_2 + 2\omega_2^2\right)}{12\omega_1^2\omega_2^4(\omega_1+\omega_2)^2}G\left(\frac{\hbar(\omega_1+2\omega_2)}{2|\mu|}\right)\end{aligned} \tag{3.43}$$

$\chi^{(3)}$ of graphene in difference-frequency mixing (DFM) process ($\omega_t = -\omega_1 + 2\omega_2$) can be expressed as [18]

$$\begin{aligned}\chi^{(3)}_{xxxx}(-\omega_1 + 2\omega_2) = &-\frac{(\hbar v_F e^2)^2}{4\pi\hbar^5\varepsilon_0 d_{gr}(-\omega_1 + 2\omega_2)}\left[\frac{-\omega_1\left(2\omega_1^2 - 7\omega_1\omega_2 + 8\omega_2^2\right)}{12\omega_2^2(\omega_2-\omega_1)^2(2\omega_2-\omega_1)}G\left(\frac{-\hbar\omega_1}{2|\mu|}\right)\right.\\ &+\frac{10\omega_1^2 - 20\omega_1\omega_2 + 4\omega_2^2}{12\omega_1^2\omega_2(-\omega_1+\omega_2)^2(-\omega_1+2\omega_2)}G\left(\frac{\hbar\omega_2}{2|\mu|}\right)\\ &-\frac{(-\omega_1+\omega_2)^3}{3\omega_1^2\omega_2^4(-\omega_1+2\omega_2)}G\left(\frac{\hbar(-\omega_1+\omega_2)}{2|\mu|}\right)\\ &\left.-\frac{4\omega_2{}^3}{3\omega_1^2\omega_2^4(-\omega_1+2\omega_2)}G\left(\frac{\hbar(2\omega_2)}{2|\mu|}\right)\right]\\ &+\frac{(-\omega_1+2\omega_2)^2\left(2\omega_1^2 - \omega_1\omega_2 + 2\omega_2^2\right)}{12\omega_1^2\omega_2^4(-\omega_1+\omega_2)^2}G\left(\frac{\hbar(-\omega_1+2\omega_2)}{2|\mu|}\right)\end{aligned} \tag{3.44}$$

3.3.4 Magnetic Modulation

Magnetization-induced SHG (MSHG) can be achieved in magnets under an external magnetic field. A general theoretical description of MSHG can be found in references [91–93]. The second-order nonlinear polarization is given by [93]

$$\mathbf{P}^{nl}(2\omega) = \chi^{(2)}_{\text{crystal}}\mathbf{E}(\omega)\mathbf{E}(\omega) + \chi_m\mathbf{E}(\omega)\mathbf{E}(\omega)\mathbf{M} \tag{3.45}$$

where $\chi^{(2)}_{\text{crystal}}$ and χ_m are the intrinsic and magnetization-induced second-order susceptibility. The first term is the intrinsic SHG originating from the non-vanishing $\chi^{(2)}_{\text{crystal}}$, while the second term describes the magnetization-induced SHG under magnetic field $\mathbf{M}$. $\chi^{(2)}_{\text{crystal}}$ is a polar third-rank tensor, while χ_m is an axial tenser of rank 4. In a lossless material, $\chi^{(2)}_{\text{crystal}}$ is a real tensor and χ_m is an imaginary tensor. Consequently, the interference between these two terms is absent. For an isotropic interface with incident light in *xz*-plane, the total $\chi^{(2)}$ can be written as

$$\begin{aligned}\chi^{(2)} &= \chi^{(2)}_{\text{crystal}} + \chi_m \cdot \mathbf{M} \\ &= \begin{bmatrix} \chi_{xxx}(M_y) & \chi_{xyy}(M_y) & \chi_{xzz}(M_y) & \chi_{xzy}(M_z) & \chi_{xzx}^{\text{crystal}} & \chi_{xxy}(M_x) \\ \chi_{yxx}(M_x) & \chi_{yyy}(M_x) & \chi_{xzz}(M_x) & \chi_{yzy}^{\text{crystal}} & \chi_{yzx}(M_z) & \chi_{yxy}(M_y) \\ \chi_{zxx}^{\text{crystal}} & \chi_{zyy}^{\text{crystal}} & \chi_{zzz}^{\text{crystal}} & \chi_{zzy}(M_x) & \chi_{zzx}(M_y) & \chi_{zxy}(M_z) \end{bmatrix}\end{aligned} \tag{3.46}$$

The recent discovery of 2D van der Waals magnets extends MSHG to layered materials. MSHG effect is achieved by breaking both spatial inversion and time-reversal symmetries simultaneously, which is termed as *c*-type SHG. *c*-type SHG is different with conventional *i*-type SHG with only spatial-inversion symmetry broken. For instance, SHG is forbidden in bilayer CrI_2 in the presence of inversion symmetry. However, *c*-type SHG can be induced by an external magnetic field attributing to the layered antiferromagnetic order [94].

3.3.5 Perspectives

Although TMDs and graphene can be integrated with Si wafer, current fabrication methods mainly rely on mechanical exfoliation. The technique of wafer scale fabrication of 2D materials with high quality is the key to practical applications. Currently, wafer scale fabrication of single crystalline graphene, TMDs and h-BN are successfully demonstrated [95–97], and it is critical to promote the transition from laboratory research to massive industrial production. Secondly, integration of 2D materials with Si waveguide is promising for on-chip devices. However, the problem of lattice distortion and strain effects should be addressed for practice applications. Recently, quantum interference of SHG in TMDs is discovered [98]. It will be of great interest to explore the quantum nonlinear effects of 2D materials.

Author Contributions

The manuscript was written through contributions of both authors. Both authors have given approval to the final version of the manuscript.

Notes

The authors declare no competing financial interest.

Acknowledgment

This work is supported by the National Natural Science Foundation of China (62005091).

References

1 Franken, P., Hill, A.E., Peters, C.E. et al. (1961). Generation of optical harmonics. *Physical Review Letters* 7 (4): 118.

2 Novoselov, K.S., Geim, A.K., Morozov, S.V. et al. (2004). Electric field effect in atomically thin carbon films. *Science* 306 (5696): 666–669.

3 Kumar, V. (2021). Linear and nonlinear optical properties of graphene: a review. *Journal of Electronic Materials* 50 (7): 3773–3799.

4 Liu, Z.B., Zhang, X.L., Yan, X.Q. et al. (2012). Nonlinear optical properties of graphene-based materials. *Chinese Science Bulletin* 57 (23): 2971–2982.

5 Cheng, J.L., Sipe, J.E., Vermeulen, N. et al. (2019). Nonlinear optics of graphene and other 2D materials in layered structures. *Journal of Physics: Photonics* 1 (1): 015002.

6 Hendry, E., Hale, P.J., Moger, J. et al. (2010). Coherent nonlinear optical response of graphene. *Physical Review Letters* 105 (9): 097401.

7 You, J.W., Bongu, S.R., Bao, Q. et al. (2019). Nonlinear optical properties and applications of 2D materials: theoretical and experimental aspects. *Nanophotonics* 8 (1): 63 97.

8 Wen, X.L., Gong, Z.B., and Li, D.H. (2019). Nonlinear optics of two-dimensional transition metal dichalcogenides. *InfoMat* 1 (3): 317–337.

9 Autere, A., Jussila, H., Dai, Y.Y. et al. (2018). Nonlinear optics with 2D layered materials. *Advanced Materials* 30 (24): 1705963.

10 Clark, D.J., Senthilkumar, V., Le, C.T. et al. (2015). Strong optical nonlinearity of CVD-grown MoS_2 monolayer as probed by wavelength-dependent second-harmonic generation. *Physical Review B* 92 (15): 121409.

11 Song, Y., Tian, R.J., Yang, J.L. et al. (2018). Second harmonic generation in atomically thin $MoTe_2$. *Advanced Optical Materials* 6 (17): 1701334.

12 Janisch, C., Wang, Y.X., Ma, D. et al. (2014). Extraordinary second harmonic generation in tungsten disulfide monolayers. *Scientific Reports* 4 (1): 1–5.

13 Zeng, H.L., Liu, G.B., Dai, J.F. et al. (2013). Optical signature of symmetry variations and spin-valley coupling in atomically thin tungsten dichalcogenides. *Scientific Reports* 3 (1): 1–5.

14 Kumar, N., Najmaei, S., Cui, Q.N. et al. (2013). Second harmonic microscopy of monolayer MoS_2. *Physical Review B* 87 (16): 161403.

15 Saynatjoki, A., Karvonen, L., Rostami, H. et al. (2017). Ultra-strong nonlinear optical processes and trigonal warping in MoS_2 layers. *Nature Communications* 8 (1): 1–8.

16 Rüning, M.G. and Attaccalite, C. (2014). Second harmonic generation in h-BN and MoS_2 monolayers: role of electron-hole interaction. *Physical Review B* 89 (8): 081102.

17 Wu, H.Y., Yen, Y., and Liu, C.H. (2016). Observation of polarization and thickness dependent third-harmonic generation in multilayer black phosphorus. *Applied Physics Letters* 109 (26): 261902.

18 Jiang, T., Huang, D., Cheng, J.L. et al. (2018). Gate-tunable third-order nonlinear optical response of massless Dirac fermions in graphene. *Nature Photonics* 12 (7): 430–436.

19 Woodward, R.I., Murray, R.T., Phelan, C.F. et al. (2017). Characterization of the second- and third-order nonlinear optical susceptibilities of monolayer MoS_2 using multiphoton microscopy. *2D Materials* 4 (1): 011006.

20 Wang, R., Chien, H.C., Kumar, J. et al. (2014). Third-harmonic generation in ultrathin films of MoS_2. *ACS Applied Materials & Interfaces* 6 (1): 314–318.

21 Yuan, C.Q., Li, X.Y., Semin, S. et al. (2018). Chiral lead halide perovskite nanowires for second-order nonlinear optics. *Nano Letters* 18 (9): 5411–5417.

22 Saouma, F.O., Stoumpos, C.C., Wong, J. et al. (2017). Selective enhancement of optical nonlinearity in two-dimensional organic-inorganic lead iodide perovskites. *Nature Communications* 8 (1): 1–8.

23 Yoshikawa, N., Tamaya, T., and Tanaka, K. (2017). High-harmonic generation in graphene enhanced by elliptically polarized light excitation. *Science* 356 (6339): 736–738.

24 Liu, H.Z., Li, Y.L., You, Y.S. et al. (2017). High-harmonic generation from an atomically thin semiconductor. *Nature Physics* 13 (3): 262–265.

25 Zhang, S.F., Dong, N.N., Mcevoy, N. et al. (2015). Direct observation of degenerate two-photon absorption and its saturation in WS_2 and MoS_2 mono layer and few-layer films. *ACS Nano* 9 (7): 7142–7150.

26 Xiao, J., Ye, Z., Wang, Y. et al. (2015). Nonlinear optical selection rule based on valley-exciton locking in monolayer WS_2. *Light: Science & Applications* 4 (12): e366.

27 Yang, H.Z., Feng, X.B., Wang, Q. et al. (2011). Giant two-photon absorption in bilayer graphene. *Nano Letters* 11 (7): 2622–2627.

28 Liu, W.W., Xing, J., Zhao, J.X. et al. (2017). Giant two-photon absorption and its saturation in 2D organic-inorganic perovskite. *Advanced Optical Materials* 5 (7): 1601045.

29 Zhang, H., Lu, S.B., Zheng, J. et al. (2014). Molybdenum disulfide (MoS_2) as a broadband saturable absorber for ultra-fast photonics. *Optics Express* 22 (6): 7249–7260.

30 Popa, D., Sun, Z., Torrisi, F. et al. (2010). Sub 200 fs pulse generation from a graphene mode-locked fiber laser. *Applied Physics Letters* 97 (20): 203106.

31 Ono, M., Hata, M., Tsunekawa, M. et al. (2020). Ultrafast and energy-efficient all-optical switching with graphene-loaded deep-subwavelength plasmonic waveguides. *Nature Photonics* 14 (1): 37–43.

32 Lim, G.K., Chen, Z.L., Clark, J. et al. (2011). Giant broadband nonlinear optical absorption response in dispersed graphene single sheets. *Nature Photonics* 5 (9): 554–560.

33 Sar, H., Gao, J., and Yang, X.D. (2021). 2D layered SiP as anisotropic nonlinear optical material. *Scientific Reports* 11 (1): 1–8.

34 Ye, C.Y., Yang, Z.Q., Dong, J.H. et al. (2021). Layer-tunable nonlinear optical characteristics and photocarrier dynamics of 2D $PdSe_2$ in broadband spectra. *Small* 17 (50): 2103938.

35 Hsu, W.T., Zhao, Z.A., Li, L.J. et al. (2014). Second harmonic generation from artificially stacked transition metal dichalcogenide twisted bilayers. *ACS Nano* 8 (3): 2951–2958.

36 Fan, X.P., Jiang, Y., Zhuang, X.J. et al. (2017). Broken symmetry induced strong nonlinear optical effects in spiral WS_2 nanosheets. *ACS Nano* 11 (5): 4892–4898.

37 Le, C.T., Kim, J., Ullah, F. et al. (2020). Effects of interlayer coupling and band offset on second harmonic generation in vertical $MoS_2/MoS_{2(1-x)}Se_{2x}$ structures. *ACS Nano* 14 (4): 4366–4373.

38 He, C., Zhao, Q.Y., Huang, Y.Y. et al. (2019). Nonlinear optical response in graphene/WX_2 (X = S, Se, and Te) van der Waals heterostructures. *The Journal of Physical Chemistry Letters* 10 (9): 2090–2100.

39 Yang, D., Hu, X.Z., Zhuang, M.H. et al. (2018). Inversion symmetry broken 2D 3R-$MoTe_2$. *Advanced Functional Materials* 28 (26): 1800785.

40 Shi, J., Yu, P., Liu, F.C. et al. (2017). 3R MoS_2 with broken inversion symmetry: a promising ultrathin nonlinear optical device. *Advanced Materials* 29 (30): 1701486.

41 He, C., Zhu, L.P., Huang, Y.Y. et al. (2022). Accurately controlling angle-resolved second harmonic generation by stacking orders from a MoS_2 homobilayer. *Journal of Physical Chemistry C* 126 (25): 10584–10592.

42 Shan, Y.W., Li, Y.G., Huang, D. et al. (2018). Stacking symmetry governed second harmonic generation in graphene trilayers. *Science Advances* 4 (6): eaat0074.

43 Steel, W.H. and Walker, R.A. (2003). Measuring dipolar width across liquid-liquid interfaces with 'molecular rulers'. *Nature* 424 (6946): 296–299.

44 Shen, Y. (1989). Surface properties probed by second-harmonic and sum-frequency generation. *Nature* 337 (6207): 519–525.

45 Enaldiev, V. (2016). Resonant enhancement of second harmonic generation by edge states in transition metal dichalcogenide monolayers. *arXiv:1609.08946*.

46 Yin, X.B., Ye, Z.L., Chenet, D.A. et al. (2014). Edge nonlinear optics on a MoS_2 atomic monolayer. *Science* 344 (6183): 488–490.

47 Lin, K.I., Ho, Y.H., Liu, S.B. et al. (2018). Atom-dependent edge-enhanced second-harmonic generation on MoS_2 monolayers. *Nano Letters* 18 (2): 793–797.

48 Surbhi, K., Bhakta, S., Sahoo, P.K. et al. (2022). Impact of defects on the $\chi^{(3)}$ optical nonlinearity of sputtered WSe_2 thin films in the optical communication band. *Journal of Applied Physics* 132 (24): 243101.

49 Dai, X., Zhang, X., Kislyakov, I.M. et al. (2019). Enhanced two-photon absorption and two-photon luminescence in monolayer MoS_2 and WS_2 by defect repairing. *Optics Express* 27 (10): 13744–13753.

50 Zhang, X.Y., Zhang, S.F., Xie, Y.F. et al. (2018). Tailoring the nonlinear optical performance of two-dimensional MoS_2 nanofilms via defect engineering. *Nanoscale* 10 (37): 17924–17932.

51 Lucking, M.C., Beach, K., and Terrones, H. (2018). Large second harmonic generation in alloyed TMDs and boron nitride nanostructures. *Scientific Reports* 8 (1): 1–13.

52 Wagner, H.P., Kuhnelt, M., Langbein, W. et al. (1998). Dispersion of the second-order nonlinear susceptibility in ZnTe, ZnSe, and ZnS. *Physical Review B* 58 (16): 10494–10501.

53 Bergfeld, S. and Daum, W. (2003). Second-harmonic generation in GaAs: experiment versus theoretical predictions of $\chi^{(2)(xyz)}$. *Physical Review Letters* 90 (3): 036801.

54 Lafrentz, M., Brunne, D., Rodina, A.V. et al. (2013). Second-harmonic generation spectroscopy of excitons in ZnO. *Physical Review B* 88 (23): 235207.

55 Minami, F., Inoue, K., Kato, Y. et al. (1991). Observation of the forbidden second-harmonic generation in resonance with 2P Wannier exciton in ZnSe thin films. *Physical Review Letters* 67 (26): 3708.

56 Wang, G., Marie, X., Gerber, I. et al. (2015). Giant enhancement of the optical second-harmonic emission of WSe_2 monolayers by laser excitation at exciton resonances. *Physical Review Letters* 114 (9): 097403.

57 Trolle, M.L., Seifert, G., and Pedersen, T.G. (2014). Theory of excitonic second-harmonic generation in monolayer MoS_2. *Physical Review B* 89 (23): 235410.

58 Trolle, M.L., Tsao, Y.C., Pedersen, K. et al. (2015). Observation of excitonic resonances in the second harmonic spectrum of MoS_2. *Physical Review B* 92 (16): 161409.

59 Balla, N.K., O'brien, M., Mcevoy, N. et al. (2018). Effects of excitonic resonance on second and third order nonlinear scattering from few-layer MoS_2. *ACS Photonics* 5 (4): 1235–1240.

60 LE, C.T., Clark, D.J., Ullah, F. et al. (2017). Impact of selenium doping on resonant second-harmonic generation in monolayer MoS_2. *ACS Photonics* 4 (1): 38–44.

61 XU, C.-Q., Kondo, T., Sakakura, H. et al. (1991). Optical third-harmonic generation in layered perovskite-type material $(C_{10}H_{21}NH_3)_2PbI_4$. *Solid State Communications* 79 (3): 245–248.

62 Abdelwahab, I., Grinblat, G., Leng, K. et al. (2018). Highly enhanced third-harmonic generation in 2D perovskites at excitonic resonances. *ACS Nano* 12 (1): 644–650.

63 Zhou, F., Abdelwahab, I., Leng, K. et al. (2019). 2D perovskites with giant excitonic optical nonlinearities for high-performance sub-bandgap photodetection. *Advanced Materials* 31 (48): 1904155.

64 Vitale, S.A., Nezich, D., Varghese, J.O. et al. (2018). Valleytronics: opportunities, challenges, and paths forward. *Small* 14 (38): 1801483.

65 Xiao, D., Liu, G.B., Feng, W.X. et al. (2012). Coupled spin and valley physics in monolayers of MoS_2 and other group-VI dichalcogenides. *Physical Review Letters* 108 (19): 196802.

66 Seyler, K.L., Schaibley, J.R., Gong, P. et al. (2015). Electrical control of second-harmonic generation in a WSe_2 monolayer transistor. *Nature Nanotechnology* 10 (5): 407–411.

67 HU, G.W., Hong, X.M., Wang, K. et al. (2019). Coherent steering of nonlinear chiral valley photons with a synthetic Au-WS_2 metasurface. *Nature Photonics* 13 (7): 467–472.

68 Zhang, D.L., Zeng, Z.X.S., Tong, Q.J. et al. (2020). Near-unity polarization of valley-dependent second-harmonic generation in stacked TMDC layers and heterostructures at room temperature. *Advanced Materials* 32 (29): 1908061.

69 Zhang, Y., Wang, Y., Dai, Y. et al. (2022). Chirality logic gates. Science. *Advances* 8 (49): eabq8246.

70 Hubert, C., Billot, L., Adam, P.M. et al. (2007). Role of surface plasmon in second harmonic generation from gold nanorods. *Applied Physics Letters* 90 (18): 181105.

71 Rodriguez, A., Soljačić, M., Joannopoulos, J.D. et al. (2007). $\chi^{(2)}$ and $\chi^{(3)}$ harmonic generation at a critical power in inhomogeneous doubly resonant cavities. *Optics Express* 15 (12): 7303–7318.

72 Fan, S., Suh, W., and Joannopoulos, J.D. (2003). Temporal coupled-mode theory for the Fano resonance in optical resonators. *Journal of the Optical Society of America A* 20 (3): 569–572.

73 Lin, Z., Liang, X., Lončar, M. et al. (2016). Cavity-enhanced second-harmonic generation via nonlinear-overlap optimization. *Optica* 3 (3): 233–238.

74 Hong, P.L., Xu, L., and Rahmani, M. (2022). Dual bound states in the continuum enhanced second harmonic generation with transition metal dichalcogenides monolayer. *Opto-Electronic Advances* 5 (7): 200097.

75 Chen, H.T., Corboliou, V., Solntsev, A.S. et al. (2017). Enhanced second-harmonic generation from two-dimensional $MoSe_2$ on a silicon waveguide. *Light: Science & Applications* 6 (10): e17060.

76 Zuo, Y.G., Yu, W.T., Liu, C. et al. (2020). Optical fibres with embedded two-dimensional materials for ultrahigh nonlinearity. *Nature Nanotechnology* 15 (12): 987–991.

77 Han, X.B., Wang, K., Persaud, P.D. et al. (2020). Harmonic resonance enhanced second-harmonic generation in the monolayer WS_2-Ag nanocavity. *ACS Photonics* 7 (3): 562–568.

78 Li, C.T., Lu, X., Srivastava, A. et al. (2021). Second harmonic generation from a single plasmonic nanorod strongly coupled to a WSe_2 monolayer. *Nano Letters* 21 (4): 1599–1605.

79 Fryett, T.K., Seyler, K.L., Zheng, J.J. et al. (2017). Silicon photonic crystal cavity enhanced second-harmonic generation from monolayer WSe_2. *2D Materials* 4 (1): 015031.

80 Shi, J.W., Wu, X.X., Wu, K.M. et al. (2022). Giant enhancement and directional second harmonic emission from monolayer WS_2 on silicon substrate via fabry-perot micro-cavity. *ACS Nano* 16 (9): 13933–13941.

81 Calafell, I.A., Rozema, L.A., Iranzo, D.A. et al. (2021). Giant enhancement of third-harmonic generation in graphene-metal heterostructures. *Nature Nanotechnology* 16 (3): 318–324.

82 Klimmer, S., Ghaebi, O., Gan, Z.Y. et al. (2021). All-optical polarization and amplitude modulation of second-harmonic generation in atomically thin semiconductors. *Nature Photonics* 15 (11): 837–842.

83 Zhang, Y., Bai, X.Y., Munoz, J.A. et al. (2022). Coherent modulation of chiral nonlinear optics with crystal symmetry. *Light: Science & Applications* 11 (1): 1–7.

84 Wen, X.L., Xu, W.G., Zhao, W.J. et al. (2018). Plasmonic hot carriers-controlled second harmonic generation in WSe_2 bilayers. *Nano Letters* 18 (3): 1686–1692.

85 Terhune, R., Maker, P., and Savage, C. (1962). Optical harmonic generation in calcite. *Physical Review Letters* 8 (10): 404.

86 Chen, S.M., Li, K.F., Li, G.X. et al. (2019). Gigantic electric-field-induced second harmonic generation from an organic conjugated polymer enhanced by a band-edge effect. *Light: Science & Applications* 8 (1): 1–6.

87 Klein, J., Wierzbowski, J., Steinhoff, A. et al. (2017). Electric-field switchable second-harmonic generation in bilayer MoS_2 by inversion symmetry breaking. *Nano Letters* 17 (1): 392–398.

88 Gan, X.T., Zhang, M.W., Liu, Y. et al. (2022). Electrically tunable second harmonic generation in atomically thin ReS_2. *ACS Nano* 16 (4): 6404–6413.

89 Yu, H.K., Talukdar, D., Xu, W.G. et al. (2015). Charge-induced second-harmonic generation in bilayer WSe_2. *Nano Letters* 15 (8): 5653–5657.

90 Soavi, G., Wang, G., Rostami, H. et al. (2018). Broadband, electrically tunable third-harmonic generation in graphene. *Nature Nanotechnology* 13 (7): 583–588.

91 Hubner, W. and Bennemann, K. (1989). Nonlinear magneto-optical Kerr effect on a nickel surface. *Physical Review B* 40 (9): 5973.

92 Gridnev, V.N., Pavlov, V.V., Pisarev, R.V. et al. (2001). Second harmonic generation in anisotropic magnetic-films. *Physical Review B* 63 (18): 184407.

93 Kirilyuk, A. and Rasing, T. (2005). Magnetization-induced-second-harmonic generation from surfaces and interfaces. *Journal of the Optical Society of America B* 22 (1): 148–167.

94 Sun, Z.Y., Yi, Y.F., Song, T.C. et al. (2019). Giant nonreciprocal second-harmonic generation from antiferromagnetic bilayer CrI_3. *Nature* 572 (7770): 497–501.

95 Lee, J.H., Lee, E.K., Joo, W.J. et al. (2014). Wafer-scale growth of single-crystal monolayer graphene on reusable hydrogen-terminated germanium. *Science* 344 (6181): 286–289.

96 Wang, L., Xu, X.Z., Zhang, L.N. et al. (2019). Epitaxial growth of a 100-square-centimetre single-crystal hexagonal boron nitride monolayer on copper. *Nature* 570 (7759): 91–95.

97 Li, T.T., Guo, W., Ma, L. et al. (2021). Epitaxial growth of wafer-scale molybdenum disulfide semiconductor single crystals on sapphire. *Nature Nanotechnology* 16 (11): 1201–1207.

98 Lin, K.Q., Bange, S., and Lupton, J.M. (2019). Quantum interference in second-harmonic generation from monolayer WSe_2. *Nature Physics* 15 (3): 242–246.

4

Characterizing the Nonlinear Optical Properties of 2D Materials by Double 4f Nonlinear Imaging System with Phase Object and Four-Wave-Mixing Microscopy

Zhongguo Li[1] and Yinglin Song[2]

[1] *Changshu Institute of Technology, School of Electronic and Information Engineering, Changshu 215500, China*
[2] *Harbin Institute of Technology, Department of Physics, Harbin 150001, China*

4.1 Introduction

The field of photonics has emerged into an indispensable part of our economy and society. The explosive growth of Internet traffic leads to next-generation microelectronic and nanophotonic devices with better performance and smaller footprint [1–3]. 2D materials with thickness regime between several to a few hundreds of nanometers represent unique building blocks of functional devices that could reshape our future life through bringing light down to the nanometer scale [4–7]. Understanding the fundamental physical properties of 2D materials is essential to the development of novel electronic and photonic devices based on these nanostructures. In the past two decades, significant effects have been devoted to revealing the mechanical, electronical, and optical responses of 2D materials and their heterostructures [8–10]. Previous results demonstrate that the nonlinear optical (NLO) properties of 2D materials can be used to create sub-wavelength photon sources, photodetectors, optical modulators, optical antennas, and integrated waveguides [11–13]. However, while the linear optical response of 2D materials is well understood, the origin and magnitude of NLO response of 2D materials remain elusive.

NLO measurement technique is a valuable tool for investigating the properties of 2D materials. Experimental data with different parameters (laser wavelength, pulse duration, polarization, etc.) can provide unique information about exciton, inter-band transition and charge-transfer processes in 2D materials that occur under laser excitation [14, 15]. NLO measurement technique have also made a major contribution to understanding the impact of defect, stoichiometry, and interface on photophysical properties of 2D materials [16, 17]. At present, a wide range of NLO measurement techniques such as harmonic generation (HG), four-wave mixing (FWM), two-photon-induced photoluminescence (TPL), Z-scan, and self-phase-modulation (SPM) have been used to investigate the NLO response

Two-Dimensional Materials for Nonlinear Optics: Fundamentals, Preparation Methods, and Applications, First Edition. Edited by Qiang Wang and Hao-Li Zhang.

of 2D materials. Therefore, it is necessary to highlight the recent progress on NLO measurement technique in 2D material research.

In this chapter, we provide a brief review on fundamentals, advancements, and potential applications of third-order NLO measurement technique in 2D material research. We mainly focus on two third-order NLO measurement techniques with microscopy: double 4f nonlinear imaging system with phase object (NIT-PO) and FWM. Their experimental setup and theoretical models are described. A few selected studied cases are highlighted. Also, we provide a brief outlook on the future development of third-order NLO measurement technique to continue the advancement of 2D material devices. This chapter is designed for both specialists in the field and general physics and chemistry readership.

4.2 Principle of NLO Measurement Technique

4.2.1 Nonlinear Optical Interaction in 2D Materials

Nonlinear optics requires nonlinear interactions between optical field and matter. In linear optics, the macroscopic polarization of materials depends linearly on the electric field:

$$P = \varepsilon_0 \chi^{(1)} E \tag{4.1}$$

where $\chi^{(1)}$ is the linear susceptibility. All of the optical properties of a material (absorption, refraction, scattering, etc.) are related to the linear polarization. When the incident light field is strong enough (such as laser irradiation), the macroscopic polarization of materials could be expressed as a power series of the incident electric field [18, 19]:

$$P = P_{\mathrm{L}} + P_{\mathrm{NL}} = \varepsilon_0 \chi^{(1)} E + \varepsilon_0 \chi^{(2)} EE + \varepsilon_0 \chi^{(3)} EEE + \dots \tag{4.2}$$

where $\chi^{(2)}$ and $\chi^{(3)}$ are the second-order and third-order NLO susceptibilities, respectively. It is well-known that the NLO susceptibility is a tensor with multiple elements, and a variety of NLO interactions that have been found in 2D materials is illustrated in Figure 4.1 [20].

The aim of NLO measurement technique is to unravel the sign, magnitude, temporal evolution, and spectral–spatial–angular characteristics of nonlinear light–matter interactions in a material. These techniques can be performed in either time domain or frequency domain [14, 15]. Note that the second-order NLO measurement method (such as second HG or sum frequency generation) has also been widely used in 2D material research. Recent reviews on this topic can be found in previous literatures [21, 22]. For the purpose of this chapter, we simply consider the third-order NLO measurement technique.

One key issue in studying the NLO response of 2D materials is unraveling the inhomogeneous effects. These effects in 2D materials can arise from film thickness, lattice orientation, grain heterogeneity, surface/bulk defects, interfacial strain, non-stoichiometry, etc. Therefore, it is common to found different values of NLO coefficients for the same 2D material in the previous reports [23–26].

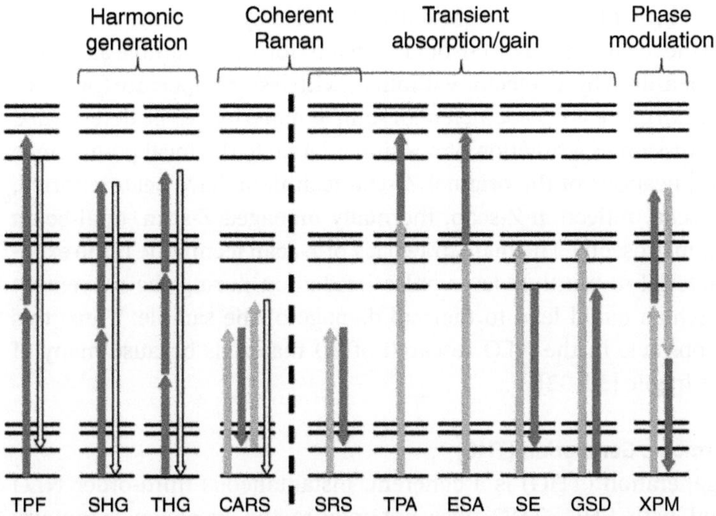

Figure 4.1 Nonlinear optical interactions available for 2D material research. Source: Reproduced from Fischer et al. [20], with permission from AIP Publishing, 2016.

4.2.2 Typical NLO Measurement Method in 2D Material Research

In the following, we highlight four NLO measurement techniques that are currently used in 2D materials research. Their advantages and drawbacks are briefly discussed.

4.2.2.1 Self-Phase Modulation (SPM)

SPM is a third-order NLO phenomenon that is induced by the optical Kerr effect. SPM can be observed experimentally by focusing a Gaussian laser beam into the sample cuvette with dispersion of 2D materials. The reorientation of 2D materials under laser excitation will lead to a change of refractive index; therefore, a phase shift will add to the transmitted light wave. According to the Kirchhoff diffraction theory, the far-field diffraction pattern is the diffraction integral of the optical field transmitted through the 2D material dispersion. As a result, a symmetrical phase shift with Gaussian profile will lead to the cone-shaped diffraction rings in the far field. The nonlinear refractive index n_2 of 2D material dispersion can be determined by analyzing the number of diffraction rings with different incident laser intensity [27–29].

4.2.2.2 Z-Scan

The Z-scan technique, reported by Sheik-Bahae et al. in 1989, is currently the most popular third-order NLO measurement method in 2D material research. This technique can simultaneously determine the sign and magnitude of nonlinear absorption coefficient and nonlinear refraction index (open-aperture and closed-aperture Z-scan). The experimental layout is simple, and the measurement sensitivity is high. The principle of Z-scan method is straightforward: a convex lens focuses the input Gaussian laser beam to the focal point. The sample is moved along the beam axis;

therefore, the input laser intensity is continuously variable by changing the sample position relative to the focal point. The far-field transmission of the sample as a function of position is measured by a detector without or with a small aperture (open- or closed-aperture Z-scan). The NLO coefficient of sample material can be determined by fitting the transmission as a function of position relative to the focal point. Since 1990, multiple modifications of the original Z-scan technique have been reported, such as top-hat Z-scan, reflection Z-scan, thermally managed Z-scan, dual-beam Z-scan, and Z-scan microscopy. The main drawback of Z-scan technique is the exposure of sample material to multiple laser pulse irradiation during the movement along beam axis, which could lead to thermal damage of the sample. Thus, this becomes a major obstacle in the NLO research of 2D materials because many of these materials are fragile [30–32].

4.2.2.3 Third-Harmonic Generation (THG)

Third-harmonic generation (THG) is a coherent, instantaneous third-order NLO process. Compared with SHG, THG does not require the inversion symmetry breaking condition; therefore, THG process can occur in both centrosymmetric and non-centrosymmetric materials. In THG measurement, the sample 2D material is excited by an ultrafast laser pulse. Three incident photons with identical photon energy are converted into a new photon whose photon energy is three times higher than the input photon. The output photons are detected by a spectrometer or a photodetector with filter. This technique is highly sensitive because THG photon has zero background. THG signal can be enhanced by exciton resonance or interfacial effect; thus, THG technique is suitable for material analysis of 2D materials. The main drawback of this technique is that THG signal is related to the total value of third-order NLO susceptibility. As a result, THG technique cannot distinguish the real and image part of NLO susceptibility [33, 34].

4.2.2.4 Two-Photon-Induced Photoluminescence (TPL)

The TPL method is a widely used technique to characterize the nonlinear absorption response of 2D materials. Compared with the three methods above, TPL method rely on an observable photoluminescence signal rather than output laser beam. Therefore, this method is an indirect measurement method and has zero background that is beneficial to increase the sensitivity. In TPL experiments, the sample material is excited by an ultrafast laser source. The strong laser field leads to the two-photon absorption (TPA) process in the sample, resulting in the inter-band transition cross the bandgap. The photo-generated carriers initially relax to the band-edge or trap states within a few picoseconds and then relax radiatively back to the valence band. The intensity of photoluminescence signal is related to the population of photo-generated carriers, which is proportional to the square of peak laser intensity. In most cases, TPL signal is detected by a photomultiplier tube (PMT) and lock-in amplifier. Therefore, TPL technique is highly sensitive. However, nonradiative sample cannot be measured by this technique, and the indirect nature of TPL signal means the data interpretation should be treated with a benchmark reference material [35, 36].

4.3 Characterizing NLO Response of 2D Materials via Double 4f Nonlinear Imaging System with Phase Object (NIT-PO) Microscopy

The 4f NIT-PO, introduced by G. Boudebs et al. in 2004 [37], is a simple yet sensitive method to determine the third-order NLO response of sample material. The principle of NIT-PO technique relies on the 4f correlator optical system used in *Fourier Optics*. The main path of the NIT-PO layout consists of two convex lenses (4f correlator). The input plane is the front focal plane of Lens 1 while the output plane is located on the back focal plane of Lens 2. The Fourier plane is located on the confocal plane between the two lenses. The sample material is placed at the Fourier plane of the 4f system, which acts as a spatial frequency filter to the incident laser beam. In the case of low incident laser intensity (linear regime), the optical nonlinearity of sample material is absent; hence, nothing changes to the output image. However, when the laser intensity is high enough (nonlinear regime), the optical nonlinearity of sample material will take place, and the NLO response will change the spatial frequency components of incident laser beam. Therefore, a change in the output light field will take place in the nonlinear regime. A CCD camera is arranged at the output plane of the 4f system, which is used to collect the output beam profile. During the measurement, an aperture with a phase object (PO) is arranged at the input plane of the 4f system. The PO can transform the nonlinear phase shift of the sample material into the amplitude change of the output image (similar to a Zernike phase contrast microscopy in *Fourier Optics*). By fitting the intensity profile of the output image numerically, the nonlinear refraction index of the sample material can be obtained. Meanwhile, the nonlinear absorption coefficient of the sample can also be determined by numerically fitting the whole output image recorded by the CCD camera. Compared with Z-scan technique, the major advantages of NIT-PO method are simple experimental layout, single-shot detection, and unrequired movement sampling. Therefore, NIT-PO technique can be easily integrated with microscopy to characterize the NLO response of 2D materials.

The experimental arrangement of the NIT-PO system and configuration of PO is shown schematically in Figure 4.2a,b, respectively. The PO is made by depositing a layer of transparent dielectric in the center of an aperture on a flat glass plate. The central dielectric layer will induce a phase retardation to the input laser beam. Moreover, the PO is illuminated at normal incidence by a linearly polarized laser beam: $E = E_0(t) \exp[-i(\omega t - kz)] + c.c$, where ω is the angular frequency, k is the wave vector in free space, and $E_0(t)$ is the amplitude of electric field of light wave. If the transmittance function of PO is defined as $t_p(x, y)$, the laser field at the rear focal plane of the convex lens L_1 is the spatial Fourier transform of $O(x, y, t) = E(x, y, t)t_p(x, y)$ [14, 38]:

$$S(u, v) = \frac{1}{\lambda f_1} \mathrm{FT}[O(x, y)] = \frac{1}{\lambda f_1} \iint O(x, y) \exp[-2\pi i(ux + vy)] \mathrm{d}x\mathrm{d}y \tag{4.3}$$

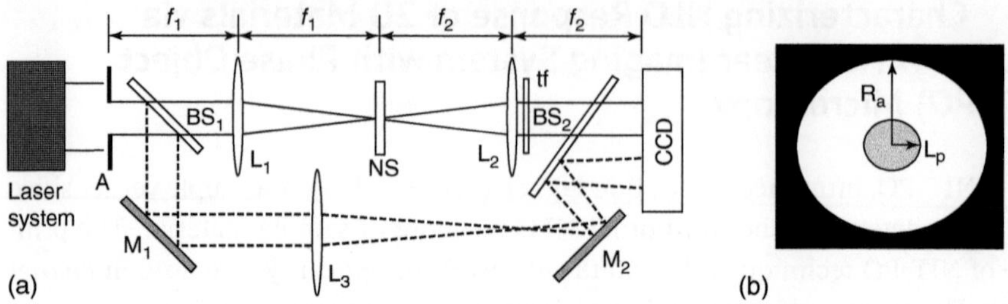

Figure 4.2 Schematic diagram of (a) the system setup of the 4f coherent imaging system with a phase object (NIT-PO) technique. (b) Configuration of the phase object with circular aperture.

where FT denotes the Fourier transform operation; u and v are the spatial frequencies in the focal plane, $u = x/\lambda f_1, v = y/\lambda f_1; f_1$ is the focal length of the lens L_1; and λ is the wavelength of the input laser beam.

The incident laser field before the sample is $S(u, v)$; therefore, the peak laser intensity at the Fourier plane is proportional to $|S(u, v)|^2$. Because the thickness of 2D material is on the range of nanometer scale, the slowly varying envelope approximation (SVEA) is suitable to describe the propagation of the electric field in 2D material:

$$\begin{cases} \dfrac{dI}{dz'} = -(\alpha_0 + \beta I) \\ \dfrac{d\Delta\phi}{dz'} = kn_2 I \end{cases} \tag{4.4}$$

where $\Delta\phi$ is the nonlinear phase change induced by Kerr effect of 2D material and z' is the propagation length within the sample. α_0, β, and n_2 are the linear absorption coefficient, TPA coefficient, and third-order nonlinear refractive index of the sample, respectively. The field at the exit surface of the 2D material can be written as

$$S_L(u, v, t) = S(u, v, t)\exp(-\alpha_0 L/2)[1 + q(u, v, t)]^{(ikn_2/\beta - 1/2)} \tag{4.5}$$

where $q(u, v, t) = \beta I(u, v, t)L_{\text{eff}}$, $L_{\text{eff}} = [1 - \exp(-\alpha_0 L)]/\alpha_0$ and L is the thickness of the sample.

At the output plane of the 4f system, the image intensity recorded by CCD camera can be expressed as

$$I_{\text{im}}(x, y, t) = |U(x, y, t)|^2 = \left|\text{FT}^{-1}\left[S_L(u, v, t)H(u, v)\right]\right|^2 \tag{4.6}$$

where FT^{-1} denotes the inverse spatial Fourier transform and $H(u, v)$ = circ $[(u^2 + v^2)^{1/2}\lambda G/N_A]$ is the coherent optical transfer function of the 4f system. The function circ(ρ) is defined to be 1 if the radius $\rho(u, v) \leq 1$ and 0 otherwise. N_A and G are the numerical aperture and magnification of the 4f nonlinear imaging system, respectively.

Because the image profile recorded by CCD camera is the fluence distribution of the laser beam intensity, this image can be expressed by the integration of the intensity:

$$F(x,y) = \int_{-\infty}^{+\infty} I_{\text{im}}(x,y,t)\text{d}t \tag{4.7}$$

Because the size of the aperture in PO is small compared with the incident laser beam, the output beam after PO could be considered as a top-hat beam. Therefore, the transmittance of the aperture is defined as

$$t_{\text{a}}(x,y) = \text{circ}[(x^2+y^2)^{1/2}/R_{\text{a}}] \tag{4.8}$$

where R_a is the radius of the aperture. So, the overall transmittance function of the PO can be given by

$$t_{\text{p}}(x,y) = t_{\text{a}}(x,y)\exp\{i\phi_{\text{L}}\text{circ}[(x^2+y^2)^{1/2}/L_{\text{P}}]\} \tag{4.9}$$

Using equations above, the third-order NLO parameters of the 2D material can be determined by numerical simulating the images recorded by CCD camera. In the NIT-PO experiment, three images are needed to be recorded for evaluating the sample's optical nonlinearities: (i) linear image, (ii) nonlinear image, and (iii) image without sample. The numerical simulated beam profile of the image at the output plane of 4f system is shown in Figure 4.3. Similar to the Zernike phase contrast microscopy, a positive nonlinear phase shift will lead to the increase of normalized transmittance, while a negative nonlinear phase shift will result in the decrease of normalized transmittance.

Many modifications of the original NIT-PO technique have been reported, such as NIT with a positive–negative annular or bar PO, time-resolved NIT-PO, and reflection NIT-PO [39–42]. One key advantage of NIT-PO technique is the exposure of sample material to only one shot of strong laser pulse irradiation (nonlinear image acquirement); therefore, the accumulated thermal effect is negligible. This intriguing feature is highly promising for NLO research of 2D material because many of

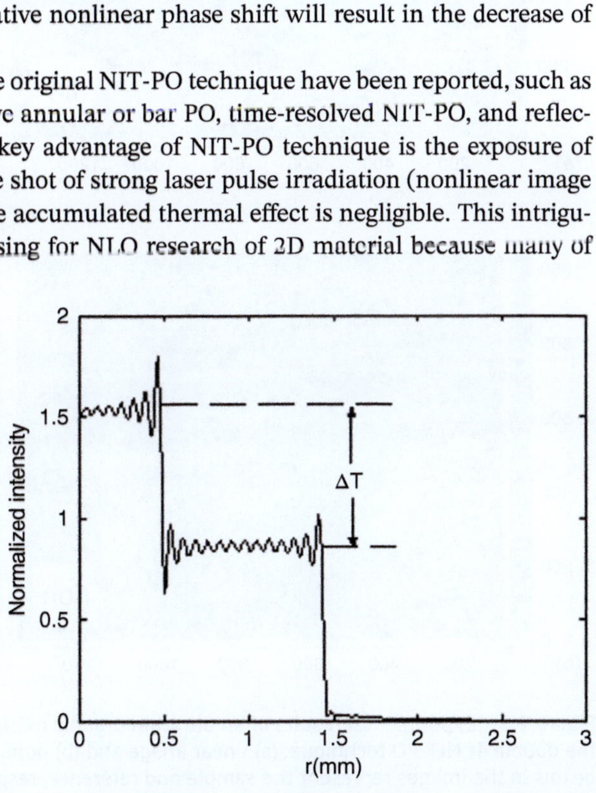

Figure 4.3 Numerical simulation of the beam profile at the output plane of NIT-PO system. The normalized transmittance is defined as the difference between the intensity value of PO area and outside ring.

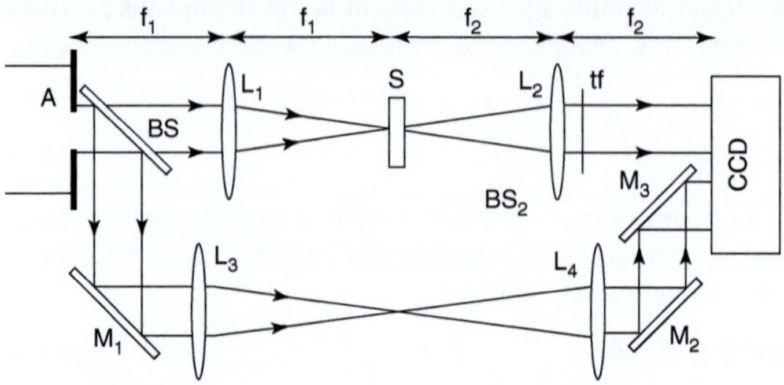

Figure 4.4 Schematic diagram of the double 4f coherent imaging system with phase object.

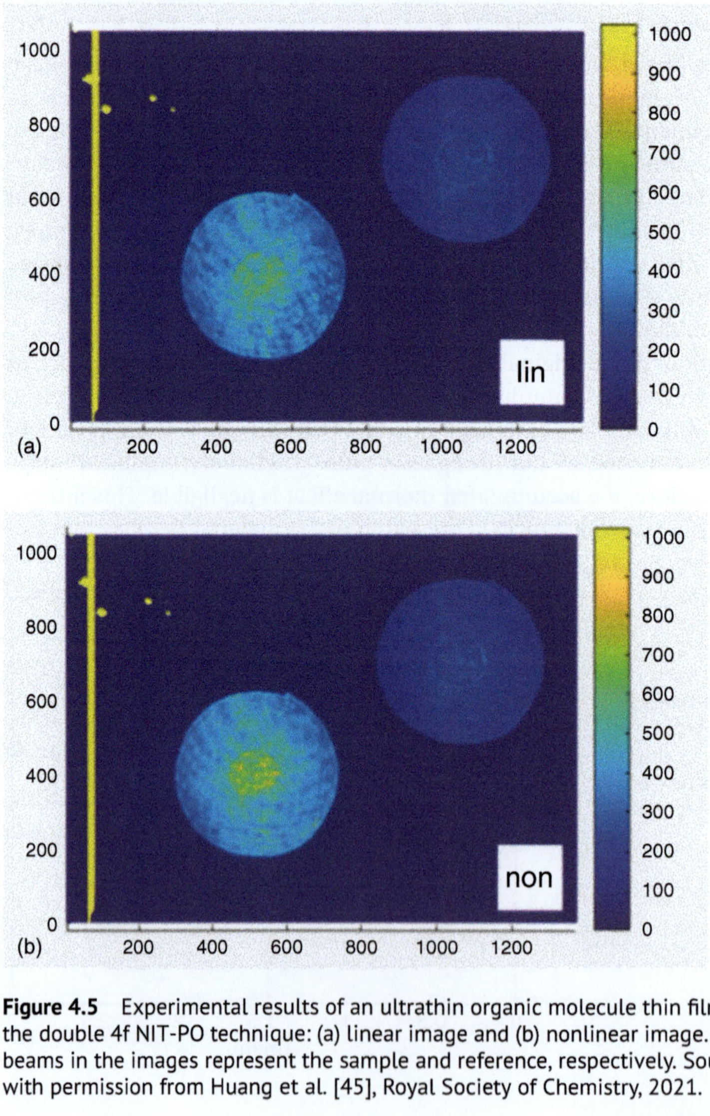

Figure 4.5 Experimental results of an ultrathin organic molecule thin film measured using the double 4f NIT-PO technique: (a) linear image and (b) nonlinear image. The left and right beams in the images represent the sample and reference, respectively. Source: Reproduced with permission from Huang et al. [45], Royal Society of Chemistry, 2021.

this material is fragile and high-intensity laser irradiation could lead to photoinduced phase transition. However, one drawback of the original NIT-PO technique is the high requirement of laser beam quality and sample homogeneities, because both distortion of the beam profile and sample-induced scattering could result in a pseudo phase shift. To address this problem, our group have developed a modification of the original NIT-PO method: double 4f NIT-PO method. In this method, both the sample path and the reference path are settled into a 4f correlator optical system [43], as shown in Figure 4.4. In this case, any fluctuations of input laser energy or beam profile are recorded simultaneously, which significantly eliminate the issue of laser beam fluctuations. As a result, the robustness of NIT-PO measurement is significantly enhanced, making the double 4f NIT-PO method a promising tool for NLO research of 2D materials [44–47]. The experimental demonstration of double 4f NIT-PO technique is shown in Figure 4.5.

4.4 Characterizing NLO Response of 2D Materials via Four-Wave-Mixing Microscopy

The double 4f NIT-PO technique described above belongs to the category of beam distortion method, in which the spatial profile of input laser beam is characterized after the nonlinear sample. In many cases, the sample 2D material is highly inhomogeneous or has strong scattering. Thus, the transmitted or reflected light through the sample is a speckle field. In this scenario, the spectrum distortion method is a better option for measuring the NLO response of highly inhomogeneous 2D material. The FWM technique is a third-order NLO measurement method that has been used in the 2D material research. Combined with a microscope, the FWM microscopy is a highly promising tool for both grain heterogeneity analysis and characterization of ultrafast light–matter interactions in 2D materials.

FWM is a typical third-order NLO process that is sensitive to both electronic and vibrational properties of sample materials. Because the third-order NLO susceptibility $\chi^{(3)}$ has many terms, there are many kinds of FWM method available today in 2D material research, such as coherent anti-Stokes Raman scattering (CARS), stimulated Raman scattering (SRS), degenerate FWM (DFWM), and TPA-enhanced difference-frequency mixing. Compared with conventional linear optical microscopy (such as PL mapping or absorption microscopy), the FWM microscopy has several advantages: (i) The wavelength of emitted photon can be chosen to spectrally separated from the incident photon; therefore, the detection sensitivity of FWM signal is very high. (ii) By tuning the temporal delay between the incident light waves, FWM can probe the ultrafast coherence dynamics and the carrier/exciton relaxation processes in 2D materials. (iii) The intensity of FWM signal scales squarely with the incident light intensity; therefore, it is possible to probe a spot smaller than diffraction limit. (iv) FWM signal is detectable in non-fluorescence sample, and the blinking issue is not a problem for FWM microscopy [48].

Despite these advantages, there are also some limitations for the FWM microscopy. The first one is the high cost of laser source and the complexity of experimental layout. Many FWM experiments use a femtosecond (fs) laser source with sufficient bandwidth to generate the FWM signal [49, 50]. Some literatures even use a dual-color light source (such as an fs laser and a synchronized pumped optical parametric oscillator) to generate the pump and stokes pulses in FWM process [51]. The key factors in a scanning optical microscope (scanner, high-numerical-aperture lens, and detector/spectrometer system) are that the microscope system is bulky and proper alignment is required. The second drawback of FWM microscopy is that the FWM signal cannot distinguish the real and image part of third-order NLO susceptibility directly. There are some reports about the phase-sensitive FWM

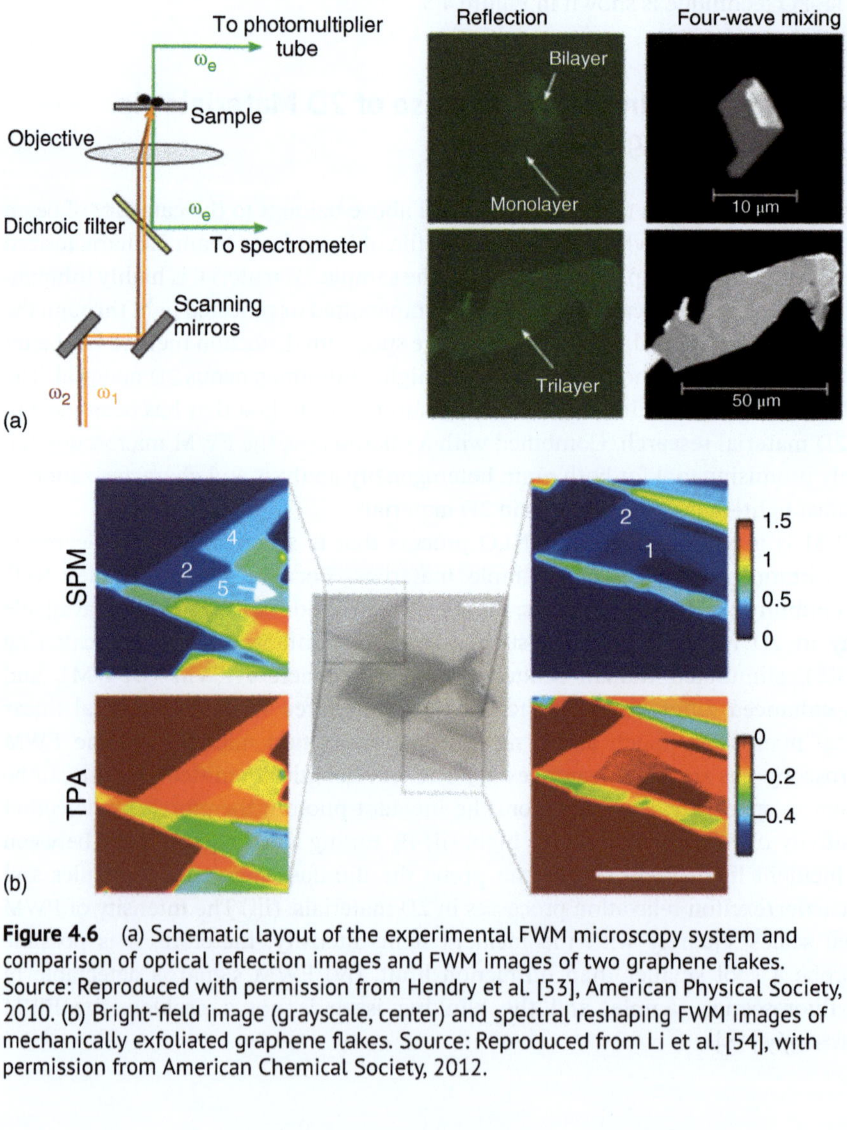

Figure 4.6 (a) Schematic layout of the experimental FWM microscopy system and comparison of optical reflection images and FWM images of two graphene flakes. Source: Reproduced with permission from Hendry et al. [53], American Physical Society, 2010. (b) Bright-field image (grayscale, center) and spectral reshaping FWM images of mechanically exfoliated graphene flakes. Source: Reproduced from Li et al. [54], with permission from American Chemical Society, 2012.

experiments with heterodyne detection configuration that can measure the real part of dielectric permittivity [52]. However, the experimental layout of these system is highly complex.

Several examples of FWM microscopy in 2D material research are shown in Figure 4.6. The image in Figure 4.6a shows the experimental scheme of FWM experiment in graphene. The FWM results indicate that the graphene has excellent third-order NLO susceptibility (on the order of 10^{-7} esu) [53]. Also, the FWM microscopy image on the right panel of Figure 4.6a clearly demonstrates the thickness-dependent FWM response in graphene, which is indistinct in linear reflection optical microscopy. Some research groups have reported a modified FWM system with a pulse shaper to control the spectrum of incident fs laser pulse (spectral reshaping FWM). Using this spectral reshaping configuration, it is possible to distinguish the TPA and cross-phase-modulation (XPM) effect in graphene (Figure 4.6b). Also, a layer-dependent FWM response is clearly visible in the FWM microscopy images [54, 55]. The Cundiff group has developed the two-dimensional (2D) coherent imaging FWM microscopy, which is a combination of the two-dimensional coherent spectroscopy (2DCS) and scanning optical microscope. Using the 2D coherent imaging FWM microscopy, exciton decoherence and dark exciton in $MoSe_2$ film can be detected [56, 57]. These information about 2D materials is indistinguishable in other optical microscopy, indicating the great potential of FWM microscopy in 2D material research.

4.5 Outlook and Perspective

In this chapter, we have presented a brief overview of the third-order NLO measurement techniques in 2D material research, with special focus on two NLO techniques with microscopy. The principles, advantages, and disadvantages of these NLO characterization techniques were provided, and a few examples of the use of NLO measurement techniques in 2D material research were highlighted. As surveyed in this chapter, the third-order NLO microscopy techniques could shed new light on characterizing the ultrafast light–matter interaction on nanoscale, elucidating previously obscured mechanisms of carriers/exciton relaxation. The potential of third-order NLO measurement techniques in 2D material research is high and far beyond the results discussed in this chapter. Despite the successes, there are still several outstanding challenges that require more in-depth research.

The first one is the further improvement of spatial resolution, both axially and radially, in NLO measurement of 2D materials. The development of novel sub-diffraction-limited NLO microscopy techniques can provide radially resolution on the order of hundreds of nanometers, and further improvement is possible with novel measurement devices and structural illumination with vortex beams [58, 59]. A new axial scanning multiphoton microscopy could provide volumetric imaging in the axial direction, which is highly promising for 2D material device engineering [60].

The second area for improvement is developing novel in situ and multimode NLO microscopy. These techniques could help to provide insight in optoelectronic response mapping of 2D material correlated with bias and grain boundaries. Combining HG, PL, and transient absorption/refraction microscopy is a promising direction for the future [61–63].

Last but not least, it is of importance to extend the spectral range and measurement sensitivity of the NLO measurement techniques. With the rapid development of new laser source with frequencies ranging from THz to X-ray regime, it is possible to investigate a larger range of photophysical processes in 2D materials [64–66].

References

1 Akinwande, D., Huyghebaert, C., Wang, C.-H. et al. (2019). Graphene and two-dimensional materials for silicon technology. *Nature* 573 (7775): 507–518.

2 Liu, C., Chen, H., Shuiyuan, W. et al. (2020). Two-dimensional materials for next-generation computing technologies. *Nature Nanotechnology* 15 (7): 545–557.

3 Tan, T., Jiang, X., Wang, C. et al. (2020). 2D material optoelectronics for information functional device applications: status and challenges. *Advanced Science* 7 (11): 2000058.

4 Wu, Y., Duan, J., Ma, W. et al. (2022). Manipulating polaritons at the extreme scale in van der Waals materials. *Nature Reviews Physics* 4 (9): 578–594.

5 Zhang, Q., Hu, G., Ma, W. et al. (2021). Interface nano-optics with van der Waals polaritons. *Nature* 597 (7875): 187–195.

6 Huang, T., Zhao, X., Zeng, S. et al. (2020). Planar nonlinear metasurface optics and their applications. *Reports on Progress in Physics* 83 (12): 126101.

7 Bloch, J., Cavalleri, A., Galitski, V. et al. (2022). Strongly correlated electron–photon systems. *Nature* 606 (7912): 41–48.

8 Regan, E.C., Wang, D., Paik, E.Y. et al. (2022). Emerging exciton physics in transition metal dichalcogenide heterobilayers. *Nature Reviews Materials* 1–18.

9 Jiang, Y., Chen, S., Zheng, W. et al. (2021). Interlayer exciton formation, relaxation, and transport in TMD van der Waals heterostructures. *Light: Science & Applications* 10 (1): 1–29.

10 Wang, Y., Nie, Z., and Wang, F. (2020). Modulation of photocarrier relaxation dynamics in two-dimensional semiconductors. *Light: Science & Applications* 9 (1): 1–16.

11 Sun, Z., Martinez, A., and Wang, F. (2016). Optical modulators with 2D layered materials. *Nature Photonics* 10 (4): 227–238.

12 Ono, M., Hata, M., Tsunekawa, M. et al. (2020). Ultrafast and energy-efficient all-optical switching with graphene-loaded deep-subwavelength plasmonic waveguides. *Nature Photonics* 14 (1): 37–43.

13 Zhao, X., Jin, H., Liu, J. et al. (2022). Integration and applications of nanomaterials for Ultrafast photonics. *Laser & Photonics Reviews* 16 (11): 2200386.

14 de Araujo, C.B., Gomes, A.S.L., and Boudebs, G. (2016). Techniques for nonlinear optical characterization of materials: a review. *Reports on Progress in Physics* 79 (3): 036401.

15 You, J.W., Bongu, S.R., Bao, Q. et al. (2019). Nonlinear optical properties and applications of 2D materials: theoretical and experimental aspects. *Nano* 8 (1): 63–97.

16 Zhu, T., Snaider, J.M., Yuan, L. et al. (2019). Ultrafast dynamic microscopy of carrier and exciton transport. *Annual Review of Physical Chemistry* 70: 219–244.

17 Ginsberg, N.S. and Tisdale, W.A. (2020). Spatially resolved photogenerated exciton and charge transport in emerging semiconductors. *Annual Review of Physical Chemistry* 71: 1–30.

18 Boyd, R.W. (2007). *Nonlinear Optics*, 3e, 1–34. Academic Press.

19 Leuthold, J., Koos, C., and Freude, W. (2010). Nonlinear silicon photonics. *Nature Photonics* 4 (8): 535–544.

20 Fischer, M.C., Wilson, J.W., Robles, F.E. et al. (2016). Invited review article: pump-probe microscopy. *Review of Scientific Instruments* 87 (3): 031101.

21 Prylepa, A., Reitbock, C., Cobet, M. et al. (2018). Material characterisation with methods of nonlinear optics. *Journal of Physics D: Applied Physics* 51 (4): 043001.

22 Zhou, L., Fu, H., Lv, T. et al. (2020). Nonlinear optical characterization of 2D materials. *Nanomaterials* 10 (11): 2263.

23 Lau, K.Y., Liu, X., and Qiu, J. (2022). MXene saturable absorbers in mode-locked fiber laser. *Laser & Photonics Reviews* 16 (9): 2100709.

24 Liu, X., Guo, Q., and Qiu, J. (2017). Emerging low-dimensional materials for nonlinear optics and ultrafast photonics. *Advanced Materials* 29 (14): 1605886.

25 Zhang, M., Wu, Q., Zhang, F. et al. (2019). 2D black phosphorus saturable absorbers for ultrafast photonics. *Advanced Optical Materials* 7 (1): 1800224.

26 Zareba, J.K., Nyk, M., and Samoc, M. (2021). Nonlinear optical properties of emerging nano-and microcrystalline materials. *Advanced Optical Materials* 9 (23): 2100216.

27 Zhang, X.-j., Yuan, Z.-h., Yang, R.-x. et al. (2019). A review on spatial self-phase modulation of two-dimensional materials. *Journal of Central South University* 26 (9): 2295–2306.

28 Zhang, C., Wu, H., Xu, Y. et al. (2020). Plasmon-assisted broadband all-optical control of highly intense femtosecond laser by weak continuous-wave laser. *Advanced Optical Materials* 8 (19): 2000560.

29 Shi, B., Miao, L., Wang, Q. et al. (2015). Broadband ultrafast spatial self-phase modulation for topological insulator Bi_2Te_3 dispersions. *Applied Physics Letters* 107 (15): 151101.

30 Sheik-Bahae, M., Said, A.A., Wei, T.-H. et al. (1990). Sensitive measurement of optical nonlinearities using a single beam. *IEEE Journal of Quantum Electronics* 26 (4): 760–769.

31 Zhao, W. and Palffy-Muhoray, P. (1994). Z-scan measurement of $\chi(3)$ using top-hat beams. *Applied Physics Letters* 65 (6): 673–675.

32 Ferdinandus, M.R., Reichert, M., Ensley, T.R. et al. (2012). Dual-arm Z-scan technique to extract dilute solute nonlinearities from solution measurements. *Optical Materials Express* 2 (12): 1776–1790.

33 Soavi, G., Wang, G., Rostami, H. et al. (2018). Broadband, electrically tunable third-harmonic generation in graphene. *Nature Nanotechnology* 13 (7): 583–588.

34 Calafell, I.A., Irati, R.L.A., Alcaraz, I.D. et al. (2021). Giant enhancement of third-harmonic generation in graphene–metal heterostructures. *Nature Nanotechnology* 16 (3): 318–324.

35 Liu, W., Li, X., Yiling, S. et al. (2018). Cooperative enhancement of two-photon-absorption-induced photoluminescence from a 2D perovskite-microsphere hybrid dielectric structure. *Advanced Functional Materials* 28 (26): 1707550.

36 Lu, S., Sui, L., Liu, Y. et al. (2019). White photoluminescent Ti_3C_2 MXene quantum dots with two-photon fluorescence. *Advanced Science* 6 (9): 1801470.

37 Boudebs, G. and Cherukulappurath, S. (2004). Nonlinear optical measurements using a 4f coherent imaging system with phase objects. *Physical Review A* 69 (5): 053813.

38 Fedus, K. and Boudebs, G. (2013). Experimental techniques using 4f coherent imaging system for measuring nonlinear refraction. *Optics Communications* 292: 140–148.

39 Nie, Z., Li, Z., Zhang, X. et al. (2014). Symmetrical positive–negative annular phase object for optical nonlinearity characterization. *Physics Letters A* 378 (3): 131–138.

40 Nie, Z., Shi, G., Li, Z. et al. (2012). Investigation of the third-order nonlinear refraction using 4f coherent imaging system with positive–negative bar phase objects. *Optics and Lasers in Engineering* 50 (10): 1405–1409.

41 Li, Y., Pan, G., Yang, K. et al. (2008). Time-resolved pump-probe system based on a nonlinear imaging technique with phase object. *Optics Express* 16 (9): 6251–6259.

42 Yang, J., Song, Y., Wang, Y. et al. (2010). Reflection 4f coherent imaging technique for measurements of optical nonlinearity. *Optics Communications* 283 (2): 209–215.

43 Li, C., Wang, Y., Shui, M. et al. (2010). Characterization of signs change of nonlinear refraction in ZnSe based on a modified double 4f imaging system with a phase object. *Optics Communications* 283 (6): 1124–1128.

44 Zong, J., Li, Z., Zhou, W. et al. (2021). Ultrafast third order nonlinear optical properties of self-assembled indium phthalocyanine multilayer films. *Thin Solid Films* 735: 138890.

45 Huang, Y., Zhou, W., Li, X. et al. (2021). Highly broadband NLO response of acceptor–donor–acceptor materials with a planar conformation. *Materials Advances* 2 (6): 2097–2103.

46 Zong, J., Wu, X., Xin, J. et al. (2022). Ultrafast third-order nonlinear optical properties of self-assembled $PCPDTPhSO_3Na/C_{70}$-$(EDA)_8$ periodically overlapping film. *Journal of Modern Optics* 69 (1): 41–46.

47 Zong, J., Wu, X., Zhou, W. et al. (2021). Ultrafast third-order nonlinear optical properties of self-assembled poly [4, 4-bis (4-sulphobutyl)-4H-cyclopenta [2, 1-b:

3, 4-b′] dithiophene-2, 6-diyl]-1, 4-phenylene sodium sa multilayer films. *Journal of Modern Optics* 68 (9): 483–488.

48 Wang, Y., Lin, C.-Y., Nikolaenko, A. et al. (2011). Four-wave mixing microscopy of nanostructures. *Advances in Optics and Photonics* 3 (1): 1–52.

49 Jakubczyk, T., Delmonte, V., Koperski, M. et al. (2016). Radiatively limited dephasing and exciton dynamics in $MoSe_2$ monolayers revealed with four-wave mixing microscopy. *Nano Letters* 16 (9): 5333–5339.

50 Ko, B.A., Sokolov, A.V., Scully, M.O. et al. (2019). Enhanced four-wave mixing process near the excitonic resonances of bulk MoS_2. *Photonics Research* 7 (3): 251–259.

51 Ganikhanov, F., Carrasco, S., Xie, X.S. et al. (2006). Broadly tunable dual-wavelength light source for coherent anti-Stokes Raman scattering microscopy. *Optics Letters* 31 (9): 1292–1294.

52 Masia, F., Langbein, W., and Borri, P. (2012). Measurement of the dynamics of plasmons inside individual gold nanoparticles using a femtosecond phase-resolved microscope. *Physical Review B* 85 (23): 235403.

53 Hendry, E., Carrasco, S., Xie, X.S. et al. (2010). Coherent nonlinear optical response of graphene. *Physical Review Letters* 105 (9): 097401.

54 Li, B., Cheng, Y., Liu, J. et al. (2012). Direct optical imaging of graphene in vitro by nonlinear femtosecond laser spectral reshaping. *Nano Letters* 12 (11): 5936–5940.

55 Ciesielski, R., Comin, A., Handloser, M. et al. (2015). Graphene near-degenerate four-wave mixing for phase characterization of broadband pulses in ultrafast microscopy. *Nano Letters* 15 (8): 4968–4972.

56 Purz, T.L., Martin, E.W., Holtzmann, W.G. et al. (2022). Imaging dynamic exciton interactions and coupling in transition metal dichalcogenides. *The Journal of Chemical Physics* 156 (21): 214704.

57 Purz, T.L., Hipsley, B.T., Martin, E.W. et al. (2022). Rapid multiplex ultra-fast nonlinear microscopy for material characterization. *arXiv* preprint arXiv:2208.06633 .

58 Jahng, J., Brocious, J., Fishman, D.A. et al. (2015). Ultrafast pump-probe force microscopy with nanoscale resolution. *Applied Physics Letters* 106 (8): 083113.

59 Doherty, T.A.S., Winchester, A.J., Macpherson, S. et al. (2020). Performance-limiting nanoscale trap clusters at grain junctions in halide perovskites. *Nature* 580 (7803): 360–366.

60 He, H., Dong, X., Ren, Y.-X. et al. (2022). Tomographic-encoded multiphoton microscopy. *ACS Photonics* 9 (10): 3291–3301.

61 Nah, S., Spokoyny, B., Stoumpos, C. et al. (2017). Spatially segregated free-carrier and exciton populations in individual lead halide perovskite grains. *Nature Photonics* 11 (5): 285–288.

62 Yu, J., Li, Z., Liao, Y. et al. (2019). Probing the spatial heterogeneity of carrier relaxation dynamics in $CH_3NH_3PbI_3$ perovskite thin films with femtosecond time-resolved nonlinear optical microscopy. *Advanced Optical Materials* 7 (22): 1901185.

63 Block, A., Principi, A., Hesp, N.C.H. et al. (2021). Observation of giant and tunable thermal diffusivity of a Dirac fluid at room temperature. *Nature Nanotechnology* 16 (11): 1195–1200.

64 Cocker, T.L., Jelic, V., Hillenbrand, R. et al. (2021). Nanoscale terahertz scanning probe microscopy. *Nature Photonics* 15 (8): 558–569.

65 Nishida, J., Alfaifi, A.H., Gray, T.P. et al. (2020). Heterogeneous cation–lattice interaction and dynamics in triple-cation perovskites revealed by infrared vibrational nanoscopy. *ACS Energy Letters* 5 (5): 1636–1643.

66 Correa-Baena, J.-P., Luo, Y., Brenner, T.M. et al. (2019). Homogenized halides and alkali cation segregation in alloyed organic-inorganic perovskites. *Science* 363 (6427): 627–631.

5

Ultrafast Carrier Dynamics in Emerging 2D Materials

Jiawei Huang[1,2], *Ningning Dong*[1], *and Jun Wang*[1]

[1] *Chinese Academy of Sciences, Shanghai Institute of Optics and Fine Mechanics, Key Laboratory of Materials for High-Power Laser, Photonic Integrated Circuits Center, No. 390, Qinghe Road, Jiading, Shanghai, 201800 China*
[2] *Songshan Lake Materials Laboratory, No.1, Xuefu Road, Songshan Lake, Dongguan, Guangdong, 523808 China*

5.1 Introduction

With the further development of microelectronic science, artificial intelligence, the Internet of Things, cloud computing, and other information technologies have been widely penetrated into various fields of human society. In the meanwhile, a long-term contradiction between the increasing amount of data and insufficient processing capacity has also prominently arisen. The modern integrated computing unit industry has been approaching the limitation of "Moore's Law" and "Dennard scaling," which is a challenge for the traditional semiconductor technologies.

In contrast with the pure electronic method that is based on "electronically controlling electricity," the optoelectronic "electronically controlling light" technology (as represented by optical fiber) has greatly released the bottleneck for mass data transmission with high speed. However, this technology still needs to rely on electrical conversion for the moment, which still largely restricts the rate of communication. This inspires us that all-optical transmission and on-chip all-optical operation that "light-controlling light" route, could be a solution for these device innovation challenges. In fact, people have explored for half a century on the research of all-optical devices, and been growing consensus that: the development of high speed, high integration, and low-power cost devices must be considered in the field combining nano and ultrafast photonics.

Two-dimensional (2D) materials offer opportunities to solve the aforementioned problem. 2D materials consist of a single or few-layer atoms, while they can exhibit various interesting optical properties. Since the quantum confinement, coulomb screening, or surface effect will modify the electronic structure of low-dimensional material, which leads to the unexpected properties and phenomena compared to that in their bulk counterpart. Besides, its atomic limit nature also provides

Two-Dimensional Materials for Nonlinear Optics: Fundamentals, Preparation Methods, and Applications, First Edition. Edited by Qiang Wang and Hao-Li Zhang.

the intrinsic advantages for high-integrated applications and possible low-driving power costs.

Ultrafast time-resolved spectroscopy offers versatile, precise, and noninvasive techniques for the discovery of properties of the 2D materials on temporal scale. Ultrafast time-resolved spectroscopy is typically based on the pump–probe technique that samples the probed photons signal interacting with materials during the adjustment of the pump delay. The pump laser pulse usually generates the nonequilibrium of materials that can reflect the temporal optical and electronic properties in picosecond or even femtosecond scale.

In this chapter, we focus on the topic of ultrafast time-resolved spectroscopy in 2D materials, such as the emerging van der Waals 2D materials and the quasi-2D materials. In Section 5.2, the general principles of ultrafast time-resolved spectroscopy are discussed briefly in theoretical and experimental aspects. As an illustration, ultrafast optical properties in typical van der Waals 2D materials, such as graphene, transition metal dichalcogenides (TMDC), and black phosphorus (BP), are presented in Section 5.3. Moreover, ultrafast absorption response properties in emerging non-layered quasi-2D materials are introduced in Section 5.4. Finally, perspectives on ultrafast optics for 2D materials are made in Section 5.5.

5.2 Ultrafast Time-Resolved Spectroscopy

Ultrafast time-resolved spectroscopy offers a powerful tool to explore the time-evolutionary optical properties in materials. Thanks to the progress of ultrashort pulse generation technique, the temporal resolution of advanced ultrafast spectroscopy can reach femtoseconds or even attoseconds timescale, nowadays. This is an impossible measuring task for pure electronics, in which the electrical detection capacity limits to about 10 GHz (>100 μs). In this section, we introduce the standard pump–probe concept and its experimental installation for ultrafast spectroscopy measurements. Besides, the fundamental recombination mechanisms of excited carriers in semiconductors are also discussed briefly.

5.2.1 Principle of Pump–Probe Measurement

The realization of ultrafast time-resolved spectroscopy measurement is usually based on the framework of pump–probe technique, as shown in Figure 5.1. Figure 5.1a exhibits the principle of a typical pump–probe experiment that the pump and the probe laser pulses are focused on the sample by the lens. The pump laser pulse, which has a stronger pulsed intensity than the probe (10 or 100 times stronger in general), is employed to perturb the material from equilibrium optically. While the probe laser pulse monitors this perturbation, by sampling the transmission or reflection of material with different time delays from the pump laser pulse to reconstruct the dynamic evolution. The pump–probe time delay is usually achieved through a fine control of the optical path length. It should be noted that the laser focusing spot of the pump upon the sample is usually controlled

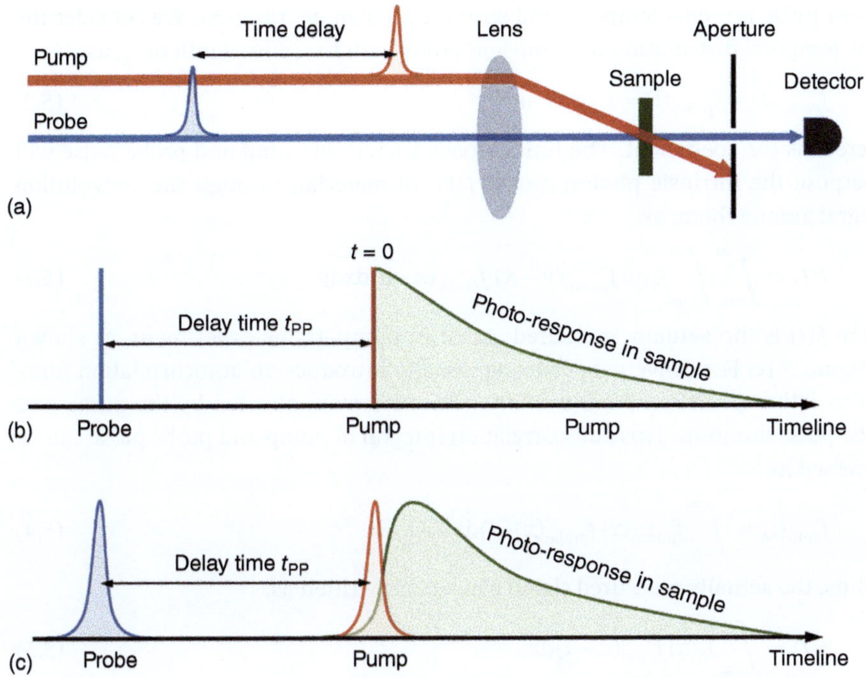

Figure 5.1 Principle of pump–probe measurement. (a) The concept of a pump–probe experiment. (b) The ideal optical response of material excited and detected by δ-functional pulse. (c) The practical optical response of material excited and detected by a finite laser pulse.

larger than that of the probe experimentally, aiming to obtain a perfect spatial overlap in the sample. Besides, one should also isolate the pump signal that arises an unwanted background from the probe signal to the detector, for example, as the aperture illustrated in Figure 5.1a. More experimental details are introduced in Section 5.2.2.

In ultrafast pump–probe experiment, the temporal resolution is mainly determined by the laser pulse duration width t_p. Assuming the pulse duration is infinitely short as a Delta function, carrier dynamic in materials responds as an instantaneous event. The dynamic properties of material can be faithfully reflected, as shown in Figure 5.1b. We suppose this event is a double exponential decay with time constant of τ_1 and τ_2. The dynamic function of this event through measuring the transmission or reflection signal can be described as,

$$f_0(t) = \begin{cases} 0, & t < 0 \\ A\exp(-t/\tau_1) + B\exp(-t/\tau_2), & t \geq 0 \end{cases} \tag{5.1}$$

where A and B are coefficients of the corresponding process. The independent variable t is experimentally equal to the pump–probe delay time t_{pp} as shown in Figure 5.1b.

The actual dynamic response of materials will be concealed in the practical measurement with a finite temporal pulse width, as shown in Figure 5.1c. In general, the fs-laser pulse presents temporal and spatial Gaussian distribution. We consider the same temporal distribution of pump and probe with the pulse width of t_p, as,

$$f_{pump}(t) = f_{probe}(t) = C exp\left(-t^2/t_p{}^2\right) \tag{5.2}$$

where C is the coefficient. The finite broaden width of pump and probe pulse will superpose the intrinsic photoresponse f_0(t) of material, through the convolution integral among them, as:

$$f(t) = \int_{-\infty}^{\infty}\int_{-\infty}^{\infty} f_0(y) f_{\text{pump}}(y - x)\, f_{\text{probe}}(x - t)\text{d}x\text{d}y \tag{5.3}$$

where $f(t)$ is the actually measured signal in pump–probe experiment, as shown in Figure 5.1c. For convenience, people usually introduce an autocorrelation function to estimate the temporal resolution in the system affected by the pump and probe pulse duration. This autocorrelation integral of pump and probe pulse can be expressed as,

$$f_{\text{auto}}(t) = \int_{-\infty}^{\infty} f_{\text{pump}}(x) f_{\text{probe}}(x - t)\text{d}x \tag{5.4}$$

Thus, the actually measured signal can be also written as,

$$f(t) = \int_{-\infty}^{\infty} f_0(x) f_{\text{auto}}(x - t)\text{d}x \tag{5.5}$$

Noted that the autocorrelation should only rely on the experimental system, and is independent of the sample. Considering the Eqs. (5.1) to (5.5), one can obtain the relationship [1]:

$$f(t) = D\int_0^{\infty}\left[A\exp\left(-\frac{x}{\tau_1}\right) + B\exp\left(-\frac{x}{\tau_2}\right)\right]\exp\left(\frac{-(x-t)^2}{2t_p{}^2}\right)\text{d}x \tag{5.6}$$

For brevity, we could introduce the error function:

$$\text{erf}\, z = \frac{2}{\sqrt{\pi}}\int_0^{z}\exp(-x^2)\text{d}x \tag{5.7}$$

or

$$\text{erfc}\, z = 1 - \text{erf}\, z = \frac{2}{\sqrt{\pi}}\int_z^{\infty}\exp(-x^2)\text{d}x \tag{5.8}$$

Therefore, the actually measured dynamic signal can be simply given as [2],

$$\begin{aligned} f(t) = {} & D_1\exp\left(-\frac{t}{\tau_1}\right)\text{erfc}\left(\frac{t_p}{\sqrt{2}\tau_1} - \frac{t}{\sqrt{2}t_p}\right) \\ & + D_2\exp\left(-\frac{t}{\tau_2}\right)\text{erfc}\left(\frac{t_p}{\sqrt{2}\tau_2} - \frac{t}{\sqrt{2}t_p}\right) \end{aligned} \tag{5.9}$$

where D_1 and D_2 are the amplitudes of the specific processes, respectively. It is worth noting that the Eq. (5.9) can be extended to contain the more exponential processes [3, 4].

5.2.2 Experiment for Ultrafast Spectroscopy

The experimental installation schematic of ultrafast pump–probe spectroscopy is illustrated in Figure 5.2. In this experiment, the pump laser (red solid line) pulse and the probe laser (blue solid line) pulse are usually configured with orthogonal light polarization to reduce their coherence and separate them before detection. This polarization control and selection can be realized through a half-wave plate or a quarter-wave plate with a polarizing film. Noted that the half-wave plate here also allows one to investigate the polarization-dependent ultrafast optical properties [5–8]. The transient dynamic signal of material usually demands an amplification through the lock-in amplifier. The chopper that modulates the light signal by means of introducing extra frequency, is usually placed on the pump laser path to modulate it, as shown in Figure 5.2. The probe signal, which is collected by the spectrograph, will also carry this modulated frequency information through the pump-induced material response. The pump signal should be isolated with the probe via setting different light path or selecting specific polarization or filters.

The probe laser beam typically changes the optical path length to control the pump–probe time delay on the sample. A fine motorized position stage with retroreflector mirror here is employed to realize it experimentally, as shown in Figure 5.2, which will help to acquire good temporal resolution. In ultrafast pump–probe spectroscopy experiment, a steady, board, and high-power white light supercontinuum (WLS) is essential to obtain a differential transmission (DT) or reflection spectroscopy. Ultrafast WLS can be generated using CaF_2, MgF_2, or sapphire crystals excited by a fs-laser. Figure 5.2 also exhibits the transmission and reflection types of ultrafast pump–probe spectroscopy, which carries the identical dynamic properties of material.

5.2.3 Fundamental Mechanism of Recombination Dynamics

For a semiconductor, the interband carrier transition and recombination between conduction band (CB) and valence bands (VB) determine the global carrier

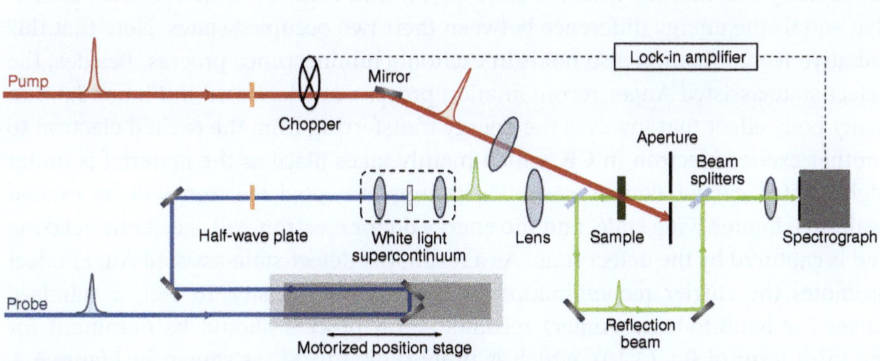

Figure 5.2 The experimental schematic configuration of ultrafast pump–probe spectroscopy. The red line denotes the pump laser beam, while the blue one is the probe counterpart. The green line indicates the white light supercontinuum excited by the probe.

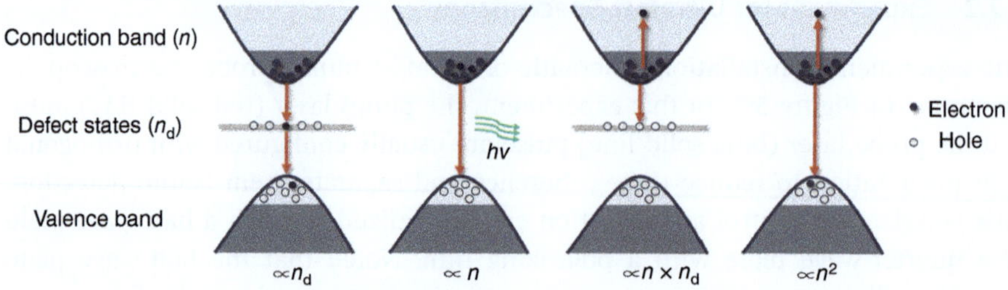

Figure 5.3 Fundamental mechanisms of recombination dynamics for semiconductor. The index n denotes the photoinduced carrier (electron) density in conduction band, while n_d stands for the carrier (electron) density captured by defect.

dynamic property of materials. Here we introduce the fundamental recombination mechanisms of a common semiconductor, as shown in Figure 5.3[9–15]. In fact, when one investigates the carrier dynamic behavior, the recombination lifetime τ of the semiconductor material is typically a function of the excited carrier density n, as illustrated in Figure 5.3.

The reciprocal of the excited carrier lifetime $1/\tau$, which is proportional to the carrier recombination rate, is dependent on the excited carrier density n as,

$$\frac{1}{\tau} = A + Bn + Cn^2 \tag{5.10}$$

where A, B, and C indicate the coefficients for the different recombination mechanisms. The first term A of Eq. (5.10) is independent of the excited carrier density n, which can be contributed from the Shockley–Read–Hall or defect-state-assisted recombination. This recombination phenomenon is proportional to the density of defect-state n_d, in which the defect-state captures the electrons from the CB and subsequently releases them to the VB with multiphonon emission, as shown in Figure 5.3. The second term of Eq. (5.10) is linearly dependent on the excited carrier density n. Both radiative and defect-state-assisted Auger recombination processes will contribute to this term. As shown in Figure 5.3, excited electron in CB directly recombines with the hole in VB and releases a photon with energy that equals the energy difference between their two occupied states. Note that this radiative recombination also holds in excitonic luminescence process. Besides, the defect-state-assisted Auger recombination process, as illustrated in Figure 5.3, is a many-body effect that involves the energy transferring from the excited electron to another excited electron in CB, which mainly takes place as the material is under high excited carrier density case. This energy-accepted electron will be excited again to a higher-lying state, and the energy donor electron will accelerate relaxing and is captured by the defect state. As a result, the defect-state-assisted Auger effect promotes the carrier recombination at high carrier density. In fact, a standard Auger (or band-to-band Auger) recombination process should be dominant for the third term of Eq. (5.10), which is proportional to n^2, as shown in Figure 5.3. This process is also common at high carrier injections since there are three particles that electron–electron–hole (or electron–hole–hole) will interact with each other.

5.3 Ultrafast Optics in van der Waals 2D Materials

5.3.1 Graphene

Graphene, which is a typical 2D material isolated from graphite, has a broadband optical response from visible to terahertz energy range. The ultrafast carrier dynamics properties in graphene are rich and attract great interest from researchers in the aspects of basic physics and advanced photonic applications. The nonequilibrium distributions of photoinduced electrons and holes in ultrafast timescale will govern the transient carrier dynamic properties of materials.

As in Figure 5.4a, M. Baudisch et al. identifies the instantaneous response of Dirac fermions, revealing the role of hot carriers in graphene on sub-100 fs timescales [16]. Similarly, P. George et al. reported the ultrafast carrier dynamic properties in intraband and interband of graphene using optical-pump THz-probe spectroscopy [19]. They found that there are three likely processes of photoinduced carriers in nonequilibrium graphene. In several tens to hundred femtoseconds after optical excitation, the photon-excited electron and hole redistribute, and rapidly thermalize and cool, subsequently. The hot carrier further cools through intraband phonon scattering in hundreds of femtoseconds to picoseconds. The electron and hole finally recombine and reach equilibrium distribution again in tens picoseconds. Although, the intraband process is rather rapid, M. Breusing et al. captures this ultrafast intraband dynamic process in freestanding graphene employing high-temporal-resolution optical pump–probe spectroscopy with 7-fs laser pulses, as shown in Figure 5.4b [17]. They have observed the ultrafast intraband carrier equilibration within 30 fs, while the phonon-assisted intraband carrier cooling process is on timescale of about 100 fs. Wang et al. fabricated multilayer graphene saturable absorber mirror (GSAM) via liquid-phase exfoliation dispersion of graphene using vacuum filtration [18]. A commercial semiconductor saturable absorber mirror (SESAM) designed for 2 μm with similar linear optical reflection was compared with the prepared GSAM, as Figure 5.4c shown. They carried out degenerate pump–probe measurements on the GSAM and the SESAM using reflection geometry with both the pump and the probe pulses at 2 μm and 800 nm. The graphene displayed carrier relaxation times of 1.14 and 1.2 ps at 800 nm and 2 μm, respectively, while the SESAM was slower at 63 ps at 2 μm, as shown in Figure 5.4d.

5.3.2 Transition Metal Dichalcogenides (TMDC)

Inspired by the isolating atomic graphene from graphite, researchers have successfully prepared diverse 2D semiconductor materials with various optical properties. Among the 2D material family, TMDC play an important role in nanophotonic functional applications due to their multiple operation routes, such as the tunable bandgap, significant excitonic effects, many-body effect, layer-twistable, heterojunction, and the like. The chemical formula of TMDC materials usually consists of one transition metal and two chalcogenide elements (as MX_2 form, like MoS_2, WSe_2, and $PtSe_2$). These TMDC molecules combine through the chemical bonds within the layer, while through van der Waals interactions between the layers.

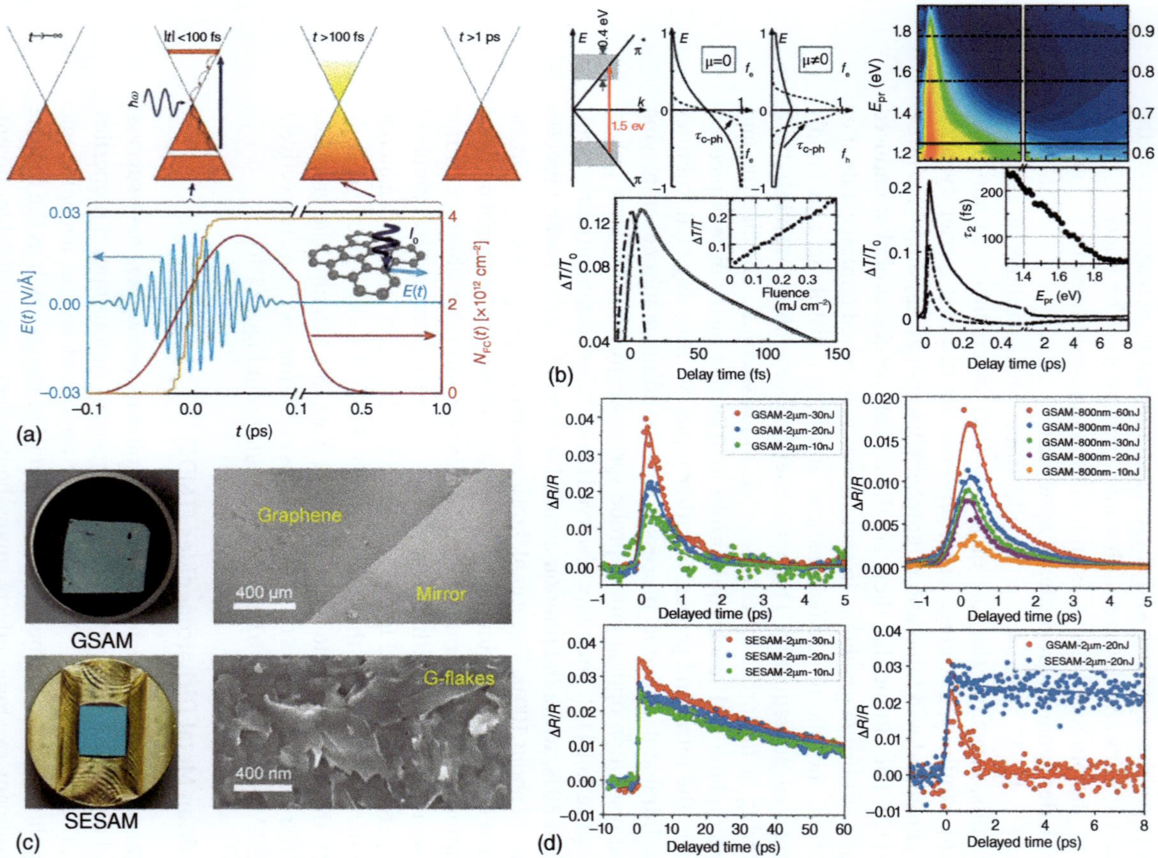

Figure 5.4 The ultrafast carrier dynamics in graphene. (a) The photoinduced electron distributions and the temporal evolution of Dirac fermions in graphene. (b) Ultrafast optical pump–probe study of graphene using an ultrashort laser pulse source with a time resolution of about 10 fs. (c) Characterizations of graphene saturable mirror (GSAM) taken by camera (top left), low magnification SEM (top right), and higher one (bottom right). The commercial SESAM (BATOP, SAM-2000-2044-2010 ps) for comparison is shown in the bottom left. (d) Time-dependent reflectivity of GSAM and the commercial SESAM with fs pulses excitation (~150 fs, 100 kHz). 1 nJ corresponds to the laser fluence of ~0.106 mJ cm^{-2} and ~ 12.0 μJ cm^{-2} for 800 nm and 2 μm, respectively. Source: (a) Reproduced with permission Baudisch et al. [16]. Copyright © 2018, Springer Nature; (b) Reproduced with permission Breusing et al. [17], © 2009, American Physical Society; (c, d) Reproduced with permission Wang et al. [18], © 2017, John Wiley & Sons.

Molybdenum/Tungsten Disulfide/Diselenide (MoS_2/$MoSe_2$/WS_2/WSe_2). In contrast to bandgap-free graphene, two crucial energy ranges within and out of the bandgap govern the global ultrafast optical property of semiconducting TMDC materials. Thereinto, the prominent exciton effect within the bandgap is essential character for low-dimensional TMDC materials. S. Dal Conte et al. employed time-resolved Faraday rotation and time-resolved circular dichroism techniques in monolayer MoS_2 as Figure 5.5a shown [20]. They found the exciton–exciton interactions of MoS_2 play a leading role in valley relaxation dynamics. A. Chernikov et al. found a giant bandgap renormalization (BGR) and population effect in few-layered WS_2, when they used the transient absorption spectra with an intense optical pump pulse, as Figure 5.5b shown [21]. Tianju zhang et al. investigated the exciton–exciton annihilation (EEA) process of four types of CVD-grown monolayer WS_2, WSe_2, MoS_2, and $MoSe_2$ films with femtosecond transient absorption spectroscopy (TAS) at higher exciton densities of $\sim 10^{12}\,cm^{-2}$ below a Mott density of $\sim 10^{13}\,cm^{-2}$, as Figure 5.5c,d shown [22]. The TA spectra were performed with the pump light of 520 nm at the temperature of 300 K. The results of TA spectroscopy demonstrate that the defect-assisted EEA process of local excitons is enhanced and plays a key role in the exciton relaxation process when the exciton density exceeds a value of approximately $10^{12}\,cm^{-2}$ due to the defect localized effect. The measured EEA rates for the atomically thin-layered WS_2, WSe_2, $MoSe_2$, and MoS_2 were 0.016, 0.026, 0.049, and $0.102\,cm^2\,s^{-1}$, respectively, implying that EEA is enhanced as the defect states increase in monolayer TMDC. Moreover, F. Rana's team found the fast and slow defect-assisted carrier recombination process in monolayer MoS_2 when they employed ultrafast optical pump–probe monitoring the mid-gap defect states [15]. Besides, the unique strong Coulomb interaction in 2D material results in a significant many-body effect like the Auger effect in monolayer MoS_2. Except for the quantum effect in monolayer, few-layer MoS_2 also undergoes remarkable surface recombination effect, which is a particularly important effect in the wafer fabrication industry [23].

Platinum Diselenide. $PtSe_2$, a rising member of group-10 TMDC family, possesses many intriguing properties including an indirect energy bandgap ranging from 0 to 1.2 eV, high carrier mobility, and good environmental stability, as shown in Figure 5.6a,b[24, 25]. L. Wang et al. prepared 4, 7, 17, and 55-layer $PtSe_2$ films through thermally assisted conversion (TAC) method, with the bandgap of 0.87, 0.59, 0.33, and 0 eV, respectively [24]. They explored the excited carrier dynamics under 380 fs pulses at 1040 nm with a repetition rate of 1 kHz at room temperature through degenerated time-resolved pump–probe technique. For the two thinner films (4 and 7 layer), transient dynamics show a two-photon absorption (2PA)-induced negative $\Delta T/T$ with a slow recovery process of more than a hundred picoseconds. Conversely, the two thicker films show an initial SA response followed by free carrier absorption process, as shown in Figure 5.6e,f. Furthermore, they proposed a four-level rate equation model to simulate the relaxation time and absorption cross section of the states in the CB, as shown in Figure 5.6g. Combining with the nonlinear optical, ultrafast carrier dynamics, and electrical property measurements, they confirmed the transition from semiconductor to semimetal in $PtSe_2$ from 17 to 55 layers.

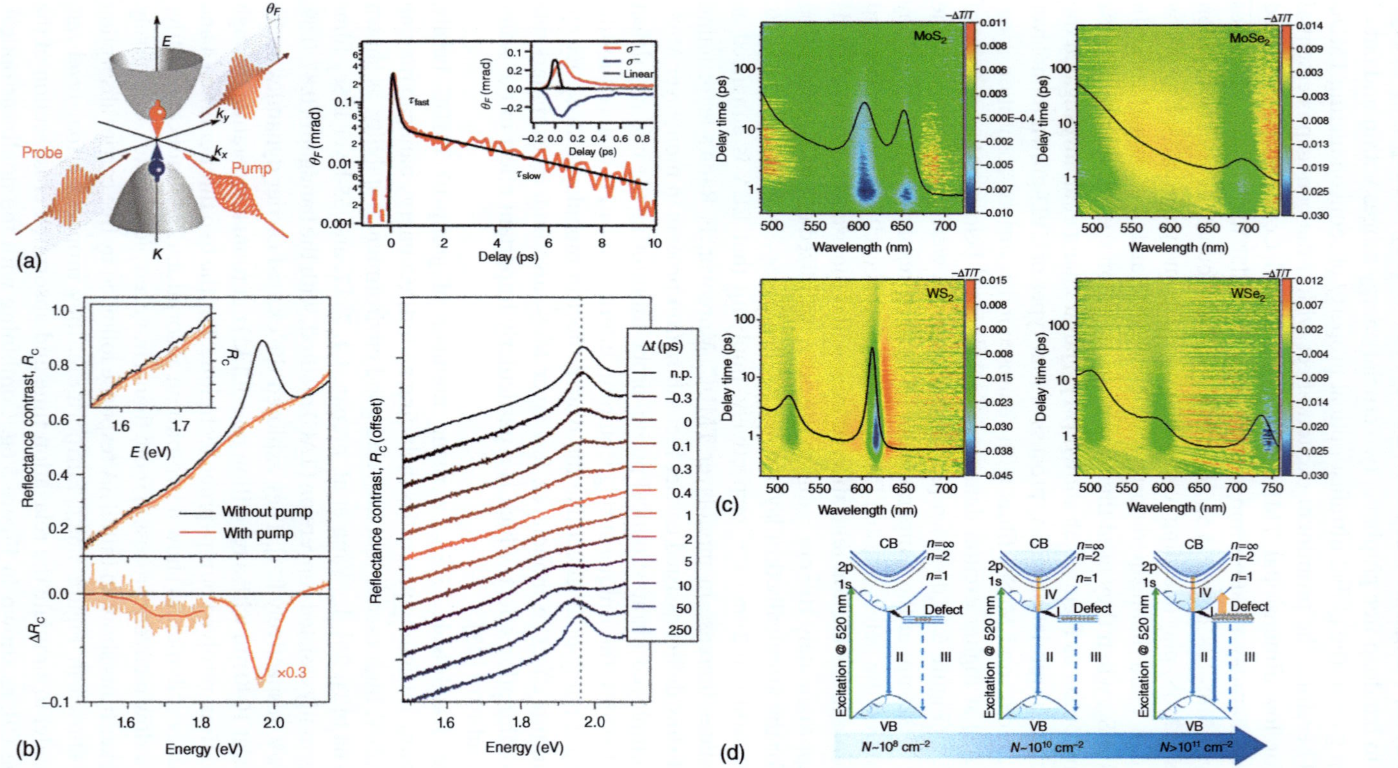

Figure 5.5 Ultrafast dynamics in 2D TMDC. (a) Ultrafast exciton valley relaxation dynamics in monolayer MoS_2. (b) The giant many-body effect in few-layered WS_2. (c) 2D pseudocolor TA spectra of the four CVD-grown monolayer TMDC obtained by pumping at a wavelength of 520 nm for MoS_2, $MoSe_2$, WS_2, and WSe_2 with exciton densities corresponding to 6.8×10^{12}, 4.2×10^{12}, 6.08×10^{12}, and 8.1×10^{12} cm^{-2}, respectively. The black solid line indicates the linear absorption spectrum. (d) Schematic of the relaxation process of excitons at different exciton concentrations *n*. Source: (a) Reproduced with permission Dal Conte et al. [20]. Copyright © 2015, American Physical Society; (b) Reproduced with permission Chernikov et al. [21]. Copyright © 2015, Springer Nature; (c,d) Reproduced with permission Zhang and Wang [22], © 2022, American Chemical Society.

As shown in Figure 5.6b, G. Wang et al. prepared three $PtSe_2$ thin films with thicknesses of 3.0, 4.6, and 16.2 nm using the same TAC method [25]. They investigated the excited-carrier dynamics of $PtSe_2$ using time-resolved spectroscopy with the pumping of 400 nm (3.1 eV) femtosecond laser pulses and the probing of super-continuum white light source from 470 nm (2.64 eV) to 720 nm (1.72 eV), as shown in Figure 5.6h,i. Positive DT signals within 1 ps near zero delay time appeared in all the three samples, and this can be attributed to the filling of states and Pauli-blocking by optically excited carriers. Negative DT signals were found around 2–4 ps which are due to BGR caused by the filling of a large amount of electron–hole plasma. This BGR results in a redshift in interband absorption and less transmission. A triple-phase decay equation was employed to obtain carrier relaxation times which follow thickness-dependent characteristics of 0.6–1.7, 2–3, and 27–250 ps for the carrier–phonon scattering, BGR, and electron–hole recombination, respectively. In addition to, they performed the chemical potential evolution analysis which implies that the carrier dynamics in few-layered $PtSe_2$ follow semiconducting behavior instead of metallic behavior, as shown in Figure 5.6j. This is consistent with Ref [25].

Palladium Diselenide. $PdSe_2$ is one kind of TMDC material. Unlike the 1 T or 2H phase in traditional TMDCs, $PdSe_2$ has a puckered pentagonal atomic structure, where one Pd atom covalently coordinates with four Se atoms and two adjacent Se atoms are bonded to each other, as Figure 5.7a shown. Besides, its bandgap can tune between 1.3 eV (monolayer) and 0 eV (bulk). Xin Chen et al. prepared a trilayer $PdSe_2$ film via chemical vapor deposition and studied its ultrafast carrier dynamics by pumping at 300 nm and probing in the range of 350–900 nm, as shown in Figure 5.7b,c[26]. The film exhibits an absolute negative absorption signal band in the 350–650 nm which is defined as photoinduced bleaching (PB), and positive signal in 650–900 nm which is known as photoinduced absorption (PIA). Besides, they also employed degenerate pump–probe measurements at 520 nm. The ultrafast dynamic process can be divided into three stages: (i) After the pump excitation, PB-induced positive differential transmittance which reaches its maximum within hundreds of femtoseconds. (ii) A fast recovery of the positive signal occurs within 3 ps with a weight of ~50%. (iii) A slow recovery lasting ~100 ps. The recombination mechanism can be explained based on the shallow and deep defect models.

Rhenium Disulfide. ReS_2, a rising member of the TMDC family, has triggered many novel studies in optoelectronic field on account of its peculiar bandgap evolution. Lei Wang et al. synthesized trilayer ReS_2 film and investigated its carrier dynamics by pump–probe spectroscopy with the pump light at 400 nm, and the probe light in the range from 430 to 640 nm and 900 to 1400 nm, as Figure 5.7d–i shown [27]. As the bandgap of trilayer ReS_2 is much smaller than the pump photon energy, the electrons in the VB can be excited into the CB effectively. For the visible range, the transient absorption dynamics show an initial PB immediately after the excitation, followed by a speedy recovery of PIA and a slow relaxation process that lasts for more than 300 ps, as shown in Figure 5.7g. For near-infrared range, the transient absorption dynamics only contain an incipient PIA component and the corresponding long recovery process. The dependence of transient transmission on pump intensity and temperature ruled out most of the recombination mechanisms like phonon-assisted

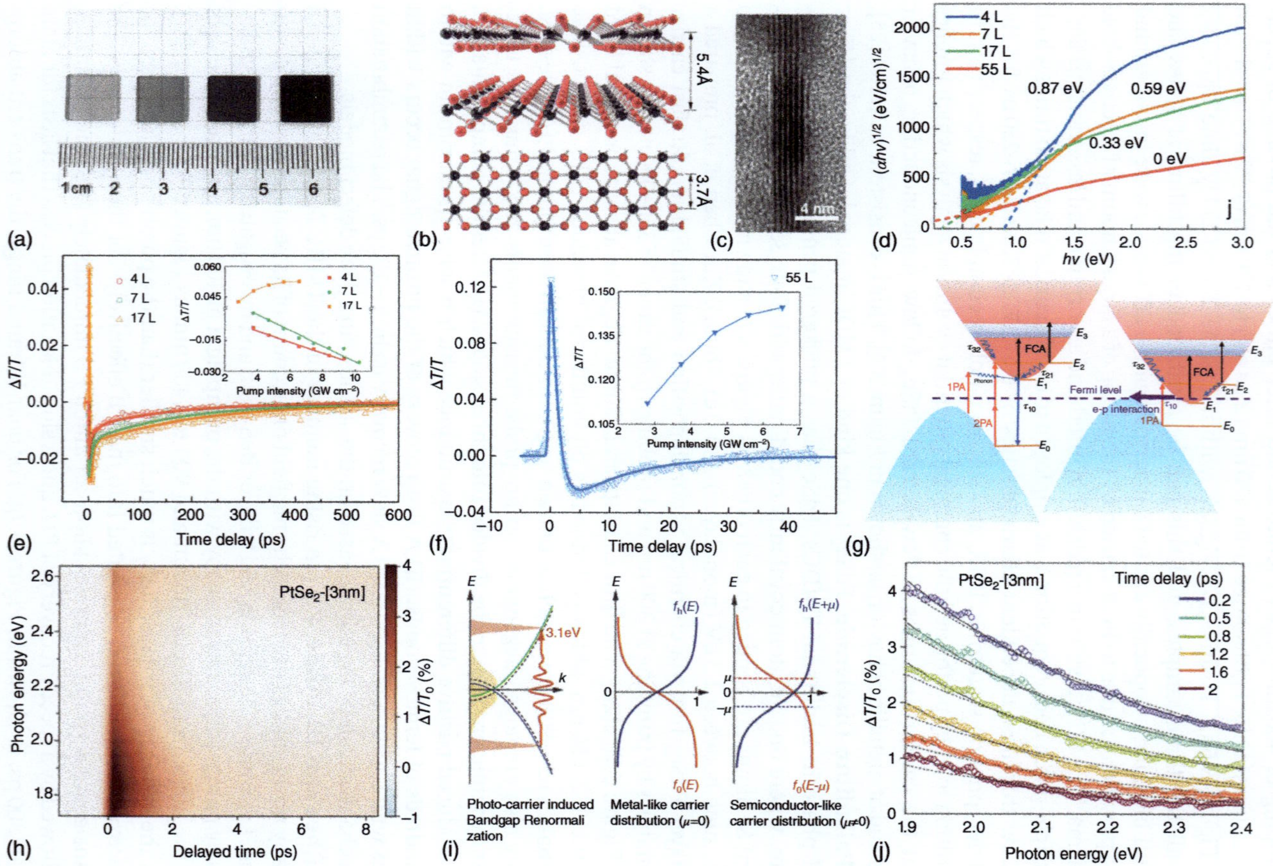

(a)
1 cm 2 3 4 5 6
(b)
5.4Å
3.7Å
(c)
4 nm
(d)
4 L
7 L
17 L
55 L
0.87 eV
0.59 eV
0.33 eV
0 eV
j
(αhν)^{1/2} (eV/cm)^{1/2}
hν (eV)
(e)
4 L
7 L
17 L
ΔT/T
Pump intensity (GW cm^{−2})
Time delay (ps)
(f)
55 L
ΔT/T
Pump intensity (GW cm^{−2})
Time delay (ps)
(g)
FCA
E_3
E_2
E_1
E_0
τ_32
τ_21
τ_10
1PA
2PA
Phonon
Fermi level
e-p interaction
(h)
PtSe_2-[3nm]
Photon energy (eV)
Delayed time (ps)
ΔT/T_0 (%)
(i)
E
k
3.1eV
f_h(E)
f_0(E)
f_h(E+μ)
f_0(E−μ)
μ
−μ
Photo-carrier induced Bandgap Renormali zation
Metal-like carrier distribution (μ=0)
Semiconductor-like carrier distribution (μ≠0)
(j)
PtSe_2-[3nm]
Time delay (ps)
0.2
0.5
0.8
1.2
1.6
2
ΔT/T_0 (%)
Photon energy (eV)

Figure 5.6 (a) Photographs of the $PtSe_2$ films synthesized by the thermally assisted conversion method with the initial Pt thicknesses of 0.5, 1, 3, and 5 nm, respectively. (b) Side- and top-view of 1 T-type octahedral $PtSe_2$ crystal structures. Dark balls represent Pt atoms while red ones indicate Se atoms. (c) TEM image of the cross section of the $PtSe_2$ film with the starting Pt metallic film thickness of 0.5 nm. (d) Tauc plots of the four $PtSe_2$ films in (a) with the layer number of 4, 7, 17, and 55, respectively. (e, f) Transient transmission dynamics of $PtSe_2$ films at 1040 nm with a pump intensity of 3.72 GW cm^{-2}, while the solid lines represent the fitting using a four-level rate equation model. The insets show $\Delta T/T_0$ versus the pump intensity at zero time delay. (g) Schematic illustration of the four-level system. (h) Time-resolved DT mapping in the visible range from 1.72 eV (720 nm) to 2.64 eV (470 nm) for the $PtSe_2$ sample with 3 nm starting Pt thickness. The pump fluence was 0.94 mJ cm^{-2}. (i) Left: schematic of the shrunken energy bands (solid lines) from original ones (dashed lines) of dE due to bandgap renormalization (BGR). Middle and right: illustrations show the metal- (zero chemical potential) and semiconductor-like carrier distributions (nonzero chemical potential) in the thermal region after optical excitation. (j) The ultrafast carrier dynamics of the $PtSe_2$ film with the starting Pt metallic film thickness of 3 nm. Source: (a, d–g) Reproduced with permission Wang et al. [24], © 2019, John Wiley & Sons; (b, c, h–j) Reproduced with permission Wang et al. [25], © 2019, John Wiley & Sons.

trapping and bandgap renormalization effect, suggesting that band-to-band Auger recombination also played an important role. With the rate equation simulations, it was concluded that Auger effect dominated the nonradiative recombination and led to the deterioration of the radiative quantum efficiency in atomic thin ReS_2 film. Their results provide a detailed understanding of recombination mechanism in visible and near-infrared range of ReS_2 and will be helpful in evaluating and enhancing the performance of ReS_2-based optoelectronic devices.

5.3.3 Black Phosphorus

BP is one of the most promising 2D materials for optoelectronic applications due to its high carrier mobility, remarkable anisotropy, broadband linear/nonlinear optical response, etc [2, 28–31]. It also maintains direct bandgap from 0.3 to ~2 eV with different number of atomic layers. BP is believed to bridge the gap between graphene and TMDCs where the former suffers the drawback of zero bandgap and the latter possesses limited response in the near- to mid-infrared region.

For example, V. Pareek et al. found the dimensionality-tunable EEA phenomenon, when they employed the micro-transient absorption spectroscopy in bilayer BP, as Figure 5.8a shown [32]. R. Suess et al. used the polarization-resolved pump–probe measurement identifying the strong anisotropic ultrafast optical absorption in fewer layer BP, as Figure 5.8b shown [6]. K. Wang et al. prepared BP nanosheet dispersions by liquid-phase exfoliation method and carried out degenerate pump–probe experiments in the range of 800–2026 nm to characterize their excited-state dynamics [2]. They also measured graphene dispersion for comparison. They found a Pauli-blocking-induced SA response in BP dispersions. They revealed BP's carrier-carrier scattering time to be 16 and 32 fs for 800 and 2026 nm, and carrier-phonon scattering time to be 0.360, 0.928, 1.358 ps at 800, 1600, and 2026 nm, decreasing with increasing wavelength. Comparative studies

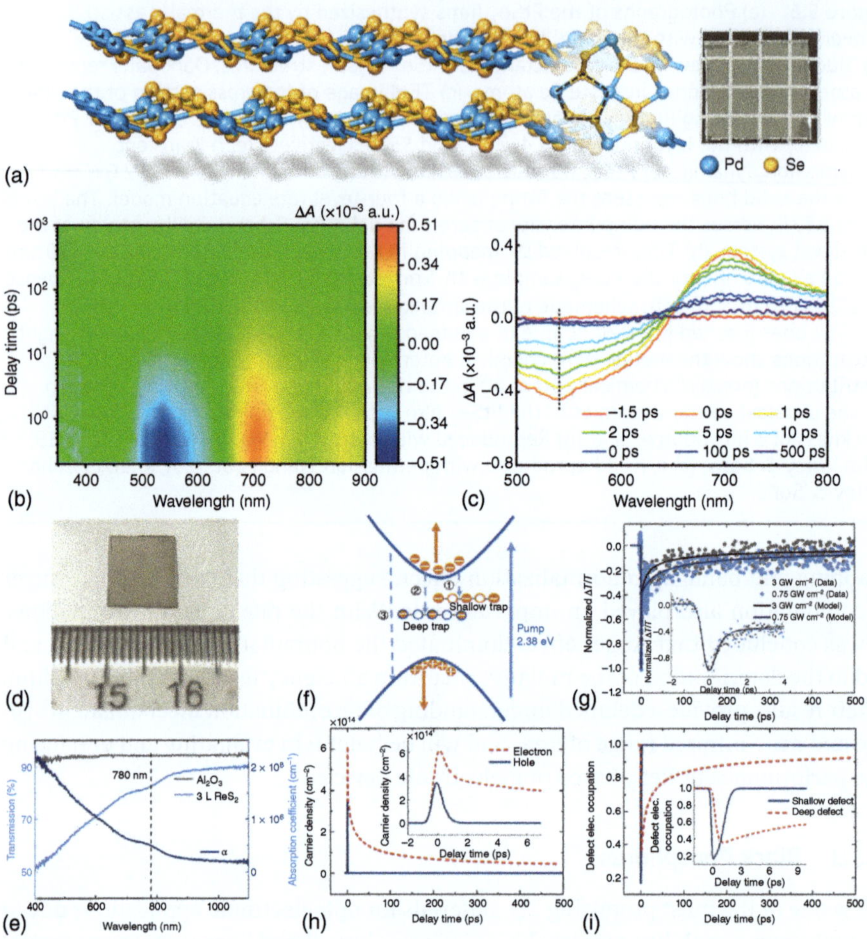

Figure 5.7 (a) Crystal structure of $PdSe_2$. The blue and yellow balls represent the Pd and Se atoms, respectively. The right is the photograph of synthesized $PdSe_2$ film. (b) TAS pumped at 300 nm and probed from 350 to 900 nm. (c) Extracted TAS with different delay time. (d) Image, and (e) linear transmission and absorption coefficient spectra of 3 L ReS_2 sample. (f) Schematic of charge carrier recombination in 3 L ReS_2 film. g The simulation and experiment results of transient dynamics are plotted versus the probe delay at two extreme values of pump intensities of 0.75 and 3 GW cm^{-2}. The inset shows the details for the first 7 ps. The simulated (h) electron and hole densities and (i) electron occupations of shallow and deep defects are plotted versus time delay with 3 GW cm^{-2} pump excitation. Source: (a–c) Reproduced with permission Chen et al. [26], © 2022, John Wiley & Sons; (d–i) Reproduced with permission Wang et al. [27], © 2020, Optica Publishing Group.

indicate much faster relaxation of excited carriers of BP than that of graphene with the carrier-carrier scattering time of 47 and 225 fs at 800 and 2026 nm, and the carrier-phonon scattering time of 0.434 and 2.05 ps at 800 and 1600 nm and a bit faster lifetime at 2026 nm. Furthermore, V. Iyer et al. identified the ultrafast optical properties of BP in energy range within and out of bandgap, using the pump–probe experiments with a board mid-infrared range, as shown in Figure 5.8c [33]. Besides,

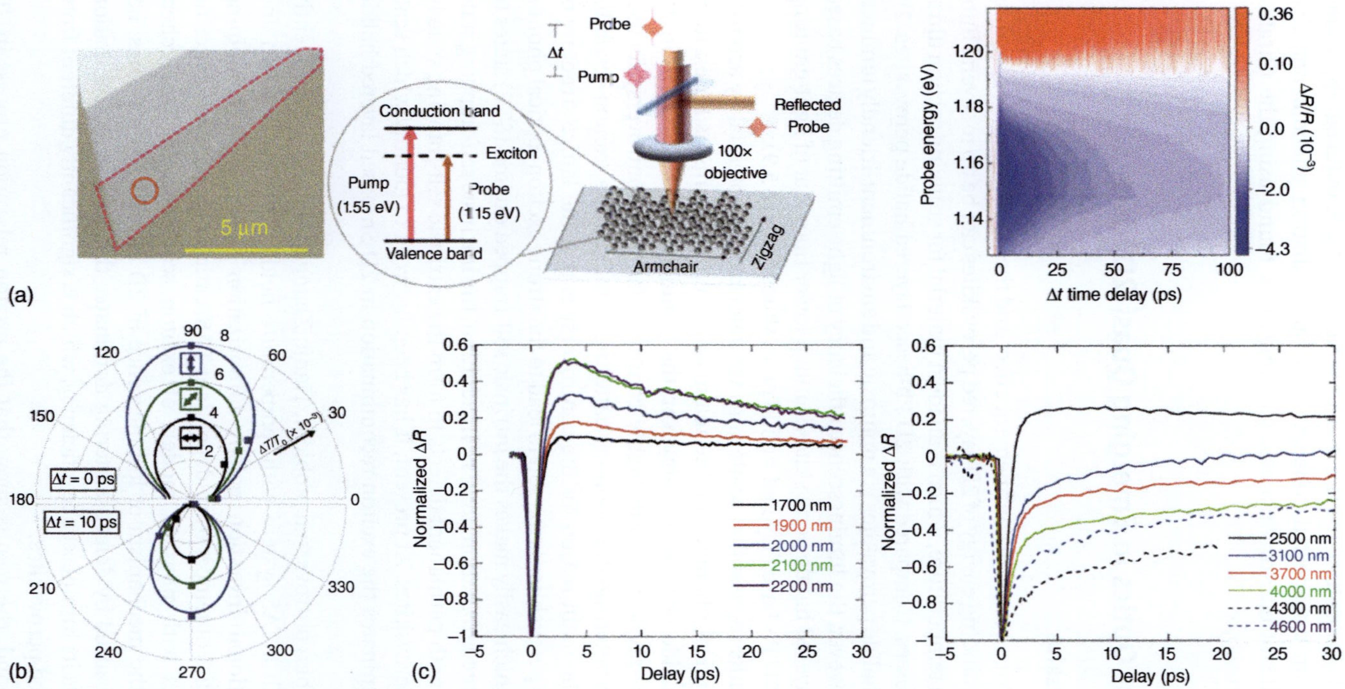

Figure 5.8 (a) Micro-transient absorption spectroscopy of bilayer BP. (b) The polarization-dependent ultrafast absorption property of few-layer BP. (c) Transient dynamics of BP in mid-infrared range. Source: (a) Reproduced with permission Pareek et al. [32]. Copyright © 2020, American Physical Society; (b) Reproduced with permission Suess et al. [6]. Copyright © 2015, AIP Publishing; (c) Reproduced with permission Iyer [33]. Copyright © 2017, IOP Publishing Ltd.

M. Ji's group employed the spatiotemporal resolved pump–probe microscopy technique on the exfoliated few-layer BP [34]. They measured the layer-dependent resonant transient absorption spectra and observed the BGR-induced absorption in BP. Moreover, an unexpected coherent phonon oscillation phenomenon that reflects the significant electron–phonon couplings was found using the ultrafast pump–probe microscopy.

5.4 Ultrafast Optics in Emerging Quasi-2D Materials

5.4.1 2D Perovskite

With the strong space quantum confinement effect and the weak dielectric shielding effect, the exciton binding energy of 2D-layered perovskites can be up to several hundred meV at room temperature, which is an ideal material for studying exciton ultrafast physical processes. Compared with 2D Pb-based layered halide perovskites, 2D Sn-based layered halide perovskites is a nontoxic and environment-friendly luminescent material. However, the luminescence efficiency of light-emitting devices based on 2D Sn-based layered halide perovskites is much lower than that of light-emitting devices based on 2D Pb-based layered halide perovskites (Figure 5.9).

The present results suggest that scatterings via deformation potential by acoustic and homopolar optical phonons are the main scattering mechanisms for excitons in 2D Pb-based halide perovskites, however, the luminescence mechanism in 2D Sn-based halide perovskites is still highly controversial. The controversy focuses on which defect scattering and phonon scattering dominate the exciton recombination process. The main reason for the controversy is that the lattice structure of 2D-layered halide perovskites is too soft to lead to the strong exciton–optical phonon coupling effect, which easily makes the intrinsic self-trapped exciton (STE) states to affect the exciton recombination process. In addition, the intrinsic self-trapping state can easily couple with permanent defects to form the extrinsic self-trapping state in Sn-based halide perovskites. At present, it has been no work to identify which scattering process dominates the exciton recombination in 2D Sn-based layered halide perovskites.

In response to this controversial problem, Tianju Zhang et al. have independently designed and built a TAS system with temperature field regulation function and obtained the national invention patent authorization [35, 36]. By combining experimental studies of temperature-dependent TAS, cation regulation engineering of 2D perovskites with physical models of exciton scattering, the researchers further revealed the mechanism of luminescence of 2D Sn-based perovskites, i.e. the deformation caused by defect scattering dominates the exciton recombination process and affects its luminescence efficiency, which is significantly different from that of 2D Pb-based perovskites.

On the other hand, the results show that the exciton relaxation process in 2D Sn-based perovskites can be divided into three components: (i) the sub-picosecond lifetime relaxation component, which is a competition between the defect-trapped

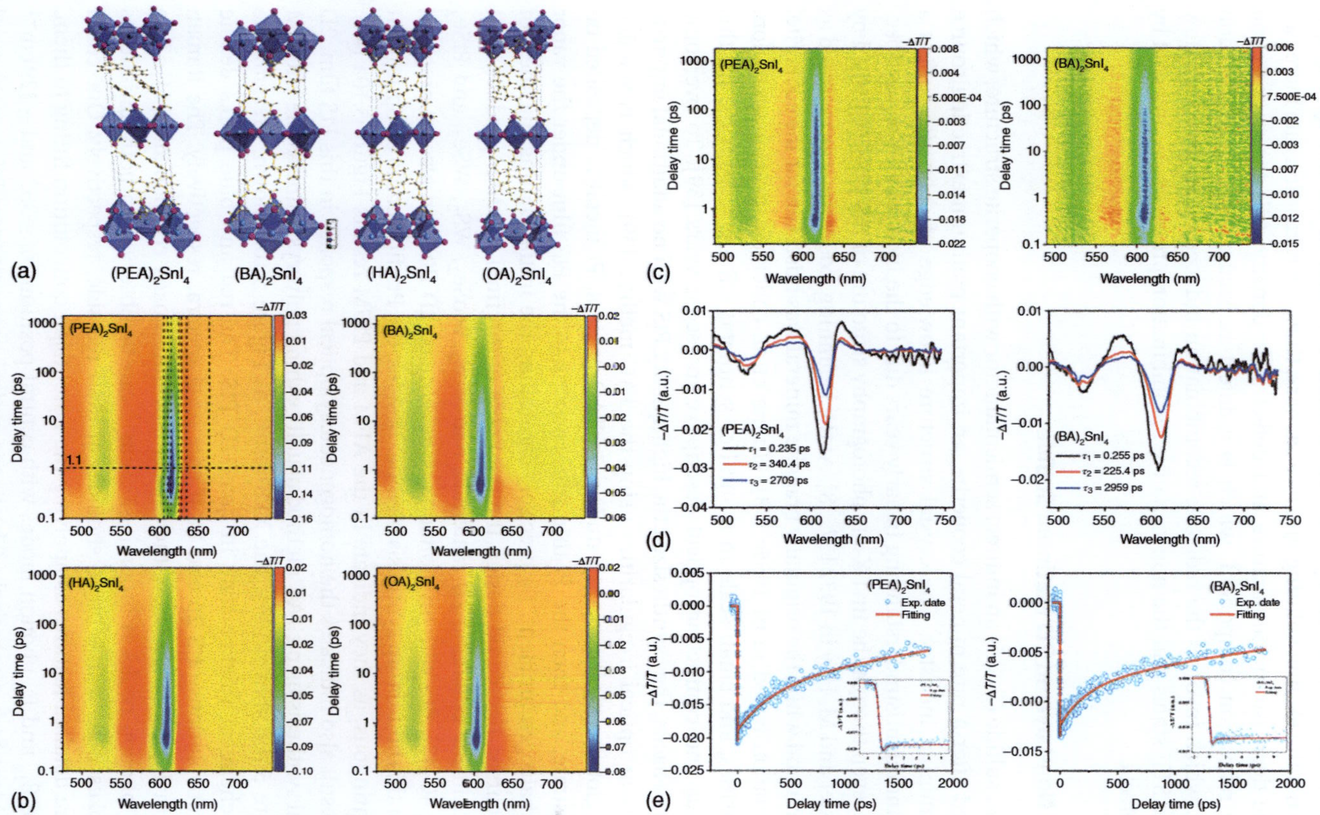

Figure 5.9 (a) Crystal structures of four types of 2D Sn-based perovskites. (b) False-color 2D TA mapping of the four perovskites with the excitation at 520 nm (2.38 eV) at a fluence of 2 μJ cm^{-2} at room temperature. The color scale represents the signal intensity of the TA spectrum, that is, the signal intensity of the differential transmittance. (c) 2D pseudocolor TA spectra of $(PEA)_2SnI_4$ and $(BA)_2SnI_4$ thin films at low exciton density. (d) The global fitting analysis was used to produce evolution-associated spectra with three varied lifetimes, and (e) three exponential fits of the relaxation process of the bandedge exciton of $(PEA)_2SnI_4$ and $(BA)_2SnI_4$. Source: (a, b): Reproduced from Zhang et al. [35], © 2022, Springer Nature/CC BY 4.0; (c–e) Reproduced with permission Zhang et al. [36], © 2022, American Chemical Society.

exciton process and the hot exciton-induced bandgap reformation process; (ii) the lifetime ~100 ps relaxation component, the interband radiation recombination process is mainly affected by the deformation potential scattering by charged defects, not the exciton splitting into free carriers; thus, the relaxation rate increases with the decrease in temperature and defect state density; (iii) the nanosecond lifetime relaxation component, which is a defect-assisted exciton recombination process, which involves the radiative recombination induced by relatively shallow extrinsic STE states and the non-radiative recombination processes are induced by deep defects.

5.4.2 Non-Layered 2D Platinum Sulfide

Platinum sulfide (PtS) is an indirect semiconductor with cooperite structure, which is named after a rare mineral cooperite, an important platinum-extracting source in the mining industry. The crystal structure knowledge of cooperite suffers a long-standing misunderstanding in early years due to the incorrect reflection spectra data, even though the first X-ray photography structure of PtS was confirmed by F. Banister and M. Hey in 1932 [37, 38]. At the beginning, cooperite was regarded as a new mineral while the originally braggite mineral was mistaken for the cooperite. This situation lasted until 1985s [38]. Huang et al. conclude the contamination, irregularities, and small scale in natural or synthetic PtS are responsible for this erroneous and conflicting optical descriptive in recent years [39]. To overcome these drawbacks, they synthesized the high-quality PtS with no additional impurity elements or optical irregularities using the TAC method [39], which is suitable for the global optical property studies. In TAC process, Pt metal is deposited on the substrate, and then the Pt thin film converts to the PtS film under the vapor phase chalcogen ambient. By precisely controlling the initial coating thickness of the precursor, one could design the thickness of the final reaction product, as it employs in the emerging 2D materials, like MoS_2, $MoSe_2$, WS_2, WSe_2, and $PtSe_2$ [24, 25, 40–43]. PtS with an atomic thickness (quasi-2D PtS) can also be prepared through this bottom-up synthesis method. Besides, the PtS film could be optical flat and continuous, as they indicate in the AFM and PiFM results Figure 5.10a [44]. PiFM result also indicates the homogeneously optical response in this PtS film. 2D material synthesis using TAC method usually exhibit polycrystalline as the HRTEM in Figure 5.10b shown [44]. The XPS result, as Figure 5.10c shown, presents the steady chemical valence state in quasi-2D PtS. The red and blue peaks denote the deconvolutions of S 2p and Pt 4f core level spectra, respectively. The Raman of PtS in Figure 5.10d show the Stokes and anti-Stokes components [44]. Two intense phonon peaks at around $\pm 332\,cm^{-1}$ can be observed in both anti-Stokes and Stokes sides with a broad shoulder at their red side, respectively. One could see more detailed features as applying the Lorentzian convolution fitting to them. And the B_{1g} and E_g phonon modes with central positions of ± 333 and $\pm 329\,cm^{-1}$ can be identified, respectively. The linear optical property from PtS is performed in Figure 5.10e. The Tauc plot is extracted from the reflection and transmission spectra for the optical bandgap estimating. It is reported that the quasi-2D PtS

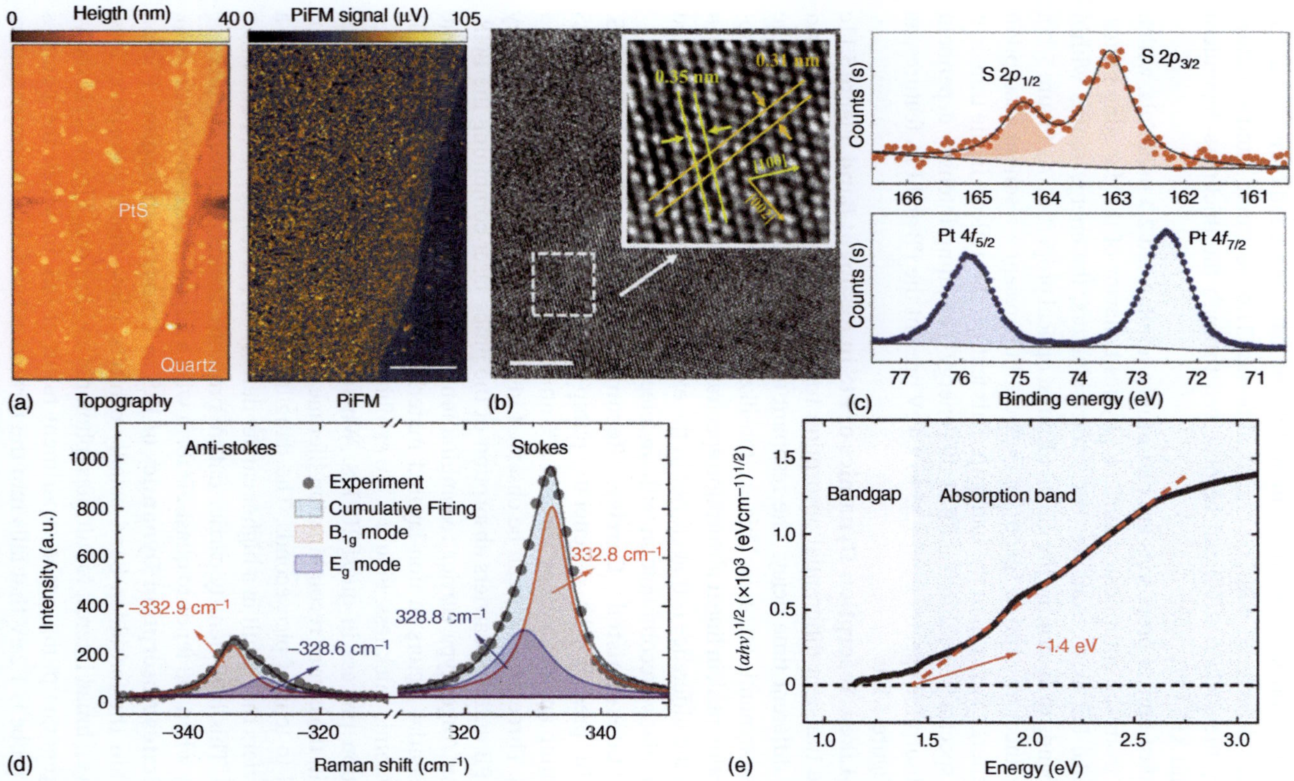

Figure 5.10 Synthesis and Characterizations of PtS. (a) Topography and PiFM images on quartz. (b) HRTEM image and its zoom-in counterpart of a selected area in PtS. (c) XPS characterization and (d) Raman spectrum of PtS. (e) The Tauc plots of PtS. Source: (a–e): Reproduced with permission Huang et al. [44], © 2022, John Wiley & Sons.

exhibit the bandgap narrowing when the thickness approaches the atomic limit, which is a different phenomenon from the conventional 2D layered van der Waals material [39]. This bandgap narrowing effect in non-layered 2D PtS probably results from the significant surface effect, in which the dangling bonds are typical for the non-layered material with 3D chemical bonds. The electronic band structure modified by the surface effect can be observed through the thickness-dependent valence-band XPS [39].

The ultrafast optical properties are plentiful in non-layered 2D materials, which may result from the complicated surface physics behavior of the dangling bonds. Here, we take PtS as an example. For semiconductors, the energy ranges within and out of bandgap determine the global ultrafast optical properties. As Figure 5.11a shown, the linear absorption spectra of the bulk PtS present a resonant absorption peak with energy larger than the bandgap. This feature also reflects in the transient absorption spectra that exhibits a PB at this energy. The pump-intensity dependent transient absorption spectra from 1.35 to 2.4 eV present the PB and PIA features, as shown in Figure 5.11b.

Exciton-Liked Absorption Dynamics of PtS in Resonant Band. Figure 5.11c exhibits the transient differential absorption traces at different probe energy and the TAS at different time delay. The resonant absorption of PtS at CB shows the exciton-liked dynamic features. This exciton-liked behavior is reflected in the resonant absorption peak in linear absorption spectra and the photoinduced bleaching in TAS. While the difference is that carrier in the exciton-liked state of PtS is governed by the nonradiative recombination with undetectable photoluminescence signal.

Surface-States-Assisted Carrier Recombination in Quasi-2D PtS. Figure 5.11d presents the degenerate pump–probe experiment result of PtS film. The pump and probe laser are both 520 nm, which is in the high-lying excited state of PtS. Three processes can be observed in this dynamic curve, in which they undergo a PB and free-carriers absorption of the probe laser combining the BGR, following an Auger-type carrier recombination process. The surface effect is one of most essential factors in non-layered material, especially for that in quasi-2D form with atomic thickness. Figure 5.11e exhibits the surface-states assisted carrier recombination process in quasi-2D PtS. When the thickness of PtS approaches atomic limit, the carrier recombination lifetime dramatically decreases from about a thousand to tens of picoseconds. The dangling bonds on the surface capture a mass of carriers that result in a higher carrier density on the surface than that inner the sample. This high-density carrier nature promotes the recombination process, particularly when PtS tends to quasi-2D form with a larger surface-to-volume ratio.

Defect-States Absorption Dynamics of PtS in Transparent Region. Photon energy within the forbidden band usually exhibits a low linear absorption loss, which draws a broad interest for ultrafast photonic applications. Figure 5.11f shows the nondegenerate pump–probe experiment result of PtS, in which the pump is 2.4 eV and probe is 1.2 eV that falls into the forbidden band. In this configuration, PtS undergoes an intensity-independent PB due to the ionization of mid-gap defect states, and a PA result from the FCA of probe laser, as the schematic shown in Figure 5.11g.

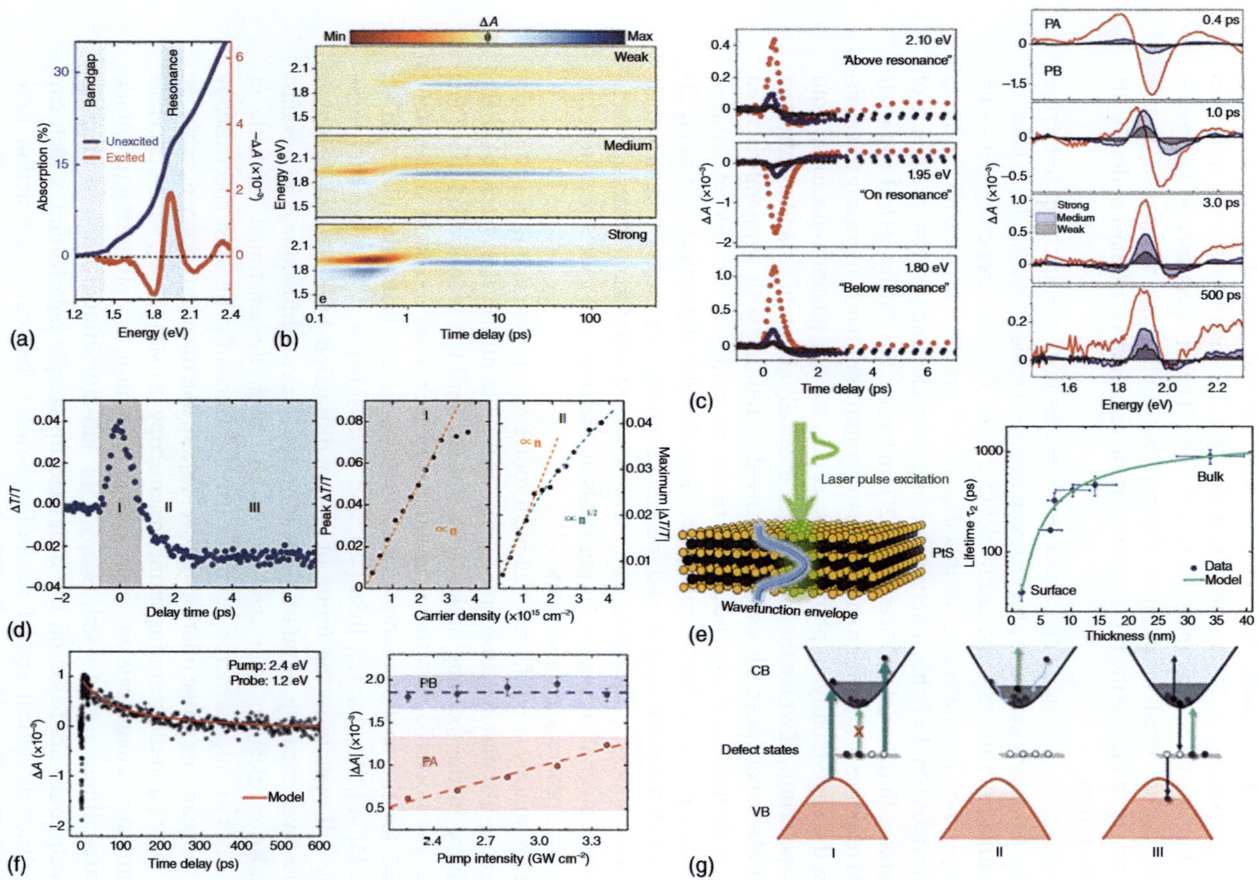

Figure 5.11 Ultrafast carrier dynamics of PtS. (a) Transient and steady absorption spectrum of PtS. (b) TAS of PtS from 1.35 to 2.4 eV. (c) Exciton-liked resonance and resonant detuning dynamic of PtS. (d) Ultrafast carrier dynamic properties of PtS in the high-lying energy band. (e) Surface-state assisted carrier recombination from quasi-2D to bulk PtS. (f) Mid-gap absorption dynamic and (g) its schematic mechanism in PtS. Source: (a–c, f, g): Reproduced with permission Huang et al. [44], © 2022, John Wiley & Sons; (d, e) Reproduced with permission Huang et al. [39]. Copyright © 2019, American Chemical Society.

5.5 Perspectives on Ultrafast Optics for 2D Material

In this chapter, we have discussed the principles and experiments of ultrafast time-resolved spectroscopy for 2D materials. Several typical ultrafast carrier dynamic properties in 2D layered (van der Waals) and non-layered materials are also introduced as examples, in which the ultrafast spectroscopy technique presents great universal, informative, damage-free, and precise advantages. In this section, we try to discuss the future development tendency of ultrafast optics for 2D materials studies.

The conventional ultrafast spectroscopy technique does well in the characterization of light-matter interactive properties on temporal scale. Besides, a large spectral range source from the extreme ultraviolet (XUV) to THz can also coordinate with the pump–probe equipment up to this day, which provides a powerful detecting tool for various physical phenomena associated with electron, exciton, phonon, and spin coupling in materials. However, for 2D materials, the locally optical and their temporal response in micro or nanoscale, can contain abundant interesting phenomena, like the edge-modifying photoluminescence, surface plasmon–polariton, and phonon–polariton, of which the information is usually absent in a standard ultrafast pump–probe experiment. This spatial-resolved combining with temporal-resolved abilities is desirable and also become a trend, no matter in the field of ultrafast optics or nanophotonics for 2D materials. Since the globally optical studies on 2D materials with temporal, spatial, and spectral knowledge are the key to realize their advanced all-optical applications. The spatial-resolved ultrafast spectroscopy technique, although challenging for 2D materials in micro or nanoscale dimension, has been ongoing progress in these years.

There are mainly two different technical routes to improve the ultrahigh spatial resolution in ultrafast spectroscopy. Firstly, one could minish the electromagnetic wave source in the focusing optics system to obtain high spatial resolution. For example, the microscopic ultrafast spectroscopy technique has been successfully employed for the spatiotemporal optical studies in 2D materials, like TMDCs [15, 23, 45, 46], BP [34, 47], InSe [48], and perovskite[49, 50]. Noted that these methods still operate in the frame of Abbe's diffraction limit. To obtain higher spatial-resolved capacity, ultrafast X-ray, ultrafast electron diffraction, and ultrafast photoemission electron microscopy have also been demonstrated available in 2D materials or microstructure samples [51–57]. Secondly, ultrafast spectroscopy merges the scanning probe microscope (SPM) [58], like the atomic force microscope (AFM)[59–66] and the scanning tunnel microscopy (STM) [67–69], which provides the advantages of nanoscale and even atomic resolution, is a promising technique for ultrafast time-resolved spectroscopy with sub-diffractional resolved imaging. These techniques are based on the near-field tip-sample polarization, which can obtain ultrahigh optical spatial resolution beyond the optical diffraction limit in far-field. For example, the ultrafast time-resolved scanning near-field optical microscope (tr-SNOM), which is a developing spatiotemporal imaging technique in these days, has shown great potential for addressing ultrafast dynamics in 2D van der Waals nano-materials [70–76]. Besides, the ultrafast STM that offers the capacity for atomic-level detection with temporal resolution, has also made significant advances in these years [67, 69, 77–79]. Other techniques that combine ultrafast spectroscopy with SPM technique, like ultrafast photoinduced force microscope[80, 81] and ultrafast tip-enhanced Raman spectroscopy[82, 83], are still developing for observing their selective and special optical response.

Ultrafast temporal and ultrahigh spatial spectroscopy provide new microscopic vision to unveil the nanoscale interactions of 2D materials, including the layered and non-layered member, and their heterostructures counterparts, which may contain unexpected and programmable functionality due to their microscopic interacting systems. The global knowledge of 2D materials in temporal, spatial, and spectral dimensions is an essential stepping-stone toward their advanced ultrafast nanophotonic applicatinons.

Author Contributions

The manuscript was written through contributions of all authors. All authors have given approval to the final version of the manuscript.

Notes

The authors declare no competing financial interest.

Acknowledgment

This work is supported by the National Natural Science Foundation of China (No. 12174414 and 61975221), the Youth Innovation Promotion Association, Chinese Academy of Sciences (CAS) (2019249). J.W.H. acknowledges the fund by Guangdong Basic and Applied Basic Research Foundation (No. 2020A1515110433).

References

1 Prasankumar, R.-P. and Taylor, A.-J. (2016). *Optical Techniques for Solid-state Materials Characterization*, 330–336. Florida: CRC Press.

2 Wang, K., Szydlowska, B., Wang, G. et al. (2016). Ultrafast nonlinear excitation dynamics of black phosphorus nanosheets from visible to mid-infrared. *ACS Nano* 10 (7): 6923–6932.

3 Zhang, X.-Y., Zhang, S.-F., Xie, Y.-F. et al. (2018). Tailoring the nonlinear optical performance of two-dimensional MoS_2 nanofilms via defect engineering. *Nanoscale* 10 (37): 17924–17932.

4 Dai, X.-R., Zhang, X.-Y., Kislyakov, I.-M. et al. (2019). Enhanced two-photon absorption and two photon luminescence in monolayer MoS_2 and WS_2 by defect repairing. *Optics Express* 27 (10): 13744–13753.

5 Lagarde, D., Bouet, L., Marie, X. et al. (2014). Carrier and polarization dynamics in monolayer MoS_2. *Physical Review Letters* 112 (4): 047401.

6 Suess, R., Jadidi, M., Murphy, T. et al. (2015). Carrier dynamics and transient photobleaching in thin layers of black phosphorus. *Applied Physics Letters* 107: 081103.

7 Zhuo, X., Lai, J.-W., Yu, P. et al. (2021). Dynamical evolution of anisotropic response of type-II Weyl semimetal $TaIrTe_4$ under ultrafast photoexcitation. *Light-Science & Applications* 10: 101.

8 Ge, S.-F., Li, C.-K., Zhang, Z.-M. et al. (2015). Dynamical evolution of anisotropic response in black phosphorus under ultrafast photoexcitation. *Nano Letters* 15 (7): 4650–4656.

9 Linnros, J. (1998). Carrier lifetime measurements using free carrier absorption transients. I. Principle and injection dependence. *Journal of Applied Physics* 84 (1): 275–283.

10 Landsberg, P. (1987). Trap-auger recombination in silicon of low carrier densities. *Applied Physics Letters* 50 (12): 745–747.

11 Shockley, W. (1952). Transistor electronics: Imperfections, unipolar and analog transistors. *Proceedings of the Institute of Radio Engineers* 40 (11): 1289–1313.

12 Shockley, W. and Read, W. (1952). Statistics of the recombinations of holes and electrons. *Physical Review* 87 (5): 835–842.

13 Vanroosbroeck, W. and Shockley, W. (1954). Photon-radiative recombination of electrons and holes in germanium. *Physical Review* 94 (6): 1558–1560.

14 Strait, J., Nene, P., and Rana, F. (2014). High intrinsic mobility and ultrafast carrier dynamics in multilayer metal-dichalcogenide MoS_2. *Physical Review B* 90: 245402.

15 Wang, H.-N., Zhang, C.-J., and Rana, F. (2015). Ultrafast dynamics of defect-assisted electron hole recombination in mono layer MoS_2. *Nano Letters* 15 (1): 339–345.

16 Baudisch, M., Marini, A., Cox, D. et al. (2018). Ultrafast nonlinear optical response of Dirac fermions in graphene. *Nature Communications* 9: 1080.

17 Breusing, M., Ropers, C., and Elsaesser, T. (2009). Ultrafast Carrier Dynamics in Graphite. *Physical Review Letters* 102 (8): 086809.

18 Wang, G.-Z., Wang, K.-P., Szydlowska, B. et al. (2017). Ultrafast nonlinear optical properties of a graphene saturable mirror in the 2 μm wavelength region. *Laser & Photonics Reviews* 11: 1700166.

19 George, P.-A., Strait, J., Dawlaty, J. et al. (2008). Ultrafast optical-pump terahertz-probe spectroscopy of the carrier relaxation and recombination dynamics in epitaxial graphene. *Nano Letters* 8 (12): 4248–4251.

20 Dal Conte, S., Bottegoni, F., Pogna, E.-A.-A. et al. (2015). Ultrafast valley relaxation dynamics in monolayer MoS_2 probed by nonequilibrium optical techniques. *Physical Review B* 92 (23): 235425.

21 Chernikov, A., Ruppert, C., Hill, H.-M. et al. (2015). Population inversion and giant bandgap renormalization in atomically thin WS_2 layers. *Nature Photonics* 9 (7): 466–469.

22 Zhang, T.-J. and Wang, J. (2021). Defect-enhanced exciton-exciton annihilation in monolayer transition metal dichalcogenides at high exciton densities. *ACS Photonics* 8 (9): 2770–2780.

23 Wang, H.-N., Zhang, C.-J., and Rana, F. (2015). Surface recombination limited lifetimes of photoexcited carriers in few-layer transition metal dichalcogenide MoS_2. *Nano Letters* 15 (12): 8204–8210.

24 Wang, L., Zhang, S.-F., Mcevoy, N. et al. (2019). Nonlinear optical signatures of the transition from semiconductor to semimetal in $PtSe_2$. *Laser & Photonics Reviews* 13: 1900052.

25 Wang, G.-Z., Wang, K.-P., Mcevoy, N. et al. (2019). Ultrafast carrier dynamics and bandgap renormalization in layered $PtSe_2$. *Small* 15: 1902728.

26 Chen, X., Huang, J.-W., Chen, C.-D. et al. (2022). Broadband nonlinear photoresponse and ultrafast carrier dynamics of 2D $PdSe_2$. *Advanced Optical Materials* 10: 2101963.

27 Wang, L., Zhang, S.-F., Huang, J.-W. et al. (2020). Auger-type process in ultrathin ReS_2. *Optical Materials Express* 10 (4): 1092–1104.

28 Churchill, H.-O. and Jarillo-Herrero, P. (2014). Two-dimensional crystals: phosphorus joins the family. *Nature Nanotechnology* 9 (5): 330–331.

29 Xia, F., Wang, H., and Jia, Y. (2014). Rediscovering black phosphorus as an anisotropic layered material for optoelectronics and electronics. *Nature Communications* 5: 4458.

30 Huang, J.-W., Dong, N.-N., Zhang, S.-F. et al. (2017). Nonlinear absorption induced transparency and optical limiting of black phosphorus nanosheets. *ACS Photonics* 4 (12): 3063–3070.

31 Liu, H., Neal, A.-T., Zhu, Z. et al. (2014). Phosphorene: an unexplored 2D semiconductor with a high hole mobility. *ACS Nano* 8 (4): 4033–4041.

32 Pareek, V., Madeo, J., and Dani, K.-M. (2020). Ultrafast control of the dimensionality of exciton-exciton annihilation in atomically thin black phosphorus. *Physical Review Letters* 124 (5): 057403.

33 Iyer, V., Ye, P., and Xu, X. (2017). Mid-infrared ultrafast carrier dynamics in thin film black phosphorus. *2D Materials* 4 (2): 021032.

34 Miao, X.-C., Zhang, G.-W., Wang, F.-J. et al. (2018). Layer-dependent ultrafast carrier and coherent phonon dynamics in black phosphorus. *Nano Letters* 18 (5): 3053–3059.

35 Zhang, T.-J., Zhou, C.-C., Feng, X.-Z. et al. (2022). Regulation of the luminescence mechanism of two-dimensional tin halide perovskites. *Nature Communications* 13 (1): 60.

36 Zhang, T.-J., Zhou, C.-C., Lin, J. et al. (2022). Regulating the auger recombination process in two-dimensional sn-based halide perovskites. *ACS Photonics* 9 (5): 1627–1637.

37 Bannister, F. and Hey, M. (1932). Determination of minerals in platinum concentrates from the Transvaal by X-ray methods. *Mineralogical Magazine* 2: 118–206.

38 Criddle, A. and Stanley, C. (1985). Characteristic optical-data for cooperite, braggite and vysotskite. *Canadian Mineralogist* 23: 149–162.

39 Huang, J.-W., Dong, N.-N., Mcevoy, N., et al. (Unpublished). Surface-state assisted carrier recombination and optical nonlinearities in bulk to 2D non-layered PtS.

40 Zhan, Y.-J., Liu, Z., Najmaei, S. et al. (2012). Large-area vapor-phase growth and characterization of MoS_2 atomic layers on a SiO_2 substrate. *Small* 8 (7): 966–971.

41 Zhang, S.-F., Dong, N.-N., Mcevoy, N. et al. (2015). Direct observation of degenerate two-photon absorption and its saturation in WS_2 and MoS_2 monolayer and few-layer films. *ACS Nano* 9 (7): 7142–7150.

42 Dong, N.-N., Li, Y.-X., Zhang, S.-F. et al. (2018). Saturation of two-photon absorption in layered transition metal dichalcogenides: experiment and theory. *ACS Photonics* 5 (4): 1558–1565.

43 Wagner, S., Yim, C., Mcevoy, N. et al. (2018). Highly sensitive electromechanical piezoresistive pressure sensors based on large-area layered $PtSe_2$ films. *Nano Letters* 18 (6): 3738–3745.

44 Huang, J.-W., Dong, N.-N., Mcevoy, N. et al. (2022). Exciton-like and mid-gap absorption dynamics of PtS in resonant and transparent regions. *Laser & Photonics Reviews* 16: 2100654.

45 Sun, D.-Z., Rao, Y., Reider, G. et al. (2014). Observation of rapid exciton-exciton annihilation in monolayer molybdenum disulfide. *Nano Letters* 14 (10): 5625–5629.

46 Wei, K., Sui, Y.-Z., Xu, Z.-J. et al. (2020). Acoustic phonon recycling for photocarrier generation in graphene-WS_2 heterostructures. *Nature Communications* 11: 3876.

47 Wu, S.-M., Lu, Z.-X., Hu, A.-Q. et al. (2021). Dichroic photoelasticity in black phosphorus revealed by ultrafast coherent phonon dynamics. *Journal of Physical Chemistry Letters* 12 (25): 5871–5878.

48 Chen, C.-D., Dong, N.-N., Huang, J.-W. et al. (2021). Nonlinear optical properties and ultrafast carrier dynamics of 2D indium selenide nanosheets. *Advanced Optical Materials* 9: 2101432.

49 Guo, Z., Wan, Y., Yang, M.-J. et al. (2017). Long-range hot-carrier transport in hybrid perovskites visualized by ultrafast microscopy. *Science* 356 (6333): 59–62.

50 Delport, G., Macpherson, S., and Stranks, S. (2020). Imaging carrier transport properties in halide perovskites using time-resolved optical microscopy. *Advanced Energy Materials* 10: 1903814.

51 Sie, E., Nyby, C., Pemmaraju, C. et al. (2019). An ultrafast symmetry switch in a Weyl semimetal. *Nature* 565 (7737): 61–66.

52 Wang, K.-P., Dahan, R., Shentcis, M. et al. (2020). Coherent interaction between free electrons and a photonic cavity. *Nature* 582 (7810): 50–54.

53 Lin, M.-F., Kochat, V., Krishnamoorthy, A. et al. (2017). Ultrafast non-radiative dynamics of atomically thin $MoSe_2$. *Nature Communications* 8: 1745.

54 Gaffney, K. and Chapman, H. (2007). Imaging atomic structure and dynamics with ultrafast X-ray scattering. *Science* 316 (5830): 1444–1448.

55 Man, M., Madeo, J., Sahoo, C. et al. (2021). Experimental measurement of the intrinsic excitonic wave function. Science. *Advances* 7 (17): eabg0192.

56 Madeo, J., Man, M., Sahoo, C. et al. (2020). Directly visualizing the momentum-forbidden dark excitons and their dynamics in atomically thin semiconductors. *Science* 370 (6521): 1199–1203.

57 Wang, L., Xu, C., Li, M.-Y. et al. (2018). Unraveling spatially heterogeneous ultrafast carrier dynamics of single-layer WSe_2 by femtosecond time-resolved photoemission electron microscopy. *Nano Letters* 18 (8): 5172–5178.

58 Atkin, J., Berweger, S., Jones, A. et al. (2012). Nano-optical imaging and spectroscopy of order, phases, and domains in complex solids. *Advances in Physics* 61 (6): 745–842.

59 Chen, J.-N., Badioli, M., Alonso-Gonzalez, P. et al. (2012). Optical nano-imaging of gate-tunable graphene plasmons. *Nature* 487 (7405): 77–81.

60 Fei, Z., Rodin, A., Andreev, G. et al. (2012). Gate-tuning of graphene plasmons revealed by infrared nano-imaging. *Nature* 487 (7405): 82–85.

61 Dai, S., Fei, Z., Ma, Q. et al. (2014). Tunable phonon polaritons in atomically thin van der waals crystals of boron nitride. *Science* 343 (6175): 1125–1129.

62 Ma, W.-L., Alonso-Gonzalez, P., Li, S.-J. et al. (2018). In-plane anisotropic and ultra-low-loss polaritons in a natural van der Waals crystal. *Nature* 562 (7728): 557–562.

63 Li, P.-N., Dolado, I., Javier, A. et al. (2018). Infrared hyperbolic metasurface based on nanostructured van der Waals materials. *Science* 359 (6378): 892–896.

64 Hu, G.-W., Ou, Q.-D., Si, G.-Y. et al. (2020). Topological polaritons and photonic magic angles in twisted α-MoO_3 bilayers. *Nature* 582 (7811): 209–213.

65 Basov, D., Fogler, M., Garcia, D.E., and Abajo, F. (2016). Polaritons in van der Waals materials. *Science* 354 (6309): aag1992.

66 Huang, J.-W., Tao, L., Dong, N.-N. et al. (2023). In-plane hyperbolic phonon-polaritons in van der Waals nanocrystals. *Advanced Optical Materials* 11 (5): 2202048.

67 Cocker, T., Peller, D., Yu, P. et al. (2016). Tracking the ultrafast motion of a single molecule by femtosecond orbital imaging. *Nature* 539 (7628): 263–267.

68 Cocker, T., Jelic, V., Gupta, M. et al. (2013). An ultrafast terahertz scanning tunnelling microscope. *Nature Photonics* 7 (8): 620–625.

69 Terada, Y., Yoshida, S., Takeuchi, O. et al. (2010). Real-space imaging of transient carrier dynamics by nanoscale pump-probe microscopy. *Nature Photonics* 4 (12): 869–874.

70 Sternbach, A., Chae, S., Latini, S. et al. (2021). Programmable hyperbolic polaritons in van der Waals semiconductors. *Science* 371 (6529): 617–620.

71 Mrejen, M., Yadgarov, L., Levanon, A. et al. (2019). Transient exciton-polariton dynamics in WSe_2 by ultrafast near-field imaging. *Science Advances* 5: eaat9618.

72 Siday, T., Sandner, F., Brem, S. et al. (2022). Ultrafast nanoscopy of high-density exciton phases in WSe_2. *Nano Letters* 22 (6): 2561–2568.

73 Wagner, M., Fei, Z., Mcleod, A. et al. (2014). Ultrafast and nanoscale plasmonic phenomena in exfoliated graphene revealed by infrared pump-probe nanoscopy. *Nano Letters* 14 (2): 894–900.

74 Ni, G., Wang, L., Goldflam, M. et al. (2016). Ultrafast optical switching of infrared plasmon polaritons in high-mobility graphene. *Nature Photonics* 10 (4): 244–247.

75 Huber, M., Mooshammer, F., Plankl, M. et al. (2017). Femtosecond photo-switching of interface polaritons in black phosphorus heterostructures. *Nature Nanotechnology* 12 (3): 207–211.

76 Sternbach, A., Latini, S., Chae, S. et al. (2020). Femtosecond exciton dynamics in WSe_2 optical waveguides. *Nature Communications* 11: 3567.

77 Cocker, T., Jelic, V., Hillenbrand, R. et al. (2021). Nanoscale terahertz scanning probe microscopy. *Nature Photonics* 15 (8): 558–569.

78 Ammerman, S., Jelic, V., Wei, Y. et al. (2021). Lightwave-driven scanning tunnelling spectroscopy of atomically precise graphene nanoribbons. *Nature Communications* 12: 6794.

79 Plankl, M., Faria, J., Mooshammer, F. et al. (2021). Subcycle contact-free nanoscopy of ultrafast interlayer transport in atomically thin heterostructures. *Nature Photonics* 15 (8): 594–600.

80 Jahng, J., Brocious, J., Fishman, D. et al. (2015). Ultrafast pump-probe force microscopy with nanoscale resolution. *Applied Physics Letters* 106: 083113.

81 Jahng, J., Fishman, D., Park, S. et al. (2015). Linear and nonlinear optical spectroscopy at the nanoscale with photoinduced force microscopy. *Accounts of Chemical Research* 48 (10): 2671–2679.

82 Gruenke, N., Cardinal, M., Mcanally, M. et al. (2016). Ultrafast and nonlinear surface-enhanced Raman spectroscopy. *Chemical Society Reviews* 45 (8): 2263–2290.

83 Sonntag, M., Pozzi, E., Jiang, N. et al. (2014). Recent advances in tip-enhanced raman spectroscopy. *Journal of Physical Chemistry Letters* 5 (18): 3125–3130.

6

Transient Terahertz Spectroscopy for 2D Materials

Jingyin Xu, Kai Zhang, Hong Li, and Tianwu Wang

Chinese Academy of Sciences, GBA branch of Aerospace Information Research Institute, 11 Kai-Yuan Blvd, Huangpu District, Guangzhou, Guangdong Province 510700, P.R.China

6.1 Introduction

The class of two-dimensional (2D) materials has recently emerged as an excellent platform for designing materials on demand for next-generation optoelectronic devices [1, 2]. Bulk crystals can be thinned down to the ultimate limit of only a single atomic layer. The reduction of dimensionality results in robust Coulomb correlations in monolayers and various charge carrier properties, which endowed the 2D materials as the promising building blocks for optoelectronic devices [2–4]. Charge carriers can have a variety of properties in 2D materials, depending on morphology, temperature, and optoelectronic properties such as the crystal structure, bandgap, dielectric function, and electron–phonon coupling strength. In contrast with the bulk material with high dielectric function in which charge carriers are efficiently screened from one another, electrons and holes will be present as free carriers and relatively mobile. For the 2D materials with atomic layer, the carrier–lattice interactions will lead to the formation of polarons, carriers that are dressed with local lattice deformations [5, 6]. Light trapped at the nanoscale in 2D materials, deep below the optical wavelength, exhibits an increase in the associated electric field strength, which results in enhanced light–matter interaction. This leads to strong nonlinearities, large photonic forces, and enhanced emission and absorption probabilities.

The characteristic of charge carriers in 2D materials can be presented in any of the forms described above, such as free carriers, polarons, and excitons. However, these charge quasiparticles all exhibit a distinct response in the low-frequency range of the electromagnetic spectrum in the meV range and are optically active. The response of charge quasiparticles typically occurs on (sub)picosecond timescales, operating dispersion in the dielectric response on meV energy scales. Pulsed THz spectroscopy that covers the energy range of meV possesses the capacity to probe the characterization of charge carriers through the distinct spectral signatures in the THz frequency [7, 8]. The contact-free probe of the frequency-dependent

Two-Dimensional Materials for Nonlinear Optics: Fundamentals, Preparation Methods, and Applications, First Edition. Edited by Qiang Wang and Hao-Li Zhang.

conductivity allowed THz spectroscopy to characterize the critical parameters of 2D materials such as carrier density and mobility in detail. As a type of electromagnetic spectrum with low photon energies (4.2 meV), THz radiation is characterized by sub-mm wavelengths (300 μm), corresponding to less-than-thermal energies at room temperature (1 THz corresponds to 48 K). THz time-domain spectroscopy makes use of sub-picosecond pulses of freely propagating electromagnetic radiation in the THz range and takes the measurement of the amplitude and phase of the transmitted or reflected pulse as a function of time. By Fourier-transforming the time-domain data, the frequency spectrum of the sample can be obtained, providing information about its physical and chemical properties.

In this chapter, we review the body of work on transient THz spectroscopy for 2D materials. We start with a description of fundamental technical details and recent advances and the development of 2D materials for time-resolved THz spectroscopy. We discuss the study of free carriers, polarons, and excitons in 2D materials via THz techniques, and then turn our focus to the central concept of coupling THz radiation for near-scanning probe microscopy, which is a powerful technique used to study the properties of 2D materials with time and spatial resolutions. Finally, we finish with a brief outlook on the future of THz spectroscopy.

6.2 Generation and Detection of THz Radiation

Commonly used technologies for generating THz radiation include photoconductive antennas (PCAs) described by the current surge model, optical rectification (OR) based on the second-order nonlinear process [9], air plasma based on the third-order nonlinear process and photocurrent model [10–12], and difference frequency generation (DFG) implemented by optical parametric amplifiers (OPA) [13]. The realization of OR is based on electro-optic (EO) crystals such as ZnTe [14], GaAs, GaSe, and $LiNbO_3$ and organic crystals such as DAST [15]. The most popular method is $LiNbO_3$ crystals based on tilted pulse front (TPF) rectification technology [16, 17]. The methods used to detect the THz wave include PCAs [18], EO sampling technique [19], air-based THz detection technique, and spectral-domain interferometry (SDI) technique based on Michelson interference principle. The principles of the three major technologies for THz radiation generation and detection such as PCAs, EO crystals, and air plasma are described in detail below. Each method has its own advantages and disadvantages, as shown in Table 6.1.

6.2.1 Generation of THz Radiation Based on Nonlinear Optical Processes

As elucidated in Section 6.2.2, the utilization of terahertz spectroscopy techniques, which seamlessly bridge the realms of photonics and electronic frequencies, has emerged as a formidable means for elucidating the characteristics of electrons and excitons in 2D materials. Commonly used technologies for generating THz radiation include PCAs described by the current surge model, OR based on the second-order

Table 6.1 The advantages and disadvantages of the three primary THz generation and detection technologies.

THz generation and detection technologies	Advantages	Disadvantages
PCAs	1. Obtain energy from bias voltage 2. High stability	1. Low damage threshold 2. High electromagnetic noise 3. THz radiation saturates easily at low optical fluence
OR	1. High conversion efficiency 2. High stability	1. Low damage threshold 2. Narrow spectral detection range 3. Thermal and multi-photon effects 4. Complicated phase-matching conditions for some crystals
Air plasma	1. No damage threshold 2. Wide frequency spectrum 3. Whole THz region covered without gap	1. Low THz energy 2. Complicated phase-matching conditions between 800- and 400-nm pulses

nonlinear process [9], air plasma based on the third-order nonlinear process and photocurrent model [10–12], and DFG implemented by OPA [13]. The realization of OR is based on EO crystals and organic crystals, EO crystals such as ZnTe [14], GaAs, GaSe, LiNbO3, and organic crystals such as DAST [15]. The most popular method is LiNbO3 crystals based on TPF rectification technology [16, 17]. In terms of THz wave detection, several methods are utilized, including PCAs [18], EO sampling techniques [19], air-based THz detection techniques, and SDI techniques based on the Michelson interference principle. The advantages and principles of the three leading technologies for THz radiation generation and detection, namely PCAs, EO crystals, and air plasma, are elaborated in detail below, as summarized in Table 6.1.

6.2.1.1 Photoconductive Antennas (PCAs)

The mechanism for generating THz radiation from PCAs is described by a current surge model [20]: a femtosecond laser pulse is applied to the semiconductor material to generate free carriers, and a pulse current is driven by an applied DC electric field. A PCA is typically composed of a high resistivity semiconductor substrate and two electrodes fabricated on one face of the substrate. The schematic of a typical PCA is shown in Figure 6.1.

Figure 6.1 Schematic of a typical PCA.

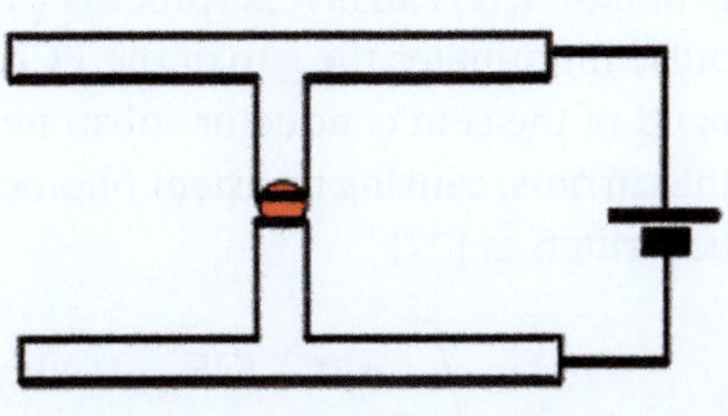

Table 6.2 Substrate materials.

Materials	Bandgap (eV)	Laser	Description
GaAs	1.44	800 nm	Low breakdown field for LAPCAs with gap sizes of few mm
Diamond, ZnO, GaN	3.1	Third harmonic	—
ZnSe	2.7	Second harmonic, two-photon absorption	Thermal properties similar to GaAS
6H-SiC	3.03	Second harmonic, one-photon absorption	Superior thermal quality compared with ZnSe crystals
4H-SiC	3.26	Second harmonic, two-photon absorption	

The THz electric field generated by PCAs is proportional to the time derivative of the transient current [21, 22]:

$$E(t) \propto \frac{\mathrm{d}J(t)}{\mathrm{d}t} \propto E_\mathrm{b}\frac{\mathrm{d}n(t)}{\mathrm{d}t} \tag{6.1}$$

where $J(t)$ is the current density, E_b is the bias field, and $n(t)$ is the carrier density. The THz electric field is linearly proportional to the applied bias field. To generate intense THz pulses from PCAs, there are several methods:

Increase the aperture of the PCAs, the large aperture photoconductive antennas (LAPCAs). These are defined to have an optically illuminated area (the gap size is a few millimeters) of dimensions much greater than the center wavelength of the emitted radiation (1 THz ~ 0.3 mm).

1. Choose a proper semiconductor material as the substrate. A list of several substrate materials is provided in Table 6.2 [15].
2. Apply voltage as high as possible between the electrodes. However, high-voltage source is hard to use. Other methods such as interdigitated LAPCA [23], phase mask [24], and plasmonic electrodes [25] are proposed.

Detecting THz Radiation by PCAs PCA is the first technology to detect coherent THz radiation [26]. The mechanism of THz radiation detection by PCAs is similar to PCA generation, it is an inverse process without a bias voltage. A femtosecond laser probe pulse illuminates the gap of the PCAs and excites the carriers in to the conduction band of the semiconductor substrate. The THz electric field to be measured drives the carriers, causing transient photocurrents in the PCAs. The photocurrent $I(t)$ can be written as [27]

$$I(t) \propto \int_{-\infty}^{t} \sigma_\mathrm{s}(t-t')E_\mathrm{THz}(t')\mathrm{d}t' \tag{6.2}$$

where E_{THz} is the incident THz field and $\sigma_s(t)$ is the transient surface conductivity. The transient surface conductivity is a time-dependent quantity, which means that the photocurrent cannot simultaneously respond to the transient changes of the THz electric field. It can be seen from Eq. (6.2) that the photocurrent is a convolution of the transient surface conductivity with the THz electric field. Therefore, the THz waveform is affected by both the laser pulse and the PCA substrate. To increase the maximum detection bandwidth, a laser pulse with a short pulse width and a substrate with a short carrier lifetime should be selected. The most commonly used material is low-temperature GaAs (LT-GaAs), which has a carrier lifetime shorter than 0.5 ps [28, 29].

6.2.1.2 Electro-optic (EO) Crystals

Due to the second-order nonlinear effect, when an EO crystal is excited by a femtosecond laser pulse, a time-dependent nonlinear polarization is generated, which contains different frequency components, resulting in ultrafast broadband THz radiation. This is an OR process. The THz electric field is proportional to the second derivative of the second-order polarization of the material ${P_{OR}}^{(2)}$ with respect to time t:

$$E_{THz}^{rad}(t) \propto \frac{\partial^2 P_{OR}^{(2)}(t)}{\partial t^2} \tag{6.3}$$

$$P_{OR}^{(2)}(\omega) = \varepsilon_0 \chi^{(2)} E(\omega_1) E^*(\omega_2) \tag{6.4}$$

where ω is the frequency difference between the two components of the pump light ω_1 and ω_2, ε_0 is the vacuum dielectric constant, and $\chi^{(2)}$ is the second-order susceptibility tensor of the material. Here, three typical kinds of EO crystals are introduced:

1. ZnTe. It has good phase-matching conditions, but the EO coefficient and THz absorption coefficient are relatively small. Larger-aperture OR emitters [30] help to reduce the saturation effect in ZnTe.
2. $LiNbO_3$. It has a high damage threshold but also has a large mismatch between the group velocity of the excited pulse and the phase velocity of the THz wave. TPF technology [31, 32] is widely used to overcome the mismatch problem in $LiNbO_3$. A simple schematic diagram is shown in Figure 6.2.
   The phase-matching condition is $v_{THz}^{ph} = v_{pump}^{gr} \cos(\gamma)$. When $\gamma \approx 63°$, $LiNbO_3$ obtains the best phase-matching condition.
3. DAST. It is an organic EO crystal with high efficiency but low laser damage threshold [33].

Free-space EO crystal sampling technique based on Pockels effect is used to detect THz electric field [27, 34, 35]. A schematic of EO crystal detection is shown in Figure 6.3.

Without THz radiation, EO crystals are not induced to birefringence. When the linear polarization probe light passes through the quarter wave plate (λ/4), the polarization state changes from linear polarization to circular polarization. The Wollaston prism (WP) divides the polarized light into S and P components and are detected by two balanced photodiodes (BPDs). Since the two components of the circularly

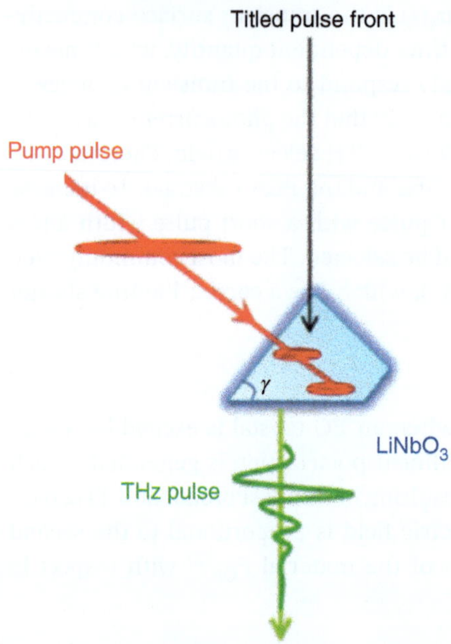

Figure 6.2 Schematic of titled pulse front technology of $LiNbO_3$. Source: Adapted from Hafez et al. [15]. Copyright © 2016, IOP Publishing.

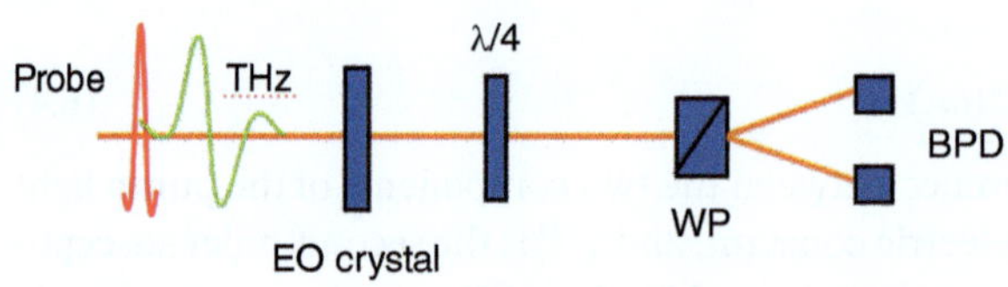

Figure 6.3 Schematic of EO crystal detection.

polarized light have the same light intensity, the difference is zero. Due to the Pockels effect, THz radiation causes birefringence in the EO crystal, and the polarization state of the probe pulse changes as it passes through the EO crystal. After λ/4, the polarized light changes from linearly polarized light to elliptically polarized light. Finally, the difference in the light intensity detected by the two balance detectors is not zero. Any change in polarization caused by the THz pulse results in an imbalance between the S and the P polarized probe beams. The THz electric field is proportional to the difference in electric field detected by BPDs. In this way, the THz electric field can be obtained.

6.2.1.3 Air Plasma

In simple terms, the principle of THz radiation generated by air plasma is that the focused fs laser pulses cause gas ionization, and the oscillating electrons of the plasma generate THz radiation. Laser-induced air plasma is first demonstrated by H. Hamster et al. in 1993 [36]. They explained that the ponderomotive forces at the focal point of the intense laser caused the electrons to accelerate and expel from the focus. The expelled charges then resulted in intense and broadband THz radiation. A significant progress was made in 2000. D.J. Cook et al. provided an efficient way to increase the THz intensity by using the fundamental wave (FW) and second harmonic (SH) wave of fs pulses to generate plasma, and the peak THz field is estimated to be 2 kV cm^{-1} [12]. The mechanism of two-color mixing to generate THz radiation is explained on the basis of four-wave rectification (FWR) [37] (also called four-wave-mixing (FWM) or four-wave OR), which is a third-order nonlinear process. The THz field is proportional to

$$E_{\text{THz}} \propto \chi^{(3)} E_{2\omega} E_{\omega}^{2} \cos(\varphi) \tag{6.5}$$

where $\chi^{(3)}$ is the third-order nonlinear susceptibility; $E_{2\omega}$ and E_{ω} are the electric field amplitudes of the SH and FW, respectively; and φ the is the relative phase difference between fundamental and SH frequencies. As seen from Eq. (6.5), the THz amplitude generated by the FWR process strongly depends on the relative phase of the fundamental and SH pulses. However, the problem here is that the third-order nonlinearity caused by the bound or free electrons in the plasma is relatively small. Intense THz radiation generated by the high laser intensity is not suitable for interpretation with FWM [38].

To describe the intense THz emission from an air plasma source with two-color laser fields, a photocurrent model (also called transient electron current model) has been provided and developed by K.Y. Kim et al. since 2007 [10, 11]. The FW and SH wave are superimposed to generate a nonuniform electric field; the total laser field is symmetry-broken (asymmetric). The laser field frees the electrons, the bound electrons undergo rapid tunnel ionization, and then the free electrons and positively charged ions are generated to form a plasma. A nonvanishing electron current surge occurs during the optical field ionization of air, which forms an instantaneous photocurrent. Since the laser pulse is on the fs timescale (<100 fs), the current surge can emit THz electromagnetic radiation in the far field. The mixed laser field with the fundamental and SH laser fields can be expressed as

$$E_{\text{L}}(t) = E_{\omega} \cos(\omega t) + E_{2\omega} \cos(2\omega t + \varphi) \tag{6.6}$$

when the relative phase between the colors $\varphi = \pi/2$, a nonzero net current generates and gradually accumulates throughout the whole laser pulse duration to form a directional current surge. The nonvanishing transverse plasma current can be described as

$$J = eN_{\text{e}}v_{\text{e}} \tag{6.7}$$

where e is the electron charge, N_{e} is the electron density, and v_{e} is the electron velocity. The THz field is proportional to the derivative of the electron density, which can be written as

$$E_{\text{THz}} \propto \frac{\mathrm{d}J(t)}{\mathrm{d}t} = e\frac{\mathrm{d}N_{\text{e}}(t)}{\mathrm{d}t} v_{\text{d}}(t) \tag{6.8}$$

where $v_{\text{d}}(t)$ is the drift velocity of the electrons. Through a series of assumptions and derivations [39], the THz field can be expressed in the following form:

$$E_{\text{THz}} \propto f(E_{\omega}) E_{2\omega} \sin(\varphi) \tag{6.9}$$

where $f(E_{\omega})$ is the function related to E_{ω}. It can conclude that E_{THz} scales linearly with $E_{2\omega}$ and the magnitude and polarity of the THz field are strongly dependent on the relative phase φ. Nowadays, it is believed that THz radiation generated by air plasma is a mixed contribution of FWR and transient photocurrent [40]. When the laser pulse is weak (below or near the air ionization threshold), the FWR nonlinear process of the neutral atoms dominates. When the laser intensity is strong (above the ionization threshold), the mechanism of transient photocurrent with free plasma electrons becomes dominant.

In addition, many other researches have contributed to the generation and understanding of THz radiation generated by air plasma. Since 2000, THz radiation from photoionized air under strong external electric bias field has been studied [41–46]. The bias field accelerates the electrons of the plasma and increases the THz amplitude. However, DC bias increases the complexity of the system and is not suitable for remote sensing applications. In 2007, researchers studied plasma induced by few-cycle laser pulses based on the theory of FWM [47, 48]. If the pulse duration is short enough (<400 fs) and only contains a few cycles in the pulse envelope, the few-cycle laser pulses can directly create spatial charges asymmetry in the plasma, producing nonzero current and radiating THz wave. Moreover, the intensity of the THz radiation from the air plasma sources is proportional to the length of the plasma in the transmission direction, which is called laser filamentation. Also in 2007, C.D' Amico et al. achieved THz radiation from air filaments with a one-color laser for the first time [49]. The THz pulse emitted from the filament is highly directional. The filamentation process can be described as a self-action process, in which the intense fs laser pulses undergo strong spatial and temporal reshaping. In the filamentation, THz pulses do not experience natural diffraction but are spatially confined to a narrow region of micron diameter and can remain almost constant over several Rayleigh lengths, leaving ionization channels [50]. The cone-shaped distributed THz radiation was observed from the air filaments, which was explained by the theory of transition–Cherenkov radiation (TCR) [49]. This technology was then realized by the two-color laser filaments [51], which has been explained by the off-axis phase matching.

In the detection part of THz-TDS system, no matter what detection method is used, the final signal detected by the PD of PMT is the electric field of the SH caused by the THz field and the additional bias voltage field. Since the intensity of the air plasma in the detection part is very weak compared with that of in the generation part, the principle used to detect THz radiation by air plasma is the FWR process in third-order nonlinearity. The first steps for THz detection with gases were realized by J.M. Dai in 2006 [52]. Since there was no additional bias voltage field and only the laser-induced ionized gas was used for the detection portion, the THz-field-induced second harmonic (TFISH) strongly depended on the probe laser intensity. When the probe intensity was smaller than $1.8 \times 10^{14}\ \mathrm{W\,cm^{-2}}$, it was incoherent detection based on multi-photon ionization (MI). The TFISH can be described as

$$E_{2\omega}^{\mathrm{THz}} \propto \chi^{(3)} E_{\omega} E_{\omega} E_{\mathrm{THz}} = \chi^{(3)} I_{\omega} E_{\mathrm{THz}} \tag{6.10}$$

$$I_{2\omega} = \left(E_{2\omega}^{\mathrm{THz}}\right)^2 \propto \left(\chi^{(3)} I_{\omega}\right)^2 I_{\mathrm{THz}} \tag{6.11}$$

where $E_{2\omega}$, E_{ω}, and E_{THz} are the electric field amplitudes of the SH wave, FW, and THz wave, respectively; $I_{2\omega}$, I_{ω}, and I_{THz} are the intensities of the SH wave, FW, and

THz wave, respectively; and $\chi^{(3)}$ is the third-order susceptibility. From the Eq. (6.10), we can see that the TFISH electric field is proportional to the THz electric field. Because the phase information cannot be extracted in this process, it is called incoherent detection. When the probe laser increased to higher than 5.5×10^{14} W cm^{-2}, it was coherent detection with the effect of tunnel ionization (TI). This is due to the laser-induced air plasma causing another SH component $E_{2\omega}{}^{\text{LO}}$. LO is stands for local oscillator. So, the TFISH in this way was generated by the interaction of the THz field and the fundamental laser field; the intensity of the TFISH can be written as

$$I_{2\omega} \propto (E_{2\omega})^2 = \left(E_{2\omega}^{\text{THz}} + E_{2\omega}^{\text{LO}}\right)^2 = \left(E_{2\omega}^{\text{THz}}\right)^2 + \left(E_{2\omega}^{\text{LO}}\right)^2 + 2E_{2\omega}^{\text{THz}}E_{2\omega}^{\text{LO}}\cos(\varphi) \quad (6.12)$$

where φ is the phase difference between $E_{2\text{w}}^{\text{THz}}$ and $E_{2\text{w}}^{\text{LO}}$. This was a coherent detection process due to the phase information included. The intensity of the $E_{2\text{w}}^{\text{LO}}$ depended on the plasma density, which depended on the probe laser intensity. Due to the optical heterodyne detection, the DC contribution from the LO $\left(E_{2\text{w}}^{\text{LO}}\right)$ can be filtered out by the lock-in amplifier. Combined with Eq. (6.10), the equation can be simplified as

$$I_{2\omega} \propto (\chi^{(3)}I_{\omega})^2 I_{\text{THz}} + 2\chi^{(3)}I_{\omega}E_{2\omega}^{\text{LO}}E_{\text{THz}}\cos(\varphi) \quad (6.13)$$

Furthermore, the intensity between the incoherent detection and the coherent detection was a hybrid detection. By increasing the probe beam intensity, the detection changed from incoherent detection to coherent detection. However, the intensity of THz radiation was very weak.

To obtain intense THz radiation, a bias voltage field was added at the overlap focus point of the THz and probe laser [53], which is called air-biased coherent detection (ABCD). The bias voltage field is used as a local oscillator to provide an additional SH. The repetition rate of the AC bias voltage was 500 Hz when the fs laser repetition rate was 1 KHz. In this situation, $E_{2\text{w}}^{\text{LO}}$ in Eq. (6.12) means the SH field is induced by the bias voltage field:

$$E_{2\omega}^{\text{LO}} \propto \chi^{(3)}I_{\omega}E_{\text{bias}} \quad (6.14)$$

Since the phase difference between $E_{2\text{w}}^{\text{THz}}$ and $E_{2\text{w}}^{\text{LO}}$ is 0 or π by changing the direction of the bias voltage field, combined with Eqs. (6.10) and (6.14), Eq. (6.12) can be written in the following form:

$$I_{2\omega} \propto (\chi^{(3)}I_{\omega})^2\left[(E_{\text{THz}})^2 \pm 2E_{\text{THz}}E_{\text{bias}} + (E_{\text{bias}})^2\right] \quad (6.15)$$

Only the cross term can be measured when the AC bias voltage is synchronized with the lock-in amplifier; therefore, Eq. (6.15) is eventually written as

$$I_{2\omega} \propto 4(\chi^{(3)}I_{\omega})^2 E_{\text{bias}}E_{\text{THz}} \quad (6.16)$$

This was a coherent detection process. Compared with the first method, the local oscillator was changed from the gas plasma generated by the probe laser to the external bias voltage field. The probe pulse energy required for coherent detection was reduced, and the signal-to-noise ratio (SNR) was significantly improved.

6.2.1.4 THz Emission of 2D Semiconductor Materials

The THz emission of 2D semiconductor materials is basically based on the THz generation principles described above. There are two competing mechanisms for THz generation in 2D semiconductors, which can be described by [54–56]

$$\vec{E}_{\mathrm{THz}} \propto \frac{\partial^2 \vec{P}}{\partial t^2} + \frac{\partial \vec{J}}{\partial t} + \frac{\partial}{\partial t}(\nabla \times \vec{M}) \tag{6.17}$$

The first term on the righthand side is nonlinear polarization (P), derived from OR. When the photon energy of the pump laser is below the bandgap, the rectification current plays a dominant role in THz radiation. The second term represents the contribution from transient photocurrent (J). When the optical excitation is above the bandgap, either transient photocurrent or a combination of the two mechanisms dominates the THz emission. If it is a magnetic material, the third term magnetization (M) is considered.

Take ultraefficient THz emission in a thin-layered gallium telluride (GaTe) film as an example [57]. The results are shown in Figure 6.4. In general, THz radiation

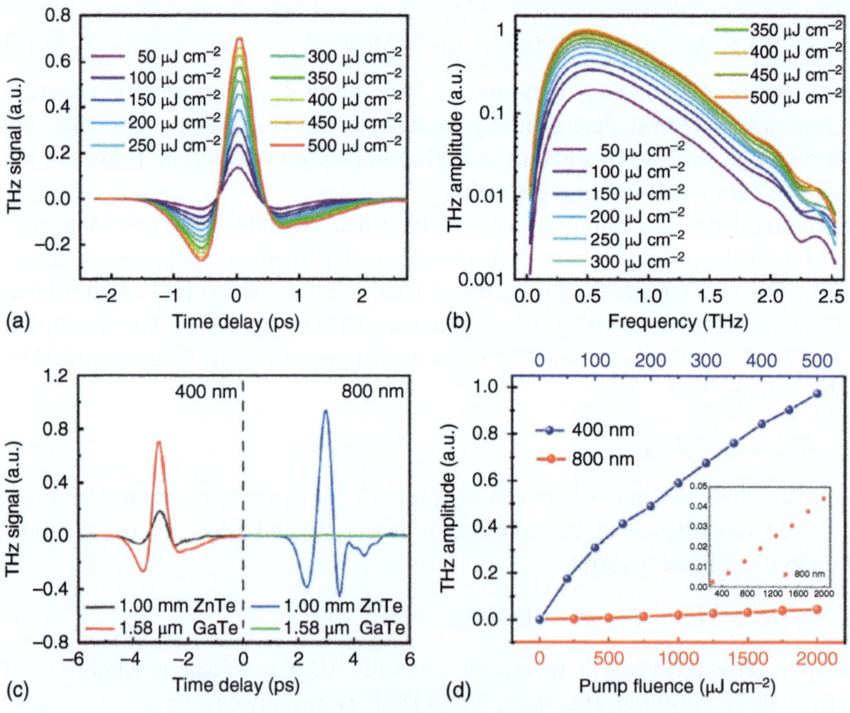

Figure 6.4 Pump-fluence- and wavelength-dependent THz generation in GaTe. (a) Time-domain THz waveforms at different photoexcitation fluences in response to 400-nm excitation. (b) Frequency-domain spectra of the THz signals that are shown in panel (a). (c) THz signals generated by GaTe and ZnTe in response to different pump lasers of different wavelengths with fluence fixed at 500 μJ cm^{-2}. (d) Peak-to-peak values of THz signals as a function of the pump fluence with wavelength centered around 400 and 800 nm, respectively. The inset shows an enlarged plot of the THz amplitudes under 800 nm excitation. Source: Adapted from Tong et al. [57].

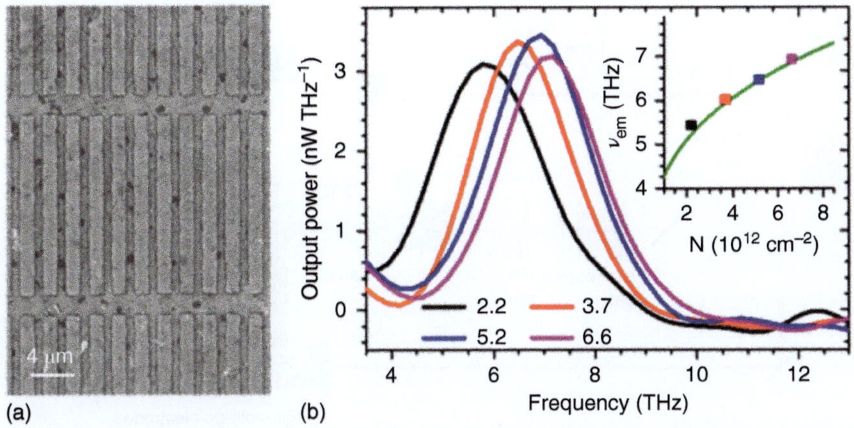

Figure 6.5 Graphene THz light-emitting metasurfaces. (a) Top-view scanning electron microscopy (SEM) image of an experimental sample. (b) Measured emission spectra for Tbase = 80 K, Pin = 0.2 W, and different values of carrier density N (listed in the legend in units of 10^{12} cm^{-2}). Each trace is normalized to the spectrum measured under the same conditions at charge neutrality. Inset: Peak emission frequencies extracted from the spectra (symbols) and calculated plasmonic resonance frequency versus carrier density (solid line). Source: Reproduced with permission from Li et al. [58]/Optica Publishing Group.

increases as pump power increases. The wavelengths of 800 and 400 nm corresponds to energies of 1.55 and 3.1 eV, respectively. Also, the bandgap of GaTe is 1.65 eV. Therefore, the relationship between excitation energy and energy band of the sample determines its radiation mechanisms and the magnitude of radiation intensity.

There are a number of advanced 2D THz emission materials out there, such as graphene, transition metal dichalcogenides (TMDs) like MoS_2 and WS_2, topological insulators, hybrid perovskites, heterostructures, and so on. Graphene is a promising material with a variety of unique properties, including its gap-free, linear electronic energy dispersion, ultrahigh intrinsic mobility, and long-wavelength plasmonic resonances. The use of metasurfaces is an attractive way to enhance light–matter interaction, which is not limited by the ultrasmall thickness of monolayer graphene. One example is shown in Figure 6.5 [58]. THz is emitted from the graphene metasurfaces. This device operation relies on the generation of hot carriers in the ribbons under direct current injection and their subsequent energy relaxation through the excitation of graphene plasmon polaritons. Narrowband free-space THz radiation is then emitted by the resulting collective oscillations of the graphene electron gas. The novel 2D materials used as THz emitting devices can not only promote the development of new and compact THz sources but also provide a nondestructive optical method for the study of the carrier transport mechanism and ultrafast dynamics of materials.

6.2.2 Typical THz Time-Domain Spectroscopy Setup

In this part, we first introduce the basic structure of a THz time-domain spectroscopy system. Figure 6.6 shows a schematic diagram of a THz-TDS system that generates

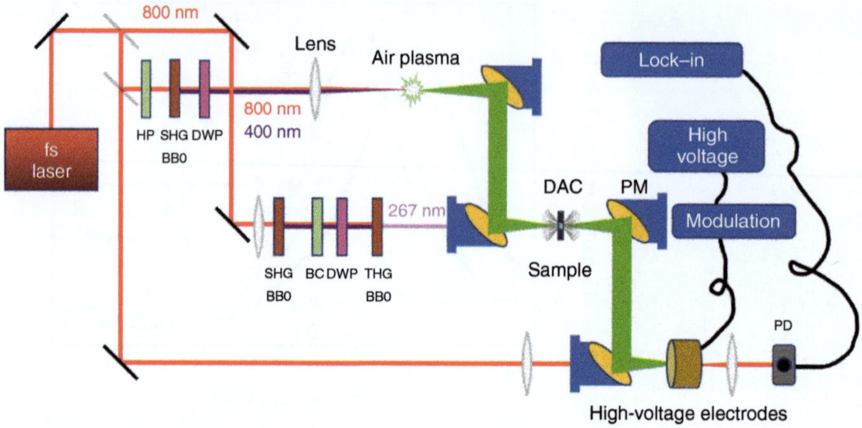

Figure 6.6 Schematic diagram of a THz-TDS system for THz radiation generated and detected by air plasma.

THz based on air plasma and detects THz based on ABCD. Here, only fs laser amplifier is considered as the excited sources.

Usually, 800-nm fs pulses from an amplified laser are split into two beams by the beam splitter (BS): one beam is used to generate THz radiation, and another is used for detection. The generate pulses with higher laser energy transmit a lens and focus to the point where the air plasma (filament) is generated due to the air or other gas ionization. A 100-nm-thick type-I β-barium borate (BBO) crystal is inserted between the lens and the air plasma to generate SH at 400 nm. It is called an SHG (second-harmonic generation) BBO. The combination of the fundamental pulse laser (ω) and its SH (2ω) enhances the amplitude of the THz field, which is the most common way to generate THz radiation by air plasma now. Previous study has shown that the optimum ratio is 1/9 between the fundamental and SH laser fields to obtain the maximal THz radiation [59]. The focal length of the lens, the angle of the BBO crystal axis with respect to the polarization direction of the incident light, and the distance from BBO to the focus point all have a significant influence on the THz radiation intensity. The FW (800 nm) and SH (400 nm) wave are orthogonally polarized via type-I phase matching. To generate maximal THz radiation, the polarization of 800 and 400-nm two-color lasers is changed by half wave plate (HP) and double-wavelength wave plate (DWP), respectively, and the polarization is finally consistent (collinearly polarized) [60]. In addition, the thickness of the HP is as thin as possible to reduce the dispersion and the optical path difference between the two-color pulses [61]. At the location of the air plasma, THz radiation is generated and transmitted along the plasma filament. A Si plate is placed behind to block the laser and transmit THz wave. For maximal THz generation, the Si plate is placed at a Brewster angle (about 74°) with respect to the beam path [61]. The emitted THz pulses are then collimated and focused by two pairs of off-axis parabolic mirrors (PM). The sample is placed in the focal position. In the detection part, the probe pulses reach a lens through a delay line and then pass through

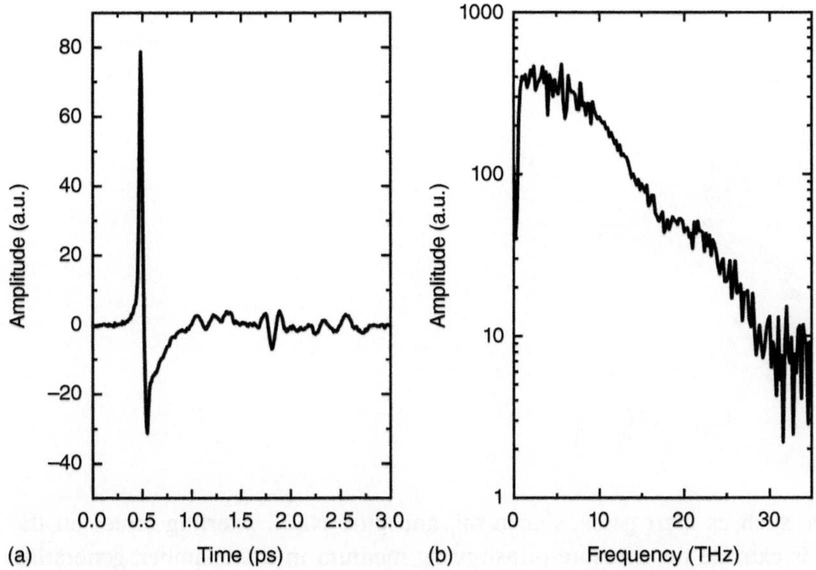

Figure 6.7 (a) THz-TDS and (b) corresponding Fourier transform spectroscopy based on air plasma generation and ABCD detection system.

the aperture on the back of the last PM and converge on the same point as the THz beam focused by the last PM. An additional bias voltage field with a pair of electrodes is applied at the focus point, which is vertical to the direction of laser propagation. Therefore, the SH wave induced by the THz field and the additional voltage bias voltage field is generated. The FW is filtered, and the SH wave signal is detected by the balance photodiode (PD) or photomultiplier tube (PMT). The THz field is proportional to the SH field; in this way, the THz field strength is obtained. The THz-TDS and corresponding Fourier transform spectroscopy are shown in Figure 6.7. In this system, the frequency range can reach to 30 THz.

If we add one pump laser to excite the sample in this system, an optical pump–THz detection spectroscopy setup can be achieved. The original laser wavelength is 800 nm. The wavelength of 400 nm can be obtained by adding SHG BBO, and the wavelength of 267 nm can be obtained by adding third-harmonic generation (THG) BBO. A birefringence crystal (BC) is used here as a phase compensator.

6.2.2.1 Time-Domain THz Spectroscopy Under Pressure

Pressure is an effective routine to alter the lattice parameters of materials and thereby modify the electron and phonon dispersions, leading to electron or even lattice structural transitions. A diamond anvil cell (DAC) is a frequently used apparatus to generate a wide range of hydrostatic or quasi-hydrostatic pressure conditions [63]. The typical schematic diagram of DAC is presented in Figure 6.8. It consists of two diamonds and a gasket with a columnar sample chamber in the middle. The sample is sealed in the chamber with filling pressure-transmitting

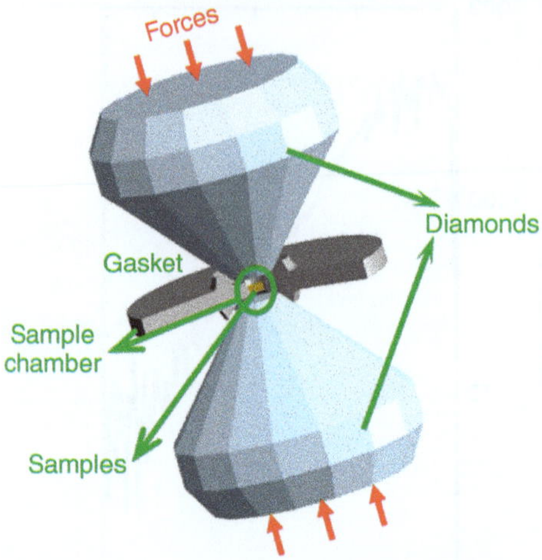

Figure 6.8 Schematic diagram of the DAC setup. Source: Reproduced with permission from Xu et al. [62]. © 2021, Optica Publishing Group.

medium, such as inert gases, silicon oil, and pure NaCl. Exerting forces on the diamonds extrude the pressure-transmitting medium in the chamber, generating hydrostatic pressure around the sample.

Generally, the natural hardest diamond exhibits excellent transmittance properties in an ultrabroad wavelength range from 225 nm (the bandgap of the pure diamond) to several hundreds of μm, covering X-ray (mostly above 5 keV), visible light, infrared, and THz waves except several phonon absorptions between 4 and 10 μm. In addition, one can also exert low/high temperature or electric/magnetic field on the DAC. Thus, DAC is always employed in the in situ measurements and studies of the high-pressure optical observation, infrared and Raman scattering spectra, X-ray diffraction, optical pump–probe spectra, electromagnetic transport, etc. [63]. In theory, DAC can also be applied in the detection of the THz transmission or reflection spectra. However, due to the diffraction limit, the practical spot size of the focused THz pulse can reach to several hundred μm or even mm scale, which is bigger than the culet diameter of the frequently used diamond anvil. This will dramatically attenuate the THz intensity, especially the low-frequency range, and thus the application of in situ time-resolved THz spectroscopy under high pressure is currently limited. In addition, with increasing pressure, the pressure-transmitting medium is also compressed, i.e. its properties also change with pressure. The spectrum analysis will be more complex.

To avoid this, diamond anvil with big culet diameter should be employed. In 2021, we investigated the photocarrier dynamics in GaAs under hydrostatic pressure using time-resolved THz spectroscopy [62]. GaAs is a typical direct gap semiconductor that hosts intriguing ultrafast photophysics such as THz nonlinear absorption [64–66] and phonon decoupling [67] as dictated by hot carriers. With the application of hydrostatic pressure, the Γ and L valley extrema move upward. However, the X valley minimum moves downward [68, 69]. Above 4 GPa, a crossover between the

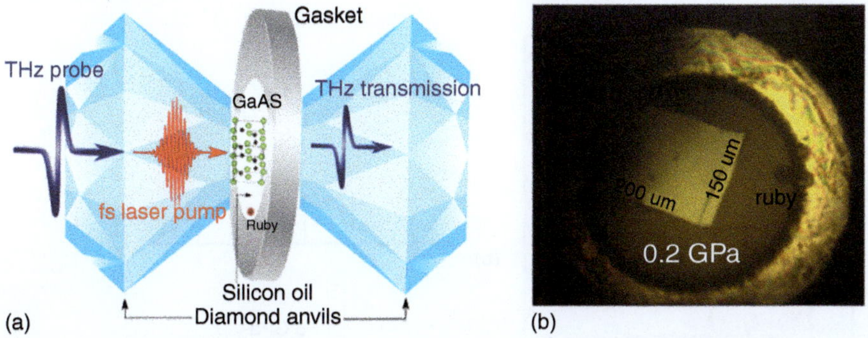

Figure 6.9 (a) Schematic of the DAC for in situ TRTS. (b) Morphology of the sample in the pressure chamber at 0.2 GPa. Source: Reproduced with permission from Xu et al. [62]/Optica Publishing Group.

Γ and X valley extrema occurs, forming an indirect bandgap at the X point. Such a shift intrigues versatile phenomena.

We used a DAC with a culet size of 500 μm to generate hydrostatic pressures up to 10 GPa. The sample chamber of the drilled gasket was about 300 μm in diameter. A piece of undoped GaAs single crystal wafer with <100> orientation and the size of about $200 \times 150 \times 10\,\mu m^3$ was loaded in the DAC chamber together with a ruby particle, as shown in Figure 6.9b. The transparent liquid silicon oil was used for the pressure-transmitting medium. The aperture diameter is obviously smaller than the spot diameter of the THz pulse, and it will hugely attenuate the transmitted THz intensity. The intense THz pulses from a $LiNbO_3$ crystal are used as the probe beam in time-resolved THz spectroscopy [16, 70]. Two beams with 600- and 800-nm wavelengths were used as the pump beams, and the applied pump fluences were kept below $60\,\mu J\,cm^{-2}$ to avoid the multi-photon absorption from diamonds. In addition, the signal of DAC with only the pressure-transmitting medium under high pressure was also examined, and no signal was observed.

The bandgap of the GaAs increases with pressure and exceeds 1.55 eV (800 nm) above 2 GPa. However, the GaAs still shows observable signal under 800-nm pump configuration, and the peak intensity nonlinearly increases with increasing pump fluence, indicating that a two-photon absorption effect occurs. The decay times of the hot carriers show significantly decrease from about 4 ns at ambient conditions to 65 ps under a pressure of 10 GPa, and a fast decay component emerges at high pressures. Such a phenomenon is suggested to stemming from the formation of surface defects due to the external pressure [71, 72], which can increase the density of nonradiative recombination centers and lead to the fast trapping of photocarriers under high pressure [73]. Furthermore, the intraband photocarrier dynamics was also investigated. The initial signal in the time range of 20 ps for photoexcitation at 800 nm and 600 nm is presented in Figure 6.10. For the 800-nm pump configuration, the initial rising process firstly prolongs with increasing pressure and then shortens

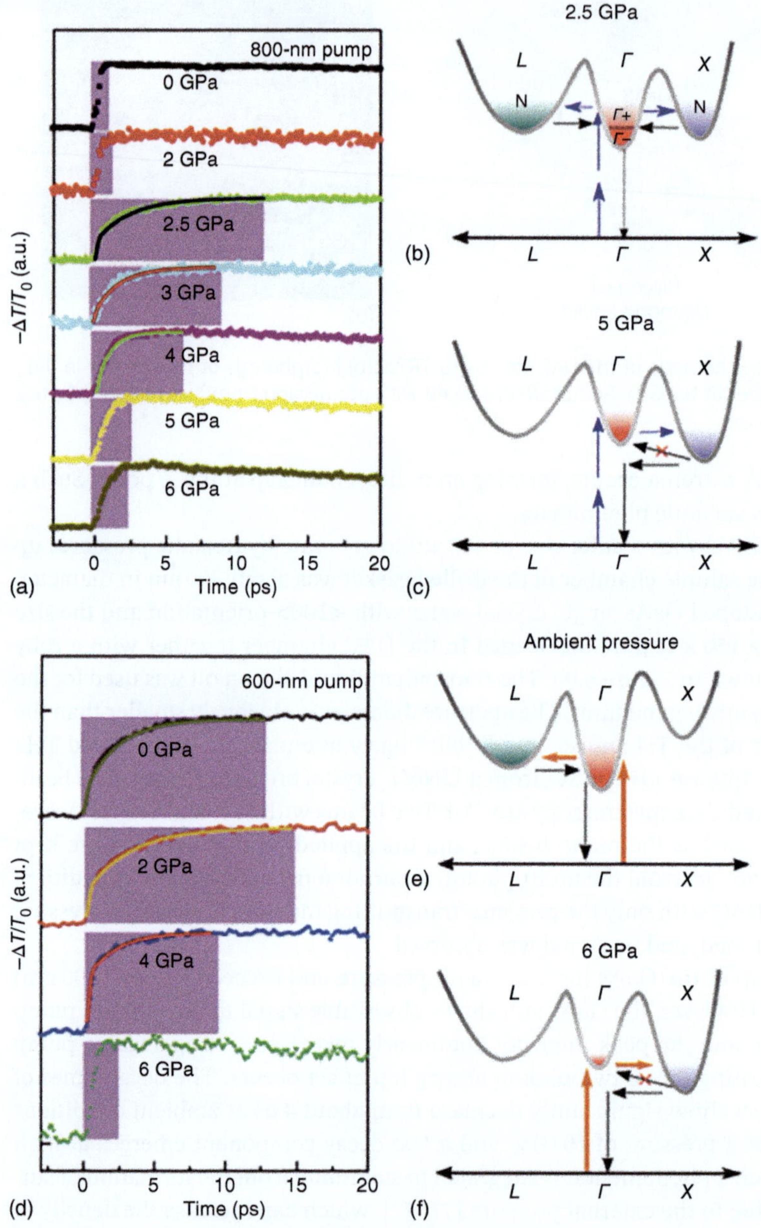

Figure 6.10 (a) and (d) Normalized $-\Delta T/T_0$ in the short time range of 20 ps for photoexcitation at 800 and 600 nm, respectively. The curves are shifted vertically for comparison. The solid circles and solid lines represent the experimental data and fitting results using the three-state model, respectively. The shaded area indicates the rising process of the signal. (b), (c), (e), and (f) Schematic of the electronic band structure of GaAs and related intervalley scattering mechanism. (b) and (c) represent the photoexcitation condition with 800 nm at 2.5 and 5 GPa, respectively. (e) and (f) correspond to the photoexcitation using 600 nm at ambient pressure and 6 GPa, respectively. Source: Reproduced with permission from Xu et al. [62]/Optica Publishing Group.

above 2.5 GPa. Pressure can effectively modify the intervalley scattering processes, i.e. the fraction of the electrons in the Γ valley can transfer into the neighboring valley via LO phonon scattering [64]. Hence, the rising time prolongs. In addition, the three-state rate model can also be used to extract the time constants at different pressures.

2D layered materials always attract a lot of attention due to their exclusive properties in scientific research and application. The neighboring layers of those materials bond by weak van der Waals (vdW) force. Therefore, 2D materials can be dissociated into thin films or even a few layers. Due to this unique feature, high-pressure technology is very applicable for the property investigation of the 2D materials because the thickness of the sample chamber for the DAC is typically less than 40 μm. A lot of experimental technologies is applied to explore the intriguing properties of 2D materials under an extreme pressure condition, such as Raman and infrared spectroscopies, electromagnetic transport, X-ray diffraction, and optical pump–probe spectroscopy. However, time-resolved THz spectroscopy is now rarely employed on it and still faces tremendous challenges.

6.2.3 Photoinduced Changes of the Dielectric Function in 2D via THz Spectroscopy

The emerging vdW heterostructures via the stacking of multiple 2D materials, breaking through the restriction of lattice matching in traditional heterostructures [74–76], are promising to tailor the physical properties for individual layers and further extend their application prospects in electronics [77], optoelectronics [77–79], photocatalysts [80, 81], and others [82]. Usually, fantastic features surpassing the separate materials can be reached for the heterostructure. For example, due to its gapless nature, graphene always shows large dark current, low light absorption, and low quantum efficiency, impeding its direct application in nanodevices. Combining graphene and other 2D material, such as MoS_2 and WS_2, as heterostructures can effectively overcome those drawbacks and broaden the application of graphene-based devices. In addition, the heterostructure also breeds plenty of novel physical phenomena, such as charge transfer and interlayer exciton, providing a versatile platform for scientific research and application [83].

THz spectroscopy can be served as an ideal noncontact probe method for carrier dynamics with a sub-picosecond temporal resolution, benefiting from simultaneous access to the frequency- and time-resolved photoconductivities. A lot of researches has been focused on the ultrafast photocarrier dynamics of layered heterostructures, including interband recombination, surface recombination, spatial diffusion, etc. As shown in Figure 6.11, the carrier dynamics of the large-scale vertically stacked heterostructures with MoS_2 and $ReSe_2$ monolayers were investigated using ultrafast spectroscopies [84]. An 800-nm fs beam was pumped on the surface of the $MoS_2/ReSe_2$ and $ReSe_2/MoS_2$ heterostructures with an incidence angle of 45°. The incident 800-nm proton energy exceeds the low-lying excitonic energy of the monolayer $ReSe_2$ (~1.45 eV), while it lies well below with that of monolayer MoS_2 (~1.9 eV). It also can be seen that the radiated THz pulses show an obvious

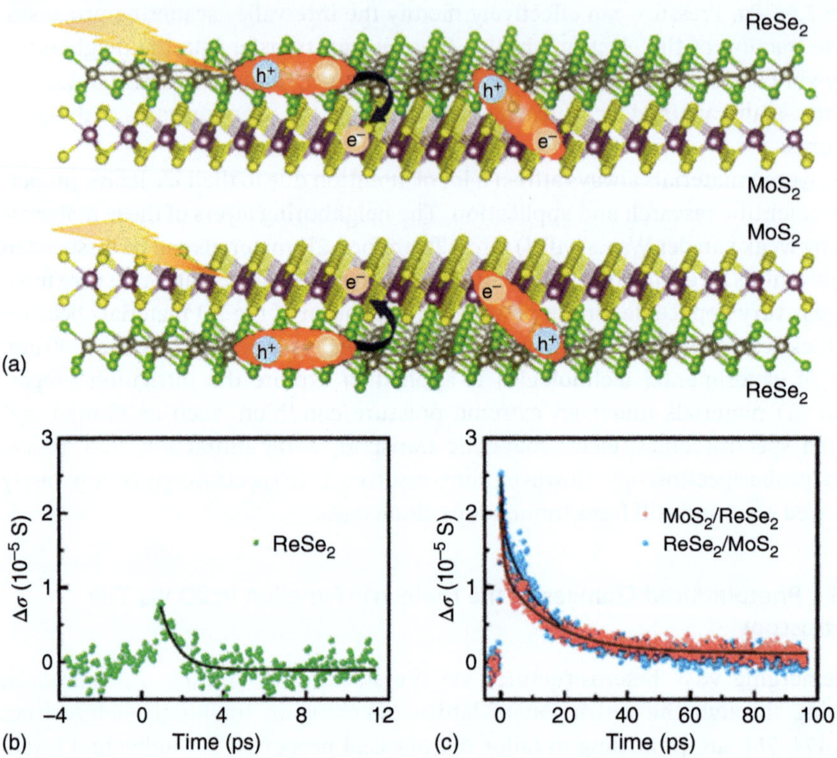

Figure 6.11 (a) Time-resolved and frequency-resolved THz photoconductivities excited with 800-nm pump pulses. (b) Time-resolved THz photoconductivity in $ReSe_2$ monolayer. The solid line represents a single exponential fitting result. (b) Time-resolved THz photoconductivities in $MoS_2/ReSe_2$ (red) and $ReSe_2/MoS_2$ (blue) heterostructures. Source: Reproduced with permission from Yang et al. [84]/American Chemical Society.

π phase shift for the two stacking configurations. All the phenomena implies that the emitted THz pulse stems from the burst of the out-of-plane current since the direction of the current is exclusively determined by the stacking orders. In addition, the photoinduced changes in the real part of the conductivity, Re $\Delta\sigma(t)$, of the heterostructures show typical two-exponential decay, which is obviously different to the isolated $ReSe_2$, demonstrating the charge transfer process in the heterostructures.

Integration of graphene and monolayer TMDs enables high-performance photodetectors by combining the strong optical absorption at exciton resonances in TMDs, with ultrafast charge transfer across vdW interfaces and ultrahigh charge mobility in graphene. Combining ultrafast THz and transient absorption spectroscopy, Fu et al. investigated the interfacial hot electron transfer and recombination processes in graphene–WS_2 heterostructures following optical excitation [85]. As the typical 2D semiconductor, the frequency-resolved THz photoconductivity can be well-fitted by a Drude–Smith model. Figure 6.12 shows the charge carrier dynamics in MXenes using THz laser spectroscopy [85].

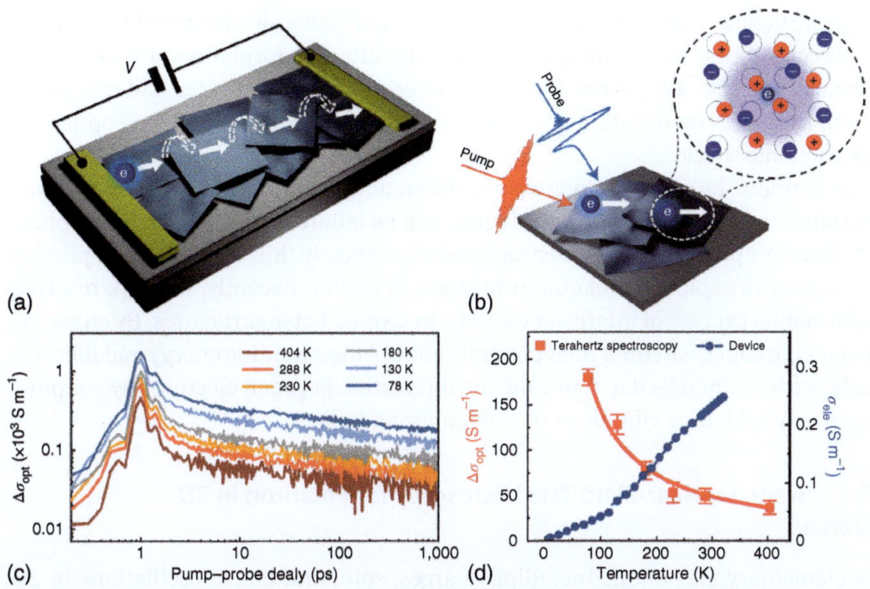

Figure 6.12 (a) Schematic of charge transport within the network of MXene flakes. (b) Scheme for ultrafast OPTP experiments to characterize the intra-MXene flake transport of charge carriers. (c) The time-resolved terahertz photoconductivity dynamics at temperatures ranging from 78 to 404 K under vacuum conditions. (d) A direct comparison of T-dependent charge carrier conductivity determined optically by terahertz spectroscopy (by taking the averaged photoconductivity from 600 to 800 ps following excitations) and electrically via the van der Pauw four-probe method. Source: Zheng et al. [85]/with permission of Springer Nature.

6.3 Nanoscale THz Scanning Probe Microscopy of 2D Materials

The class of 2D vdW materials has recently emerged as a great platform to design materials on demand. Bulk crystals can be thinned down to the ultimate limit of only a single atomic layer. The reduction of dimensionality results in exceptionally strong Coulomb correlations in monolayers. Light trapped at the nanoscale, deep below the optical wavelength, exhibits an increase in the associated electric field strength, which results in enhanced light–matter interaction. This leads to strong nonlinearities, large photonic forces, and enhanced emission and absorption probabilities. A practical approach toward nanoscale light trapping and manipulation is offered by interfaces separating media with permittivities of opposite signs. Such interfaces sustain hybrid light–matter modes involving collective oscillations of polarization charges in matter, hence the term polaritons.

We have focused our research recently on the semiconducting 2D materials made from TMDs. There, optically excited electrons and holes are bound to each other by their strong Coulomb attraction – just like how electrons are bound to the nucleus in the hydrogen atom – forming the so-called excitons. By observing their atom-like energy structure with subsequent ultrashort light pulses in the mid infrared, we can directly access the binding energy and the ultrafast formation of these quasi-particles. Thereby, we are not limited by interband selection rules and can probe exciton dynamics on ultrafast timescales including the typically hidden "optically" dark excitonic species.

Furthermore, by stacking and twisting different monolayers in a targeted manner, band structures and Coulomb correlations can be tailored to achieve unique optical and electronic properties. Heterostructures of atomically thin materials thus provide a laboratory to explore novel quantum states of matter. Recently, we have resolved the formation process of interlayer excitons in layered heterostructures. By changing the stacking angle, we could then precisely control their binding energy and lifetime. Lately, we have managed to fine-tune the interaction between electrons by coupling to polar crystal lattice vibrations of a neighboring layer.

6.3.1 Scanning Near-Field THz Microscopy for Polariton in 2D Materials

The elementary excitations including charges, spin, and lattice oscillations in 2D materials are covering the THz range (from 0.1 to 30 THz). Therefore, ultrafast THz spectroscopy is being used for insight into the understanding of materials for next-generation technologies for tracking emergent phenomena in low-dimensional materials. In particular, for 2D materials, the electronic and optical properties are dominated by the surface and interfaces due to the quantum confinement effect. As a result, light can be trapped and confined within a thin layer near the surface, leading to an enhanced light–matter interaction. Such coupling between photon and matter modes involves collective oscillations of polarization charges in the matter, so-called polariton. Several types of polaritons including plasmon–polariton, exciton–polaritons, and phonon–polaritons have been reported by scientists recently. The dispersion of polariton in 2D materials can be described in the following equation when the thickness of materials d is much smaller than the wavelength of polaritons λ_p:

$$\lambda_\text{p} = \frac{2\pi}{K_\text{p}} = 4\pi^2 \text{Im}\left\{\frac{\sigma}{\omega\varepsilon_\text{a}}\right\}, d \ll \lambda_\text{p} \tag{6.18}$$

where the σ is the conductivity, ω is the frequency, ε_a is the permittivity of the environment, and K_p is the in-plane polariton wave vector. The conductivity σ in semiconductor is commonly described as

$$\sigma(\omega) = \frac{i}{\pi}\frac{S_\text{f}}{\omega + i\tau_\text{f}^{-1}} + \frac{i}{\pi}\frac{\omega S_\text{b}}{\omega^2 - \omega_\text{b}^2 + i\omega\tau_\text{f}^{-1}} \tag{6.19}$$

where the first part of the Drude model and the second part of the Lorentz model represent the contributions of free (f) and bond (b) charges, respectively.

Unlike the limitations of traditional far-field characterization methods on both time and spatial resolutions for 2D materials, scattering scanning near-field optical microscopy (SNOM) is a powerful technique that can reveal novel physical phenomena and dispersion behaviors of 2D polaritons. In SNOM, a sharp tip is used to

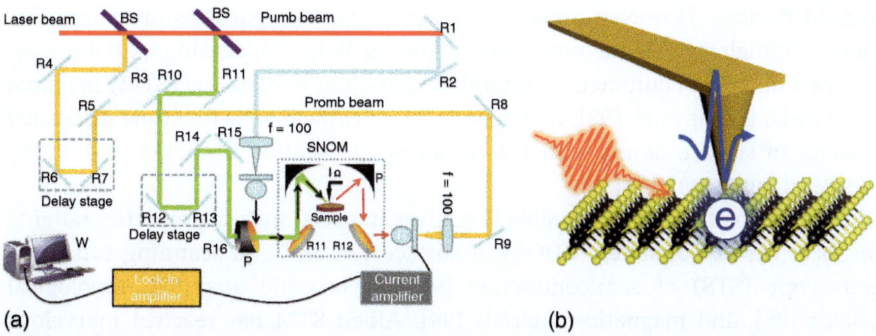

Figure 6.13 (a) Typical scanning near-field THz microscopy setup. (b) A metallized tip scans over 2D materials, and terahertz laser light illuminates the tip and split.

scan the surface of the sample, and the near-field signal is detected using a detector located at the tip apex for measuring the near-field intensity and phase information of 2D polaritons. For s-SNOM at THz frequencies, a sharp tip is also used to focus the THz pulse onto the sample and collect the scattered or transmitted light for high-resolution imaging of the local electric field distribution. Figure 6.13 showed the setup of the scanning near-field THz microscopy.

The evanescent THz fields coupled to an s-SNOM tip are polarized normally to the surface of the tip and oriented longitudinally at its apex [86]. When the s-SNOM tip is brought close to the sample surface, the evanescent THz fields interact with the sample and induce an additional polarization in the tip [87, 88]. This happens because the near fields interact with the charges in the sample, causing them to oscillate and emit THz radiation. This radiation then interacts with the tip, inducing a polarization in the opposite direction to the incident THz field. The induced polarization in the s-SNOM tip can be described as a net dipole that radiates THz light into the far field, where it can be detected by a detector. The THz radiation is scattered by the sample, and the dielectric properties of the sample affect how the THz radiation scatters. Specifically, the near-field reflection coefficient of the sample affects how the scattered THz field is imprinted with information about the sample's dielectric properties. The local Fresnel reflection coefficient describes the reflection of the THz radiation at the interface between the s-SNOM tip and the sample, and it plays an important role in determining the contrast in the scattered THz field. By analyzing this scattered THz field, it is possible to create an image of the sample's dielectric properties at the nanoscale level.

6.3.2 THz-STM

In the recent years, people's understanding about material has been broadly stretched especially in the microscopic field. A number of research investigating microscopic dynamics emerged that prompted the development of nanoscale material science. One of the most popular topics is surface science that engages in exploring the dynamics of quasiparticles such as phonons and excitons on the nm and fs scales [89–91]. These studies will play fundamental roles in developing advanced devices. However, tools used to study surface particles are scant that require ultrahigh spatial and temporal resolutions. [90]. Considering spatial resolution, one important equipment is scanning tunneling microscope (STM) invented in 1982 by G. Binnig et al. [92]. It was since then people for the first time unraveled the secret of surface atoms with the initial demonstration of Si 111 in 1990 by R. Wiesendanger et al. [93].

With the help of STM, it is possible to analyze material surface properties ranging from local density of states (LDOS) of surface [94, 95] and scanning tunneling spectroscopy (STS) of semiconductors [96], superconductors [97], topological insulator [98], and magnetic materials [99]. Albeit STM has reached marvelous resolution in space, time resolution remains problematic. This situation continued until 2013 and was broken by the appearance of THz-combined scanning tunneling microscopy (THz-STM), a creative system proposed by T.L. Cocker [100]. THz is one kind of electromagnetic wave whose frequency ranges from 0.3 to 10 THz. Unlike visible and near-infrared light, it has many advantages when combined with STM. Its wavelength is 300um, and this derives a Keldysh parameter [101] value far less than 1, indicating that the role of THz on STM is more or less like a bias proportional to its electrical field strength. In addition, Zheltikov's calculation [102] pointed out that THz's electrical field established within a few femtoseconds meaning that the tunneling currents emerge with incident THz field simultaneously. Typically, THz-induced instant tunneling current lasts for a few picoseconds coherent to its electric field; then it is normalized by the current amplifier due to its limited bandwidth. Therefore, THz current acts as a DC part of the total currents characterized by the unit e/pulse [103] (Figure 6.14).

Owing to its role as a bias, THz is able to map material surface atomically like the DC. Actually, in 2017, Jelic et al. [104] managed to picture the 7*7 atom structure of Si 111 surface attaining a lateral resolution less than 0.3 nm (Figure 6.15). They proved and utilized the bulk current coming from belt bending due to THz to enhance tunneling current.

Later in 2020, Luo Yang et al. [105] pictured Cu 111 with THz-driven STM imaging. They distinguished the steps on the Cu 111 surface and further developed one useful model for characterizing THz tunneling current on metal surface. Apart from crystal surface, the imaging ability of THz-STM even reached molecule level as was shown in Cocker's experiment mapping the molecule orbital states of Pentacene [106] and Peller's study measuring the switch probability of MgPc molecule to come up with its LDOS. [107]. Additionally, the imaging ability of THz-STM can even excess the DC as was evidenced with the resolution of 7-AGNR lobe structure carried out by Ammermann et al. [108].

In contrast to visible and near-infrared light, photon energy of THz wave is quite small with only a few meV. Benefiting from this, THz has subtle impact on STM tip and is suitable for pump–probe experiment. It usually works as probe pulse but can also use as pump pulse as illustrated in the experiment by Cocker in 2016 [106].

In Cocker's experiment, he manipulated one THz pulse as pump pulse to activate the vibration of Pentacene while probing the vibrating movement with the other one. Through this approach, he successfully obtained the vibrational frequency

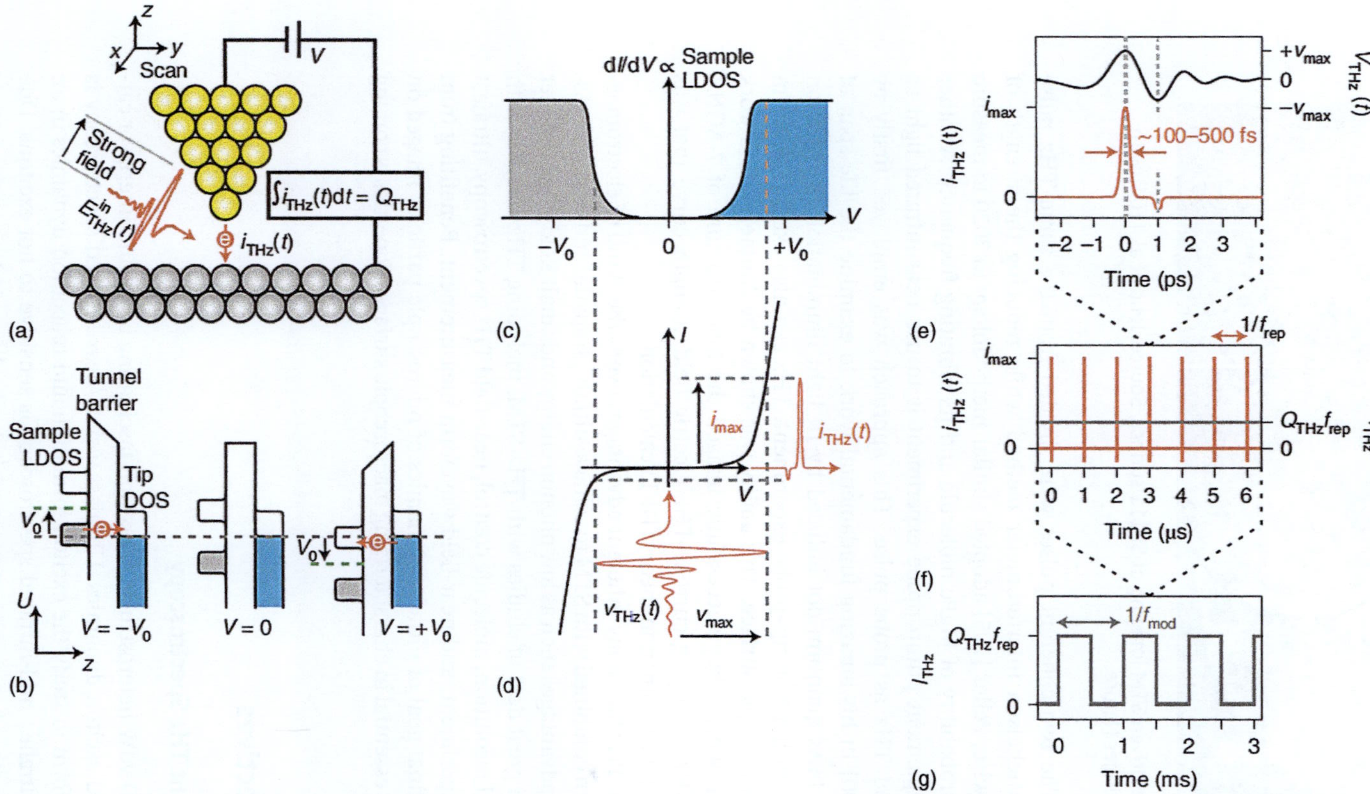

Figure 6.14 Operational concept of ultrafast light wave-driven THz-STM. (a) A single-cycle, phase-stable terahertz pulse is coupled into a scanning tunneling microscope tip–sample junction. (b) A static b as applied to the tunnel junction shifts the Fermi level of the sample (green dashed line) relative to that of the tip (grey dashed line), such that a net current flows through the tunnel barrier. (c) A conventional STS experiment reveals the differential conductance (dI/dV) of the junction. (d) Features in the LDOS correspond to nonlinearities in the I–V characteristic of the junction. (e) The full-width at half-maximum (FWHM) of the main peak of $i_{THz(t)}$ is sub-cycle on the terahertz oscillation timescale due to the nonlinear generation process. (f) The terahertz pulse focused onto the tip is part of a pulse train, as it is generated by a train of ultrafast near-infrared pulses from a laser amplifier. (g) THz from the normal tunnel current and improved signal-to-noise ratio of the detection. The THz pulse train is modulated at a frequency f_{mod} within the bandwidth of the preamplifier but above the bandwidth of the STM feedback loop. Source: Reproduced with permission from Cocker et al. [103]/Springer Nature.

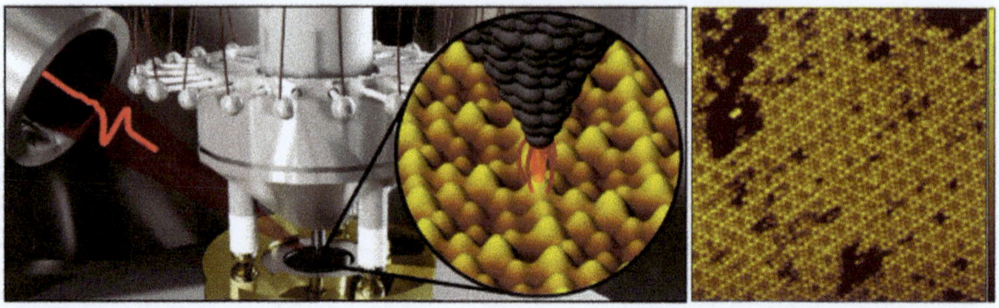

Figure 6.15 THz-driven STM imaging of Si 111 surface. Source: Jelic et al. [104]/with permission of Springer Nature.

of pentacene. The two coherent pulses were divided from the same THz pulse following the Michelson interferometer method while reducing the strength of one pulse. Likewise, Peller [105] adopted similar methodology in 2020 to measure the switching probability of MgPc molecule and its vibrating frequency. Another approach to implement pump probe experiment is to use near-infrared light as pump pulse and THz as probe pulse. This approach was employed firstly by T.L. Cocker [100] in his amazing fundamental work to examine the life span of carriers within InAs quantum dot induced by IR light illumination. In addition to imaging and ultrafast pump–probe experiment, THz can also extract spectrum information from sample surface. This ability was shown by Ammermann's work in 2020 [108] in which they successfully imaged the lobe structure of 7-AGNR and obtained THz tunneling current–THz electric field strength curve that was consistent with density function theory (DFT) calculation.

In conclusion, THz has many inherent advantages over other kind of electromagnetic waves when combined with STM to achieve ultrahigh spatial and timescale resolutions. These advantages such as low phonon energy and small Keldysh parameter brought about a great deal of studies with THz-STM, including THz imaging with ultrahigh spatial resolution, molecule control, near-field THz spectroscopy, ultrafast pump–probe experiment, and near-field waveform measurement. Benefiting from THz-STM, the final goal of ultrafast dynamics of microscopic particles imaged on nm and fs levels essential in characterizing macroscopic surface properties is coming into reality soon.

6.4 Perspectives

6.4.1 Transient THz Spectroscopy

Monolayer TMD vdW heterostructures exhibit fascinating intra- and interlayer carrier transfer and exciton dynamics. Ultrafast visible/near-infrared spectroscopy is a standard platform to study the exciton formation and relaxation dynamics in an indirect way. Ultrafast mid-infrared spectroscopy is sensitive to hot excitons. Due to the weak coupling between electrons and holes in different layers, the feature energy of interlayer excitons stays in the multi-THz range, directly observing that the dynamics has been realized. Transient THz spectroscopy that is sensitive to free carriers and partially localized excitons further provides complementary information. Combining all these ultrafast spectroscopies in wavelengths from visible to THz range would comprehensively investigate the carrier transfer and exciton dynamics in vdW heterostructures.

6.4.2 High-Resolution Spatial and Temporal THz Spectroscopies

Atomically thin 2D materials or vdW heterostructures ranging from graphene to TMDs have been extensively studied with various pump–probe spectroscopy techniques. Ultrafast optical pump–probe spectroscopy provides the possibility of tracing the carrier dynamics with femtosecond temporal and hundreds of nm spatial resolutions. Time-resolved THz time-domain spectroscopy (TRTS) is an ideal tool to probe the carrier processes due to the low THz photon energy. However, the spatial resolution is limited to mm scale because of the diffraction limit. Combing THz spectroscopy with nanometer probe tip pushes the spatial resolution to tens of nm scale, like THz s-SNOM. Recently, It has been demonstrated that by replacing the DC bias of STM with ps THz pulse, THz-STM has fs temporal and atomical scale spatial resolutions. Since many interesting physical phenomena happen at nanometer scale, the emerging techniques with high spatial and temporal resolution offer new opportunities to study the ultrafast dynamics of 2D materials in situ.

Acknowledgment

This work is supported by the National Natural Science Foundation of China (NSFC. 61988102, 12274424), the National Key Research and Development Program of China (2022YFA1203500), Key Research and Development Program of Guangdong Province (2019B090917007), Science and Technology Planning Project of Guangdong Province (2019B090909011).

References

1 Low, T., Chaves, A., Caldwell, J.D. et al. (2017). Polaritons in layered two-dimensional materials. *Nature Materials* 16: 182–194.
2 Basov, D.N., Fogler, M.M., and Garcia de Abajo, F.J. (2016). Polaritons in van der Waals materials. *Science* 354: aag1992.
3 Gopalan, P. and Sensale-Rodriguez, B. (2020). 2D materials for terahertz modulation. *Advanced Optical Materials* 8: 190055.
4 Yagodkin, D., Nádvorník, L., Gueckstock, O. et al. (2021). Ultrafast photocurrents in $MoSe_2$ probed by terahertz spectroscopy. *2D Materials* 8: 025012.
5 Rivera, N. and Kaminer, I. (2020). Light–matter interactions with photonic quasiparticles. *Nature Reviews Physics* 2: 538–561.
6 Kasprzak, J., Richard, M., Kundermann, S. et al. (2006). Bose–Einstein condensation of exciton polaritons. *Nature* 443: 409–414.

7 Katzenellenbogen, N. and Grischkowsky, D. (1992). Electrical characterization to 4 THz of N- and P-type GaAs using THz time-domain spectroscopy. *Applied Physics Letters* 61: 840–842.

8 Grischkowsky, D., Keiding, S., van Exter, M., and Fattinger, C. (1990). Far-infrared time-domain spectroscopy with terahertz beams of dielectrics and semiconductors. *Journal of the Optical Society of America B* 7: 2006–2015.

9 Yeh, K.L., Hoffmann, M.C., Hebling, J., and Nelson, K.A. (2007). Generation of 10 μJ ultrashort terahertz pulses by optical rectification. *Applied Physics Letters* 90: 171121.

10 Kim, K.-Y. (2009). Generation of coherent terahertz radiation in ultrafast laser-gas interactions. *Physics of Plasmas* 16: 056706.

11 Kim, K.Y., Glownia, J.H., Taylor, A.J., and Rodriguez, G. (2007). Terahertz emission from ultrafast ionizing air in symmetry-broken laser fields. *Optics Express* 15: 4577–4584.

12 Cook, D.J. and Hochstrasser, R.M. (2000). Intense terahertz pulses by four-wave rectification in air. *Optics Letters* 25: 1210–1212.

13 Sell, A., Leitenstorfer, A., and Huber, R. (2008). Phase-locked generation and field-resolved detection of widely tunable terahertz pulses with amplitudes exceeding 100 MV/cm. *Optics Letters* 33: 2767–2769.

14 Sanjuan, F., Gaborit, G., and Coutaz, J.-L. (2018). Influence of two-photon absorption anisotropy on terahertz emission through optical rectification in zinc-blende crystals. *Journal of Infrared, Millimeter, and Terahertz Waves* 39: 378–386.

15 Hafez, H.A., Chai, X., Ibrahim, A. et al. (2016). Intense terahertz radiation and their applications. *Journal of Optics* 18: 093004.

16 Hirori, H., Doi, A., Blanchard, F., and Tanaka, K. (2011). Single-cycle terahertz pulses with amplitudes exceeding 1 MV/cm generated by optical rectification in $LiNbO_3$. *Applied Physics Letters* 98: 091106.

17 Hirori, H. and Tanaka, K. (2012). Single-cycle terahertz pulses with amplitudes exceeding 1 MV/cm generated by optical rectification in LiNbO3 and applications to nonlinear optics[C]. In: *Nonlinear Frequency Generation and Conversion: Materials, Devices, and Applications XI*, vol. 8240, 48–53. SPIE.

18 Nguyen, T.K., Kim, W.T., Kang, B.J. et al. (2017). Photoconductive dipole antennas for efficient terahertz receiver. *Optics Communications* 383: 50–56.

19 Nahata, A., Weling, A.S., and Heinz, T.F. (1996). A wideband coherent terahertz spectroscopy system using optical rectification and electro-optic sampling. *Applied Physics Letters* 69: 2321–2323.

20 Darrow, J.T., Zhang, X.-C., Auston, D.H., and Morse, J.D. (1992). PCA_Saturation properties of large-aperture photoconducting antennas. *IEEE Journal of Quantum Electronics* 28: 1607–1616.

21 Auston, D.H., Cheung, K.P., and Smith, P.R. (1984). Picosecond photoconducting Hertzian dipoles. *Applied Physics Letters* 45: 284–286.

22 Stone, M.R., Naftaly, M., Miles, R.E. et al. (2004). Electrical and radiation characteristics of semilarge photoconductive terahertz emitters. *IEEE Transactions on Microwave Theory and Techniques* 52: 2420–2429.

23 Dreyhaupt, A., Winnerl, S., Dekorsy, T., and Helm, M. (2005). High-intensity terahertz radiation from a microstructured large-area photoconductor. *Applied Physics Letters* 86: 121114.

24 Ropagnol, X., Morandotti, R., Ozaki, T., and Reid, M. (2011). THz pulse shaping and improved optical-to-THz conversion efficiency using a binary phase mask. *Optics Letters* 36: 2662–2664.

25 Berry, C.W., Wang, N., Hashemi, M.R. et al. (2013). Significant performance enhancement in photoconductive terahertz optoelectronics by incorporating plasmonic contact electrodes. *Nature Communications* 4: 1622.

26 Auston, D.H., Johnson, A.M., Smith, P.R., and Bean, J.C. (1980). Picosecond optoelectronic detection, sampling, and correlation measurements in amorphous semiconductors. *Applied Physics Letters* 37: 371–373.

27 Lee, Y.-S. (2009). *Principles of Terahertz Science and Technology*. Berlin: Springer.

28 Kono, S., Tani, M., and Sakai, K. (2001). Ultrabroadband photoconductive detection: comparison with free-space electro-optic sampling. *Applied Physics Letters* 79: 898–900.

29 Singh, A., Pal, S., Surdi, H. et al. (2015). Carbon irradiated semi insulating GaAs for photoconductive terahertz pulse detection. *Opt Express* 23: 6656–6661.

30 Blanchard, F., Razzari, L., Bandulet, H.-C. et al. (2007). Generation of 1.5 μJ single-cycle terahertz pulses by optical rectification from a large aperture ZnTe crystal. *Optics Express* 15: 13212–13220.

31 Blanchard, F., Ropagnol, X., Hafez, H. et al. (2014). Effect of extreme pump pulse reshaping on intense terahertz emission in lithium niobate at multimilli-Joule pump energies. *Optics Letters* 39: 4333–4336.

32 Hebling, J., Stepanov, A.G., Almási, G. et al. (2004). Tunable THz pulse generation by optical rectification of ultrashort laser pulses with tilted pulse fronts. *Applied Physics B: Lasers and Optics* 78: 593–599.

33 Vicario, C., Ovchinnikov, A.V., Ashitkov, S.I. et al. (2014). Generation of 0.9-mJ THz pulses in DSTMS pumped by a Cr:Mg(2)SiO(4) laser. *Optics Letters* 39: 6632–6635.

34 Jepsen, P.U., Jacobsen, R.H., and Keiding, S.R. (1996). Generation and detection of terahertz pulses from biased semiconductor antennas. *Journal of the Optical Society of America B* 13: 2424–2436.

35 Gallot, G. and Grischkowsky, D. (1999). Electro-optic detection of terahertz radiation. *Journal of the Optical Society of America B* 16: 1204–1212.

36 Hamster, H., Sullivan, A., Gordon, S. et al. (1993). Subpicosecond, electromagnetic pulses from intense laser-plasma interaction. *Physical Review Letters* 71: 2725–2728.

37 Xie, X., Dai, J., and Zhang, X.C. (2006). Coherent control of THz wave generation in ambient air. *Physical Review Letters* 96: 075005.

38 Kress, M., Ler, T.L., Eden, S. et al. (2004). Terahertz-pulse generation by photoionization of air with laser pulses composed of both fundamental and second-harmonic waves. *Optics Letters* 29: 1120–1122.

39 Luk, K.M., Zhou, S.F., Li, Y.J. et al. (2017). A microfabricated low-profile wideband antenna array for terahertz communications. *Scientific Reports* 7: 1268.

40 Shkurinov, A.P., Sinko, A.S., Solyankin, P.M. et al. (2017). Impact of the dipole contribution on the terahertz emission of air-based plasma induced by tightly focused femtosecond laser pulses. *Physical Review E* 95: 043209.

41 Löffler, T., Jacob, F., and Roskos, H.G. (2000). Generation of terahertz pulses by photoionization of electrically biased air. *Applied Physics Letters* 77: 453–455.

42 Löffler, T. and Roskos, H.G. (2002). Gas-pressure dependence of terahertz-pulse generation in a laser-generated nitrogen plasma. *Journal of Applied Physics* 91: 2611–2614.

43 Bhasin, L. and Tripathi, V.K. (2011). Terahertz generation from laser filaments in the presence of a static electric field in a plasma. *Physics of Plasmas* 18: 123106.

44 Wang, T.-J., Marceau, C., Chen, Y. et al. (2010). Terahertz emission from a dc-biased two-color femtosecond laser-induced filament in air. *Applied Physics Letters* 96: 211113.

45 Houard, A., Liu, Y., Prade, B. et al. (2008). Strong enhancement of terahertz radiation from laser filaments in air by a static electric field. *Physical Review Letters* 100: 255006.

46 Chen, Y., Wang, T.J., Marceau, C. et al. (2009). Characterization of terahertz emission from a dc-biased filament in air. *Applied Physics Letters* 95: 101101.

47 Kreß, M., Löffler, T., Thomson, M.D. et al. (2006). Determination of the carrier-envelope phase of few-cycle laser pulses with terahertz-emission spectroscopy. *Nature Physics* 2: 327–331.

48 Dai, J., Xie, X., and Zhang, X.C. (2007). Terahertz wave amplification in gases with the excitation of femtosecond laser pulses. *Applied Physics Letters* 91: 211102.

49 D'Amico, C., Houard, A., Franco, M. et al. (2007). Conical forward THz emission from femtosecond-laser-beam filamentation in air. *Physical Review Letters* 98: 235002.

50 Zhao, J., Chu, W., Wang, Z. et al. (2016). Strong spatial confinement of terahertz wave inside femtosecond laser filament. *ACS Photonics* 3: 2338–2343.

51 You, Y.S., Oh, T.I., and Kim, K.Y. (2012). Off-axis phase-matched terahertz emission from two-color laser-induced plasma filaments. *Physical Review Letters* 109: 183902.

52 Dai, J., Xie, X., and Zhang, X.C. (2006). Detection of broadband terahertz waves with a laser-induced plasma in gases. *Physical Review Letters* 97: 103903.

53 Karpowicz, N., Dai, J., Lu, X. et al. (2008). Coherent heterodyne time-domain spectrometry covering the entire "terahertz gap". *Applied Physics Letters* 92: 011131.

54 Lei, Z., Huang, Y., Du, W. et al. (2020). Nonlinear optical response on the surface of semiconductor SnS_2 probed by terahertz emission spectroscopy. *The Journal of Physical Chemistry C* 124: 21559–21567.

55 Huang, Y., Zhu, L., Yao, Z. et al. (2018). Terahertz surface emission from layered MoS_2 crystal: competition between surface optical rectification and surface photocurrent surge. *The Journal of Physical Chemistry C* 122: 481–488.

56 Fan, Z., Xu, M., Huang, Y. et al. (2020). Terahertz surface emission from $MoSe_2$ at the monolayer limit. *ACS Applied Materials & Interfaces* 12: 48161–48169.

57 Tong, M., Hu, Y., He, W. et al. (2021). Ultraefficient terahertz emission mediated by shift-current photovoltaic effect in layered gallium telluride. *ACS Nano* 15: 17565–17572.

58 Li, Y., Krisshnamurthi, M.R., Luo, W. et al. (2022). Graphene metasurfaces for terahertz wavefront shaping and light emission [Invited]. *Optical Materials Express* 12: 4528–4546.

59 Massaouti, M., Manceau, J.M., Selimis, A., and Tzortzakis, S. (2011). An intense tunable femtosecond gas-plasma THz source: application in spectroscopic studies of polycyclic aromatic hydrocarbons. *Journal of Molecular Structure* 1006: 28–33.

60 Yoo, Y.-J., Kuk, D., Zhong, Z., and Kim, K.-Y. (2017). Generation and characterization of strong terahertz fields from kHz laser filamentation. *IEEE Journal of Selected Topics in Quantum Electronics* 23: 8501007.

61 Oh, T.I., Yoo, Y.J., You, Y.S., and Kim, K.Y. (2014). Generation of strong terahertz fields exceeding 8 MV/cm at 1 kHz and real-time beam profiling. *Applied Physics Letters* 105: 041103.

62 Xu, S., Huang, D., Liu, Z. et al. (2021). Hydrostatic pressure effect of photocarrier dynamics in GaAs probed by time-resolved terahertz spectroscopy. *Optics Express* 29: 14058–14068.

63 Mao, H.-K., Chen, X.-J., Ding, Y. et al. (2018). Solids, liquids, and gases under high pressure. *Reviews of Modern Physics* 90: 015007.

64 Su, F.H., Blanchard, F., Sharma, G. et al. (2009). Terahertz pulse induced intervalley scattering in photoexcited GaAs. *Optics Express* 17: 9620–9629.

65 Sharma, G., Al-Naib, I., Hafez, H. et al. (2012). Carrier density dependence of the nonlinear absorption of intense THz radiation in GaAs. *Optics Express* 20: 18016–18024.

66 Shin, H.J., Nguyen, V.L., Lim, S.C., and Son, J.-H. (2018). Ultrafast nonlinear travel of hot carriers driven by high-field terahertz pulse. *Journal of Physics B: Atomic, Molecular and Optical Physics* 51: 144003.

67 Bowlan, P., Kuehn, W., Reimann, K. et al. (2011). High-field transport in an electron-hole plasma: transition from ballistic to drift motion. *Physical Review Letters* 107: 256602.

68 Yu, P.Y. and Welber, B. (1978). High pressure photoluminescence and resonant Raman study of GaAs. *Solid State Communications* 25: 209–211.

69 Boucenna, M. and Bouarissa, N. (2004). Predicted electronic properties of GaAs under hydrostatic pressure. *Materials Chemistry and Physics* 84: 375–379.

70 Hirori, H. and Tanaka, K. (2016). Dynamical nonlinear interactions of solids with strong terahertz pulses. *Journal of the Physical Society of Japan* 85: 082001.

71 Misiuk, A. (1991). *Evolution of Process-Induced Defects in Silicon Under Hydrostatic Pressure, Solid State Phenomena*, 387–392. Trans. Tech. Publ.

72 Gray, G.T. (1997). Shock-induced defects in bulk materials. *MRS Online Proceedings Library (OPL)* 499: 87.

73 Ceballos, F., Cui, Q., Bellus, M.Z., and Zhao, H. (2016). Exciton formation in monolayer transition metal dichalcogenides. *Nanoscale* 8: 11681–11688.

74 Geim, A.K. and Grigorieva, I.V. (2013). Van der Waals heterostructures. *Nature* 499: 419–425.

75 Novoselov, K.S., Mishchenko, A., Carvalho, A., and Castro Neto, A.H. (2016). 2D materials and van der Waals heterostructures. *Science* 353: aac9439.

76 Jin, C., Ma, E.Y., Karni, O. et al. (2018). Ultrafast dynamics in van der Waals heterostructures. *Nature Nanotechnology* 13: 994–1003.

77 Liu, Y., Weiss, N.O., Duan, X. et al. (2016). Van der Waals heterostructures and devices. *Nature Reviews Materials* 1: 16042.

78 Shim, J., Kang, D.-H., Kim, Y. et al. (2018). Recent progress in Van der Waals (vdW) heterojunction-based electronic and optoelectronic devices. *Carbon* 133: 78–89.

79 Paik, E.Y., Zhang, L., Burg, G.W. et al. (2019). Interlayer exciton laser of extended spatial coherence in atomically thin heterostructures. *Nature* 576: 80–84.

80 Guo, J., Zhang, Y., Shi, L. et al. (2017). Boosting hot electrons in hetero-superstructures for plasmon-enhanced catalysis. *Journal of the American Chemical Society* 139: 17964–17972.

81 Bai, L., Wang, X., Tang, S. et al. (2018). Black phosphorus/platinum heterostructure: a highly efficient photocatalyst for solar-driven chemical reactions. *Advanced Materials* 30: 1803641.

82 Barati, F., Grossnickle, M., Su, S. et al. (2017). Hot carrier-enhanced interlayer electron–hole pair multiplication in 2D semiconductor heterostructure photocells. *Nature Nanotechnology* 12: 1134–1139.

83 Hong, X., Kim, J., Shi, S.-F. et al. (2014). Ultrafast charge transfer in atomically thin MoS_2/WS_2 heterostructures. *Nature Nanotechnology* 9: 682–686.

84 Yang, J., Jiang, S., Xie, J. et al. (2021). Identifying the intermediate free-carrier dynamics across the charge separation in monolayer $MoS_2/ReSe_2$ heterostructures. *ACS Nano* 15: 16760–16768.

85 Zheng, W., Sun, B., Li, D. et al. (2022). Band transport by large Fröhlich polarons in MXenes. *Nature Physics* 18: 544–550.

86 Valk, N.C.J.V.D. and Planken, P.C.M. (2002). Electro-optic detection of subwavelength terahertz spot sizes in the near field of a metal tip. *Applied Physics Letters* 81: 1558–1560.

87 Knoll, B. and Keilmann, F. (1999). Near-field probing of vibrational absorption for chemical microscopy. *Nature* 399: 134–137.

88 Hillenbrand, R. and Keilmann, F. (2000). Complex optical constants on a subwavelength scale. *Physical Review Letters* 85: 3029–3032.

89 Wu, S.W. and Ho, W. (2010). Two-photon-induced hot-electron transfer to a single molecule in a scanning tunneling microscope. *Physical Review B* 82: 085444.

90 Lee, J., Perdue, S.M., Rodriguez Perez, A., and Apkarian, V.A. (2014). Vibronic motion with joint angstrom–femtosecond resolution observed through Fano progressions recorded within one molecule. *ACS Nano* 8: 54–63.

91 Li, S., Chen, S., Li, J. et al. (2017). Joint space-time coherent vibration driven conformational transitions in a single molecule. *Physical Review Letters* 119: 176002.

92 Binnig, G., Rohrer, H., Gerber, C., and Weibel, E. (1982). Surface studies by scanning tunneling microscopy. *Physical Review Letters* 49: 57–61.

93 Wiesendanger, R., Güntherodt, H.J., Güntherodt, G. et al. (1990). Observation of vacuum tunneling of spin-polarized electrons with the scanning tunneling microscope. *Physical Review Letters* 65: 247–250.

94 Hasegawa, Y. and Avouris, P. (1993). Direct observation of standing wave formation at surface steps using scanning tunneling spectroscopy. *Physical Review Letters* 71: 1071–1074.

95 Crommie, M.F., Lutz, C.P., and Eigler, D.M. (1993). Imaging standing waves in a two-dimensional electron gas. *Nature* 363: 524–527.

96 Giessibl, F.J. (1995). Atomic resolution of the silicon (111)-(7×7) surface by atomic force microscopy. *Science* 267: 68–71.

97 Hess, H.F., Robinson, R.B., Dynes, R.C. et al. (1989). Scanning-tunneling-microscope observation of the abrikosov flux lattice and the density of states near and inside a fluxoid. *Physical Review Letters* 62: 214–216.

98 Fu, Y.-S., Kawamura, M., Igarashi, K. et al. (2014). Imaging the two-component nature of Dirac–Landau levels in the topological surface state of Bi_2Se_3. *Nature Physics* 10: 815–819.

99 Loth, S., Etzkorn, M., Lutz, C.P. et al. (2010). Measurement of fast electron spin relaxation times with atomic resolution. *Science* 329: 1628–1630.

100 Cocker, T.L., Jelic, V., Gupta, M. et al. (2013). An ultrafast terahertz scanning tunnelling microscope. *Nature Photonics* 7: 620–625.

101 Keldysh, L. (1965). Ionization in the field of a strong electromagnetic wave. *Soviet Physics—JETP* 20: 1307–1314.

102 Zheltikov, A.M. (2016). Keldysh parameter, photoionization adiabaticity, and the tunneling time. *Physical Review A* 94: 043412.

103 Cocker, T.L., Jelic, V., Hillenbrand, R., and Hegmann, F.A. (2021). Nanoscale terahertz scanning probe microscopy. *Nature Photonics* 15: 558–569.

104 Jelic, V., Iwaszczuk, K., Nguyen, P.H. et al. (2017). Ultrafast terahertz control of extreme tunnel currents through single atoms on a silicon surface. *Nature Physics* 13: 591–598.

105 Luo, Y., Jelic, V., Chen, G. et al. (2020). Nanoscale terahertz STM imaging of a metal surface. *Physical Review B* 102: 205417.

106 Cocker, T.L., Peller, D., Yu, P. et al. (2016). Tracking the ultrafast motion of a single molecule by femtosecond orbital imaging. *Nature* 539: 263–267.

107 Peller, D., Kastner, L.Z., Buchner, T. et al. (2020). Sub-cycle atomic-scale forces coherently control a single-molecule switch. *Nature* 585: 58–62.

108 Ammerman, S.E., Jelic, V., Wei, Y. et al. (2021). Lightwave-driven scanning tunnelling spectroscopy of atomically precise graphene nanoribbons. *Nature Communications* 12: 6794.

7

Graphene Glass for Nonlinear Optics

Qi Xiao[1], *Jingyu Sun*[2], *and Hao-Li Zhang*[1]

[1] *Lanzhou University, State Key Laboratory of Applied Organic Chemistry, College of Chemistry and Chemical Engineering, Lanzhou, 730000, P. R. China*
[2] *Soochow University, College of Energy, Soochow Institute for Energy and Materials Innovations, Light Industry Institute of Electrochemical Power Sources, Key Laboratory of Advanced Carbon Materials and Wearable Energy Technologies of Jiangsu Province, Suzhou 215006, P. R. China*

7.1 Light Absorption of Graphene

Graphene is a piece of thin film made up of an atom-thick single layer of carbon atoms. The carbon atoms in graphene are arranged in a hexagonal honeycomb lattice through sp^2 hybridization (Figure 7.1a). As a zero-gap semiconductor, the valance band and conduction band of graphene touch one another at the Dirac points (K and K') in the first Brillouin zone (Figure 7.1b). The importance of the Dirac point is that the charge carriers (electrons and holes) in graphene behave like massless Dirac-like elementary particles around the Dirac point, following the relativistic massless Dirac equation. Thus, they are known as the Dirac fermions. These Dirac fermions in graphene exhibit peculiar linear band structures at the Dirac point in the first Brillouin zone [1].

As for the inter-band transitions excited by the incident light of different frequencies, graphene generates charge carriers with the same possibility because of the linear dispersion at the Dirac point. From visible to mid-infrared frequency ranges, single-layer graphene absorbs approximately $A = \pi\alpha \approx 2.3\%$ of the incident light (Figure 7.1c), where $\alpha = e^2/\mathrm{hc} \approx 1/137$ is the fine structure constant. Meanwhile, the reflectivity of graphene $R < 0.1\%$, which is significantly less than the transmittance $T = 1 - A$. Thus, the absorbance of multilayer graphene is $A = N\pi\alpha$, where N is the layer of graphene [2]. This signifies that graphene could be seen with the naked eye. For example, on a Si substrate with a certain thickness of SiO_2 coating layer, owing to the diffraction of the optical path and the interference effect, graphene possessing different layers shows different colors and contrasts under an optical microscopy (Figure 7.1d) [3].

Two-Dimensional Materials for Nonlinear Optics: Fundamentals, Preparation Methods, and Applications, First Edition. Edited by Qiang Wang and Hao-Li Zhang.

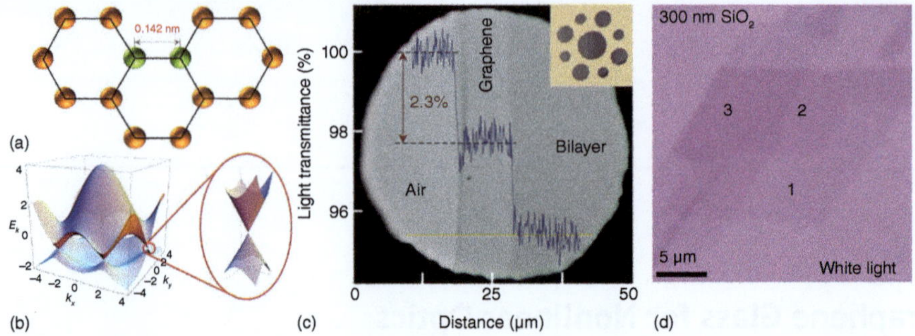

Figure 7.1 (a) Crystal structure of graphene. (b) Graphene band structure and Dirac point. Source: Adapted from Castro Neto et al. [1]. (c) Light transmittance of graphene and its bilayer. Source: Reproduced with permission Nair et al. [2]. © 2008, American Association for the Advancement of Science - AAAS. (d) Optical image of graphene with different layers. Source: Reproduced with permission Blake et al. [3]. © 2007, AIP Publishing LLC.

7.2 Nonlinear Optical Properties of Graphene

7.2.1 Third-Order Nonlinear Optical Properties of Graphene

The graphene crystal lattice possesses D6h point group symmetry. This means that the graphene crystal lattice is a centrosymmetric 2D crystal lattice. Therefore, some of the optical processes are absent in this material due to its centrosymmetric crystal lattice structure, such as second-harmonic generation (SHG), second-order optical susceptibility, and nonlinear optical conductivity. However, researchers from the Free University of Brussels in Belgium measured the third-order nonlinear parameters of graphene earlier in 2009. They were surprised to find that the nonlinear refractive index of graphene at 1550 nm was 10^{-7} cm^2 W^{-1}, which was about 5 orders of magnitude higher than that of ordinary dielectric materials. Later, the third-order nonlinear susceptibility ($\chi^{(3)}$) of graphene in the near-infrared range was measured by a four-wave mixing method with the result of 1.5×10^{-7} esu, also 5 orders of magnitude higher than that of ordinary dielectric materials [4].

The low dimension of graphene means its nonlinear optical properties can be easily modulated. Under the excitation of a specific light, due to the limitation of the Pauli exclusion principle, gradually adjusting the Fermi level can successively prohibit the one-photon, two-photon, and three-photon inter-band transitions. In the FET devices prepared by graphene, Jiang et al. used ionic gel as the gate electrode to control the Fermi level of graphene (Figure 7.2a) [5]. Using linearly polarized femtosecond pulsed laser as the incident light source, they observed the third-order nonlinear effect of graphene was adjusted by the chemical potential and incident photon energy. The experimental results show a shoulder-shaped rising curve in a specific region, reflecting the resonance-type electronic transition mechanism behind the third harmonic generation (Figure 7.2b). The third-order nonlinear optical response in graphene is an overall effect caused by the cooperative competition among multiple quantum resonance transition channels with

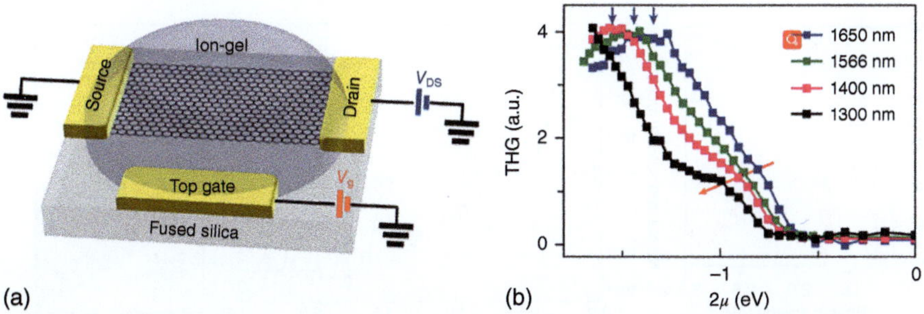

Figure 7.2 (a) Schematic of ion-gel-gate controlled graphene field effect transistor. (b) THG signal as a function of chemical potential (2 μ) generated by different input wavelengths. Source: (a) and (b) Reproduced with permission Jiang et al. [5]. Copyright © 2018, Springer Nature.

different phases and intensities, which is strongly influenced by carrier density. The switching of correlated quantum resonant transition channels can be controlled by the modulation of the carrier density and chemical potential in graphene, ultimately leading to the interesting results. Similarly, by adjusting the Fermi level of single-layer graphene by the back-gate effect, Ferrari et al. from the University of Cambridge also realized the wide-spectrum electric field adjustment of the third-order nonlinear effect of graphene [6].

7.2.2 Saturable Absorption of Graphene

Saturable absorption is a universal property of any material related to the imaginary part of the third-order nonlinear susceptibility ($\chi^{(3)}$). Graphene shows superior saturable absorption properties. On the one hand, the remarkably high absorption of single-layer graphene is 50 times higher than GaAs with the same layers. Therefore, under illumination in the visible to near-infrared wavelengths, graphene's light absorption easily reaches saturation. On the other hand, the zero-bandgap enables graphene to be used as a saturable absorber (SA) in various wavelengths without processing. The broadband saturable absorption properties of graphene have essential applications in the field of laser and optical communications. Since graphene was first used in the field of pulsed lasers in 2009, passive mode-locked fiber lasers based on graphene materials have appeared successively, covering 1, 1.25, 1.5, 2, 3 μm, and other bands [7]. In particular, Jaroslaw et al. used graphene SA in a cascaded fiber laser, which was first pumped with a diode laser to achieve 1.5 μm mode locking, and then pumped with the output 1.5 μm laser to achieve 2.0 μm locking [8]. It fully verifies that the excellent properties of graphene-SAs can be used in multiple wavelengths at the same time.

7.2.3 High-Order Harmonic Generation of Graphene

Optical harmonic generation was first investigated intensively in atomic gases for the generation of deep ultraviolet and attosecond pulsed lasers. Using high-order

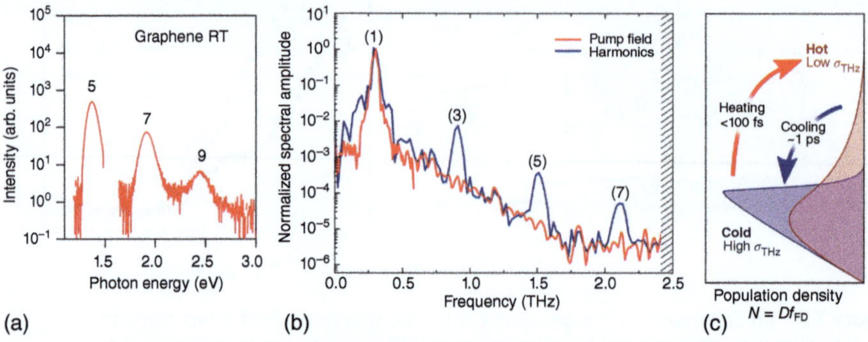

Figure 7.3 (a) High-harmonic radiation spectrum of graphene. Source: Reproduced with permission Yoshikawa et al. [9]. Copyright © 2017, American Association for the Advancement of Science. (b) Red line, amplitude spectrum of the incident pump THz wave at the fundamental frequency. Blue line, the spectrum of the same THz wave transmitted through graphene, with clearly visible generated harmonics of third, fifth, and seventh order. (c) Illustration of the mechanism of THz harmonic generation in graphene. Source: (b) and (c) Reproduced with permission Hafez et al. [10]. Copyright © 2018, Springer Nature.

harmonics to generate photons in different frequencies can cover most of the electromagnetic spectrum and even extend to the terahertz range, enriching and developing spectroscopy detection technology. Due to the zero bandgap and linear energy dispersion of graphene, its group velocity only depends on the direction of the $\boldsymbol{k}$-vector, making its nonlinear electromagnetic response very strong. Its optical harmonics are expected to have very high efficiency. Both semiclassical models and quantum mechanical theories predict the high-harmonic generation in monolayer graphene. For the first time, Yoshikawa et al. used a mid-infrared pulsed laser to observe the generation of higher harmonics, even up to the ninth harmonic in graphene at room temperature (Figure 7.3a) [9].

In the near-infrared region, the generation of graphene high harmonics mainly depends on the inter-band transitions of graphene in strong electric fields. While in the terahertz region, the efficient high-order harmonic of graphene at room temperature is also predicted. Initial investigation reported the third and fifth harmonic generation of 45-layer graphene in a low-temperature environment, but the experimental third harmonic efficiency was much lower than the theoretical value. After that, Gensch et al. used a new accelerator-based superconducting radio frequency terahertz radiation source, whose pulse frequency was a hundred times higher than that of the laser terahertz source. For the first time, they demonstrated the third, fifth, and seventh harmonic in graphene terahertz range at room temperature, with the efficiency as high as 10^{-3}, 10^{-4}, and 10^{-5} respectively (Figure 7.3b) [10]. Another key factor is the used graphene, which contains many free electrons that originate from the doping of substrate or the environment. These electrons act as a nonlinear intermediary to transfer energy from the absorbed terahertz field to the graphene lattice, making the electrical conductivity of graphene change over time. Finally, the nonlinear current oscillations in graphene were driven, resulting in the emission of high odd harmonics (Figure 7.3c).

7.3 Nonlinear Optical Properties of Graphene Glass

7.3.1 Graphene-Doped Glass

So far, the combination of graphene and glass mainly includes four methods. The first one is graphene-doped glass, whether by gelation of the glass in a solution with dispersed graphene sheets or polymerization of graphene-dispersed methyl methacrylate (MMA) (Figure 7.4a). Zheng et al. investigated the nonlinear optical properties of graphene-doped gel glass using ns and ps Z-scan techniques at 532 nm [11]. The observed nonlinear absorption effects were affected by pulse width of laser. Reverse saturable absorption was observed in the ns regime, while saturable absorption dominated in the ps regime. Ouyang et al. reported graphene/PMMA organic glasses hybrid with various semiconductors, such as CdS, Ag_2S, ZnO, and MoS_2 [12–15]. The addition of graphene can enhance the RSA properties of semiconductors/PMMA (Figure 7.4b) due to which the interfacial charge transfer between graphene and those semiconductors results in an extra 2PA process (Figure 7.4c). However, affected by the intrinsic poor transmittance of gel or organic glass and the inevitable aggregation of graphene, graphene-doped glass shows the poor transmittance lower than ~60%. Such composite does not reflect the intrinsic properties of graphene.

7.3.2 Graphene Glass Fabricated by Spin Coating

The second is liquid-phase coating of reduced graphene oxide or liquid-phase exfoliated graphene on glass surface. For example, E.I. Anila reported that reduced graphene oxide incorporated potassium-doped zinc oxide (KZO:rGO) composite thin films were synthesized on glass substrates by spin-coating technique [16]. The composite thin films exhibit enhanced RSA behavior, which is mainly attributed to two-photon absorption. Obviously, it is due to the small size and abundant defects

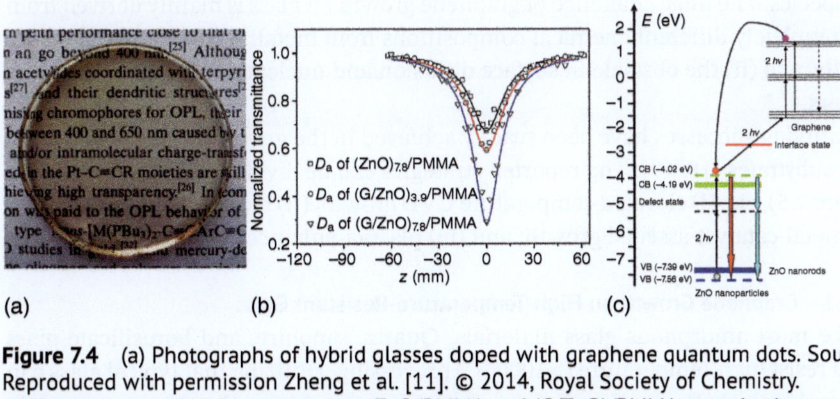

Figure 7.4 (a) Photographs of hybrid glasses doped with graphene quantum dots. Source: Reproduced with permission Zheng et al. [11]. © 2014, Royal Society of Chemistry. (b) Comparisons of NLA curves among ZnO/PMMA and (G/ZnO)/PMMA organic glasses. (c) Energy level diagram of G/ZnO composites. Source: (b) and (c) Reproduced with permission Ouyang et al. [12]. Copyright © 2013, Elsevier Ltd.

of reduced graphene oxide, the inability to achieve precise and uniform thickness control, and most importantly, a weak adhesion of graphene to Glass.

7.3.3 Graphene Glass Fabricated by Transfer Method

The third is photoresist-assisted transfer of copper foil CVD-grown graphene to the glass surface. The process of this transfer includes: First, the graphene surface is coated with a photoresist such as PMMA; then, the copper foil substrate is removed by a liquid-phase etching process and the graphene is transferred to the desired glass substrate; Finally, the surface polymer is washed away by an organic solvent. Sobon et al. fabricated bilayer graphene on Glass by this method, forming a SA [17]. Low non-saturable losses of only ~5% and modulation depth as high as 55% were obtained. Such SA was successfully used in mode-locked fiber laser to achieve nearly-transform limited 315 fs soliton pulses. However, the complicated polymer-mediated transfer process is time-consuming and lacks scalability for mass production. It also introduces the contamination of polymer residues, as well as defects and cracks in the perfect graphene structure.

7.3.4 Directly Growth Graphene on Glass

Graphene directly grown on a glass substrate using a CVD method is believed to be the most practical approach to fabricating graphene glass. Strong adhesion between graphene film and the glass surface is caused by the high-temperature CVD process. The direct CVD process has a better control on growth quality of graphene compared with the abovementioned solution-casting and mechanical transfer methods, thus able to solve the challenges, such as film continuity, thickness control, and mass producibility with low cost.

With respect to the graphene growth by CVD method, usually a metallic catalyst surface (typically Cu or Ni) is required to lower the decomposition temperature of carbon precursors (such as methane) and the graphitization temperature of the carbon species. The huge challenge of graphene growth on glass is mainly derived from the completely different chemical compositions from metal: (i) the lack of catalytic activity, and (ii) the obstacle in surface diffusion and nucleation of carbon atoms or clusters.

Gratifying successes have been readily achieved in the growth of graphene on the glass substrates [18, 19]. The reported strategies can be divided into four categories (Figure 7.5), viz. (i) elevated-temperature CVD growth, (ii) molten-bed CVD growth, (iii) metal-catalyst-assisted growth, and (iv) plasma-enhanced CVD growth.

7.3.4.1 Graphene Growth on High-Temperature-Resistant Glass

Unlike most amorphous glass materials, Quartz, sapphire, and borosilicate glass could resist high-temperature (>1000 °C) processing. Utilizing that typical glass can help tackle the softening issue of the glass substrates during the graphene growth. Under such metal-catalyst-free growth conditions, note that the carbon precursors would merely proceed with thermal decomposition at elevated growth temperatures

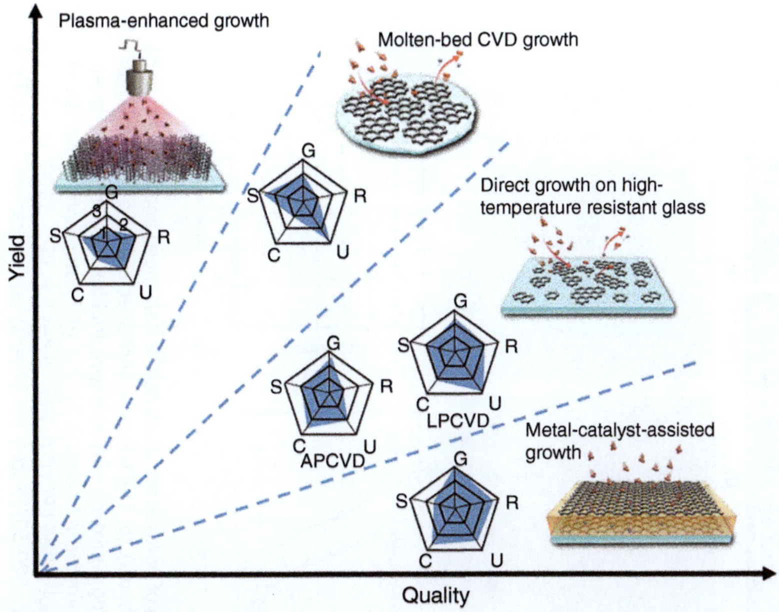

Figure 7.5 Comparison of strategies for the growth of graphene on Glass in terms of yield and quality. Each strategy was evaluated in terms of graphene crystallinity (G), growth rate (R), uniformity (U), cost (C), and scalability (S). Each ring marked by 1, 2, and 3 indicates low, medium, and high levels, respectively. Source: Reproduced with permission Chen et al. [18]. Copyright © 2018, Wiley-VCH.

rather than catalytic decomposition. In this sense, different pressure parameters were employed to adjust the diffusion of the carbon species, rendering atmospheric- and low-pressure CVD (APCVD/LPCVD) route.

By virtue of APCVD route, Liu and coworkers reported the direct growth of graphene on borosilicate, quartz, and sapphire glass substrates at 1000 °C [20]. Figure 7.6a shows the layer thickness of graphene on borosilicate glass can be controlled by varying the flow rate of the carbon precursor (CH_4). The optical transmittance was 96.3% at 550 nm (Figure 7.6b), markedly higher than the graphene-doped glass samples. Graphene films with one to three-layer thickness were found in transmission electron microscopy (TEM) observations (Figure 7.6c).

In comparison with graphene grown on Cu foil (~10 minutes to realize a full coverage) [22], a continuous graphene film grown on glass surface costs significantly longer time of over 1–7 hours, even by increasing the concentration of carbon precursor [23]. To boost the growth efficiency of graphene, a confined-flow CVD was developed by the same group [24]. A frosted quartz plate was placed on the glass substrate, rendering a gap with a 2–4 μm spacing. When CH_4 flows through the gap, the intense surface collisions of CH_4 molecules could increase the local concentration of carbon fragments and boost the growth kinetics.

Owing to the violent collisions in the gas phase, the mean-free paths of carbon species are quite short, leading to a high diffusion barrier and a low surface flux

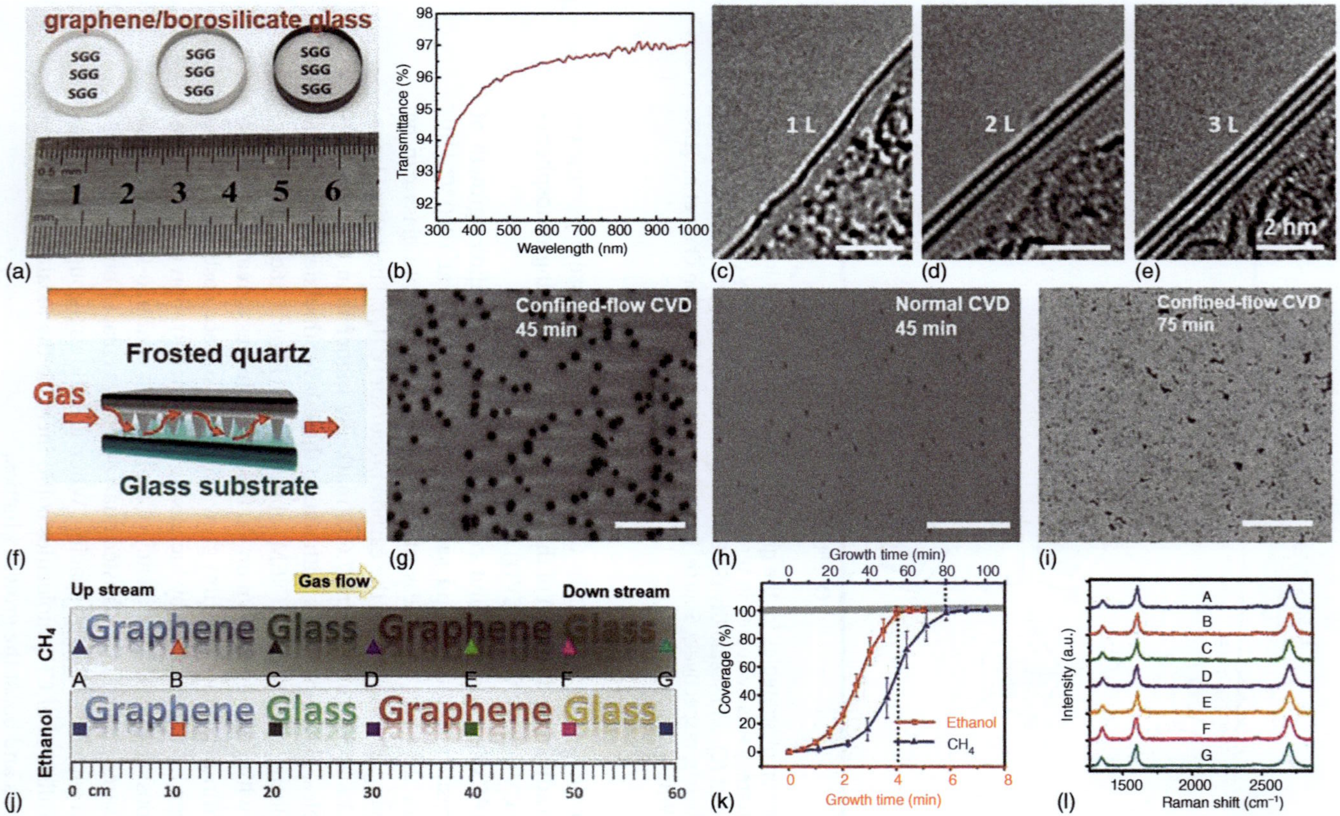

Figure 7.6 Direct growth of graphene on high-temperature-resistant Glass. (a) Photograph of borosilicate glass before (left-most) and after graphene growth with different CH_4 flow rates of 2 and 7.5 sccm (Ar/H_2:100/50 sccm at 1000 °C for 2 hours). (b) The transmittance spectrum of a graphene film directly grown on quartz glass. (c) TEM images of graphene transferred from glasses, showing a mixture of monolayer, double-layer, and three-layer graphene, respectively. Source: (a)–(c) Reproduced with permission Sun et al. [20]. © 2015, American Chemical Society. (d) Photographs of the derived 60 cm long graphene on quartz glass obtained by methane-based APCVD (upper) and ethanol-based LPCVD (lower) routes (LPCVD: ~1100 °C, Ar/H_2/ethanol: 1000/1000/500 sccm at 1250 Pa for 10 minutes; APCVD: ~1100 °C, $Ar/H_2/CH_4$: 500/50/15 sccm for 2 hours). (e) Comparison of graphene growth time via ethanol-based LPCVD and methane-based APCVD routes toward a full coverage of monolayer on quartz glass. (f) Raman spectra of graphene glass produced using an ethanol-based LPCVD route, collected from positions labeled A-G in (d). Source: (d)–(f) Chen et al. [21]/Reproduced with permission of John Wiley & Sons.

of carbon fragments to glass substrate. This is so-called mass-transport limited reaction, which is one of the main reasons for the low growth rate of graphene on glass substrates [25]. In addition, large-area uniform growth becomes very difficult, which is affected by the nonuniform concentration of carbon precursors from upstream to downstream in the gas flow direction (Figure 7.6d) [21]. To overcome these obstacles, LPCVD was adopted. In a typical LPCVD process, carbon species have fewer intermolecular collisions and longer diffusion mean-free paths, leading to a shorter retention time in the growth furnace. Hence, the surface reaction becomes the rate-determining step, readily confirmed by theoretical calculations [26].

Taking advantages of the LPCVD route, Liu and coworkers successfully achieved fast and uniform growth of graphene on glass substrates using ethanol as the carbon precursor. Full-coverage monolayer graphene was obtained for only 4 minutes, which was 20 times faster than that of APCVD based on methane precursor (Figure 7.6e) [21]. In this regard, uniform graphene film with a length of 60 cm was realized on quartz glass (Figure 7.6d). Along the gas flow direction, a constant I_G/I_D ratio of graphene was found from the Raman spectra (Figure 7.6f), showing the improved crystalline quality. The fewer gas-phase collisions of carbon species also reduce the formation of amorphous carbon. In the same time, the OH radicals decomposed by ethanol might help the removal of the byproducts. Collectively, controlling the transport, collision, and reaction of carbon precursors by lower the system pressure and/or alter the carbon sources would offer a practical route for the controllable production of high-quality "super graphene glass." Nevertheless, continuous efforts are still needed to devise the related equipment to allow mass production.

7.3.4.2 Graphene Growth on Molten Glass

The ubiquitous soda-lime glass with a low softening point of ~620 °C seems not compatible with the elevated-temperature CVD route for graphene growth (>950 °C). Liu and coworkers proposed a molten-bed CVD method for graphene grown on such glasses [27]. During the molten-bed CVD, the furnace was heated over the softening point of the glass substrate to create a molten glass surface, where graphene was grown on such a surface with CH_4 serving as the carbon precursor (Figure 7.7a). A high-temperature-resistant container, e.g. a graphite crucible, was required to support the molten glass. Uniformly distributed graphene disks appeared and then merged into a continuous graphene film (Figure 7.7b–d). The final product exhibited little change in optical transparency (Figure 7.7a) and enabled a sharp 2D peak with a high I_{2D}/I_G ratio, indicating the formation of single-layer graphene on soda-lime glass (Figure 7.7e). This was further verified by atomic force microscopy (AFM) and TEM characterizations (Figure 7.7f–h). It is noted that a stronger graphene–substrate interaction existed as compared with that obtained by a transfer method (Part 3.3), as evidenced by the Raman shift at 14 and 16 cm^{-1} of the G and 2D band, respectively (Figure 7.7e).

The advantage of graphene grown on molten glass is summarized as follows. First, on the molten glass surface, nucleation centers are uniformly distributed,

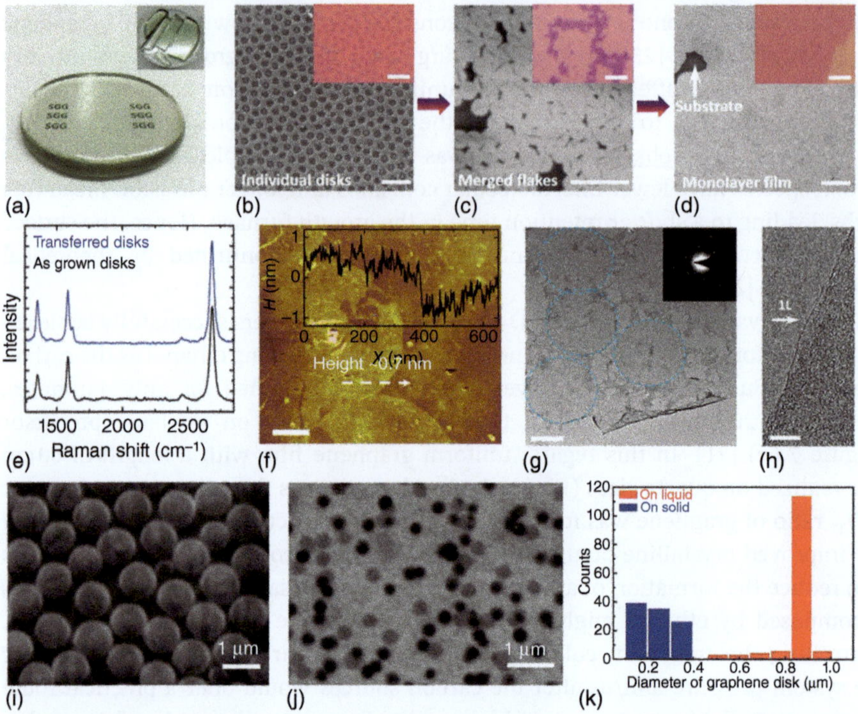

Figure 7.7 Graphene grown on a molten soda lime glass surface. (a) Photograph of a graphene glass with full graphene coverage. Inset: Various shaped graphene glass after cutting. (b)–(d) The morphology evolution of graphene grown on molten Glass from individual disks to a continuous film with the experimental conditions. (b) ~970 °C, $Ar/H_2/CH_4$: 150/20/8 sccm for 1 hour, (c) ~1000 °C, $Ar/H_2/CH_4$: 150/15/5 sccm for 2 hours and (d) ~1020 °C, $Ar/H_2/CH_4$: 150/15/6 sccm for 2 hours. (e) Comparison of Raman spectra of the as-grown and transferred graphene disk samples. (f) AFM image of transferred graphene disks. Scale bar: 200 nm. Inset: The height profile along the white dashed line showing a thickness of ≈0.7 nm. (g) TEM image of transferred graphene film. Scale bar: 500 nm. Inset: corresponding SAED pattern. (h) A high-resolution TEM image of continuous graphene indicating the monolayer feature. Scale bars: 10 nm. (i), (j) SEM images of graphene growth on (i) molten glass and (j) solid glass, respectively. (k) Size distributions of graphene grown on molten Glass (red column) and solid quartz glass (blue column). Source: Reproduced with permission Chen et al. [27]. © 2015, John Wiley & Sons.

which eventually evolve into nearly identical diameters of graphene flakes (Figure 7.7i and k). In contrast, on the solid glass surface, the nucleation centers are randomly scattered (Figure 7.7j and k). Such a difference is derived from the molten feature of the growth substrate since the high-energy active sites for graphene nucleation such as defects, steps, and scratches have been mostly eliminated. Second, the growth rate of graphene on the molten glass is an order of magnitude higher than that on solid quartz glass. This is because the ultrasmooth surface of molten glass exhibits a low diffusion barrier of carbon species, which is in favor of the fast formation of graphene. Third, graphene grown on molten glass is predominantly single-layer, while bilayer and few-layer counterparts occur concurrently on the growth over solid glass. The reason lies in the fact that the graphene growth is a self-limiting process on molten glass surface, which is speculated to follow a surface-oxygen-assisted growth mechanism [28]. Moreover, the isotropic molten glass surface allows for a homogenous growth of graphene, as shown by the nearly circular graphene domains (Figure 7.7i).

The molten-bed CVD approach for producing "super graphene glass" is potentially compatible with the industrial production line of glass. With an optimized glass production process, the graphene growth could be completed during the solidification process of glass. Nevertheless, the mismatch of thermal expansion coefficients between glass and graphene could introduce stress and even result in mechanical breakage of the graphene film.

7.3.4.3 Metal-Catalyst-Assisted Graphene Growth

Metals such as Ga, Cu, Ni, and Pt are perfect catalysts to render high-quality and large-area growth of graphene [20, 29–32]. Therefore, to grow graphene on the surface of glass with catalytic inertness, incorporating metal catalysts to improve the reaction activity is one of the meaningful tactics in producing super graphene glass. There have been several strategies for graphene growth assisted by metal catalysts, mainly encompassing metal vapor catalyst [33, 34], interfacial segregation technique [35, 36], sacrificial metal coating method [37], and use of metal-containing precursors [38].

Metal Vapor Catalyst Metal vapor is generated by heating the metal in a CVD growth furnace. The metal vapor contains gasified metal atoms, which can be used as a catalyst to promote the decomposition of carbon precursors, thus facilitating the growth of graphene on the insulators. On the other hand, metal atoms can also participate in the surface growth process. It could be concluded that metal catalysts can improve the growth quality and expedite the growth rate of graphene over nonmetallic glass substrates.

In the report by Chiu et al. [34], copper foil was placed upstream of the glass substrate to produce vaporized catalyst atoms. Still, this method is highly dependent on the distance of the copper foil, which strongly affects the uniformity of graphene along the gas flow direction on the glass surface. To solve this problem, Choi et al. placed a copper foil directly above the glass substrate to provide Cu vapors [33], with no physical contact between the two. In this manner, sufficient supply of uniform vaporized Cu atoms is realized to catalyze the growth of graphene. The quality and uniformity of graphene on glass were greatly improved as compared to graphene grown directly on Cu foil. This is due to the relatively high ratio of copper vapor to methane gas. For further improvement, Liu et al. used copper foil around the glass substrate to improve growth quality and uniformity [39]. A large-area graphene/quartz glass sample ($6 \times 4\,cm^2$) harvested good optical transparency (97.1% at 550 nm). The derived graphene glass shows relatively high electrical conductivity, making it a perfect candidate for transparent electrode applications. In another attempt, low melting point Ga metal was employed to generate catalytic metal vapor to grow graphene on glass. Under the same conditions, the vapor

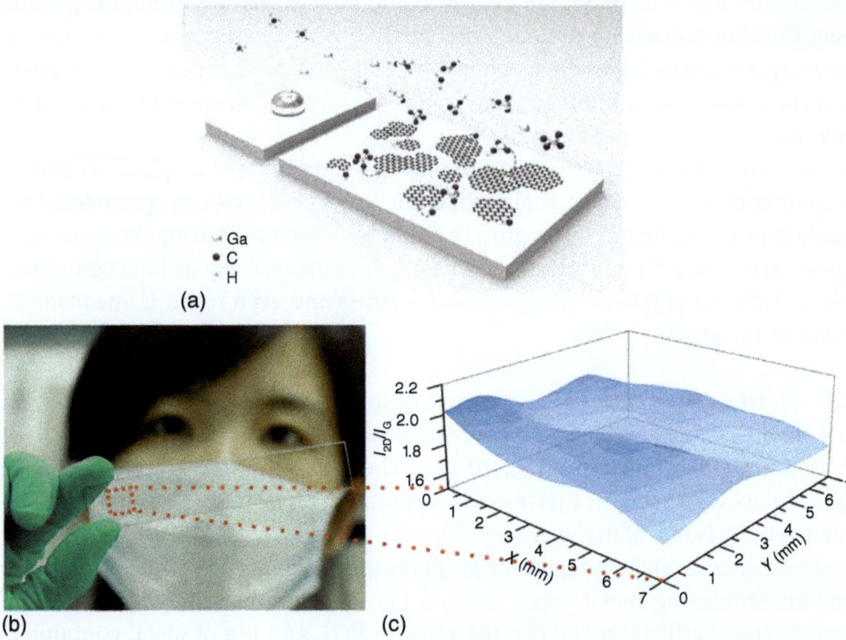

Figure 7.8 Graphene grown with Metal vapor. (a) Schematic illustrations of the remote catalyzation process for direct growth of single-layer graphene with Ga vapor-assisted. (b) Photograph of a 1.8 cm × 10 cm graphene film directly grown on a quartz substrate. (c) A 3D-colormap surface derived from Raman mapping result acquired from the dotted area in (b). Source: Tan et al. [41]/John Wiley & Sons, Inc.

pressure of Ga was nearly 10 times higher than that of Cu [40]. Fu and colleagues demonstrated the Ga-assisted rapid growth of graphene on quartz glass, as illustrated in Figure 7.8f [41]. The as-grown $1.8 \times 10\,\text{cm}^2$ graphene film on a quartz substrate shows excellent optical transparency (Figure 7.8b) and perfect uniformity (Figure 7.8c).

Interfacial Segregation Growth Li and colleagues proposed the interfacial segregation growth of graphene on quartz glass [35]. The designing principle is shown in Figure 7.9a. Firstly, 300 nm copper as a catalyst was deposited on quartz glass, and the precursor molecules were added. The precursor molecules decomposed into carbon clusters on the copper surface. With the increase in processing temperature, the carbon clusters diffused into the copper phase along the grain boundary, then segregated and grew into graphene at the interface between copper and quartz glass. The TEM evidence for the formation of graphene at the copper/glass interface is provided in Figure 7.9b. The growth temperature was set at 900 °C to avoid dehumidification or evaporation of Cu. After the top-sided graphene and copper film were removed by chemical etching, the graphene film was obtained on the quartz glass. Kwon and colleagues demonstrated the growth of graphene at a temperature quite close to room temperature (25–160 °C) [36], using polycrystalline nickel layers deposited on glass substrate as the catalyst. When the graphite powder was coated

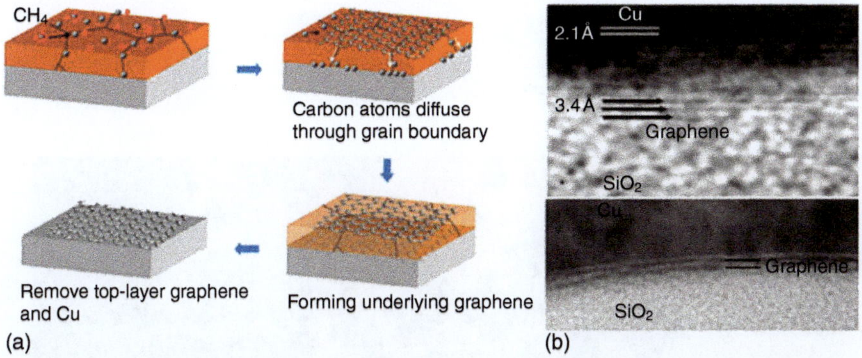

Figure 7.9 Segregation growth of graphene at the interface of a pre-deposited metal layer and glass. (a) A schematic of the interfacial segregation growth. (b) A high-magnification cross-section TEM image of the Cu–graphene–SiO_2 interface. Growth conditions: H_2/CH_4: 15/75 sccm, 900 °C, 800 mTorr, 5 minutes. Source: (a) and (b) Reproduced with permission Su et al. [35]. © 2011, American Chemical Society.

on the nickel layer as a solid carbon source, the nickel layer possessed the capability to crack carbon–carbon bonding [42–44].

Low-temperature growth on glass is of paramount importance toward practical applications. Considering the low softening point of glass, the interfacial segregation growth technique offers a satisfactory solution for graphene to grow on arbitrary glass using suitable metal catalysts. It is also cost-effective because of the low energy consumption required. Despite these conspicuous merits, there still exists problematic issues: how to remove the metal layer after growth – this is a technical challenge that needs to be solved in future work.

Sacrificial Metal Coating Approach In the case of the interface separation method discussed above, the graphene layer is formed at the metal/glass interface. Therefore, the technical challenge is to remove the metal coating after CVD growth to obtain glass with mere graphene coating. Zhang et al [37]. designed a sacrificial metal method to solve the problem. After growing graphene, the copper layer was gradually evaporated by heating with the growth time from 15 through 60 to 420 minutes. Then a monolayer of graphene remained on quartz glass, as shown in Figure 7.10a. The Raman spectra collected from three different locations (marked in Figure 7.10b–d) confirm the evolution of monolayer graphene (Figure 7.10e). There are still copper residues on the graphene/quartz surface, which can be seen in the optical microscope image (Figure 7.10f) and energy-dispersive X-ray spectroscopic profile (Figure 7.10g). The electrical conductivity of synthesized samples was significantly improved, accompanied by good optical transparency. Such a sacrificial metal coating method does not include the time-consuming chemical etching process upon CVD growth. This would be beneficial to practical application.

Metal Element-Containing Precursors Liu and colleagues directly employed nickelocene sulfonate as the growth precursor [21]. The precursor was decomposed into

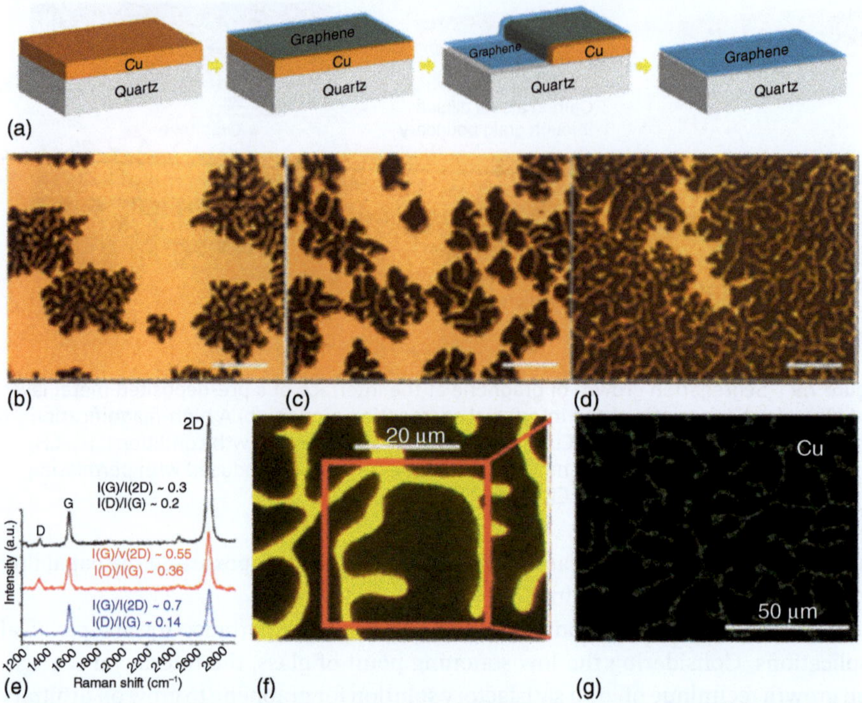

Figure 7.10 Graphene growth on glass using the sacrificial metal coating approach. (a) A schematic representation of graphene growth on quartz with a copper sacrificial layer. (b)–(d) Optical microscopy images showing the morphology evolution for a 450 nm thick Cu film on the quartz substrate with increasing growth time from 15, 60 to 420 minutes. (e) Raman spectra from the three different points marked in (d). (f) Optical image of a quartz sample with a 450 nm Cu coating layer after a 2 hours CVD growth. Growth conditions: H_2/CH_4: 2 : 35 sccm, 1000 °C, 100–500 mTorr. (g) Cu map from the EDX analysis. Source: Reproduced with permission Ismach et al. [37]. © 2010, American Chemical Society.

nickelocene and carbon five-membered rings at 150 °C. The decomposed nickel species was used as a catalyst to catalyze the growth of carbon five-membered rings into graphene on the surface of h-BN (Figure 7.11a). This method has a high reaction rate at low temperatures, which is analogous to that of metal-catalyzed growth scenarios at 1000 °C [45–47]. The theoretical calculation also showed that the reaction energy barrier was greatly reduced in the presence of Ni catalyst (Figure 7.11d). Upon growth, the synthesized graphene afforded good electrical conductivity and crystallinity (Figure 7.11b,c).

7.3.4.4 Plasma-Enhanced Low-Temperature Graphene Growth

Owing to the softening phenomenon of glass products, reducing the growth temperature of graphene represents one of the technical considerations. In this sense, PECVD technology has advantages in meeting this requirement. In a PECVD system, a mixture of high-energy plasma, high-energy electrons, charged particles, free radicals, and photons can be produced at a lower temperature under a strong electric field [48–51]. Taking methane precursors as an example, CH_4 molecules carried

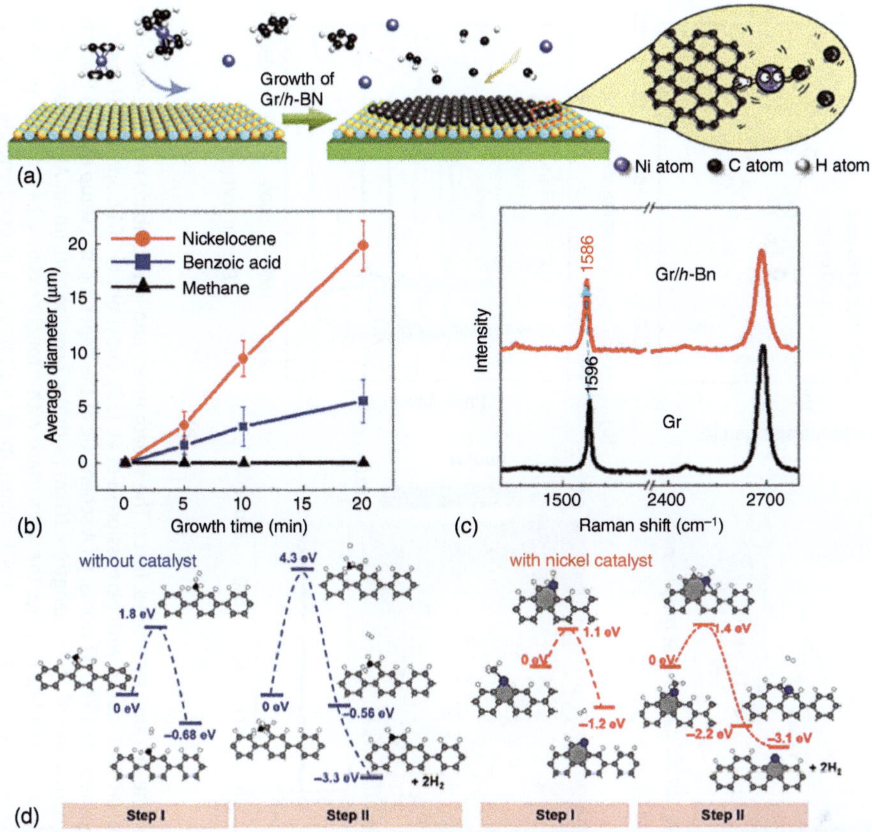

Figure 7.11 (a) A schematic of graphene growth on h-BN/Cu foil using a nickelocene precursor. (b) A plot of graphene domain size as a function of growth time with nickelocene (red), benzoic acid (blue), and methane (black) precursors. (c) Raman spectra of graphene/h-BN and pure graphene transferred to SiO_2/Si. (d) DFT calculations for the minimum energy paths for two key dehydrogenation processes for graphene growth, without (blue), and with nickel catalysts (red), respectively. Gray, blue, and white spheres represent carbon, nickel, and hydrogen atoms. Source: Reproduced with permission Li et al. [38]. Copyright © 2017, Wiley-VCH.

by argon (Ar) and hydrogen (H_2) are decomposed into highly active substances in a plasma generator, which are mainly H derivatives, argon derivatives, and C derivatives (CH, CH_2.CH_3, and $C_xH_y^+$) [52–54]. By properly controlling the growth conditions, the high-energy carbon intermediates can be rearranged into graphene films on the glass surface at a reasonably low temperature.

The graphene obtained by PECVD normally shows vertically erected morphology, which has key implications in applications such as biosensors [55], supercapacitors [56], catalysts [57], and energy storage and conversion [58]. Figure 7.12a schematically depicts the fundamental steps for graphene growth on the glass surface. In the first step, CH_4 molecules are decomposed into active intermediates, followed by surface carbonization under PECVD conditions. This highly complex plasma-aided

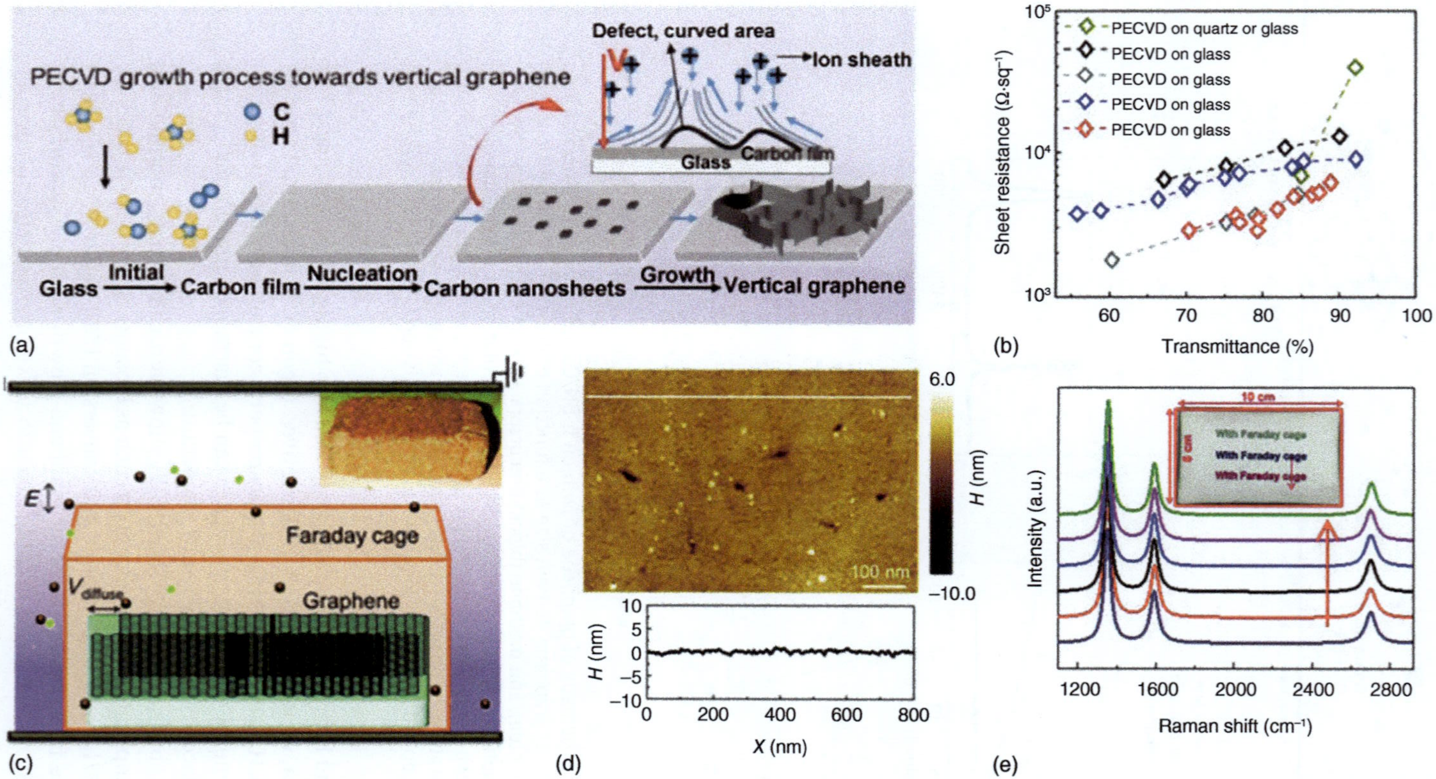

Figure 7.12 (a) A schematic for the direct growth of vertical graphene on glass substrates. Inset: The sheath effect and ion bombardment between the bulk plasma and defect/curved areas of the glass substrate. Source: (a) Reproduced with permission Ci et al. [59]. Copyright © 2017, Springer Nature. (b) Statistics for the transmittance and sheet resistance from various reported PECVD works. (c) A schematic of the experimental setup using a Faraday cage. Inset: A photograph of the copper-foam-based Faraday cage with a size of 15 cm (length) × 10 cm (width) × 2 cm (height). (d) An AFM image of a horizontal graphene film grown on glass. The bottom height profile along the line displays the flat surface with a roughness of ≈0.4 nm. (e) A series of Raman spectra from graphene glass (10 cm × 6 cm) along the arrow, with a growth time of ≈60 minutes. Inset: A photograph of graphene glass showing a relatively uniform contrast. Growth conditions: 580 °C, 80 W, 40 sccm CH_4. Source: (b)–(e) Reproduced with permission Qi et al. [60]. © 2018, John Wiley & Sons.

process will generate strong internal stress and a graphite layer with many defects on the glass surface. With the continuous propagation of carbon intermediates, graphene begins to grow vertically [59]. In addition, the diffusion of positively charged carbon ions along the vertical nanowires will be enhanced by the local electric field generated in the plasma sheath, which is also conducive to the vertical growth of graphene [61]. Zhang and colleagues reported the growth of graphene on a 4-in. glass wafer with excellent uniformity [48, 50]. Figure 7.12b shows the thin layer resistance-optical transmittance correlation of PECVD-derived graphene glasses from different research groups [48, 50, 59, 60]. As a typical example given by Liu and colleagues, graphene can grow on soda-lime glass at 400 °C, showing a sheet resistance of 3700 Ω sq.$^{-1}$ at 79.0% and 5200 Ω sq.$^{-1}$ at 84.6% transmittance, respectively [50]. The vertically oriented graphene flakes on glass have shown promise in solar cells and photothermal devices [59]. However, note that the vertical growth of graphene leads to poor in-plane conductivity due to the serious scattering of electric charges on the exposed surface [62, 63]. In addition, the complex plasma derivative reaction intermediates make it difficult to control the growth quality of graphene.

Furthermore, a Faraday cage-assisted PECVD route was proposed [60] to improve the growth direction of graphene. This method has been proved to grow graphene horizontally on the glass surface with the aid of plasma. In such an attempt, the glass substrate was placed into a designed Faraday cage, as shown in Figure 7.12c. The Faraday cage was made of commercial Cu foam (inset of Figure 7.12c). Figure 7.12d shows the successful growth of horizontal graphene on glass at 580 °C using Faraday cage-assisted PECVD method. Raman spectra obtained from the glass substrate proved the formation of uniform graphene films (Figure 7.12e). The optical transmittance of 10 cm × 6 cm graphene glass reaches 80.0% (inset of Figure 7.12e). The Faraday cage blocks ion bombardment, reduces the formation of defects in the graphite layer, and ultimately promotes the horizontal growth of graphene.

7.4 Perspectives

From the fundamental optical properties of graphene to the emerging material design of "super graphene glass," this chapter mainly introduces the nonlinear properties and strategies to produce graphene glass. Graphene has ultra-broadband absorption in visible and near-infrared wavelengths, high nonlinear optical response, and sensitive and tunable optoelectronic properties, and is expected to become the key building blocks of next-generation optical devices.

Many efforts have been made for the preparation of graphene glass, and its application covering harmonic generation, optical limiting, SA, and so on. However, most of the glass components in graphene glass currently used for nonlinear optics are only served as a carrier for nonlinear optics testing. The low transmittance of graphene glass goes against the original intention of combining graphene and glass. The direct growth of graphene on glass maintains the high transmittance of graphene as well as the advantage of large-scale production. Due to the limitation of sample preparation, the corresponding nonlinear optics research is currently lacking. In general, the optical applications of graphene are still in the early stage of research. To make full use of the advantages in optical properties of graphene and introducing graphene into our daily life, graphene glass could be an excellent platform, whereas in-depth studies are needed.

Acknowledgment

This work is supported by the National Natural Science Foundation of China (NSFC. 92256202, U22A20399, 22221001), the 111 project 2.0 (BP1221004), The Science and Technology Major Program of Gansu Province (22ZD6GD060, 22ZD6FA006), and Supercomputing Center of Lanzhou University.

References

1 Castro Neto, A.H., Guinea, F., Peres, N.M.R. et al. (2009). The electronic properties of graphene. *Reviews of Modern Physics* 81: 109–162.

2 Nair, R.R., Blake, P., Grigorenko, A.N. et al. (2008). Fine structure constant defines visual transparency of graphene. *Science* 320: 1308.

3 Blake, P., Hill, E.W., Castro Neto, A.H. et al. (2007). Making graphene visible. *Applied Physics Letters* 91: 063124.

4 Hendry, E., Hale, P.J., Moger, J. et al. (2010). Coherent nonlinear optical response of graphene. *Physical Review Letters* 105: 097401.

5 Jiang, T., Huang, D., Cheng, J. et al. (2018). Gate-tunable third-order nonlinear optical response of massless Dirac fermions in graphene. *Nature Photonics* 12: 430–436.

6 Soavi, G., Wang, G., Rostami, H. et al. (2018). Broadband, electrically tunable third-harmonic generation in graphene. *Nature Nanotechnology* 13: 583–588.

7 Peng, X. and Yan, Y. (2021). Graphene saturable absorbers applications in fiber lasers. *Journal of the European Optical Society-Rapid Publications* 17.

8 Ramachandran, S., Sotor, J., Soboń, G. et al. (2014). Dual-wavelength fiber mode-locked laser based on graphene saturable absorber. In: *Fiber Lasers XI: Technology, Systems, and Applications*, 89612A. Society of Photo-Optical Instrumentation Engineers (SPIE) https://doi.org/10.1117/12.2041041.

9 Yoshikawa, N., Tamaya, T., and Tanaka, K. (2017). High-harmonic generation in graphene enhanced by elliptically polarized light excitation. *Science* 356: 736–738.

10 Hafez, H.A., Kovalev, S., Deinert, J.C. et al. (2018). Extremely efficient terahertz high-harmonic generation in graphene by hot Dirac fermions. *Nature* 561: 507–511.

11 Zheng, X., Feng, M., Li, Z. et al. (2014). Enhanced nonlinear optical properties of nonzero-bandgap graphene materials in glass matrices. *Journal of Materials Chemistry C* 2: 4121–4125.

12 Ouyang, Q., Xu, Z., Lei, Z. et al. (2014). Enhanced nonlinear optical and optical limiting properties of graphene/ZnO hybrid organic glasses. *Carbon* 67: 214–220.

13 Ouyang, Q., Yu, H., Wu, H. et al. (2013). Graphene/MoS_2 organic glasses: fabrication and enhanced reverse saturable absorption properties. *Optical Materials* 35: 2352–2356.

14 Ouyang, Q., Di, X., Lei, Z. et al. (2013). Enhanced reverse saturable absorption in graphene/Ag_2S organic glasses. *Physical Chemistry Chemical Physics* 15: 11048–11053.

15 Ouyang, Q., Yu, H., Xu, Z. et al. (2013). Synthesis and enhanced nonlinear optical properties of graphene/CdS organic glass. *Applied Physics Letters* 102.

16 Sreeja, V.G. and Anila, E.I. (2019). Studies on the effect of reduced graphene oxide on nonlinear absorption and optical limiting properties of potassium doped zinc oxide thin film by Z-scan technique. *Thin Solid Films* 685: 161–167.

17 Sobon, G., Sotor, J., Pasternak, I. et al. (2012). Er-doped fiber laser mode-locked by CVD-graphene saturable absorber. *Journal of Lightwave Technology* 30: 2770–2775.

18 Chen, Z., Qi, Y., Chen, X. et al. (2019). Direct CVD growth of graphene on traditional glass: methods and mechanisms. *Advanced Materials* 31: e1803639.

19 Sun, J., Chen, Y., Priydarshi, M.K. et al. (2016). Graphene glass from direct CVD routes: production and applications. *Advanced Materials* 28: 10333–10339.

20 Sun, J., Chen, Y., Priydarshi, M.K. et al. (2015). Direct chemical vapor deposition-derived graphene glasses targeting wide ranged applications. *Nano Letters* 15: 5846–5854.

21 Chen, X.D., Chen, Z., Jiang, W.S. et al. (2017). Fast growth and broad applications of 25-inch uniform graphene glass. *Advanced Materials* 29.

22 Sun, Z., Yan, Z., Yao, J. et al. (2010). Growth of graphene from solid carbon sources. *Nature* 468: 549–552.

23 Chen, J., Guo, Y., Jiang, L. et al. (2014). Near-equilibrium chemical vapor deposition of high-quality single-crystal graphene directly on various dielectric substrates. *Advanced Materials* 26: 1348–1353.

24 Chen, Z., Guan, B., Chen, X.-d. et al. (2016). Fast and uniform growth of graphene glass using confined-flow chemical vapor deposition and its unique applications. *Nano Research* 9: 3048–3055.

25 Bhaviripudi, S., Jia, X., Dresselhaus, M.S., and Kong, J. (2010). Role of kinetic factors in chemical vapor deposition synthesis of uniform large area graphene using copper catalyst. *Nano Letters* 10: 4128–4133.

26 Li, G., Huang, S.H., and Li, Z. (2015). Gas-phase dynamics in graphene growth by chemical vapour deposition. *Physical Chemistry Chemical Physics* 17: 22832–22836.

27 Chen, Y., Sun, J., Gao, J. et al. (2015). Growing uniform graphene disks and films on molten glass for heating devices and cell culture. *Advanced Materials* 27: 7839–7846.

28 Chen, J., Wen, Y., Guo, Y. et al. (2011). Oxygen-aided synthesis of polycrystalline graphene on silicon dioxide substrates. *Journal of the American Chemical Society* 133: 17548–17551.

29 Zhang, P., Liu, X., Fang, X., and Chen, X. (2020). Synthesis of single-layer graphene film by chemical vapor deposition with molten gallium catalyst on silicon dioxide. *Journal of Materials Science* 55: 2787–2795.

30 Yan, Z., Lin, J., Peng, Z. et al. (2012). Toward the synthesis of wafer-scale single-crystal graphene on copper foils. *ACS Nano* 6: 9110–9117.

31 Gao, T., Xie, S., Gao, Y. et al. (2011). Growth and atomic-scale characterizations of graphene on multifaceted textured Pt foils prepared by chemical vapor deposition. *ACS Nano* 5: 9194–9201.

32 Wu, T., Zhang, X., Yuan, Q. et al. (2016). Fast growth of inch-sized single-crystalline graphene from a controlled single nucleus on cu–Ni alloys. *Nature Materials* 15: 43–47.

33 Kim, H., Song, I., Park, C. et al. (2013). Copper-vapor-assisted chemical vapor deposition for high-quality and metal-free single-layer graphene on amorphous SiO_2 substrate. *ACS Nano* 7: 6575–6582.

34 Teng, P.-Y., Lu, C.-C., Akiyama-Hasegawa, K. et al. (2012). Remote catalyzation for direct formation of graphene layers on oxides. *Nano Letters* 12: 1379–1384.

35 Su, C.-Y., Lu, A.-Y., Wu, C.-Y. et al. (2011). Direct formation of wafer scale graphene thin layers on insulating substrates by chemical vapor deposition. *Nano Letters* 11: 3612–3616.

36 Kwak, J., Chu, J.H., Choi, J.-K. et al. (2012). Near room-temperature synthesis of transfer-free graphene films. *Nature Communications* 3: 645.

37 Ismach, A., Druzgalski, C., Penwell, S. et al. (2010). Direct chemical vapor deposition of graphene on dielectric surfaces. *Nano Letters* 10: 1542–1548.

38 Li, Q., Zhao, Z., Yan, B. et al. (2017). Nickelocene-precursor-facilitated fast growth of graphene/h-BN vertical heterostructures and its applications in OLEDs. *Advanced Materials* 29: 1701325.

39 Sun, J., Chen, Z., Yuan, L. et al. (2016). Direct chemical-vapor-deposition-fabricated, large-scale graphene glass with high carrier mobility and uniformity for touch panel applications. *ACS Nano* 10: 11136–11144.

40 Ding, G., Zhu, Y., Wang, S. et al. (2013). Chemical vapor deposition of graphene on liquid metal catalysts. *Carbon* 53: 321–326.

41 Tan, L., Zeng, M., Wu, Q. et al. (2015). Direct growth of ultrafast transparent single-layer graphene defoggers. *Small* 11: 1840–1846.

42 Vang, R.T., Honkala, K., Dahl, S. et al. (2005). Controlling the catalytic bond-breaking selectivity of Ni surfaces by step blocking. *Nature Materials* 4: 160–162.

43 Lamber, R., Jaeger, N., and Schulz-Ekloff, G. (1988). Electron microscopy study of the interaction of Ni, Pd and Pt with carbon. *Surface Science* 197: 402–414.

44 Fisher, J.C. (1951). Calculation of diffusion penetration curves for surface and grain boundary diffusion. *Journal of Applied Physics* 22: 74–77.

45 Wu, W., Jauregui, L.A., Su, Z. et al. (2011). Growth of single crystal graphene arrays by locally controlling nucleation on polycrystalline cu using chemical vapor deposition. *Advanced Materials* 23: 4898–4903.

46 Li, X., Magnuson, C.W., Venugopal, A. et al. (2011). Large-area graphene single crystals grown by low-pressure chemical vapor deposition of methane on copper. *Journal of the American Chemical Society* 133: 2816–2819.

47 Yu, Q., Jauregui, L.A., Wu, W. et al. (2011). Control and characterization of individual grains and grain boundaries in graphene grown by chemical vapour deposition. *Nature Materials* 10: 443–449.

48 Zhang, L., Shi, Z., Wang, Y. et al. (2011). Catalyst-free growth of nanographene films on various substrates. *Nano Research* 4: 315–321.

49 Yang, W., He, C., Zhang, L. et al. (2012). Growth, characterization, and properties of nanographene. *Small* 8: 1429–1435.

50 Sun, J., Chen, Y., Cai, X. et al. (2015). Direct low-temperature synthesis of graphene on various glasses by plasma-enhanced chemical vapor deposition for versatile, cost-effective electrodes. *Nano Research* 8: 3496–3504.

51 Shang, N.G., Papakonstantinou, P., McMullan, M. et al. (2008). Catalyst-free efficient growth, orientation and biosensing properties of multilayer graphene nanoflake films with sharp edge planes. *Advanced Functional Materials* 18: 3506–3514.

52 Do, H.T., Thieme, G., Fröhlich, M. et al. (2005). Ion molecule and dust particle formation in Ar/CH_4, Ar/C_2H_2 and Ar/C_3H_6 radio-frequency plasmas. *Contributions to Plasma Physics* 45: 378–384.

53 Vizireanu, S., Stoica, S.D., Luculescu, C. et al. (2010). Plasma techniques for nanostructured carbon materials synthesis. A case study: carbon nanowall growth by low pressure expanding RF plasma. *Plasma Sources Science and Technology* 19: 034016.

54 Denysenko, I.B., Xu, S., Long, J.D. et al. (2004). Inductively coupled $Ar/CH_4/H_2$ plasmas for low-temperature deposition of ordered carbon nanostructures. *Journal of Applied Physics* 95: 2713–2724.

55 Akhavan, O., Ghaderi, E., and Rahighi, R. (2012). Toward single-DNA electrochemical biosensing by graphene nanowalls. *ACS Nano* 6: 2904–2916.

56 Bo, Z., Zhu, W., Ma, W. et al. (2013). Vertically oriented graphene bridging active-layer/current-collector interface for ultrahigh rate supercapacitors. *Advanced Materials* 25: 5799–5806.

57 Zhao, J., Sun, M., Liu, Z. et al. (2015). Three dimensional hybrids of vertical graphene-nanosheet sandwiched by ag-nanoparticles for enhanced surface selectively catalytic reactions. *Scientific Reports* 5: 16019.

58 Mao, S., Wen, Z., Ci, S. et al. (2015). Perpendicularly oriented $MoSe_2$ graphene nanosheets as advanced electrocatalysts for hydrogen evolution. *Small* 11: 414–419.

59 Ci, H., Ren, H., Qi, Y. et al. (2018). 6-inch uniform vertically-oriented graphene on soda-lime glass for photothermal applications. *Nano Research* 11: 3106–3115.

60 Qi, Y., Deng, B., Guo, X. et al. (2018). Switching vertical to horizontal graphene growth using faraday cage-assisted PECVD approach for high-performance transparent heating device. *Advanced Materials* 30: 1704839.

61 Neyts, E.C., Van Duin, A.C.T., and Bogaerts, A. (2012). Insights in the plasma-assisted growth of carbon nanotubes through atomic scale simulations: effect of electric field. *Journal of the American Chemical Society* 134: 1256–1260.

62 Kondo, D., Sato, S., Yagi, K. et al. (2010). Low-temperature synthesis of graphene and fabrication of top-gated field effect transistors without using transfer processes. *Applied Physics Express* 3: 025102.

63 Garcia-Lekue, A., Balashov, T., Olle, M. et al. (2014). Spin-dependent electron scattering at graphene edges on Ni(111). *Physical Review Letters* 112: 066802.

8

2D Materials for Nonlinear Optical Limiting

Wen Shang, Bolong Wang, and Qiang Wang

Lanzhou University, State Key Laboratory of Applied Organic Chemistry (SKLAOC), Key Laboratory of Nonferrous Metal Chemistry and Resources Utilization of Gansu Province, College of Chemistry and Chemical Engineering, Key Laboratory of Special Function Materials and Structure Design, Ministry of Education, No. 222 South Tianshui Road, Lanzhou 730000, China

8.1 Introduction

In 1960, Maiman invented the first laser in history [1]. Lasers are now used in a broad field covering optical communication, image processing, environmental monitoring, laser processing, medicine, and other fields [2–7]. In addition, lasers have also been developed into various types of laser weapons, showing unique and important strategic and tactical value in future wars, antiterrorism, security, and rescue [8, 9]. At the same time, the wide laser applications have vigorously promoted the research of laser protection for high-value optoelectronics, military equipment, and personnel [10–13]. Studies have shown that when the energy of the continuous wave laser reaches 2 μJ, 50% of the retina of the human eyes may be damaged; 10 μJ will cause permanent retinal damage; and 20 μJ will cause retinal hemorrhage [14]. However, some traditional laser protection equipment such as glasses has significant deficiencies. As the protection principle mainly relies on linear absorption of light, they exhibit disadvantages such as limited protection wavelength, poor anti-pulse laser capability, and contradiction between light protection and transmission. Other techniques like the use of phase transition materials of VO_2 and V_2O_5 are low in response time [15, 16].

Therefore, the development of efficient optical limiters to protect human eyes as well as optoelectronic sensors is of great importance. Under this context, optical limiting based on the third-order nonlinearity has become one promising strategy. It eliminates the drawbacks of the above two routes by demonstrating fast response and highly effective advantages. More importantly, nonlinear optical limiters behave like a smart optical switch by allowing weak light to pass but block too strong light harmful for the eyes/devices, i.e. self-adaptive properties. To achieve simultaneous protection against both pulsed laser and continuous wave or quasi-continuous wave laser damage, a lot of efforts have been invested in the development of high-performance optical limiting materials in the past few decades,

Two-Dimensional Materials for Nonlinear Optics: Fundamentals, Preparation Methods, and Applications, First Edition. Edited by Qiang Wang and Hao-Li Zhang.

such as fullerenes, carbon nanotubes, graphene, porphyrins, phthalocyanines, carbon black suspensions, polymers, and organic/inorganic composites [17–23]. Organic materials have the following advantages as nonlinear optical materials: (i) Due to the high polarizability of the π electron network in organic molecules, they usually possess large nonlinear optical coefficients; (ii) rich selectivity as they can be flexibly designed and synthesized; and (iii) solution processability, roll-to-roll production, and flexibility. However, organic materials usually exhibit low damage threshold and are easily destroyed by strong incident light. By contrast, inorganic materials usually have large damage threshold. Hence, hybrid materials of organic/inorganic composites could be promising by combining the advantages but eliminating the drawbacks of different types of materials.

Particularly, among the various materials, 2D materials such as graphene, transition metal disulfides, and black phosphorus possess unique optoelectronic properties and have stimulated a wide range of research interests [24–27]. Functional small molecules or polymers can be used to modify 2D materials at molecular level to fabricate nonlinear optical composites to greatly enhance the performance and adaptability of optical limiting devices [28–37]. This chapter briefly reviews the representative 2D materials, as well as the laser protection mechanism, research progress, key scientific problems to be solved, and future development trends.

8.2 Nonlinear Optical Limiting Mechanism

From the working principle, laser protection technologies can be divided into those based on linear optics and those based on nonlinear optics. The former includes absorption filters, reflection filters, and coherence filters, while the latter mainly comprises nonlinear absorption, nonlinear scattering (NLS), and nonlinear refraction (NLR). Other methodologies like phase transformation are also feasible. Among these various routes, third-order optical nonlinearity-based methods demonstrate many advantages like broadband spectral response, ultrafast response time, and intelligent protection by allowing transmission of weaker light but blocking harmful input. Thus, these methods are intensively investigated. The mechanism mainly includes reverse saturable absorption (RSA), two-photon absorption/multiphoton absorption (TPA/MPA), free carrier absorption (FCA), NLR, and NLS.

8.2.1 Reverse Saturable Absorption

RSA refers to the phenomenon that with the increase of the incident light intensity, the light transmittance decreases and results into optical limiting effect. For example, fullerenes, phthalocyanines, and porphyrins exhibit excellent RSA properties for laser pulses in the visible region. In a molecular system, when the absorption cross section of the excited singlet (σ_S) and triplet (σ_T) states is larger than that of the ground state, RSA occurs. A simple energy level model can be used to describe or simulate the RSA process [38]. As shown in Figure 8.1a, when light irradiates the molecule, electrons transit from the ground state S_0 to the first excited singlet

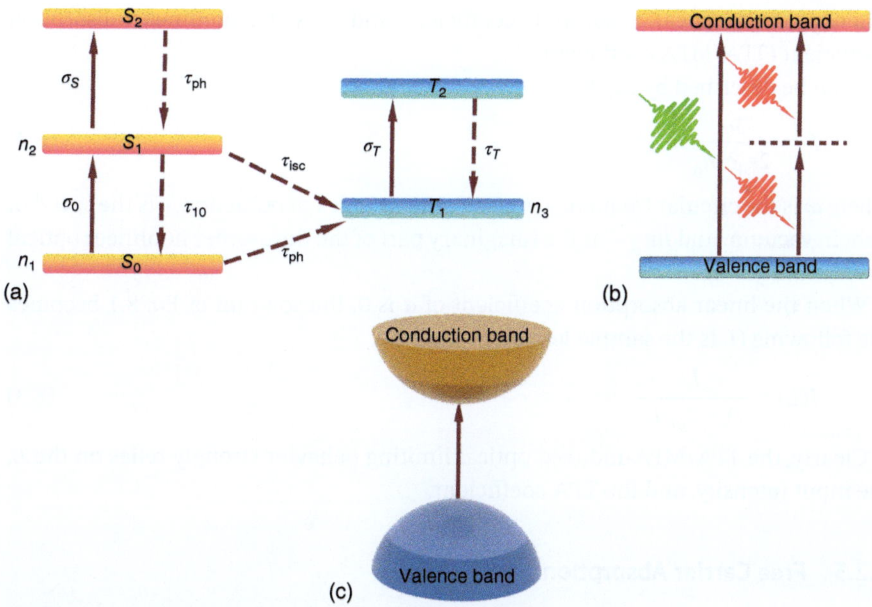

Figure 8.1 The illustration of processes of reverse saturable absorption (a), two-photon absorption/multiphoton absorption, (b) and free carrier absorption (c).

state S_1. Within the pulse width of the laser, the electrons located at S_1 continue to be excited to S_2 and then rapidly relax to S_1. Then electrons either return to the ground state from S_1 or jump to the first excited triplet state T_1 via intersystem crossing and are further excited to T_2. During the process, larger excited state absorption than the ground state absorption (i.e. $\sigma_S > \sigma_0$) will lead to reserve RSA behaviors.

8.2.2 Two-Photon/Multiphoton Absorption

TPA/MPA is the nonlinear process where two or multiple photons are simultaneously absorbed by the atoms/molecules. As shown in Figure 1b, molecules absorb two lower energy photons and transits to an excited state through a "virtual" intermediate state. TPA/MPA manifests themselves as long-wave absorption and short-wave emission, which effectively reduce Rayleigh scattering and linear absorption compared with single-photon absorption and penetrate deeply into the medium. In addition, TPA intensity is proportional to the square of the incident light intensity and occurs only near the focal point, which makes the excitation of the chromophore extremely sterically selective.

TPA/MPA can be used for optical limiting in a way similar to RSA described by a three-level RSA model: herein, the intermediate state lifetime is about zero, and the ground state absorption is very small.

The laser intensity changes as it propagates along the material [39]:

$$\partial I/\partial z = -(\alpha + \beta I)I \tag{8.1}$$

where α is the linear absorption coefficient and β is the nonlinear absorption coefficient (TPA/MPA coefficient).

β can be described by Eq. 8.2:

$$\beta = \frac{3\omega}{2\varepsilon_0 c^2 n_0^2} \mathrm{Im}\chi^{(3)} \tag{8.2}$$

where ω is the circular frequency, n_0 is the linear index of refraction, c is the speed of light in vacuum, and $Im\chi^{(3)}$ is the imaginary part of the third-order nonlinear optical susceptibility.

When the linear absorption coefficient of α is 0, the solution of Eq. 8.1 becomes the following (L is the sample length):

$$I(L) = \frac{I_0}{(1 + I_0 \beta L)} \tag{8.3}$$

Clearly, the TPA/MPA-induced optical limiting behavior strongly relies on the L, the input intensity, and the TPA coefficient.

8.2.3 Free Carrier Absorption

Upon photoexcitation, electrons are pumped from the valence band (VB) to the conduction band (CB), leaving holes in the former for semiconductors (Figure 8.1c). The optically generated free carriers are often phonon assisted and can also be used for optical limiting in a way similar to excited state absorption in molecules. Either single-photon or TPA/MPA excitation can generate free carriers and lead to optical limiting effects [39, 40]. FCA relies on the building up of carrier population in the bands. For the simplified situation where only linear absorption and FCA occur, the process can be described by the following [39]:

$$\partial \mathrm{I}/\partial \mathrm{z} = -(\alpha + \sigma N)I \tag{8.4}$$

where N is the number density of electron–hole pairs and σ is the total (electron + hole) FCA cross section.

8.2.4 Nonlinear Refraction

Nonlinear refraction includes self-focusing and self-defocusing, which stems from the refractive index change of the material caused by the incident laser [41]. Under the Gaussian beam excitation, refractive index changes in the nonlinear optical materials, and the relationship between the nonlinear refractive index and the light intensity can be described as follows:

$$n = n_0 + n_2 \mathrm{I} \tag{8.5}$$

where n_0, n_2, and I are the linear refractive index, nonlinear refractive index, and incident intensity, respectively. When the incident light intensity is smaller than the self-focusing or self-defocusing threshold, the material exhibits high transmission properties. However, once the incident light intensity reaches the threshold, the transmittance is suppressed, i.e. occurrence of optical limiting behaviors. The self-focusing effect generally occurs in materials with a positive nonlinear refractive index, whereas the self-defocusing effect takes place in the medium with a negative nonlinear refractive index.

8.2.5 Nonlinear Scattering

NLS is another route for optical limiting. Scattering reduces incident intensity by dispersing the high-intensity laser light into a larger space, thereby leading to the optical limiting effect. The occurrence of effective scattering requires scattering centers. On the one hand, nanomaterials may absorb incident laser energy, and heat is transferred to the surrounding solvent to cause bubbles. The difference in refractive index between the bubbles and the surrounding solvent is large and leads to light scattering. On the other hand, high-intensity pulsed lasers may ionize nanomaterials to form miniature plasmas, which rapidly expand to form scattering centers to attenuate the incident laser. For the ns laser, the bubble scattering efficiency is high, whereas for picosecond/femtosecond pulses, light attenuation caused by the bubbles is negligible. In addition, compared with the relatively longer time for bubble formation, plasma generation is a sub-nanosecond process and gives rise to faster optical limiting responses [42].

Our group built an NLS measurement setup to distinguish NLS signals from absorption ones [43]. The strategy is to measure the nonlinear responses at different angles. Usually, the scattering rate increases linearly at lower incidence but becomes constant at higher incidence due to NLS. The scattering signals can be subtracted by utilizing an integrating sphere in the setup.

8.3 2D Materials for Optical Limiting

For laser protection materials, they need to have a low onset threshold to effectively protect human eyes/sensors from the damage of lasers. At the same time, they should have high damage threshold to prevent the destruction on themselves. Traditional laser protection materials and devices including photochromic materials, filters, and optoelectronic shutters have shortcomings of low linear transmittance, narrow operational range, slow response, etc. Among the various newly developed materials, 2D materials with high third-order nonlinear absorption coefficients and damage thresholds demonstrate strong RSA behaviors and have been employed as effective optical limiters. Table 8.1 lists typical 2D nonlinear optical limiting materials. Based on those results, general effective strategies used to improve the optical limiting performance of 2D materials mainly include the following:

(1) Tuning bandgaps. The bandgaps of 2D materials can be tuned through different technologies, like varying the layer thickness or chemical modification. Fine-tuning bandgaps give rise to expanded light absorption ranging from UV–vis to the near-infrared and mid-infrared bands or enhances photoresponses at particular wavelength for the 2D materials;

(2) Atomic fabrication method. Atomic fabrication is a useful strategy to achieve designed materials with precise control at the atomic scale. Chemical vapor deposition, atomic layer deposition, sputtering and evaporation methods, and precise chemical synthesis as well are all typical atomic fabrication methods to obtain high-quality 2D materials.
(3) Fermi-level modulation. Through methods like chemical doping or gate control, Fermi level of 2D materials can be modulated. Through the regulation of the Fermi level to tailor the concentration of carriers, the optical nonlinearity of the 2D material can be fine-tuned.
(4) Utilization of plasmonic effect. Localized surface plasmon resonance (LSPR) occurs when the incident light causes collective oscillation of electrons in resonance on the metal nanoparticle surface. LSPR effect promotes light–matter interaction via the enhanced the electric field strength and can thus be harnessed to improve nonlinear optical responses.

In the following part, we introduce some representative 2D optical limiting materials in details.

8.3.1 Graphene and Analogs

8.3.1.1 Graphene

In 2004, Novoselov et al. used the mechanical force separation method to obtain a one-atom-thick graphene sheet from graphite [59]. Graphene is a honeycomb hexagonal monolayer 2D crystal composed of carbon atoms in the form of sp^2 hybrid orbitals, with a thickness of only 0.35 nm, which is the thinnest material known so far. The unique structure of graphene enables it to possess many excellent mechanical, electrical, thermal, and optoelectronic properties. In the visible light range, single-layer graphene is almost transparent with a transmittance of 97.7% [60].

In 2009, Wang et al. reported the optical limiting properties of high-quality graphene for the first time [61]. They used a liquid-phase exfoliation technique to prepare a graphene suspension without oxidation and lattice defects. Under visible light (532 nm) and near-infrared (1064 nm) nanosecond pulses, graphene showed a significant broadband optical limiting performance due to thermally induced nonlinear light scattering. Cheng et al. used air pressure to adjust the optical limiting of graphene and improved the performance by orders of magnitude [44]. Sun et al. used Fe nanoparticles as catalysts to prepare etched graphene by catalyzing carbon steam and tailor the graphene energy band, thereby improving its optical limiting performance [62]. The research on the optical limiting properties of graphene extended to its analogs and composites. For example, Feng and co-workers reported graphene with different nanostructures, including graphene oxide nanosheets, graphene oxide nanoribbons, graphene nanosheets, and graphene nanoribbons [63], all of which exhibited excellent broadband optical limiting performance.

Chen et al. modified reduced graphene oxide chemically with aggregation-induced emission (AIE) polymer as solid-state optical limiters [46]. The as-prepared PFTP-RGO/PMMA film showed strong RSA responses with low OL threshold (1.44 J/cm^2

Table 8.1 A brief summary of representative 2D nonlinear optical limiting materials.

Sample	State	T_0%	Laser	β_{eff}	Im$\chi^{(3)}$	Reference
Graphene	NMP	51	532 nm, 6 ns, 2 Hz	2.6 ± 0.13 cm GW^{-1}	—	[44]
GO-Pt-2	DMF	69.3	532 nm, 4 ns, 10 Hz	77.3 ± 0.23 cm GW^{-1}	$(26.6 \pm 0.08) \times 10^{-12}$ esu	[45]
Annealed PFTP-RGO/PMMA	Film	83.61/92.03	532/1064 nm, 6 ns, 2 Hz	296.79/369.89 cm GW^{-1}	$(126.73/316.26) \times 10^{-12}$ esu	[46]
MoS_2-PVK/PMMA	Film	64.42/67.78	532/1064 nm, 6 ns, 2 Hz	917.57/461.63 cm GW^{-1}	$(392.13/197.28) \times 10^{-12}$ esu	[47]
BP:C_{60} (annealed)	Film	61.72	532 nm, 6 ns, 2 Hz	241.73 cm GW^{-1}	103.22×10^{-12} esu	[48]
BP	IPA	30.3/31.0/36.0	355/532/1064 nm, 20 ps/30 ps/40 ps	5/16/57 cm GW^{-1}	$(2.6/5.5/9.7) \times 10^{17}$ (m^3 [s W] $-^1$)	[49]
h-BNNS/PMMA	Film	—	532/1064 nm, 4 ns, 10 Hz,	1.52 /1.05×10^6 cm GW^{-1}	—	[50]
Antimonene nanosheets	Ethanol	77.24/73.04	532, 30 ps, 10 Hz/1064 nm, 40 ps, 10 Hz	0.43/0.48 cm GW^{-1}	$(3.3/7.4) \times 10^{-6}$ esu	[51]
Tellurene nanosheets	Ethanol	—	1064 nm, 6 ns	72.35 cm GW^{-1}	1.78×10^{-2} esu	[52]
Te NSs	NMP	70	532 nm, 1 kHz, <110 fs	$(9.0 \pm 4.3\%) \times 10^{-9}$ m W^{-1}	—	[53]
Por-COF-HH	2-Propanol	—	532 nm, 6 ns, 10 Hz	1040 cm GW^{-1}	—	[54]
Por-COF-ZnNi	2-Propanol	—	532 nm, 6 ns, 10 Hz	4170 cm GW^{-1}	—	[54]
Por-COF-ZnCu	2-Propanol	—	532 nm, 6 ns, 10 Hz	4470 cm GW^{-1}	—	[54]
Ag-f-HEG	Ethylene glycol	—	400/800 nm, 60 fs, 1 K Hz	$(16/1.7) \times 10^{-13}$ m W^{-1}	$(36.8/7.8) \times 10^{-23}$ m^2 V^{-2}	[55]
HAgNC-630	Colloidal solutions	—	808 nm, 150 fs, 1 K Hz	0.0053 cm GW^{-1}	—	[56]
BPCSA 2D PVK	Film	—	800 nm, 100 fs, 1 kHz	11.25 ± 2.0 cm GW^{-1}	—	[57]
$(PEA)_2PbI_4$	Flake	—	800 nm, 100 fs, 1 kHz	211.5 cm MW^{-1}	—	[58]

at 532 nm and 1.08 J/cm^2 at 1064 nm) and large nonlinear coefficient (296.79 cm/GW at 532 nm and 369.89 cm/GW at 1064 nm).

The modification of graphene is also effective to improve its optical limiting properties. Girisun et al. reported the optical limiting of reduced graphene oxide (RGO) nanocomposites of Au-Fe_2O_3-RGO doped with different ratios of Au-Fe_2O_3 (15, 25, 40 wt%) under 700- to 900-nm femtosecond pulsed laser irradiation, as shown in Figure 8.2a [64]. The materials all exhibited typical RSA signals when irradiated with 700-, 800-, and 900-nm lasers, and the one doped with a ratio of 15 wt% had the best optical limiting performance under 700- and 800-nm laser excitations, with β values of 6.5×10^{-10} and 9.3×10^{-10} m/W, respectively. However, under the excitation of 900-nm laser, Au-Fe_2O_3-RGO doped with 15 and 25 wt% ratio had similar optical limiting performance. This study verified the enhancement of optical limiting through surface plasmon resonance effect of inorganic metal nanoparticles. Moreover, by loading different proportions of inorganic nanoparticles onto the graphene surface, the tuning of the nonlinear optical properties of the material to the near-infrared region was thus achieved. Stavrou et al. [65] doped graphene oxide (GO) by annealing with boric acid and melamine as boron and nitrogen sources, respectively. After B and/or N doping, the chemically modified GO exhibited significantly enhanced nonlinear optical limiting responses under 4-ns and 532- and 1064-nm excitations (Figure 8.2b [I]). The β values of GO, nitrogen (N-GO) doped, boron (B-GO) doped, B and N domain doped (BN-GO), and B, N individually doped B, N-GO at 532 nm were $43.5\pm4.0\times10^{-11}$, $453\pm70\times10^{-11}$, $524\pm52\times10^{-11}$, $595\pm68\times10^{-11}$—, and $648\pm65\times10^{-11}$ m W^{-1}, respectively. Clearly, B-GO and B, N-GO even demonstrated stronger NLO responses than those of N-GO and BN-GO, respectively. The micro-Raman spectra indicated that defects in the samples caused stronger NLO responses. The doped graphene also exhibited nonlinear refraction response, manifested itself as a valley–peak shape in Figure 8.2b (II). The work strongly proved defect engineering by chemical doping as one effective way to fine-tune NLO properties of 2D materials.

Compared with graphene loaded with inorganic nanoparticles, graphene derivative materials modified by organic small molecules and polymers usually had better solubility, stability, and processability. Liu et al. modified GO surface with Pt complexes through cyclization reaction and electrostatic adsorption, and systematically studied the synergistic effect on optical limiting [45]. In Figure 8.2c, compared with GO, Pt-1, Pt-2, and C_{60}, the optical limiting performance of Pt complex modified materials, GO-Pt-1 and GO-Pt-2, was greatly improved. Interestingly, in traditional cognition, the nonlinear optical properties of non-covalently modified graphene materials are generally slightly inferior to those of covalently modified graphene materials due to their poor solubility and dispersion stability. However, in this work, the non-covalently modified GO material GO-Pt-2 using electrostatic adsorption had better optical limiting performance than the covalently modified GO-Pt-1. As shown in Figure 8.2c (I), the normalized transmittances of GO-Pt-1 and GO-Pt-2 were 45% and 28%, respectively, under the test condition of 532-nm laser wavelength. At the same time, GO-Pt-2, a non-covalently modified GO material with Pt complexes, had the lowest critical value (0.85 J/cm^2; Figure 8.2c (II)). On the same

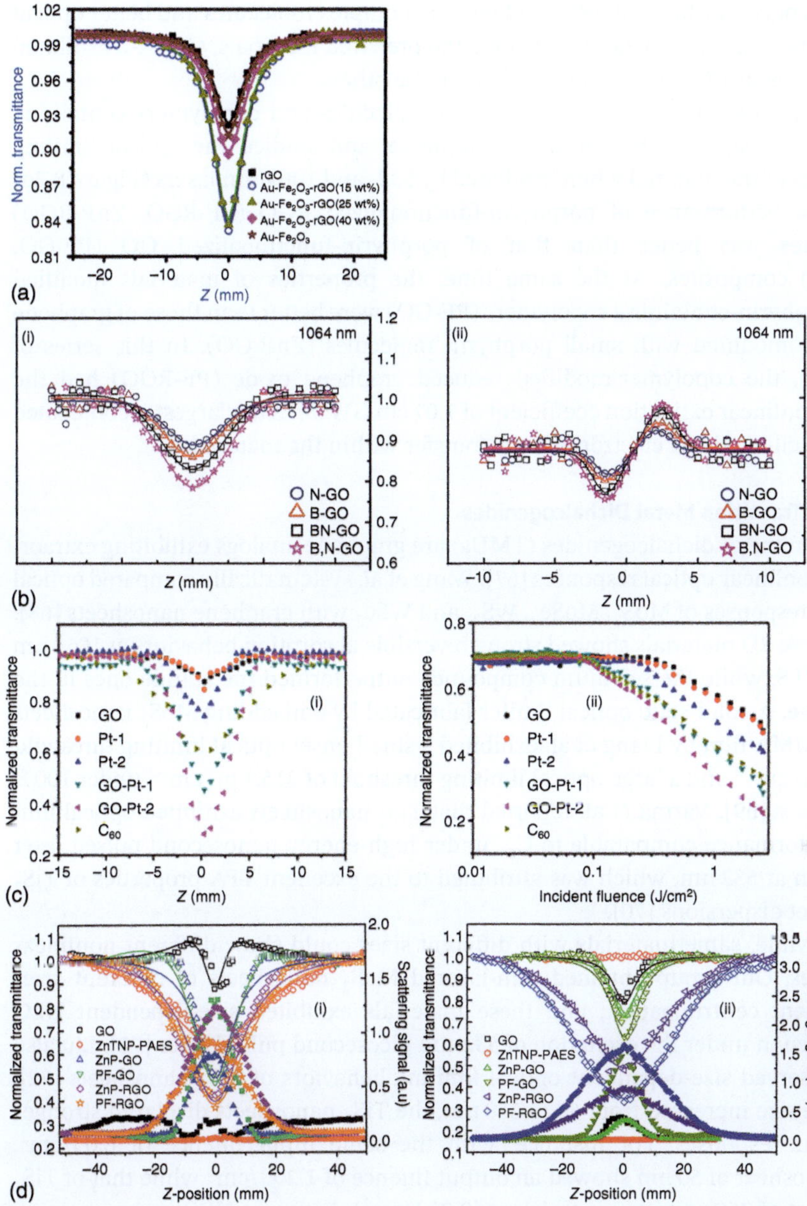

Figure 8.2 (a) Open-aperture Z-scan curves of pure rGO, Au-Fe_2O_3, and Au-Fe_2O_3-(15, 25, and 40 wt%) rGO nanocomposite at 900 nm fs laser excitation. Source: Adapted from Girisun et al. [64]. Copyright © 2018, American Chemical Society. (b) OA (I) and closed-aperture (CA) (II) Z-scans of some N-GO, B-GO, BN-GO, and B, N-GO dispersions under 1064-nm excitation. Source: Adapted from Stavrou et al. [65]. Copyright © 2021 American Chemical Society. (c) OA Z-scan curves (I) and optical limiting performance of GO, Pt-1, Pt-2, GO–Pt-1, GO-Pt-2, and C_{60} with the same linear transmittance of 70% under 4 ns, 532-nm laser excitation (II). Source: Adapted from Liu et al. [45]. Copyright © 2017, American Chemical Society. (d) OA Z-scan curves with normalized transmittance (open symbols) and scattering signal (solid symbols) for the samples at 532-nm (I) and 1064-nm (II) excitations. Source: Adapted from Du et al. [66]. Copyright ©2017, Royal Society of Chemistry.

unit area of GO surface, GO-Pt-2 had more Pt complex molecules and better optical limiting response than GO-Pt-1. Among the prepared materials, GO-Pt-2 exhibited larger β_{eff} value (77.3 ± 0.23 cm GW^{-1}) than the others.

Du et al. covalently modified porphyrin molecules and copolymers containing porphyrin groups on the surface of graphene, and studied the optical limiting properties of the material when irradiated by 532- and 1064-nm lasers (Figure 8.2d) [66]. The performance of porphyrin-functionalized RGO (PF-RGO, ZnP-RGO) composites was better than that of porphyrin-functionalized GO (PF-GO, ZnP-GO) composites. At the same time, the properties of materials modified with porphyrin-containing copolymers (PF-GO) were better than those of graphene materials modified with small porphyrin molecules (ZnP-GO). In this series of materials, the copolymer-modified reduced graphene oxide (PF-RGO) had the largest nonlinear extinction coefficient of 7.07 cm GW^{-1}, as the largest π-conjugated system facilitated fast electron/energy transfer within the material.

8.3.1.2 Transition Metal Dichalcogenides

Transition metal dichalcogenides (TMDs) are graphene analogs exhibiting extraordinary nonlinear optical responses [67]. Wang et al. systematically compared optical limiting responses of MoS_2, $MoSe_2$, WS_2, and WSe_2 with graphene nanosheets [68]. All of these 2D materials showed strong reversible absorption behaviors at 1064 nm due to NLS, while the selenium compounds outperformed the sulfide ones in the NIR range. A solid-state optical limiter fabricated by embedding MoS_2 nanosheets into a PMMA film by Liang et al. exhibited a small onset optical limiting threshold of 21.5 mJ cm^{-2} and a large optical limiting threshold of 315.1 mJ cm^{-2} under 100 fs laser pulses [69]. Varma et al. reported that TiS_2 nanosheets exhibited optical limiting performance comparable to C_{60} under high-energy nanosecond pulsed laser excitation at 532 nm, which was attributed to the excellent TPA properties of TiS_2 nanosheet dispersions [70].

Meanwhile, same materials with different sizes could show different nonlinear properties. Our group obtained thin-layered MoS_2 nanosheets of different sizes by gradient centrifugation, and these materials exhibited size-dependent NLO phenomenon under the excitation of 532-nm picosecond pulsed laser [71]. Liu also et al. observed size-dependent optical limiting behaviors of TiS_2 nanosheets [72]. With the size increased from 50 to 350 nm, the TiS_2 nanosheets displayed stronger nonlinear responses. For instance, with the same input fluence of 6.0 J/cm^2, TiS_2 nanosheet of 50 nm showed an output fluence of 1.20 J/cm^2 while that of TiS_2 nanosheet of 350 nm was much lower (0.73 J/cm^2). Different NLO responses were expected to stem from different bandgaps of the nanosheets.

Covalent modification of MoS_2 nanosheets by chemical synthesis could effectively prevent the agglomeration of MoS_2 nanomaterials that significantly reduces the nonlinear optical effects. The intramolecular charge transfer effect between the electron-donating polymer and MoS_2 would greatly improve optical limiting performance of the material. At the same time, the material was soluble in common organic solvents, which brought great technical convenience for the processing of low-cost materials and devices. Based on these ideas, Cheng et al. synthesized

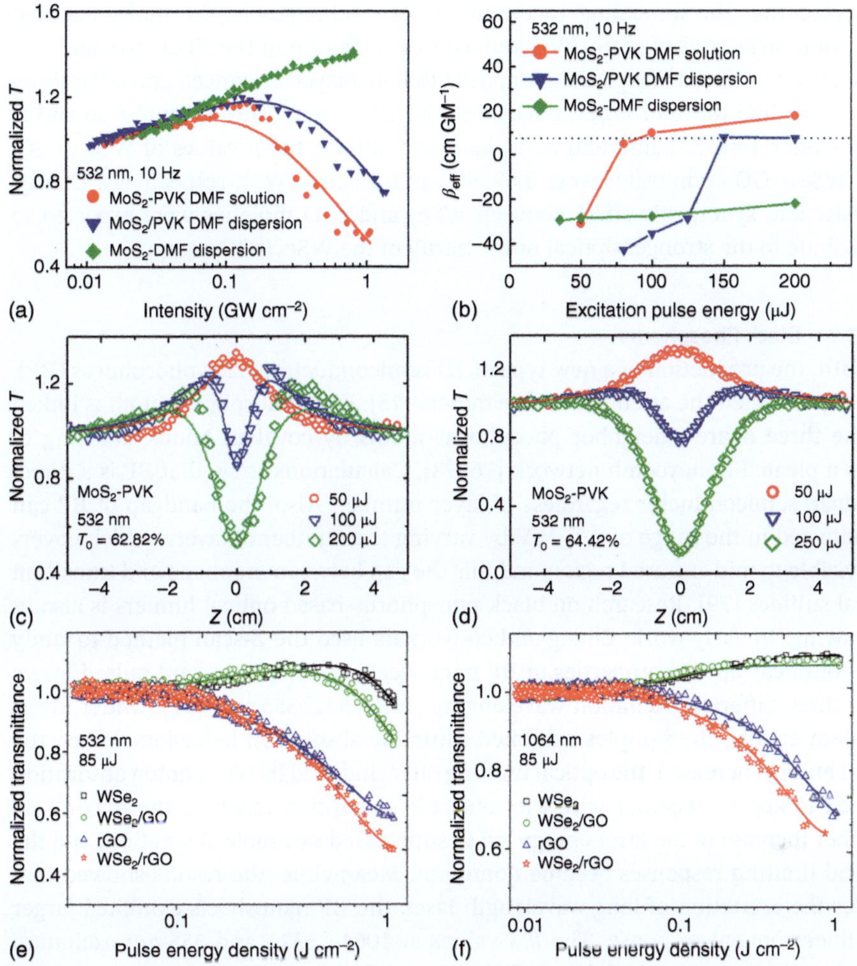

Figure 8.3 (a) The normalized transmittance as a function of input laser intensity. (b) β_{eff} value of the samples in DMF. (c) Typical OA Z-scan data for samples dispersed in DMF. (d) Typical OA Z-scan data for the PMMA-based films. Source: Adapted from Cheng et al. [47]. Copyright © 2016, WILEY-VCH Verlag GmbH & Co. KGaA, Weinheim. (e, f) The optical limiting of the different sample dispersions at 532 and 1064 nm. Source: Adapted from Zu et al. [73]. Copyright © 2021 Elsevier Ltd.

in situ poly(N-vinyl carbazole) (PVK) covalently grafted MoS_2 (MoS_2-PVK) as laser-protective functional material (Figure 8.3a-d) [47, 74]. Compared with MoS_2 and MoS_2/polymer blends, the polymer covalently grafted MoS_2 exhibited strong and broad laser protection properties in both *N*, *N*-dimethylformamide solution and PMMA solid film. Interestingly, MoS_2-PVK/PMMA solid film exhibited saturation absorption properties under 50-μJ laser irradiation, which however transformed into reverse saturation absorption under 250-μJ irradiation. This phenomenon was most likely related to the thermal effect of the laser on the thin film. These observations indicated that the annealing treatment induced changes in the molecular and electronic structures of MoS_2-PVK and led to a transition in the NLO response.

Zu et al. fabricated WSe_2/rGO composite that displayed enhanced optical limiting responses than those of WSe_2, rGO, and WSe_2/GO composites (Figure 8.3e and f) [73]. Under 4-ns, 532-nm, and 85-μJ laser excitations, the β values of WSe_2, rGO, and WSe_2/rGO composites were 0.69, 4.3, and 5.1 cm GW^{-1}, respectively. Charge transfer and synergistic effect between WSe_2 and rGO moieties were expected to contribute to the stronger optical nonlinearity of the WSe_2/rGO composite.

8.3.1.3 Black Phosphorus

In 2014, the production of a new type of 2D semiconductor, black phosphorus (BP), quickly attracted the attention of researchers [75]. Each phosphorus atom is linked to the three nearest neighbor phosphorus atoms by covalent bonds, stacking to form a pleated honeycomb network [76–78]. Calculations show that BP is a direct bandgap semiconductor regardless of layer number. Also, the bandgap of BP can be adjusted in the range of 0.3–2 eV by varying the number of layers, which covers the visible to mid-infrared region and fills the gap between graphene and transition metal sulfides [79]. Research on black phosphorus-based optical limiters is also in full swing. In early work, Zhang and co-workers used the Z-scan method to study the nonlinear optical properties of BP nanosheets under picosecond pulsed lasers with three different excitation wavelengths (1064, 532, 355 nm) [49]. Under lower incident energy, the samples exhibited saturable absorption behaviors. When the input energy increased, the optical limiting effect induced by two-photon absorption started to appear together with the saturable absorption reaching the maximum. Further increase of the laser energy led to suppressed saturable absorption, and the optical limiting responses became dominant. Meanwhile, the results showed that under the excitation of long-wavelength laser, the BP nanosheets exhibited larger nonlinear optical response. The β_{eff} values at 1064-, 532-, and 355-nm excitation wavelengths were 57, 16, and 5 cm GW^{-1}, respectively.

Huang et al. reported a suspension of BP nanosheets prepared by a liquid-phase exfoliation method [80]. Using femtosecond pulses of 1030/515 nm and nanosecond pulses of 1064/532 nm, the wavelength and pulse-width-dependent nonlinear optical responses were observed. The authors applied OA Z-scan technique and corrected the scattering light. The BP nanosheets exhibited SA at a relatively weak excitation (0.64 J/cm^2), but with the input energy increase to 1.46 J/cm^2, thermally induced NLS became obvious. With even higher excitation fluence (4.5 J/cm^2), even stronger optical limiting occurred, accompanied by the enhanced scattering signals. The result indicated NLS played important roles in the observed optical limiting effects.

Chen et al. fabricated a donor–acceptor (D-A)-type blends with BP as an electron donor and C_{60} as an electron acceptor [48]. The blend was incorporated into the poly(methylmethacrylate) (PMMA) matrix for Z-scan measurements, which exhibited an enhanced optical limiting response at 532 nm, with a large nonlinear coefficient of 241.73 cm/GW, a low limiting threshold of 4.5 J/cm^2, and a large damage threshold of 19.54 J/cm^2. Thermal-induced intermolecular charge transfer between BP and C_{60} was attributed to the observed excellent optical limiting performance.

8.3.1.4 Hexagonal Boron Nitride (h-BNS)

Hexagonal boron nitride (h-BN), i.e. the "white graphene," is also a graphene analog but unlike the zero bandgap of graphene has a wide bandgap of ~6.1 eV [81]. h-BN also demonstrates remarkable nonlinear optical features in addition to its unique properties of high stability, insulation, strong mechanical strength, and anti-oxidation capacity [82–84].

Ajayan et al. studied the optical limiting responses of h-BN [85], which stemmed from TPA with a TPA coefficient of 74.84 cm/GW. Unlu et al. introduced C and O atoms into the hexagonal boron nitride powder by ultrasonic and heat treatment [50]. Then, they dispersed hexagonal boron nitride with heteroatom doping in water and embedded it into polymethylmethacrylate films (h-BN/PMMA). The Z-scan results showed that boron nitride with heteroatom doping exhibited obvious nonlinear absorption characteristics at 532 nm and 1064 nm (Figure 8.4). The β_{eff} values of h-BN/PMMA at 532 nm and 1064 nm were 1.52×10^6 cm GW^{-1} and 1.05×10^6 GW^{-1}, respectively. Theoretical models implied that the nonlinear optical responses were mainly associated with the presence of many deep-level defects in hexagonal boron nitride with heteroatom doping.

8.3.2 Novel 2D Materials

8.3.2.1 Antimonene

The aforementioned material of BP has unique properties and great potential as functional materials. However, they are extremely sensitive to the environment and cannot exist stably in air, which limits their applications. In this regard, antimony of the same main group is a very useful complementary [75]. Studies showed that antimonene possessed various remarkable physical and chemical properties, such as tunable bandgap (0–2.28 eV) and high electrical conductivity (10 000 S m^{-1})

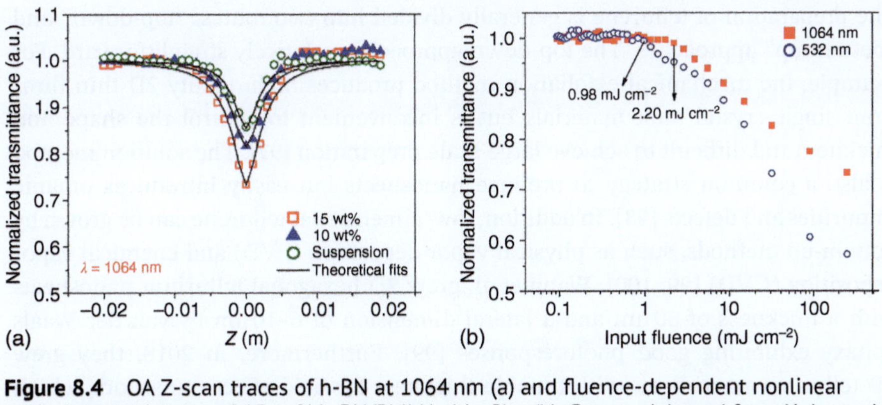

Figure 8.4 OA Z-scan traces of h-BN at 1064 nm (a) and fluence-dependent nonlinear optical transmittance of 15 wt% h-BN/PMMA thin film (b). Source: Adapted from Unlu et al. [50]. Copyright © 2021, Elsevier.

[86]. Antimonene had been successfully applied in optoelectronics, energy storage, electrocatalysis, cancer treatment, etc. [87–90]. It also showed great potential for applications in nonlinear optics, such as all-optical Kerr switches and optical limiting [91, 92].

Zhang et al. fabricated β-antimonene nanosheets using an ultrasound-assisted liquid-phase exfoliation technique [51]. The nonlinear absorption properties of β-antimonene were investigated using a Z-scan method. TPA processes accounted for the observed optical limiting responses, with the extracted TPA coefficients being 0.48 and 0.43 cm GW^{-1} at 1064 and 532 nm, respectively.

Xing et al. also used the ultrasonic-assisted liquid phase exfoliation technique to produce antimonene nanosheets, which were subsequently functionalized by 3-glycidoxypropyltrimethoxysilane to obtain Si-antimonene [93]. The as-prepared Si-antimonene was then composited with hybrid gel glass by sol–gel method, which showed strong RSA properties at 532 and 1064 nm (Figure 8.5a, b). The nonlinear refractive properties of the antimonene and Si-antimonene dispersion were also evidenced by closed-aperture Z-scan measurements. Figure 8.5c, d shows the typical self-focusing phenomenon of Si-antimonene. Moreover, the functionalized antimonene was found more uniformly dispersed in the glass (Figure 8.5e). The above results showed that antimonene was a promising candidate as optical confinement materials.

8.3.2.2 Tellurene

Low-dimensional materials have made great progress in the field of photonics research [94]. With the advancement of fabrication technology, the preparation of many carbons, nitrogen, and oxo single-element alkene materials, such as arsenene, antimonene, and tellurene, has been realized [52, 88, 95].

Among them, tellurene belongs to the chalcogens. Figure 8.6a and b shows the atomic structure model of tellurium. The tellurium atoms are connected by covalent bonds along the c-axis to form a helical chain, and the other directions are combined by van der Waals forces and arranged in a hexagonal structure [96]. The preparation of tellurene is generally divided into two routes: "top-down" and "bottom-up" approaches. The top-down approach is relatively straightforward. For example, the mechanical exfoliation method produces high-quality 2D thin films from single-crystal bulk materials but is inconvenient to control the shape and thickness and difficult to achieve large-scale preparation [97]. The solution method is also a common strategy to produce nanosheets but easily introduces organic impurities and defects [98]. In addition, low-dimensional tellurene can be grown by bottom-up methods, such as physical vapor deposition (PVD) and chemical vapor deposition (CVD) [99, 100]. Wang et al. grew 2D hexagonal tellurium nanosheets with a thickness of 30 nm and a lateral dimension of 6–10 μm by van der Waals epitaxy exhibiting good photoresponses [99]. Furthermore, in 2018, they grew 2D tellurium nanosheets with tunable thickness for field-effect transistors, which demonstrated excellent electronic properties and environmental stability [98]. Therefore, tellurene was an excellent material for optoelectronic applications [101].

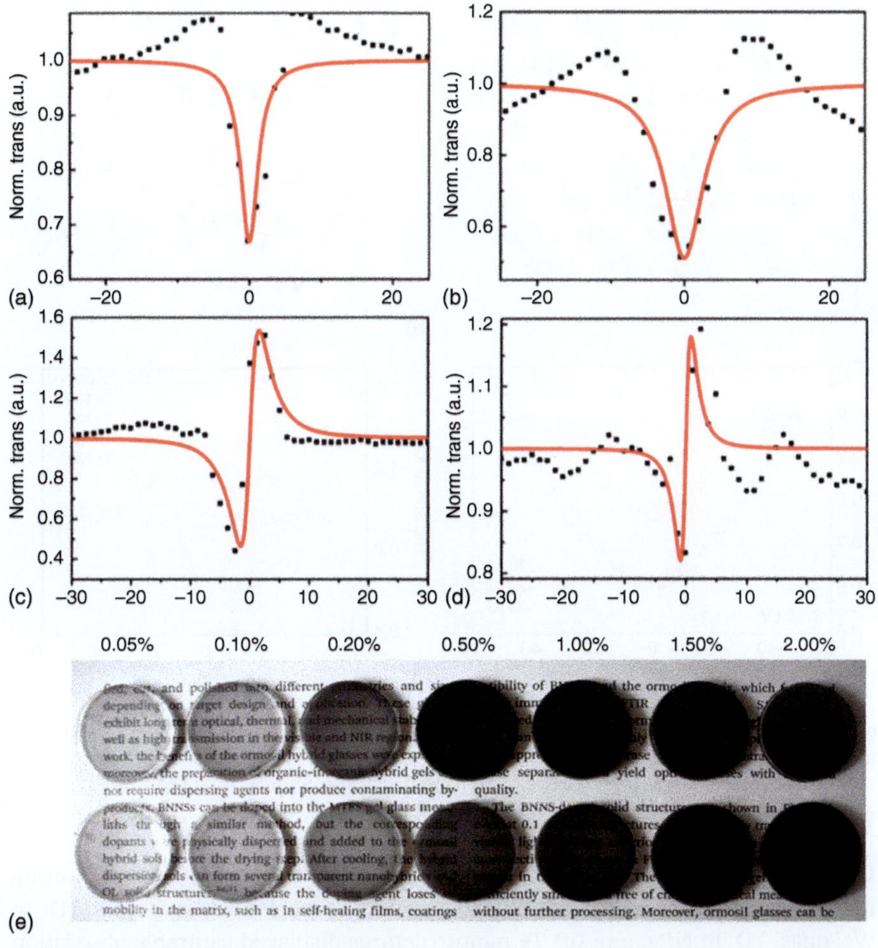

Figure 8.5 OA Z-scan curves at 532 (a) and 1064 nm (b) and closed-aperture Z-scan curves at 532 (c) and 1064 nm (d) of Si-antimonene dispersed in ethanol. (e) Photographs of the antimonene hybrid glasses with different doped wt%: doped glasses of antimonene (top) and Si-antimonene (bottom). Source: Xing et al. [93]. © 2021, American Chemical Society.

Tellurene also showed extraordinary nonlinear optical properties. Zhang et al. studied broadband NLA behaviors of 2D tellurene nanosheets fabricated by liquid-phase exfoliation [52]. Figure 8.6c displays their typical optical limiting behaviors at the excitation of photon energy of 2.76 eV with different laser intensities. At the excitation intensities of 533 μJ, a modulation depth of 40% was obtained. In this experiment, no apparent SA behavior was observed at these three wavelengths, and the OL behavior dominated the NLA process. Our group also prepared a series of Te nanostructures by electrochemical exfoliation and liquid-phase exfoliation methods, including one-dimensional (1D) Te nanowires (NWs), quasi-1D Te nanorods (NRs), zero-dimensional (0D) Te nanodots (NDs), and two-dimensional (2D) Te nanosheets (NSs) [53]. Femtosecond Z-scan measurements disclosed dimension-dependent

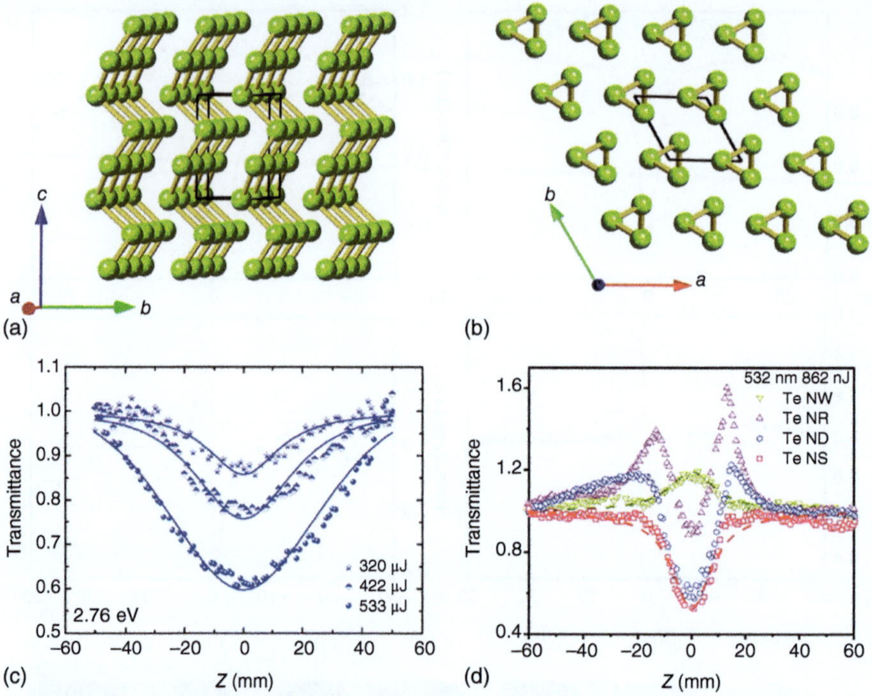

Figure 8.6 (a) and (b) Lattice structure of Te. (c) OA Z-scan results of tellurene nanosheets. Source: Adapted from Zhang et al. [52] Copyright ©2019, Royal Society of Chemistry. (d) Femtosecond (fs) OA Z-scan results of Te nanostructures. Source: Adapted from Xiao et al. [53] Copyright ©2021, Royal Society of Chemistry.

NLO properties for these Te structures. 2D Te NSs showed efficient optical limiting with a large β value of $9.0 \pm 4.3\% \times 10^{-9}\,\mathrm{m\,W^{-1}}$ (Figure 8.6d), whereas 1D Te NWs, quasi-1D Te NRs, and 0D Te nanostructures displayed saturable absorption behaviors. Transient absorption measured indicated that faster exciton relaxation and stronger excited state absorption led to the observed RSA of 2D Te NSs. This work paved the way for the applications of 2D Te in nonlinear optical fields.

8.3.2.3 2D Perovskites

In 1839, Russian mineralogist Lev Perovski first found perovskite in the metamorphic rocks of the Ural Mountains in Russia and named it after himself [102]. At first, the term "perovskite" described a type of mineral calcium titanate ($CaTiO_3$). Now, the term represents a large class of crystal materials with the same structure as that of $CaTiO_3$, usually expressed by the formula ABX_3, where A and B are cations and X are anions [103]. Compared with 3D perovskites, the advantages of novel 2D layered perovskites include high humidity resistance, excellent optical and thermal stabilities, ultralow self-doping behavior, low defect density, and other excellent photoelectric properties [27]. Due to the reduction of dimension and size and the smaller dielectric constant of the organic layer, the quantum confinement effect, electron–hole interaction, and exciton binding energy of 2D layered perovskites distinguish from their

3D counterparts [104] and have emerged as promising candidates in optoelectronic devices such as photovoltaics and laser protection [27, 104–108].

Liu et al. systematically studied the nonlinear optical properties of 2D perovskite of phenylethylamine lead iodide ($(PEA)_2PbI_4$) using a femtosecond OA Z-scan system [58]. Figure 8.7a shows characteristic reversible absorption stemming from TPA effect of the perovskite. Due to the interference of strong linear scattering, the TPA coefficient could not be obtained by directly fitting the Z-scan curves [109]. Hence, the authors put the sample at the laser beam waist and measure the incident and transmitted intensity at the same time. Figure 8.7b shows the inverse transmission the $(PEA)_2PbI_4$ flake as a function of the incident intensity. Eventually, a giant TPA coefficient of 211.5 cm MW^{-1} was extracted for the $(PEA)_2PbI_4$ flake, one order of magnitude larger than those previously reported for 3D perovskite films. The reason might be that the multi-quantum-well structure of the perovskite flake enhanced quantum and dielectric confinement. Interestingly, the optical limiting behavior also showed thickness dependence as reflected by the TPA coefficient changes with the flake thickness shown in Figure 8.7c. The TPA coefficient reduced with the increase of the sample thickness due to self-absorption of the sample.

Lim et al. synthesized a quasi-2D hybrid perovskite (BPCSA 2D PVK) with two-photon absorption characteristics, where BPCSA was (Z)-2-([1,1′-biphenyl]-4-yl)-3-(4-(3-aminopropoxy)phenyl)acrylonitrile [57]. BPCSA 2D PVK was synthesized under stoichiometric control of Pb^{2+} and BPCSA cation iodide for 2D PVK layers. The prepared BPCSA 2D PVK was spin-coated on the ITO substrate, and the nonlinear optical properties were tested. The authors applied femtosecond open- and closed-aperture Z-scans measurements, and disclosed strong optical limiting responses due to TPA for the 2D perovskite. The film thickness was determined to be 147 nm via cross-sectional SEM images (Figure 8.7d). Through data fitting, TPA coefficient of 11.25 ± 2.0 cm GW^{-1} (Figure 8.7e), nonlinear Kerr coefficient with a positive (self-focusing) sign (Figure 8.7f), and $n_2 = 1.2 \pm 0.3 \times 10^{-13}$ cm^2 W^{-1} were obtained.

8.3.2.4 2D Metal–Organic Frameworks (2D-MOFs)

Metal–organic frameworks (MOFs) are composed of metal ions or metal clusters and organic ligands [110]. MOFs have the advantages of high porosity, large specific surface area, and tunable pore structure, which make them be widely used in the fields of catalysis, drug delivery, gas separation and storage, proton transport, detection, and sensors [111–115]. Compared with 3D MOFs, the layered 2D MOFs, also known as 2D coordination polymers, have more prominent 2D spatial connectivity and unique advantages in energy storage and optoelectronic devices [116–119]. This is mainly related to their larger specific surface area, tunable bandgap, biocompatibility, controllable fabrication, and ease of assembly.

The nonlinear optical properties of 2D MOFs had been studied. For example, Liang et al. prepared few-layer Mg-MOF-74 nanosheets by liquid-phase exfoliation and investigated their nonlinear properties at different wavelengths by the OA Z-scan and I-scan methods [120]. Clearly, Mg-MOF-74 transformed from saturable absorption to RSA with the increase of input intensity at 2-μm excitation (Figure 8.8a

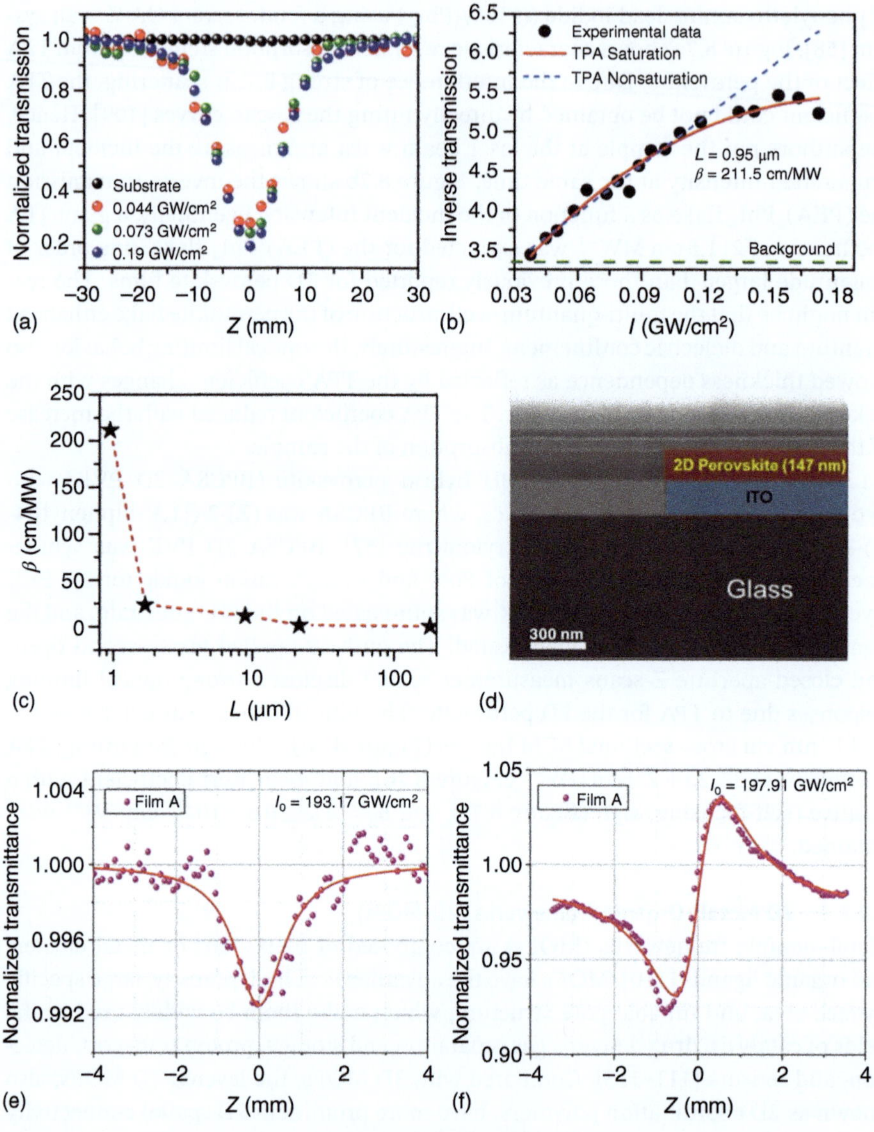

Figure 8.7 (a) OA Z-scan curve. (b) Inverse transmission as a function of peak intensity and (c) TPA coefficient versus sample thickness for the $(PEA)_2PbI_4$ flake. Source: Adapted with permission Liu et al. [58] Copyright © 2017, WILEY-VCH Verlag GmbH & Co. KGaA, Weinheim. (d) The cross-sectional SEM image of the 2D perovskite of BPCSA-coated ITO glass. (e) The OA and (f) CA Z-scan measurements of the 2D BPCSA perovskite thin film. Source: Reproduced with permission Lim et al. [57]. © 2020, John Wiley & Sons.

(I)). Through fitting the RSA curve, modulation depth of 46% and reverse saturation intensity of 1210 mJ/cm^2 were extracted (Figure 8.8a (II)). Figure 8.8a (III) shows the closed-aperture Z-scan curve of the 2D MOF, where a nonlinear refractive index of $-131 \pm 5\,cm^2\,W^{-1}$ was extracted.

Covalent organic frameworks (COFs) have also emerged as a new type of nonlinear optical materials. Biswal et al. created a series of COFs showing optical switching behaviors [54]. Using the OA Z-scan technique at 532-nm excitation (Figure 8.8b), they found the COFs transited from saturable absorption to RSA with the increase of the input pulse energy, and some of the metalated Por-COFs exhibited high nonlinear absorption coefficient of ~4500 cm GW^{-1}.

8.3.3 Materials with New Structures and Mechanism

8.3.3.1 Nonlinear Photonic Metamaterials

The emergence of optical metasurfaces has brought new opportunities for nonlinear optics, realizing multidimensional optical field modulation crucial for the fields of micro/nano-optical devices, optical information processing, high-resolution imaging, and optical limiting [121]. Photonic metamaterials are a class of ultrathin planar devices composed of spatially varying metal or dielectric metastructure functional units, enabling regulation of the polarization, phase, and amplitude of light at the subwavelength scale [122, 123]. The use of 2D metasurface design makes it easier to fabricate compact, functionally integrated, and low-loss micro/nano-optical devices. Myriad excellent optical properties had been realized based on metasurfaces, including plane lens imaging, high-efficiency holographic imaging, and generation of spin and orbital angular momentums of light [122, 124–126].

Many functional devices based on nonlinear optical metasurfaces had been developed. Specifically, beam steering devices, optical holographic imaging devices, and image encryption devices based on nonlinear optical metasurfaces are among those intensively investigated [127–130]. Figure 8.9a showed giant enhancement of the handedness-dependent optical response in planar metallic nanostructures. When illuminated by the light of opposite circular polarizations, the planar hybrid metasurface exhibits over an order of magnitude difference in nonlinear optical response, where hybridized plasmonic–photonic modes accounted for the enhancement. In another work, Wang et al. generated a third-harmonic field and achieved a highly efficient wavefront control using silicon-based metasurfaces [127]. At a designed angle, nonlinear focusing vortex beams were also successfully produced. Phase gradients with a large diffraction efficiency of 92% over a full 0–2π phase range were generated by the nonlinear metasurfaces (Figure 8.9b).

Metamaterial has also been applied to optical limiting. Guddala et al. demonstrated optical limiting effects based on a metamaterial resonant absorber (MRA) composed of a trilayer structure of gold (Au) disks on ZnS/Au thin films (Figure 8.9c) [132]. For the MRA, a significant nonlinear change of the reflectivity was observed under the 8-ps pulse excitation at 1064 nm. Figure 8.9d shows the Z-scan results of gold and ZnS thin films with the same thickness as that in the MRA. The

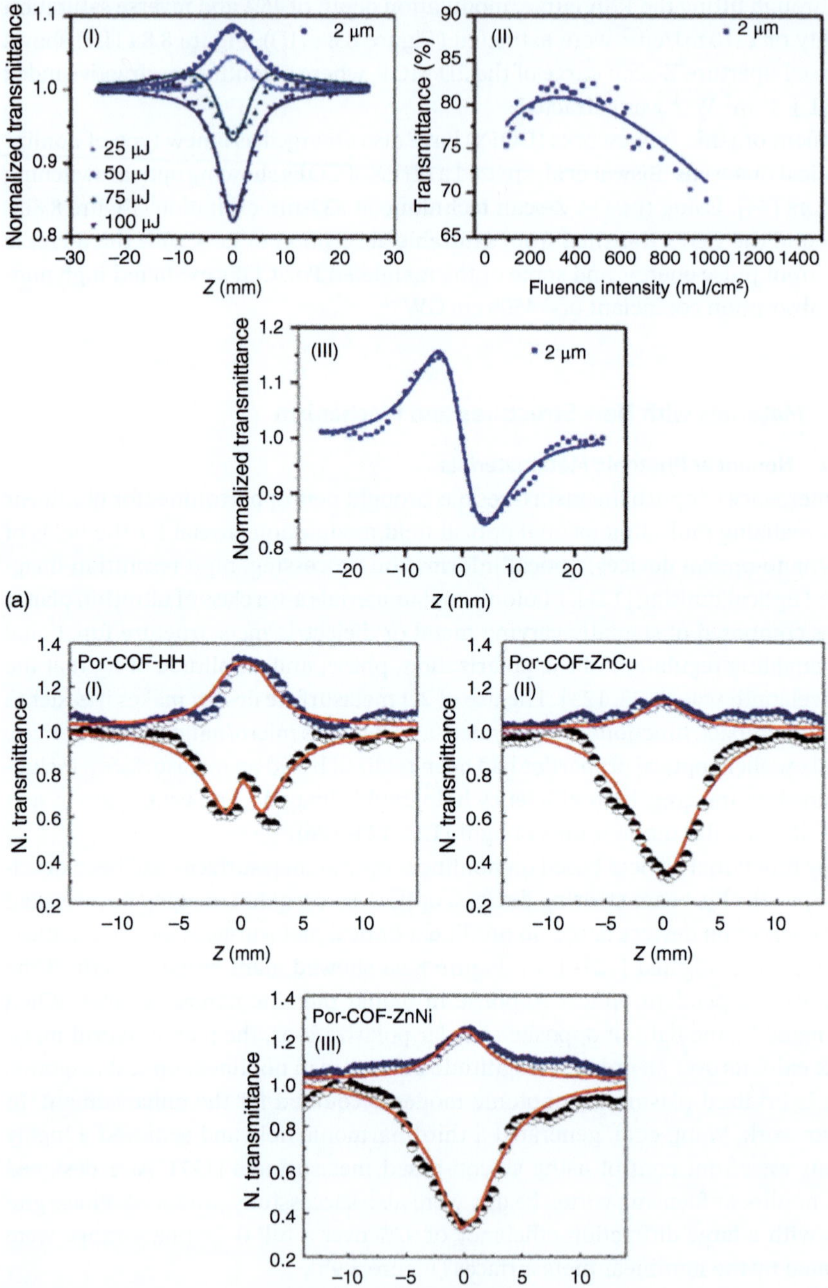

Figure 8.8 (a) OA Z-scan results (I), I-scan results (II), and CA Z-scan results (III) of the YAG-based Mg-MOF-74 nanosheet. Source: Adapted from Liang et al. [120] Copyright ©2021, Optical Society of America. (b) OA Z-scan traces for Por-COFs: (I) Por-COF-HH, (II) Por-COF-ZnCu, and (III) Por-COF-ZnNi at the excitation of 532 nm, ns laser at the input intensity of 0.25 GW/cm^2 (represented by circles) and 0.01 GW/cm^2 (represented by hexagons). Source: Adapted from Biswal et al. [54] Copyright ©2019, WILEY-VCH Verlag GmbH & Co. KGaA, Weinheim.

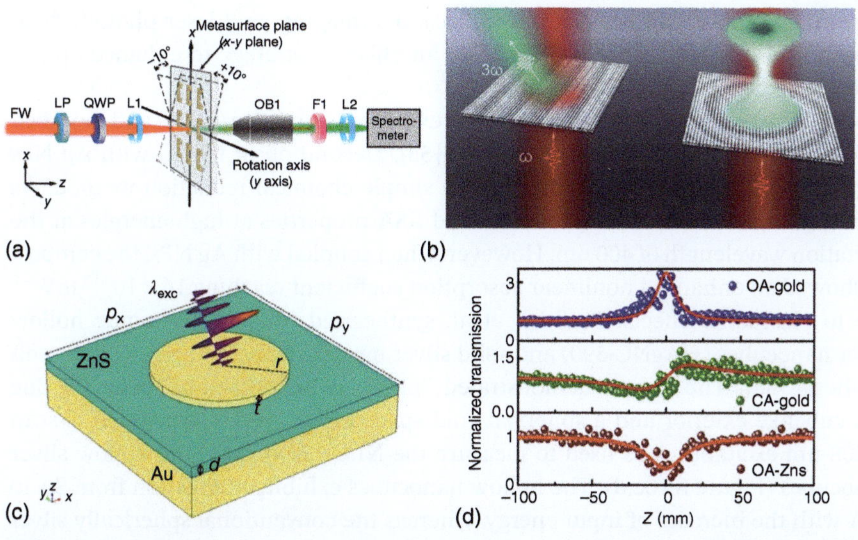

Figure 8.9 (a) Circular dichroism for second harmonic generation waves observed on the gold SRR metasurface. Source: Adapted with permission Chen et al. [131]/Reproduced with permission of John Wiley & Sons. (b) A nonlinear beam deflector and a nonlinear vortex beam generator. Source: Adapted with permission Wang et al. [127]/with permission of American Chemical Society. (c) Schematic picture of the trilayer (Au/ZnS/Au) MRA and (d) OA and CA Z-scan plots of gold and ZnS thin films. Source: Adapted with permission Guddala and Ramakrishna [132] Copyright © 2016, The Optical Society.

nonlinear susceptibility for gold ($\chi_m^{(3)}$) and ZnS thin films ($\chi_{ZnS}^{(3)}$) were found to be $-7.0\times10^{-15}+i3.70\times10^{-15}$ m^2/V^2 and $i7.59\times10^{-19}$ m^2/V^2, respectively.

Howes et al. fabricated a transmissive optical limiter through incorporating phase transition material of VO_2 with a metasurface structure [133]. The metasurfaces included a 190-nm silicon device layer on a SiO_2 wafer, with VO_2 deposited on the top. Then, electron beam lithography (EBL) and reactive ion etching (RIE) were employed to create the resonator structures. Normally, the metasurface allows high and broadband transmittance in "on" state. After VO_2 underwent phase transition from the semiconducting to the metallic state, the metasurface then became "off" state and limited the input light intensity. Additionally, by using the epsilon-near-zero (ENZ) mode of VO_2, the light transmittance was further reduced. The optical limiter thus demonstrated a large turndown ratio with remarkable transmission contrast of 25.2 dB between the on state and off state (on-state transmission was −4.8 dB). The work provided useful insight into realizing intelligent optical modulation based on metasurface architecture.

8.3.3.2 Plasmonic Effect-Enhanced Nonlinearity

Localized surface plasmon resonance (LSPR) stems from the collective oscillation of free electrons when some nanoscale metals are excited by light. As the electric field is strongly enhanced at resonance, neighboring materials usually demonstrate much enhanced optical responses including absorption, emission, Raman scattering, and nonlinear optical properties [134]. The physical phenomena thus find a large range of applications in photovoltaics, bio-imaging, and laser photothermal therapy and have also been harnessed as an effective strategy to enhance optical limiting effects [135, 136].

Bongu et al. prepared hydrogen-induced exfoliated graphene (HEG) and modified it with -OH and -COOH groups (f-HEG) [55]. Decoration of f-HEG with Ag NPs (Ag-f-HEG) was achieved by employing a simple chemical reduction method. As shown in Figure 8.10a and b, both showed RSA properties at high energies at the excitation wavelength of 400 nm. However, when coupled with Ag NPs, the composite showed an enhanced nonlinear absorption coefficient reaching 16×10^{-13} mW^{-1} due to plasmonic effects. Dadhich et al. synthesized tunable plasmonic hollow silver nanocubes (HAgNC-590) and solid silver nanoparticles (SAgNPs-400) by soft synthesis [56]. The former demonstrated "plasmon hybridization" behavior due to a cubelike exterior and a spherical void space inside. Femtosecond OA Z-scan at 808-nm excitation was used to measure the NLO response of the hollow silver nanocubes (Figure 8.10c,d). The hollow nanocubes exhibited transition from SA to RSA with the increase of input energy, whereas the conventional spherically silver NPs only showed RSA behaviors. Additionally, a much larger enhancement of β_{eff} (from 0.0002 to 0.0075 cm GW^{-1}) was observed for the former.

8.3.4 Multifunctional Optical Limiting Devices

With the increasing depletion of the Earth's resources and the excessive impact of human beings on the environment, green, environment-friendly, energy-saving, and low-carbon materials have become a research hotspot. Color-changing materials are new functional materials developed under this situation. The phenomenon of color change comes in many forms, such as photochromism, where a substance undergoes a reversible change between different colors when exposed to light [137]; thermochromism, where a substance changes color when the temperature changes [138]; or piezochromism, where pressure causes a color change, usually a change in the aggregated morphology of the substance [139]. Electrochromism refers to a special phenomenon in which a material undergoes a reversible change between a colored state with low transmittance and an achromatic state with high transmittance by injecting or extracting charges under the action of positive and negative external electric fields [140]. In terms of appearance performance, it manifests as a reversible change in color and transparency. In 1969, Deb first discovered that the amorphous WO_3 film had electrochromic properties, which paved the way of electrochromic research [141]. In the 1970s, a large number of reports on electrochromic mechanisms and inorganic electrochromic materials appeared [142]. Electrochromic devices are optoelectronic devices that combine electrochromic materials and ionic electrolytes on conductive transparent electrodes to form optoelectronic thin films. Reversible color changes can be achieved through a low external driving voltage [143].

Chen et al. designed a Fabry–Perot (F-P) cavity-type electrochromic supercapacitors with versatile color tunability. The basic structure was shown in Figure 8.11

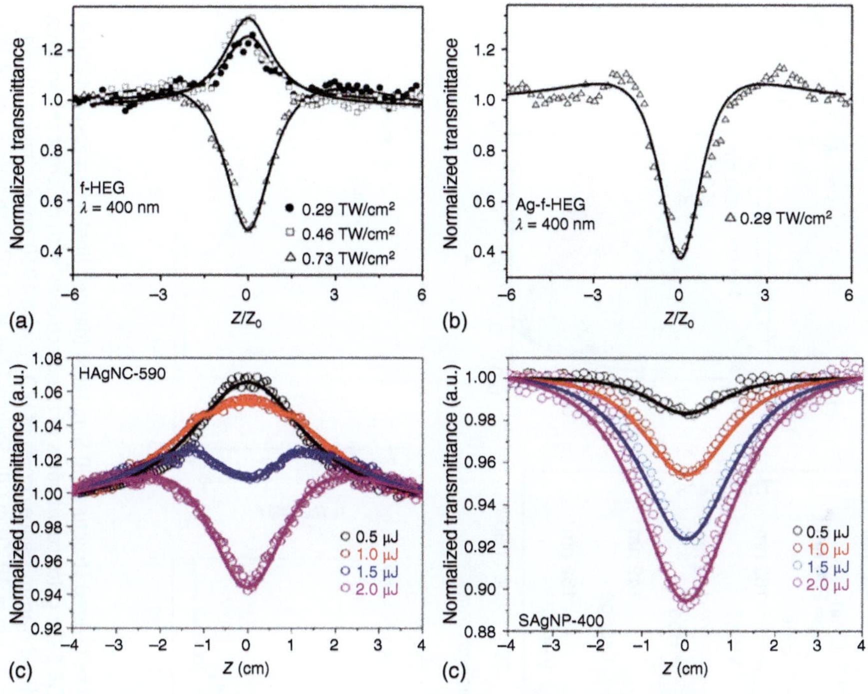

Figure 8.10 (a) and (b) The intensity-dependent Z-scan profiles of graphene (f-HEG) and Ag-f-HEG (graphene/Ag NP composites), respectively. Source: Adapted from Bongu et al. [55] Copyright © 2014, American Institute of Physics. (c) and (d) The fs OA Z-scan for SAgNPs-400 (hollow silver nanocubes) and HAgNC-590 (conventional solid silver nanoparticles) at different input energies, respectively. Source: Adapted from Dadhich et al. [56]. Copyright © 2020, American Chemical Society.

a [144]. These electrochromic supercapacitors not only exhibited high-performance electrochemical properties with high areal capacitance of ~23.4 mF cm^{-2} but also displayed high coloration efficiency and fast switching responses that could be used for intelligent light protection propose.

Electrochromic devices generally are capable of limiting light. Li et al. prepared a WO_{3-x} on a fluorine-doped tin oxide (FTO) substrate by annealing, and the film thickness was about 700 nm [145]. By adjusting the external voltage to change the color of WO_{3-x}, the author used Z-scan techniques to explore its nonlinear properties under different voltages. The experimental results showed that the sign and magnitude of the nonlinear absorption coefficient of WO_{3-x} could be adjusted by applying an external bias voltage. Figure 8.11b indicates the normalized transmittance of WO_{3-x} showed an RSA response at 1030-nm excitation with V_A in the range of 2.5 to −1.5 V. However, when V_A was negative than −2 V, the device showed strong SA behaviors. This work provided a new avenue for light modulation by combining the intrinsic third-order nonlinearity of materials and electrical control to realize multifunctional devices.

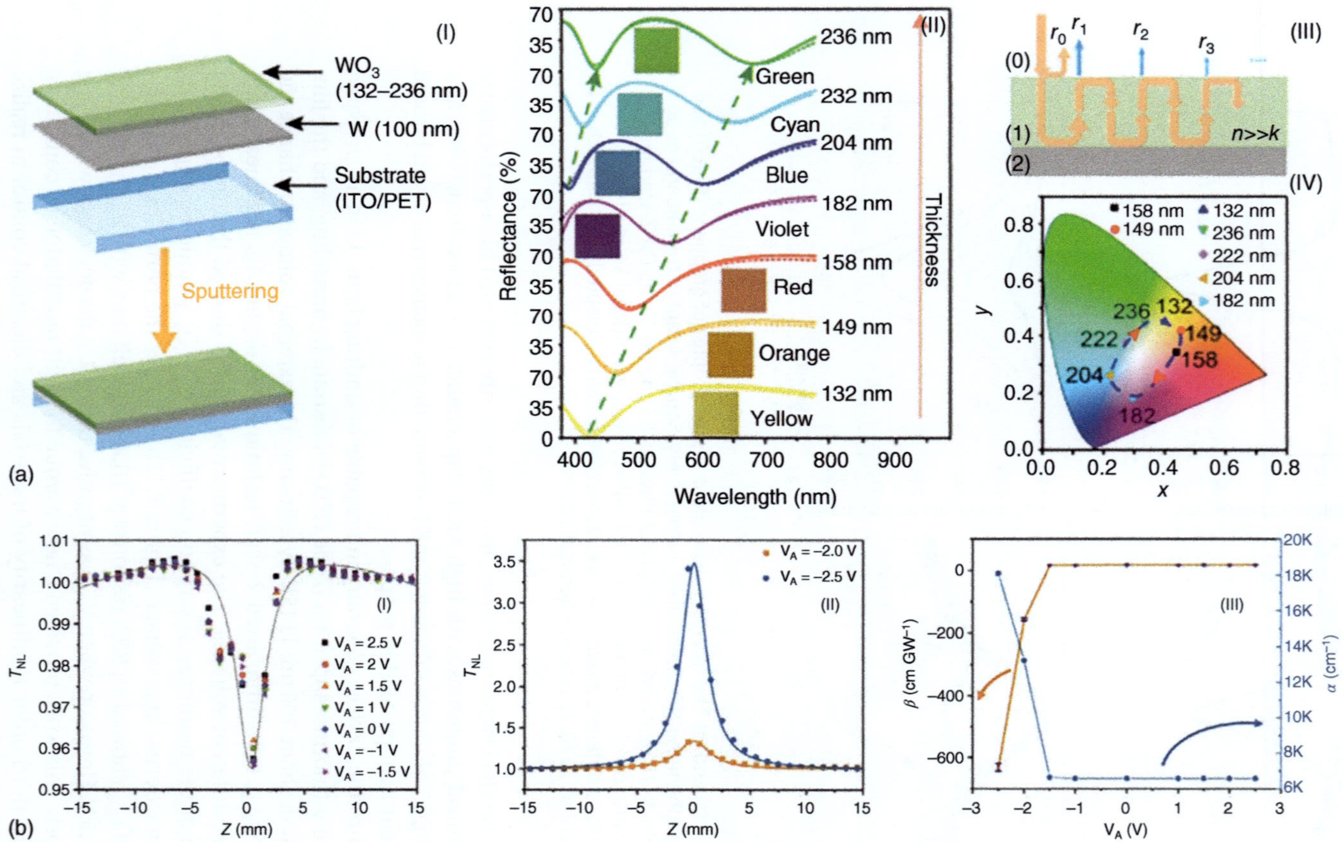

Figure 8.11 (a) The F-P cavity-type electrochromic supercapacitor electrodes. (I) The layered structure. (II) Simulated (dashed line) and measured (solid line) reflection spectra. (III) Light wave reflection processes of the electrodes. (IV) CIE 1931 color coordinates of the electrodes with different thicknesses. Source: Chen et al. [144]/with permission of American Chemical Society. (b) Normalized transmittance curves of WO_{3-x} with 1030-nm excitation (I) and varying voltage of V_A (II); $\beta_{eff} - V_A$ and $\alpha - V_A$ relationships for WO_{3-x} with laser excitation at 1030 nm (III). Source: Li et al. [145]/with permission of John Wiley & Sons.

8.4 Conclusions and Prospects

This chapter summarizes the applications of typical 2D materials in the field of optical limiting, including graphene in the early days and recently emerged novel graphene analogs such as single-elemental 2D materials from Group 15, 2D perovskites, and 2D MOFs and COFs. The following address several key requirements of future 2D materials for laser protection:

(1) Broadband responses. To cope with the development of tunable lasers with output wavelength extending from UV–vis to near IR or even beyond mid IR (>3–5 μm), optical limiting 2D materials should demonstrate broadband responses. Particularly, the operation wavelength covering mid-IR and beyond is of great value for civil and defense fields, such as laser guidance and environment monitoring and laser medicine, as the IR spectra cover two important atmosphere windows (3–5 and 8–12 μm).

(2) Even larger nonlinear nonlinearity. Larger nonlinear parameters determine the optical limiting efficiencies of 2D materials. Currently, the nonlinear absorption coefficients of most 2D materials hardly exceed the order of magnitude $\sim 10^{-6}$ m W^{-1} (at 1064 nm). Light attenuation ratio for most optical limiters based on third-order nonlinearity is less than 10 dB. Consequently, few products have reached high technological readiness level enough for practical use. Hence, screening materials with giant absorption coefficients of β or more generally third-order nonlinear coefficients of $\chi^{(3)}$ has been ever-pursing goals.

(3) Rapid response time. Mechanic optical limiters and switches respond on the time scale of second or millisecond and microsecond to the limit. Considering the development of ultrashort (ns–fs), broadband, and high-peak-power pulsed lasers for versatile use has been the tide in laser industry, optical limiters with response time as fast as femtosecond should be the future trend, which are still scant among the existing 2D materials.

(4) High linear transmittance. To effectively protect optoelectronic sensors or human eyes without interfering their normal function, optical limiters need to be intelligent enough to maintain good photon transmission in the "off" state while blocking the intense harmful input in the "on" state. Hence, optical limiting devices need to have high linear transmittance generally larger than 0.5, which again demands a high nonlinearity of 2D materials for effective strong light attenuation.

(5) High damage threshold. This is required for the optical limiting devices; otherwise, they themselves will be destroyed upon strong laser pulses. The damage threshold of most existing optical limiting materials is less than 10 J cm^{-2} (at 532 nm).

(6) Wide dynamic range and large attenuation ratios. Ideal optical limiters would start to work at a relatively small energy input while maintaining intact at a large energy input. For instance, to protect human eyes with a damage threshold of 0.5 μJ cm^{-2}, the limiter should function at very small energy start. A wide

dynamic range covering 3–4 orders of magnitude from a small onset threshold to a large damage threshold is desired.

(7) Large size. The preparation of large and uniform films of 2D materials is an important step toward practical applications. The development of large-scale film fabrication technology is thus in great need.

Other requisites include low toxicity, lightweight, easy installation, and capability to work in a wide temperature range. Considering the above statements, fundamental innovation is required to further improve the optical limiting performance of 2D materials. Researchers have been endeavoring to enhance third-order nonlinear responses of 2D materials through the atomic manufacturing strategy such as elemental doping, chemical modification at the atomic level, and full utilization of novel physical concepts including plasmonic enhancement, Fermi-level regulation, and metasurfaces. Additionally, new structural designs encompassing comprehensive laser protection methodologies including light reflection, electrochromism, and nonlinear absorption also offer new thoughts for effective optical limiting.

Acknowledgments

This work is supported by the National Natural Science Foundation of China (NSFC.22073038), Key Research and Development Program of Gansu Province (20YF3WA013), and Supercomputing Center of Lanzhou University.

References

1 Danson, C.N., White, M., Barr, J.R.M. et al. (2021). A history of high-power laser research and development in the United Kingdom. *High Power Laser Science and Engineering* 9: e18.

2 Khalighi, M.A. and Uysal, M. (2014). Survey on free space optical communication: a communication theory perspective. *IEEE Communications Surveys and Tutorials* 16: 2231–2258.

3 Richardson, D.J., Fini, J.M., and Nelson, L.E. (2013). Space-division multiplexing in optical fibres. *Nature Photonics* 7: 354–362.

4 Xiao, D.W., Huang, X.J., Cun, Y.K. et al. (2022). Large reversible upconversion luminescence modification and 3D optical information storage in femtosecond laser irradiation-subjected photochromic glass. *Science China Materials* 65: 1586–1593.

5 Yang, F.F., Mou, J., Ma, C.G., and Cao, Y.H. (2020). Dynamic analysis of an improper fractional-order laser chaotic system and its image encryption application. *Optics and Lasers in Engineering* 129: 106031.

6 Elman, M.J., Bressler, N.M., Qin, H.J. et al. (2011). Expanded 2-year follow-up of ranibizumab plus prompt or deferred laser or triamcinolone plus prompt laser for diabetic macular edema. *Ophthalmology* 118: 609–614.

7 Traxer, O. and Keller, E.X. (2020). Thulium fiber laser: the new player for kidney stone treatment? A comparison with Holmium:YAG laser. *World Journal of Urology* 38: 1883–1894.

8 Fleet, L. (2019). Lasers, death rays, and the long, strange quest for the ultimate weapon. *Nature* 565: 158–159.

9 Seet, B. and Wong, T.Y. (2001). Military laser weapons: current controversies. *Ophthalmic Epidemiology* 8: 215–226.

10 Dini, D., Calvete, M.J.F., and Hanack, M. (2016). Nonlinear optical materials for the smart filtering of optical radiation. *Chemical Reviews* 116: 13043–13233.

11 Zhou, G.J. and Wong, W.Y. (2011). Organometallic acetylides of Pt-II, Au-I and Hg-II as new generation optical power limiting materials. *Chemical Society Reviews* 40: 2541–2566.

12 Geng, X.Z., Hasi, W.L.J., Jin, C.Y. et al. (2012). Investigation of optical limiting characteristic based on the combination of stimulated Brillouin scattering and metal-phthalocyanine complex. *Laser and Particle Beams* 30: 17–21.

13 Perumbilavil, S., Sankar, P., Rose, T.P., and Philip, R. (2015). White light z-scan measurements of ultrafast optical nonlinearity in reduced graphene oxide nanosheets in the 400–700 nm region. *Applied Physics Letters* 107: 05110.

14 Barkana, Y. and Belkin, M. (2000). Laser eye injuries, Survey of. *Ophthalmology* 44: 459–478.

15 Parida, M.R., Vijayan, C., Rout, C.S. et al. (2011). Room temperature ferromagnetism and optical limiting in V_2O_5 nanoflowers synthesized by a novel method. *Journal of Physical Chemistry C* 115: 112–117.

16 Maaza, M., Hamidi, D., Simo, A. et al. (2012). Optical limiting in pulsed laser deposited VO_2 nanostructures. *Optics Communications* 285: 1190–1193.

17 Spangler, C.W. (1999). Recent development in the design of organic materials for optical power limiting. *Journal of Materials Chemistry* 9: 2013–2020.

18 Chen, Y., Hanack, M., Araki, Y., and Ito, O. (2005). Axially modified gallium phthalocyanines and naphthalocyanines for optical limiting. *Chemical Society Reviews* 34: 517–529.

19 Wang, J., Chen, Y., and Blau, W.J. (2009). Carbon nanotubes and nanotube composites for nonlinear optical devices. *Journal of Materials Chemistry* 19: 7425–7443.

20 Zhang, S.F., Dong, N.N., McEvoy, N. et al. (2015). Direct observation of degenerate two-photon absorption and its saturation in WS_2 and MoS_2 mono layer and few-layer films. *ACS Nano* 9: 7142–7150.

21 Zhao, M., Chang, M.J., Wang, Q. et al. (2015). Unexpected optical limiting properties from MoS_2 nanosheets modified by a semiconductive polymer. *Chemical, Communications* 51: 12262–12265.

22 Zhao, M., Liu, K., Zhang, Y.D. et al. (2015). Singlet fission induced giant optical limiting responses of pentacene derivatives. *Materials Horizons* 2: 619–624.

23 Chen, Y., Bai, T., Dong, N.N. et al. (2016). Graphene and its derivatives for laser protection. *Progress in Materials Science* 84: 118–157.

24 Zhai, X.P., Ma, B., Wang, Q., and Zhang, H.L. (2020). 2D materials towards ultrafast photonic applications. *Physical Chemistry Chemical Physics* 22: 22140–22156.

25 Li, N., Zhai, X.P., Yan, W.K. et al. (2021). Boosting cascade electron transfer for highly efficient CO_2 photoreduction. *Solar RRL* 5: 2100558.

26 Ma, B., Liu, R.T., Zhang, X.D. et al. (2022). Ultrafast generation of coherent phonons in two-dimensional bismuthene. *The Journal of Physical Chemistry Letters* 3072–3078.

27 Song, T., Ma, Q.X., Wang, Q., and Zhang, H.L. (2022). Design of two-dimensional halide perovskite composites for optoelectronic applications and beyond. *Materials, Advances* 3: 756–778.

28 Bentivegna, F., Canva, M., Georges, P. et al. (1993). Reverse saturable absorption in solid xerogel matrices. *Applied Physics Letters* 62: 1721–1723.

29 Deng, X.X., Zhang, X.R., Wang, Y.X. et al. (1999). Intensity threshold in the conversion from reverse saturable absorption to saturable absorption and its application in optical limiting. *Optics Communications* 168: 207–212.

30 Wei, T.H., Huang, T.H., and Wen, T.C. (1999). Mechanism of reverse saturable absorption in chloro-aluminum phthalocyanine solution studied with z-scan, Chemical. *Physics Letters* 314: 403–410.

31 Shirk, J.S., Pong, R.G.S., Flom, S.R. et al. (2000). Effect of axial substitution on the optical limiting properties of indium phthalocyanines. *Journal of Physical Chemistry A* 104: 1438–1449.

32 Dini, D., Vagin, S., Hanack, M. et al. (2005). Nonlinear optical effects related to saturable and reverse saturable absorption by subphthalocyanines at 532 nm. *Chemical Communications* 3796–3798.

33 Gao, Y.C., Zhang, X.R., Li, Y.L. et al. (2005). Saturable absorption and reverse saturable absorption in platinum nanoparticles. *Optics Communications* 251: 429–433.

34 Chen, Y., Hanack, M., Blau, W.J. et al. (2006). Soluble axially substituted phthalocyanines: synthesis and nonlinear optical response. *Journal of Materials Science* 41: 2169–2185.

35 Elim, H.I., Yang, J., Lee, J.Y. et al. (2006). Observation of saturable and reverse-saturable absorption at longitudinal surface plasmon resonance in gold nanorods. *Applied Physics Letters* 88: 083107.

36 Wang, S.X., Zhang, Y.X., Zhang, R. et al. (2015). High-order nonlinearity of surface plasmon resonance in Au nanoparticles: paradoxical combination of saturable and reverse-saturable absorption. *Advanced, Optical Materials* 3: 1342–1348.

37 Wang, G., Ma, Y.X., Shang, C. et al. (2021). Influence of reverse saturable absorption effect on conventional and dissipative solitons fiber lasers. *Optics and Laser Technology* 137: 106805.

38 Liu, Z., Zhang, B., and Chen, Y. (2019). Recent progress in two-dimensional nanomaterials for laser protection. *Chemistry* 1: 17–43.

39 Tutt, L.W. and Boggess, T.F. (1993). A review of optical limiting mechanisms and devices using organics, fullerenes, semiconductors and other materials. *Progress in Quantum Electronics* 17: 299–338.

40 Boggess, T.F., Bohnert, K.M., Mansour, K. et al. (1986). Simultaneous measurement of the two-photon coefficient and free-carrier cross section above the bandgap of crystalline silicon. *IEEE Journal of Quantum Electronic* 22: 360–368.

41 Sheikbahae, M., Hagan, D.J., and Vanstryland, E.W. (1990). Dispersion and band-gap scaling of the electronic Kerr effect in solids associated with two-photon absorption. *Physical Review Letters* 65: 96–99.

42 Francois, L., Mostafavi, M., Belloni, J. et al. (2000). Optical limitation induced by gold clusters. 1. Size effect. *Journal of Physical Chemistry B* 104: 6133–6137.

43 Wang, S.Y., Zhen, X.J., Feng, Z.Z. et al. (2022). Third-order optical nonlinearities of two-dimensional SnS under irradiation: implications for space use. *Journal of Materials Chemistry C* 10: 18025–18032.

44 Cheng, X., Dong, N., Li, B. et al. (2013). Controllable broadband nonlinear optical response of graphene dispersions by tuning vacuum pressure. *Optics Express* 21: 16486–16493.

45 Liu, R., Hu, J., Zhu, S. et al. (2017). Synergistically enhanced optical limiting property of graphene oxide hybrid materials functionalized with Pt complexes. *ACS Applied Materials & Interfaces* 9: 33029–33040.

46 Liu, Z., Dong, N., Jiang, P. et al. (2018). Reduced graphene oxide chemically modified with aggregation-induced emission polymer for solid-state optical limiter. *Chemistry A European Journal* 24: 19317–19322.

47 Cheng, H.X., Dong, N.N., Bai, T. et al. (2016). Covalent modification of MoS_2 with Poly(N-vinylcarbazole) for solid-state broadband optical limiters. *Chemistry A European Journal* 22: 4500–4507.

48 Shi, M., Huang, S., Dong, N. et al. (2018). Donor-acceptor type blends composed of black phosphorus and C_{60} for solid-state optical limiters. *Chemical Communications* 54: 366–369.

49 Zhang, F., Wu, Z., Wang, Z. et al. (2016). Strong optical limiting behavior discovered in black phosphorus. *RSC Advances* 6: 20027–20033.

50 Unlu, B.A., Karatay, A., Yildiz, E.A. et al. (2021). Defect assisted optical limiting performance of hexagonal boron nitride nanosheets in aqueous suspension and PMMA nanocomposite films. *Optical Materials* 121: 111630.

51 Zhang, F., Wang, M., Wang, Z. et al. (2018). Excellent nonlinear absorption properties of β-antimonene nanosheets. *Journal of Materials Chemistry C* 6: 2848–2853.

52 Zhang, F., Liu, G., Wang, Z. et al. (2019). Broadband nonlinear absorption properties of two-dimensional hexagonal tellurene nanosheets. *Nanoscale* 11: 17058–17064.

53 Xiao, Q., Ma, B., Fei, X. et al. (2021). Unveiling the dimension-dependence of femtosecond nonlinear optical properties of tellurium nanostructures. *Nanoscale Horizons* 6: 918–927.

54 Biswal, B.P., Valligatla, S., Wang, M. et al. (2019). Nonlinear optical switching in regioregular porphyrin covalent organic frameworks. *Angewandte Chemie International Edition in English* 58: 6896–6900.

55 Bongu, S.R., Bisht, P.B., Namboodiri, R.C.K. et al. (2014). Influence of localized surface plasmons on Pauli blocking and optical limiting in graphene under femtosecond pumping. *Journal of Applied Physics* 116: 073101.

56 Dadhich, B.K., Bhattacharya, S., Ballav, S. et al. (2020). Femtosecond-laser-induced saturable absorption and optical limiting of hollow silver nanocubes: implications for optical switching and bioimaging. *ACS Applied Nano Materials* 3: 11620–11629.

57 Lim, C.K., Maldonado, M., Zalesny, R. et al. (2020). Interlayer-sensitized linear and nonlinear photoluminescence of quasi-2D hybrid perovskites using aggregation-induced enhanced emission active organic cation layers. *Advanced Functional Materials* 30: 1909375.

58 Liu, W., Xing, J., Zhao, J. et al. (2017). Giant two-photon absorption and its saturation in 2D organic-inorganic perovskite. *Advanced Optical Materials* 5: 1601045.

59 Gerstner, E. (2010). Nobel Prize 2010: Andre Geim & Konstantin Novoselov. *Nature Physics* 6: 836–836.

60 Wang, J.G., Ma, F.C., and Sun, M.T. (2017). Graphene, hexagonal boron nitride, and their heterostructures: properties and applications. *RSC advances* 7: 16801–16822.

61 Wang, J., Hernandez, Y., Lotya, M. et al. (2009). Broadband nonlinear optical response of graphene dispersions. *Advanced Materials* 21: 2430–2435.

62 Sun, Z., Dong, N., Xie, K. et al. (2013). Nanostructured few-layer graphene with superior optical limiting properties fabricated by a catalytic steam etching process. *Journal of Physical Chemistry C* 117: 11811–11817.

63 Feng, M., Zhan, H., and Chen, Y. (2010). Nonlinear optical and optical limiting properties of graphene families. *Applied Physics Letters* 96: 033107.

64 Girisun, T.C.S., Saravanan, M., and Soma, V.R. (2018). Wavelength-dependent nonlinear optical absorption and broadband optical limiting in Au-Fe_2O_3-rGO nanocomposites. *ACS Applied Nano Materials* 1: 6337–6348.

65 Stavrou, M., Papadakis, I., Bawari, S. et al. (2021). Giant broadband (450–2300 nm) optical limiting and enhancement of the nonlinear optical response of some graphenes by defect engineering. *The Journal of Physical Chemistry C* 125: 16075–16085.

66 Du, Y., Dong, N., Zhang, M. et al. (2017). Covalent functionalization of graphene oxide with porphyrin and porphyrin incorporated polymers for optical limiting. *Physical Chemistry Chemical Physics* 19: 2252–2260.

67 Tan, C.L., Cao, X.H., Wu, X.J. et al. (2017). Recent advances in ultrathin two-dimensional nanomaterials. *Chemical Reviews* 117: 6225–6331.

68 Dong, N., Li, Y., Feng, Y. et al. (2015). Optical limiting and theoretical modelling of layered transition metal dichalcogenide nanosheets. *Scientific Reports* 5: 14646.

69 Liang, G., Tao, L., Tsang, Y.H. et al. (2019). Optical limiting properties of a few-layer MoS_2/PMMA composite under excitation of ultrafast laser pulses. *Journal of Materials Chemistry C* 7: 495–502.

70 Varma, S.J., Kumar, J., Liu, Y. et al. (2017). 2D TiS_2 layers: a superior nonlinear optical limiting material. *Advanced Optical Materials* 5: 1700713.

71 Zhou, K.G., Zhao, M., Chang, M.J. et al. (2015). Size-dependent nonlinear optical properties of atomically thin transition metal dichalcogenide nanosheets. *Small* 11: 694–701.

72 Liu, Y., Li, X., Wang, E. et al. (2021). Exceptional size-dependent property of TiS_2 nanosheets for optical limiting. *Applied Surface Science* 541: 148371.

73 Zu, Y., He, C., Liu, D. et al. (2021). Nonlinear optical properties of poly (vinyl alcohol) thin films doped with in-situ WSe_2/rGO composite. *Optics & Laser Technology* 142: 107198.

74 Fan, F., Zhang, B., Cao, Y., and Chen, Y. (2017). Solution-processable poly(*N*-vinylcarbazole)-covalently grafted MoS_2 nanosheets for nonvolatile rewritable memory devices. *Nanoscale* 9: 2449–2456.

75 Abate, Y., Akinwande, D., Gamage, S. et al. (2018). Recent progress on stability and passivation of black phosphorus. *Advanced Materials* 30: 1704749.

76 Castellanos-Gomez, A., Vicarelli, L., Prada, E. et al. (2014). Isolation and characterization of few-layer black phosphorus. *2D Materials* 1: 025001.

77 Hirsch, A. and Hauke, F. (2018). Post-graphene 2D chemistry: the emerging field of molybdenum disulfide and black phosphorus functionalization. *Angewandte Chemie International Edition* 57: 4338–4354.

78 Woomer, A.H., Farnsworth, T.W., Hu, J. et al. (2015). Phosphorene: synthesis, scale-up, and quantitative optical spectroscopy. *ACS Nano* 9: 8869–8884.

79 Cai, Y., Ke, Q., Zhang, G. et al. (2015). Giant phononic anisotropy and unusual anharmonicity of phosphorene: interlayer coupling and strain engineering. *Advanced Functional Materials* 25: 2230–2236.

80 Huang, J., Dong, N., Zhang, S. et al. (2017). Nonlinear absorption induced transparency and optical limiting of black phosphorus nanosheets. *ACS Photonics* 4: 3063–3070.

81 Zhao, G., Zhang, F., Wu, Y. et al. (2016). One-step exfoliation and hydroxylation of boron nitride nanosheets with enhanced optical limiting performance. *Advanced Optical Materials* 4: 141–146.

82 Chang, C.W., Okawa, D., Majumdar, A., and Zettl, A. (2006). Solid-state thermal rectifier. *Science* 314: 1121–1124.

83 Golberg, D., Costa, P.M.F.J., Lourie, O. et al. (2007). Direct force measurements and kinking under elastic deformation of individual multiwalled boron nitride nanotubes. *Nano Letters* 7: 2146–2151.

84 Chen, Y., Zou, J., Campbell, S.J., and Le Caer, G. (2004). Boron nitride nanotubes: Pronounced resistance to oxidation. *Applied Physics Letters* 84: 2430–2432.

85 Kumbhakar, P., Kole, A.K., Tiwary, C.S. et al. (2015). Nonlinear optical properties and temperature-dependent UV–Vis absorption and photoluminescence emission in 2D hexagonal boron nitride nanosheets. *Advanced Optical Materials* 3: 828–835.

86 Zhang, S., Xie, M., Li, F. et al. (2016). Semiconducting group 15 monolayers: a broad range of band gaps and high carrier mobilities. *Angewandte Chemie International Edition in English* 55: 1666–1669.

87 Wang, X., He, J., Zhou, B. et al. (2018). Bandgap-tunable preparation of smooth and large two-dimensional antimonene. *Angewandte Chemie International Edition in English* 57: 8668–8673.

88 Gu, J., Du, Z., Zhang, C. et al. (2017). Liquid-phase exfoliated metallic antimony nanosheets toward high volumetric sodium storage. *Advanced Energy Materials* 7: 1700447.

89 Gusmao, R., Sofer, Z., Bousa, D., and Pumera, M. (2017). Pnictogen (As, Sb, Bi) nanosheets for electrochemical applications are produced by shear exfoliation using kitchen blenders. *Angewandte Chemie International Edition in English* 56: 14417–14422.

90 Tao, W., Ji, X., Zhu, X. et al. (2018). Two-dimensional antimonene-based photonic nanomedicine for cancer theranostics. *Advanced Materials* 30: e1802061.

91 Singh, D., Gupta, S.K., Sonvane, Y., and Lukačević, I. (2016). Antimonene: a monolayer material for ultraviolet optical nanodevices. *Journal of Materials Chemistry C* 4: 6386–6390.

92 Wang, G., Higgins, S., Wang, K. et al. (2018). Intensity-dependent nonlinear refraction of antimonene dispersions in the visible and near-infrared region. *Applied Optics* 57: E147–E153.

93 Xing, F., Wang, J., Wang, Z. et al. (2021). Covalently silane-functionalized antimonene nanosheets and their copolymerized gel glasses for broadband Vis-NIR optical limiting. *ACS Applied Materials & Interfaces* 13: 897–903.

94 Zhang, Z.T., Yang, Q.Q., Zhen, X.J. et al. (2021). Two-dimensional bismuthene showing radiation-tolerant third-order optical nonlinearities. *ACS Applied Materials & Interfaces* 13: 21626–21634.

95 Yu, Q., Chen, C., Guo, K. et al. (2021). Deterministic transfer of large-scale β-phase arsenic on fiber end cap for near-infrared ultrafast pulse generation. *Frontiers in Materials* 8: 721587.

96 Du, Y., Qiu, G., Wang, Y. et al. (2017). One-dimensional van der Waals material tellurium: Raman spectroscopy under strain and magneto-transport. *Nano Letters* 17: 3965–3973.

97 Wu, L., Huang, W., Wang, Y. et al. (2019). 2D tellurium based high-performance all-optical nonlinear photonic devices. *Advanced Functional Materials* 29: 1806346.

98 Wang, Y., Qiu, G., Wang, R. et al. (2018). Field-effect transistors made from solution-grown two-dimensional tellurene. *Nature Electronics* 1: 228–236.

99 Wang, Q., Safdar, M., Xu, K. et al. (2014). Van der Waals epitaxy and photoresponse of hexagonal tellurium nanoplates on flexible mica sheets. *ACS Nano* 8: 7497–7505.

100 Peng, M., Xie, R.Z., Wang, Z. et al. (2021). Blackbody-sensitive room-temperature infrared photodetectors based on low-dimensional tellurium grown by chemical vapor deposition. *Science Advances* 7: eabf7358.

101 Calavalle, F., Suarez-Rodriguez, M., Martin-Garcia, B. et al. (2022). Gate-tuneable and chirality-dependent charge-to-spin conversion in tellurium nanowires. *Nature Materials* 21: 526–532.

102 Katz, E.A. (2020). Perovskite: name puzzle and German–Russian odyssey of discovery. *Helvetica Chimica Acta* 103: e2000061.

103 Shen, W., Chen, J., Wu, J. et al. (2020). Nonlinear optics in lead halide perovskites: mechanisms and applications. *ACS Photonics* 8: 113–124.

104 Chen, K., Schuenemann, S., Song, S., and Tueysuez, H. (2018). Structural effects on optoelectronic properties of halide perovskites. *Chemical Society Reviews* 47: 7045–7077.

105 Li, P., Chen, Y., Yang, T. et al. (2017). Two-dimensional CH3NH3PbI3 perovskite nanosheets for ultrafast pulsed fiber lasers. *ACS Applied Materials & Interfaces* 9: 12759–12765.

106 Chen, B., Zhang, Q., Shen, X. et al. (2022). Nonlinear optical and optical limiting properties of $MAPbBr_3$/polymethyl methacrylate organic glasses. *The Journal of Physical Chemistry C* 126: 5743–5750.

107 Cheng, P. and Zhan, X. (2016). Stability of organic solar cells: challenges and strategies. *Chemical Society Reviews* 45: 2544–2582.

108 Shi, E., Gao, Y., Finkenauer, B.P. et al. (2018). Two-dimensional halide perovskite nanomaterials and heterostructures. *Chemical Society Reviews* 47: 6046–6072.

109 Walters, G., Sutherland, B.R., Hoogland, S. et al. (2015). Two-photon absorption in organometallic bromide perovskites. *ACS Nano* 9: 9340–9346.

110 Li, N., Xu, J., Feng, R. et al. (2016). Governing metal-organic frameworks towards high stability. *Chemical Communications* 52: 8501–8513.

111 Adil, K., Belmabkhout, Y., Pillai, R.S. et al. (2017). Gas/vapour separation using ultra-microporous metal-organic frameworks: insights into the structure/separation relationship. *Chemical Society Reviews* 46: 3402–3430.

112 Yu, M.H., Space, B., Franz, D. et al. (2019). Enhanced gas uptake in a microporous metal–organic framework via a sorbate induced-fit mechanism. *Journal of the American Chemical Society* 141: 17703–17712.

113 Yin, H.Q., Wang, X.Y., and Yin, X.B. (2019). Rotation restricted emission and antenna effect in single metal–organic frameworks. *Journal of the American Chemical Society* 141: 15166–15173.

114 Wu, M.X. and Yang, Y.W. (2017). Metal-organic framework (MOF)-based drug/cargo delivery and cancer therapy. *Advanced Materials* 29: 1606134.

115 Bosch, M., Yuan, S., Rutledge, W., and Zhou, H.C. (2017). Stepwise synthesis of metal–organic frameworks. *Accounts of Chemical Research* 50: 857–865.

116 Dong, R., Zhang, T., and Feng, X. (2018). Interface-assisted synthesis of 2D materials: trend and challenges. *Chemical Reviews* 118: 6189–6235.

117 Zhang, H., Liu, X., Wu, Y. et al. (2018). MOF-derived nanohybrids for electrocatalysis and energy storage: current status and perspectives. *Chemical Communications* 54: 5268–5288.

118 Feng, D., Lei, T., Lukatskaya, M.R. et al. (2018). Robust and conductive two-dimensional metal-organic frameworks with exceptionally high volumetric and areal capacitance. *Nature Energy* 3: 30–36.

119 Dong, R., Han, P., Arora, H. et al. (2018). High-mobility band-like charge transport in a semiconducting two-dimensional metal-organic framework. *Nature Materials* 17: 1027–1032.

120 Liang, Y., Qiao, W., Feng, T. et al. (2021). Investigation on the optical nonlinearity of the layered magnesium-mediated metal organic framework (Mg-MOF-74). *Optics Express* 29: 23786–23798.

121 Li, G., Zhang, S., and Zentgraf, T. (2017). Nonlinear photonic metasurfaces. *Nature Reviews Materials* 2: 17010.

122 Yu, N., Genevet, P., Kats, M.A. et al. (2011). Light propagation with phase discontinuities: generalized laws of reflection and refraction. *Science* 334: 333–337.

123 Sun, S., He, Q., Xiao, S. et al. (2012). Gradient-index meta-surfaces as a bridge linking propagating waves and surface waves. *Nature Materials* 11: 426–431.

124 Chen, W.T., Zhu, A.Y., Sanjeev, V. et al. (2018). A broadband achromatic metalens for focusing and imaging in the visible. *Nature Nanotechnolgy* 13: 220–226.

125 Zheng, G., Muhlenbernd, H., Kenney, M. et al. (2015). Metasurface holograms reaching 80% efficiency. *Nature Nanotechnolgy* 10: 308–312.

126 Li, L., Liu, Z., Ren, X. et al. (2020). Metalens-array-based high-dimensional and multiphoton quantum source. *Science* 368: 1487–1490.

127 Wang, L., Kruk, S., Koshelev, K. et al. (2018). Nonlinear wavefront control with all-dielectric metasurfaces. *Nano Letters* 18: 3978–3984.

128 Schlickriede, C., Kruk, S.S., Wang, L. et al. (2020). Nonlinear imaging with all-dielectric metasurfaces. *Nano Letters* 20: 4370–4376.

129 Ye, W., Zeuner, F., Li, X. et al. (2016). Spin and wavelength multiplexed nonlinear metasurface holography. *Nature Communications* 7: 11930.

130 Walter, F., Li, G., Meier, C. et al. (2017). Ultrathin nonlinear metasurface for optical image encoding. *Nano Letters* 17: 3171–3175.

131 Chen, S., Reineke, B., Li, G. et al. (2019). Strong nonlinear optical activity induced by lattice surface modes on plasmonic metasurface. *Nano Letters* 19: 6278–6283.

132 Guddala, S. and Ramakrishna, S.A. (2016). Optical limiting by nonlinear tuning of resonance in metamaterial absorbers. *Optical Letters* 41: 5150–5153.

133 Howes, A., Zhu, Z., Curie, D. et al. (2020). Optical limiting based on huygens' metasurfaces. *Nano Letters* 20: 4638–4644.

134 Luther, J.M., Jain, P.K., Ewers, T., and Alivisatos, A.P. (2011). Localized surface plasmon resonances arising from free carriers in doped quantum dots. *Nature Materials* 10: 361–366.

135 Atwater, H.A. and Polman, A. (2010). Plasmonics for improved photovoltaic devices. *Nature Materials* 9: 205–213.

136 Jain, P.K., Huang, X., El-Sayed, I.H., and El-Sayed, M.A. (2008). Noble metals on the nanoscale: optical and photothermal properties and some applications in imaging, sensing, biology, and medicine. *Accounts of Chemical Research* 41: 1578–1586.

137 Wang, L. and Li, Q. (2018). Photochromism into nanosystems: towards lighting up the future nanoworld. *Chemical Society Reviews* 47: 1044–1097.

138 Seeboth, A., Lotzsch, D., Ruhmann, R., and Muehling, O. (2014). Thermochromic polymers--function by design. *Chemical Reviews* 114: 3037–3068.

139 Hendon, C.H., Wittering, K.E., Chen, T.H. et al. (2015). Absorbate-induced piezochromism in a porous molecular crystal. *Nano Letters* 15: 2149–2154.

140 Zhang, W., Li, H., Yu, W.W., and Elezzabi, A.Y. (2020). Transparent inorganic multicolour displays enabled by zinc-based electrochromic devices. *Light-Science & Applications* 9: 121.

141 Deb, S.K. (1969). A novel electrophotographic system. *Applied Optics* 8: 192–195.

142 Colton, R.J., Guzman, A.M., and Rabalais, J.W. (1978). Photochromism and electrochromism in amorphous transition metal oxide films. *Accounts of Chemical Research* 11: 170–176.

143 Yang, G., Zhang, Y.M., Cai, Y. et al. (2020). Advances in nanomaterials for electrochromic devices. *Chemical Society Reviews* 49: 8687–8720.

144 Chen, J., Wang, Z., Chen, Z. et al. (2020). Fabry-perot cavity-type electrochromic supercapacitors with exceptionally versatile color tunability. *Nano Letters* 20: 1915–1922.

145 Li, H., Hou, R., Sun, Y. et al. (2021). Switching the nonparametric optical nonlinearity of tungsten oxide by electrical modulation. *Advanced Optical Materials* 9: 2002188.

9

The Saturable Absorbers Based on 2D Materials

Xin-Hai Yan[1,2], Lei Zhang[3,4], and Kai-Ge Zhou[1,2]

[1] *Tianjin University, Institute of Molecular Plus, Department of Chemistry, No. 92 Weijin Road, Tianjin 300110, China*
[2] *Haihe Laboratory of Sustainable Chemical Transformations, 116 Chengdu Road, Tianjin 300192, China*
[3] *Lanzhou University, State Key Laboratory of Applied Organic Chemistry (SKLAOC), College of Chemistry and Chemical Engineering, 222 Tianshui South Road, Lanzhou 730000, China*
[4] *Lanzhou University, Office of Laboratory and Equipment Management, 222 Tianshui South Road, Lanzhou 730000, China*

9.1 Introduction

According to the Pauli exclusion principle, no two electrons would show an identical quantum state. In linear absorption, the extinction coefficient is independent of the incident light intensity as a result. However, when a large number of photons are irradiated onto a substance in a short time, all electrons at the valence band would be pumped to the conduction band, leaving insufficient time for the exciting state to decay. As a consequence, the subsequent photons are unable to further excite the electrons, showing a significant decrease in the extinction coefficient as the light intensity increases [1]. Therefore, the materials with such saturable adsorption (SA) phenomenon become more "transparent" when the intensity of incident light escalates while maintaining opaque at normal states.

Recently, many two-dimensional (2D) materials, including graphene, black phosphorus (BP), germanene, antimonene, bismuthene, transition metal dichalcogenides (TMDs), topological insulators (TIs), MXenes, Metal–organic frameworks (MOFs), and 2D Perovskites, have attracted increasing attention in the area of SA materials [2–10]. Owing to their tunable electric band structure, the family of 2D materials has exhibited abnormal SA phenomenon in broadband spectral region, e.g. large modulation depth, giant NLO coefficient, low saturation intensity, and large third-order nonlinear optics (NLO) susceptibility. Taking advantage of the extraordinary SA properties, the saturable absorbers based on 2D materials have shown great potential in the design of nonlinear optical devices. For instance, by employing 2D materials, the Q-switched component in the laser cavity can generate intense laser pulses (nJ to mJ) with a width ranging from nano to microsecond [11, 12]. In fiber or solid laser devices, the 2D SA component has been used as a mode-locked modulator in order to generate femtosecond laser [13, 14]. In addition,

Two-Dimensional Materials for Nonlinear Optics: Fundamentals, Preparation Methods, and Applications, First Edition. Edited by Qiang Wang and Hao-Li Zhang.

the SA mirror based on 2D materials has also been applied in the construction of nonlinear optical devices.

In this chapter, we mainly focus on the fundaments and applications of the 2D SA materials. First, a theoretical model describing SA phenomena in 2D materials is presented. Then, the development of 2D SA families is summarized, followed by the practical applications in NLO based on 2D materials. Finally, a roadmap toward the new avenue for the future development of 2D SA materials is given.

9.2 The Fundaments in the 2D Materials-Based Saturable Absorbers

9.2.1 SA Theory

To discuss the SA phenomenon in semiconductors, a two-level system is introduced to describe the light-matter interaction under irradiation with high intensity (I). When the incident photon energy ($h\nu$) is equal to the energy band gap, two processes exist simultaneously: the excitation from the ground states and the relaxation from excited states. To describe both processes, the following equation is employed:

$$\frac{dn_2}{dt} = -\frac{dn_1}{dt} = \frac{\sigma_0}{h\upsilon}I(n_1 - n_2) - \frac{n_2}{\tau_{21}} \tag{9.1}$$

where n_1 and n_2 represent the ground- and excited-state particle number densities, respectively, σ_0 equals to the absorption cross section from the ground state to the excited state, and τ_{21} refers to the relaxation time from the excited state back to the ground state.

When the laser pulse width (τ_L) is far beyond the relaxation time (τ_{21}), the system reaches a steady state, and Eq. (9.1) equals zero ($\frac{dn_2}{dt} = -\frac{dn_1}{dt} = 0$) as a result. Consequently, the particle number density difference between the two energy levels can be expressed as:

$$\Delta n = n_1 - n_2 = \frac{N}{1 + \frac{I}{I_s}} \tag{9.2}$$

where N refers to the total particle number density and I refers to the SA intensity, I_s can be described as:

$$I_s = \frac{h\upsilon}{2\sigma_0\tau_{21}} \tag{9.3}$$

Meanwhile, the linear and total absorption coefficients can be expressed as:

$$\alpha_0 = \sigma_0 N \tag{9.4}$$

$$\alpha = \sigma_0 \Delta n \tag{9.5}$$

By combining Eq. (9.2) with (9.4) and (9.5), the total absorption coefficient can be transformed into the following equation:

$$\alpha = \frac{\alpha_0}{1 + \frac{I}{I_s}} \tag{9.6}$$

To characterize the SA behavior of 2D materials, z-scan or I-scan technique can be employed to measure the transmitted light as a function of the incident light. More details can be found in the above chapter of "Novel nonlinear optical characterization techniques for 2D materials."

9.2.2 Slow- and Fast-Saturable Absorber

The above model based on a two-energy level system is suitable to describe the SA phenomenon under continuous-wave laser irradiation. However, in practical applications, e.g. mod-locked and Q-switched pulse laser, the relationship between the relaxation time of excited carriers and laser pulse duration should be considered. According to the difference between the width of the laser pulse and the relaxation time, SA materials can be categorized into fast and slow saturable absorbers, which can be described by the Frantz–Nodvik model and the steady-state solution of Hercher's rate equations, respectively [15]. Both slow- and fast-SA have been found in 2D nanomaterials [14]. Burshtein et al. provided a method to distinguish the slow- or fast-saturable absorber by analyzing the energy state population and photon density kinetics in Cr^{4+} doped Garnet crystals [15]. In addition, the pump–probe techniques can also be used to identify the slow- and fast-saturable absorber (see more details in chapter :Novel nonlinear optical characterization techniques for 2D materials").

In order to analyze the energy state population and photon density kinetics in saturable adsorber, the rate equation was employed. As shown in Figure 9.1a, a beam of light with photon density $E(z, t)$ propagates inside the material with some photons absorbed by the atoms. In nonlinear optical materials, both excited- and ground-states can absorb a photon, and their absorption cross sections are denoted as σ_{gs} and σ_{es}, respectively (Figure 9.1b). Meanwhile, the electron in the excited state also relaxes back to the ground state within a short time denoted as τ. Since the photons are absorbed by both the ground and excited states within a certain volume $\Delta z \cdot A$, the rate equation can be written as:

$$[E(z,t) - E(z,t+\Delta t)] \cdot \Delta z \cdot A + \Delta z \cdot A \cdot [E(z,t) - E(z+\Delta z,t)] =$$
$$(\sigma_{gs} \cdot \Delta z) \cdot N(z,t) \cdot [E(z,t) \cdot \Delta z \cdot A] + (\sigma_{es} \cdot \Delta z) \cdot [N_T - N(z,t)] \cdot [E(z,t) \cdot \Delta z \cdot A] \tag{9.7}$$

where c is the light speed, z is the propagation distance in the materials, Δz and Δt refer to the light travel distance and time, respectively. $N(z, t)$ is the density of ground-state atoms, and N_T is the total absorption atoms density. A is the interacting area of the light and material. By assuming that the photon energy absorbed by exciting states dissipates instantaneously, Eq. (9.7) can be further simplified as [15]:

$$\frac{\partial}{\partial t}E(z,t) + c\frac{\partial}{\partial z}E(z,t) = -\Delta\sigma c N(z,t)E(z,t) - \sigma_{es} c N_T E(z,t) \tag{9.8}$$

where $\Delta\sigma$ equals the difference of cross section between the ground (σ_{gs}) and excited states (σ_{es}). Considering the relaxation from the excited state to the ground state, the

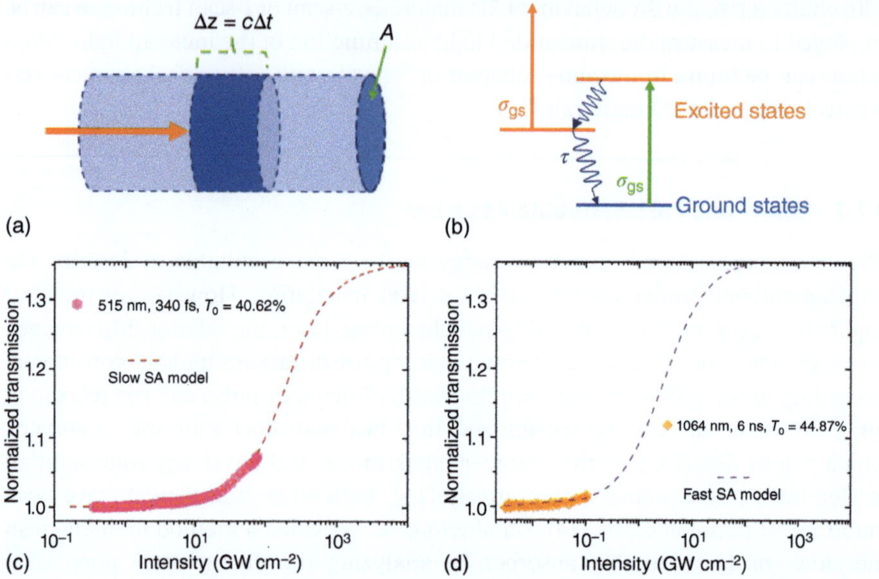

Figure 9.1 (a) Schematic diagram of a light beam traveling through the medium within the certain time element Δt. (b) Schematic diagram of the excitation and relaxation of the carriers. (c) and (d) Normalized nonlinear transmission of (c) slow- and (d) fast-SA materials as a function of the input laser intensity. Source: Adapted from Feng et al. [16] Copyright 2015, Optical Society of America.

ground states density can be expressed as:

$$\frac{\partial}{\partial t}N(z,t) = -\sigma_{gs}cN(z,t) + \frac{N_T - N(z,t)}{\tau} \tag{9.9}$$

The second term on the right side of Eqs. (9.8) and (9.9) indicates the photon absorbed by the exciting states and the relaxation back to ground states.

9.2.2.1 Slow-Saturable Absorber

In a slow saturable absorber, the light pulse duration is much shorter than the lifetime of the excited carriers, and then the second term of Eq. (9.9) can be neglected. For an ideal slow saturable absorber with $\sigma_{es} \equiv 0$, the energy transmittance, T_{FN}, of an ideal slow saturable absorber with a thickness of L can be expressed by solving Eqs. (9.8) and (9.9).

$$T_{FN} \equiv \frac{E(L)}{E(0)} = \frac{\ln\left[1 + T_0\left(e^{\sigma_{gs}E(0)-1}\right)\right]}{[\sigma_{gs}E(0)]}, T_0 = e^{-\sigma_{gs}NL} \tag{9.10}$$

where T_0 is the linear transmission of the absorber. This expression is known as the Frantz–Nodvik equation. For a practical saturable absorber, $\sigma_{es} \neq 0$ and transmissions obtained from the rate equations are given as:

$$T(F) = T_0 + \frac{T_{FN} - T_0}{1 - T_0}(T_{max} - T_0), \quad F = \hbar\omega E, T_{max} = e^{-\sigma_{es}NL} \tag{9.11}$$

where T_{max} is the maximum transmission achieved by the absorber under extremely high-intensity light irradiation.

The slow SA model can be used to obtain key parameters, such as saturation intensity, maximum transmittance, ground, and excited state absorption cross sections, NLO modulation depth, and unsaturable loss. As seen in Eqs. (9.10) and (9.11), the thickness and absorption cross section of the sample play a critical role in both NLO loss and modulation depth (Figure 9.1c).

9.2.2.2 Fast-Saturable Absorber

Compared to the slow-saturable absorber, the spontaneous decay of the excited state in a fast-saturable absorber occurs in a short time. Therefore, the second term on the right side of Eq. (9.9) cannot be neglected. Therefore, the lifetime of excited state τ has to be taken into account, and then Eq. (9.9) can be expressed as follows:

$$\frac{\partial}{\partial t}N(z) = -\sigma_{gs}N(z)I(z) + \frac{N_{\text{total}} - N(z)}{\tau} = 0 \tag{9.12}$$

where $I(z)$ is the photon flux density (photons per unit time per unit area at distance z. By combing Eq. (9.12) with Eq. (9.7), then we get:

$$\frac{\mathrm{d}I(z)}{\mathrm{d}z} = -I(z)\left\{\frac{N_{\text{total}}}{1+\sigma_{gs}\tau I(z)}\Delta\sigma + N_T\sigma_{es}\right\} \tag{9.13}$$

$$T\left[\frac{I(L)+S}{I(0)+S}\right]^D = T_0,\ \ S = \frac{1}{\tau\sigma_{gs}}, D = \frac{\Delta\sigma}{\sigma_{gs}} \tag{9.14}$$

where $I(0)$ refers to the incident intensity, S indicates the ground-state absorption (GSA), and D equals the difference of cross section between the ground states (σ_{gs}) and excited states (σ_{es}) divided by the ground states cross section (σ_{gs}).

The above expression is known as the steady-state solution of Hercher's rate equations, which can well explain the SA processes for an instantaneous transmission when the lifetime of the excited state is sufficiently fast compared to the incident pulse. By analyzing the transmittance against the incident light intensity with Eqs. (9.13) and (9.14), the corresponding parameters, such as saturation intensity, maximum transmittance, ground, and excited state absorption cross sections, NLO modulation depth, and unsaturable loss in fast-saturable absorbers can be obtained (Figure 9.1d).

9.3 The Family of 2D Material-Based Saturable Absorbers

Only four years after Maiman's demonstration of laser operation [17], saturable absorbers were used to generate intense laser pulse: SA-based pulse generation was first demonstrated in 1964, using both a dye [18] and a colored glass [19] filter for the Q-switch of a ruby laser. Since then, the development of saturable absorbers has aided in the innovation of lasers.

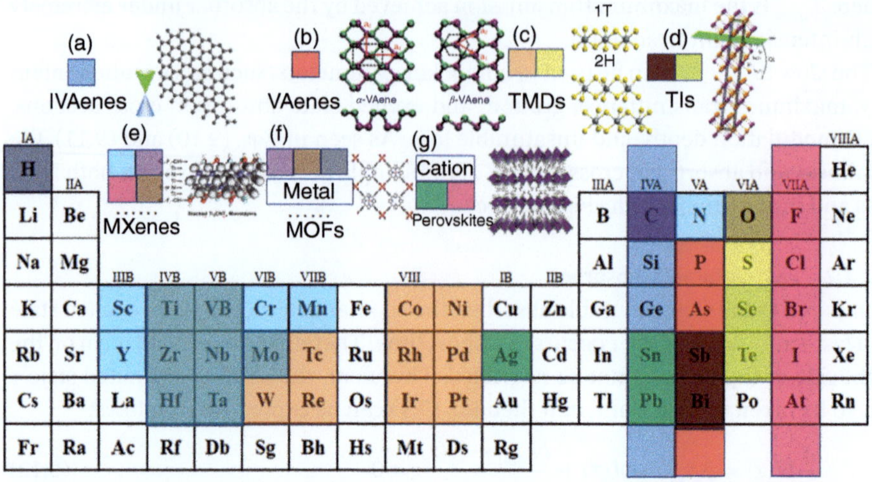

Figure 9.2 Elemental composition of 2D nanomaterials with saturable absorption properties. (a) IVAenes: graphene, silicene, germanene. (b) VAenes: phosphorene, arsenene, antimonene, bismuthene. (c) TMDs: MoS_2, WS_2, $MoSe_2$, $MoSe_2$. (d) Topological insulators (TIs): Sb_2Te_3, Bi_2Te_3, Bi_2Se_3. (e) MXenes. (f) MOFs. (g) 2D perovskites.

Since Sir Andre Geim and Kostya Novoselov successfully separated graphene from graphite in 2004 [20], the intriguing light-matter interactions of 2D materials have been extensively investigated. Many 2D materials have shown excellent SA behavior, such as low saturation intensity (I_s), high nonlinear coefficient, large modulation depth, and wide spectral region. An abundant variety of 2D materials provides various band structures, leading to a broadband response region from the near UV to the microwave region. In Figure 9.2, we collect the present 2D saturable absorbers species and their elemental composition in the periodic table of elements. Following that, their SA performance is introduced.

9.3.1 Mono-Elemental 2D Materials

9.3.1.1 Group IVA-Elemental 2D Materials

Graphene Graphene is a honeycomb lattice constructed by a single atomic layer of sp^2-hybridized carbon. Its optical and electronic properties can be described by massless Dirac fermions with linear dispersion near the Fermi energy [2, 21]. Figure 9.3a shows excitation processes responsible for the absorption of light in a graphene monolayer. Noting that the interband transition is frequency independent, making graphene responsive to an ultra-broadband incident light ranging from ultraviolet to terahertz. Shortly after photoexcitation (within 10–150 fs), these hot electrons can be thermalized and cooled down to form a hot Fermi–Dirac distribution (Figure 9.1b) with an electronic temperature (T_e) generated. At very high excitation intensities, the concentration of photogenerated carriers increased and then filled the states near the edges of the conduction and valence bands. Due to the Pauli exclusion principle, further absorption was blocked and gave transparency to the light with photon energy just above the band edge [2, 21].

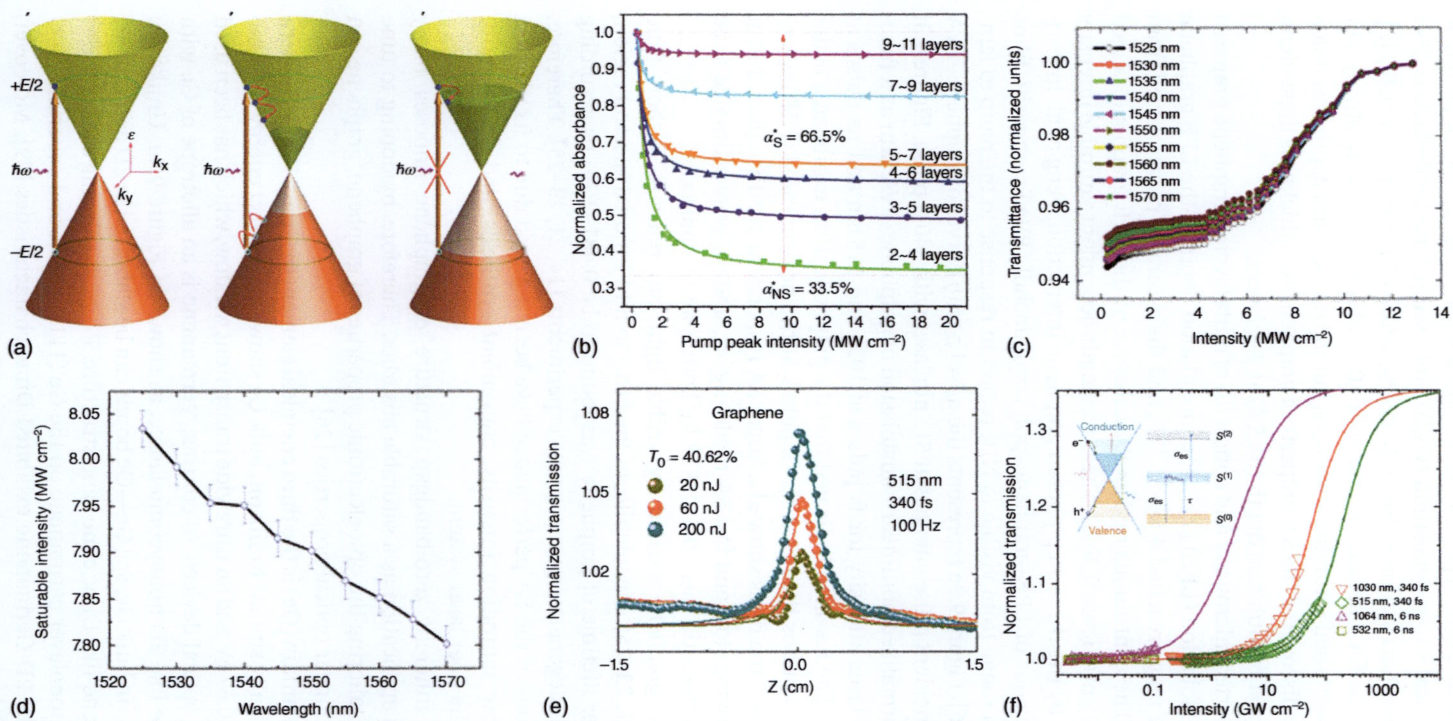

Figure 9.3 (a) Schematic diagram of the light Absorption in graphene. (b) Nonlinear absorption of graphene films with a different number of layers. Source: Bao et al. [2]/with permission of John Wiley & Sons. (c) Wavelength-dependent nonlinear transmission of graphene. (d) Wavelength-dependent saturable intensity curve of graphene. Source: Adapted with permission Zheng et al. [22] © The Optical Society. (e) The *z*-scan curves of graphene–PVA film at different incident pulse energies at 515 nm fs laser pulse. (f) Normalized nonlinear transmission of graphene–PVA film as a function of the input laser intensity for fs pulses at 1030 and 515 nm and ns pulses at 1064 and 532 nm. Source: (e) and (f) Adapted with permission Feng et al. [16] © The Optical Society.

Bao et al. for the first time reported the SA properties in atomic-layered graphene prepared by chemical vapor deposition (CVD) and used it for a mode-locked fiber to generate an ultrashort soliton pulse with a pulse width of 756 fs (Figure 9.3b) [2]. They found that the SA intensity of graphene (0.71 MW cm^{-2}) was relatively low compared to conventional saturable absorbers, but the modulation depth was much greater. They also discovered the SA property of graphene was thickness dependent, reaching the highest modulation depth (66.5%) at 2–4 layers.

Zheng et al. reported microwave and optical SA of multilayered graphene prepared by liquid phase exfoliation (LPE) [22]. The modulation depth of the LPE graphene in the microwave range reached 4.58–12.77%, and the saturable power varied from 28.8 to 79.6 μW. The continuous optical SA behavior of the LPE graphene ranged from 1525 to 1570 nm (Figure 9.3c,d). The wavelength-dependent saturable intensity can be explained by the three-step photon graphene interaction. Feng et al. investigated the SA behavior of LPE–graphene–polyvinyl alcohol (PVA) composite films over a broadband range from visible to NIR region in response to the nano to femtosecond laser [16]. Figure 9.3e represents the novel z-scan curves of graphene–PVA film at different incident pulse energies at 515 nm laser with 340 fs pulse. Figure 9.3f represents the normalized nonlinear transmission of graphene–PVA film as a function of the input laser intensity for fs pulses at 1030 and 515 nm and ns pulses at 1064 and 532 nm. The research indicated that the graphene films exhibited a better SA response for ns pulses than that for fs pulses at the near-infrared (NIR) wavelength. For fs pulses, the films showed a larger SA response at 1030 nm than that at 515 nm. By employing slow and fast SA modeling, σ_{es} and σ_{gs} were estimated to be ~10–17 cm^2, and the ratio was ~0.6 at NIR for both fs and ns pulses.

The carriers of graphene excited by visible light can relax in a short time (0.4–11 ps) [14, 21, 23–25]. Meanwhile, in the near- to mid-infrared spectral region, the excited carrier lifetime of graphene was found to be 0.481–2.291 ps according to the wavelength-degenerate pump–probe experiments [14, 21, 23–25]. Therefore, various applications of the SA performance have been reported due to its ultrafast recovery time, low saturation intensity, convenient synthesis, and flexibility of integration into the fiber laser system.

However, the inherent zero-bandgap structure of graphene imposes many limitations on its application as a saturable absorber. Therefore, by doping to tune the bandgap and tailoring the optoelectronic properties of graphene, graphene can exhibit better nonlinear optical properties [26].

Germanene Germanium (Ge) is the third chemical element in Group IVA of the periodic table. Chemically similar to silicon, bulk Ge shows an indirect energy bandgap of approximately 0.67 eV and an ultrawide transparent window, which has been used in electronic and optical devices. In contrast, germanene is an allotrope of Ge with buckled graphene-like 2D honeycomb lattice, as shown in Figure 9.4a. Unlike the sp^3 Ge—Ge bonds in bulk Ge, sp^2 Ge—Ge bonds can be stabilized in 2D germanene, implying a graphene-like Dirac cone band structure. Recently, Zhuang et al. successfully fabricated monolayer germanene on the Ge (111) surface and experimentally confirmed that the 2D Germanene exhibited Dirac characteristics [29]. Moreover, the Angle-resolved photoemission spectroscopy (ARPES) technique was used to

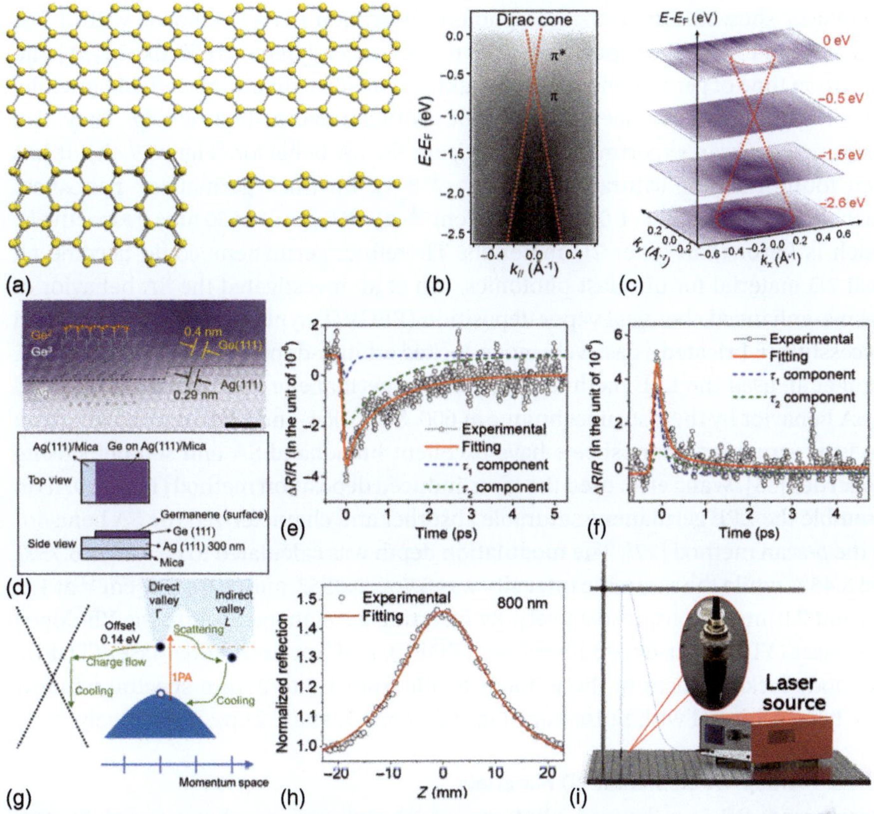

Figure 9.4 (a) The buckled honeycomb lattice of germanene nanosheet and its top and side views. (b) Energy–k dispersion obtained by the ARPES technique showing two faint linear dispersed bands cross at the Brillouin zone (BZ) center Γ point. (c) Equidistant constant energy contours demonstrating Dirac point of about 0.5 eV (error: ±0.1 eV) below the Fermi level. (d) Schematic of the germanene/Ge(111)/Ag samples and its high-resolution HAADF STEM image. (e) and (f) Transient reflective dynamics of Germanene nanosheets with a thickness of 1 nm (e) and 8–12 nm (f) pumped and probed at 800 nm. (g) Schematic energy diagram showing the optical process in the germanium sample. (h) z-scan data of germanene nanosheets with a thickness of 8–12 nm (corresponding to 20–36 l germanene) at 800 nm. (i) Experimental device diagram of laser-induced deposition method. Source: (a) and (i) were reproduced with permission from Wang et al. [27], © 2022, Walter de Gruyter GmbH/CC BY 4.0. (b)–(h) were reproduced with permission Mu et al. [28]. © 2021, John Wiley & Sons, Inc.

characterize the band structure of 2D germanene (Figure 9.4b,c) [28]. Inspired by the theoretical prediction of superior NLO in germanene [30], Mu et al. first reported the NLO properties of germanene [28]. They used a PVD technique to deposit the atomic germanene layer on Ge (111) on a 300 nm silver film supported by a mica substrate (Figure 9.4d). They investigated the carrier dynamics of germanene monolayer and a few layered germanium nanosheets by pump–probe technique. As shown in Figure 9.4e, f, the typical SA phenomenon was observed in an 8–12 nm thick nanosheet sample with a fast relaxation time in 128 fs. Meanwhile, the germanene monolayer showed a typical three-photon absorption (TPA) behavior with a fast relaxation in 135 fs. Therefore, compared to the carrier dynamics of bulk Ge [31], the relaxation time of germanene is faster, making it an ideal quantum well. Meanwhile, the few-layered Ge can act as a reservoir, as illustrated in Figure 9.4g. They also performed a z-scan experiment to investigate the SA behavior (Figure 9.4h). It has been found that the saturable intensity of 8–12 nm thick germanene nanosheet reached 16.4 ± 0.2 and $1.04 \pm 0.08\,GW\,cm^{-2}$ at 800 and 1550 nm, respectively, which is lower than other 2D materials. Therefore, germanene could become an ideal 2D material for ultrafast photonics. Lin et al. investigated the SA behavior of Plasma-enhanced chemical vapor deposition (PECVD) synthesized germanene and successfully fabricated a passively mode-locked erbium-doped fiber (EDF) laser [32]. Zhou et al. used the LPE method to synthesize germanene nanosheets and studied its SA behavior by the z-scan technique at 600, 800, 1064, and 1300 nm, showing that the LPE germanene nanosheets have excellent broadband SA and self-defocusing properties [33]. Wang et al. used the laser-induced deposition method (Figure 9.4i) to assemble the LPE germanene saturable absorber and characterized the SA behavior by the p-scan method [27]. The modulation depth was calculated to be 4.19%, 6.75%, and 8.45%, while the saturable intensity was 2.702, 0.0552, and $0.074\,GW\,cm^{-2}$ at 1.0, 1.5, and 2.0 μm regions, respectively. By inserting germanene-SA into the Yb-doped fiber laser (YDFL), Er-doped fiber laser (EDFL), and Tm-doped fiber laser (TDFL), the mode-locked pulses in three major bands from 1.0 to 2.0 μm spectral window have been achieved with pulse durations of 79.6, 0.7, and 1.21 ps, respectively.

9.3.1.2 Group VA-Elemental 2D Materials

Phosphorene BP is a layered allotrope of phosphorus, weakly bonded by van der Waals forces. Monolayer BP, also known as phosphorene, has a puckered honeycomb structure, and each phosphorus atom in the layer is connected with three neighboring phosphorus atoms by covalent bonds, as shown in Figure 9.5a,c [25, 36]. The BP is a semiconductor with a bandgap ranging from 0.3 to 2.0 eV, depending on the number of layers. The monolayer BP shows a direct bandgap of 2.0 eV, and as the number of layers increases, the bandgap turns to be indirect and gets smaller until it finally reaches 0.3 eV for bulk BP [39, 40]. Zhang et al. characterized the NLO behavior of BP dispersion by the z-scan technique [41]. Typical SA curves were observed under excitation with an 800-nm laser. A mode-locked Ti: sapphire oscillator-seeded regenerative amplifier was combined with the as-prepared BP nanosheets to generate femtosecond pulses with a peak wavelength at 800 nm, and the pulse width was narrowed down to 100 fs with a repetition rate of 1 kHz. In 2015, Lu et al. investigated the broadband NLO properties of LPE ultrathin multilayer BP by the z-scan technique [42]. The broadband SA properties have been found in layered BP from the visible (400 nm) to mid-IR (1930 nm) region. The nonlinear optical properties and the carrier relaxation dynamics of BP suspension have been reported by Wang et al. By using the pump–probe technique, they observed the ultrafast recovery time (24 fs) of BP while the exciting wavelength was 1550 nm (Figure 9.5f) [36]. Zhang et al. found that the slow SA model was

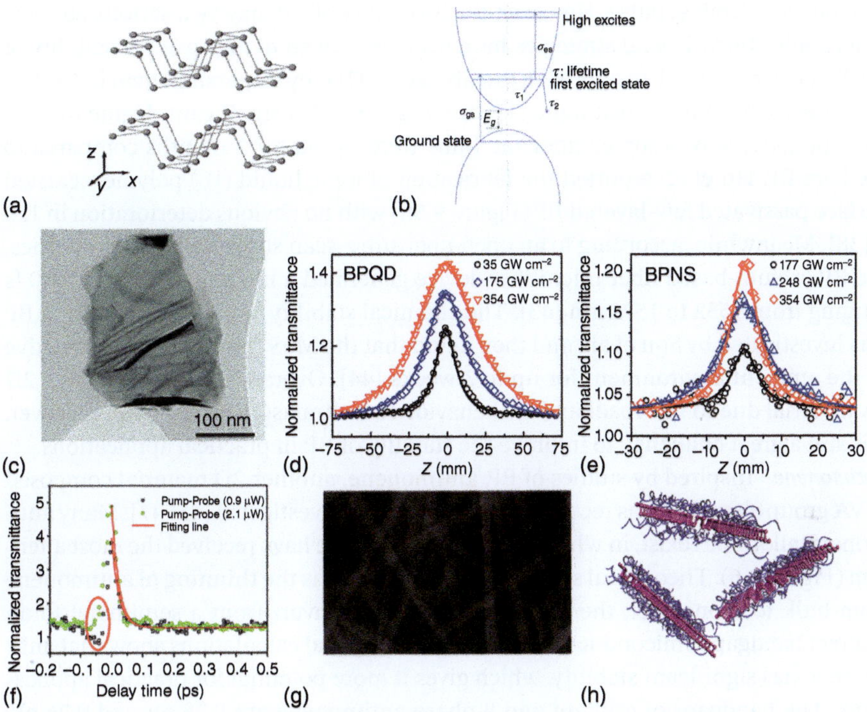

Figure 9.5 (a) Schematic diagram of the crystalline structure of black phosphorus. Source: Castellanos-Gomez et al. [34]/with permission of IOP Publishing. (b) Schematic diagram of the saturable absorption in BP NPs. (c) TEM images of BP nanosheets. Source: (b~c) Reproduced with permission Zhang et al. [24] © Royal Society of Chemistry, 2016. *z*-Scan curves of (d) BP quantum dots and (e) BP nanosheets. (d) and (e) were reprinted with permission from Xu et al. [35]/ with permission of John Wiley & Sons. (f) Fast relaxation dynamics for BP suspension with a pump wavelength of 1550 nm and pump average power of 0.9 and 2.1 μW. Source: Wang et al. [36]/with permission of AIP Publishing LLC. (g) SEM images of the electrospinning BPQDs/PMMA composite nanofiber film to stabilize BP. Source: Reproduced with permission Xu et al. [37]. © 2017, John Wiley & Sons. (h) The schematic diagram of the PILs-assisted surface passivated few-layered BP. Source: Hu et al. [38]/with permission of John Wiley & Sons.

suitable for describing the dynamic carrier process of BP [24]. The ratio of the GSA cross sections (σ_{gs}) to the excited-state absorption cross sections (σ_{es}) was estimated to be about 0.24 and 0.09 at 515 nm and 1030 nm, respectively (Figure 9.5b). The low ratios indicate that BP has a relatively large ground-state transition probability and weak loss of excited-state absorption. Xu et al. investigated the SA property of solvothermal-synthesized BP quantum dots (QDs) and revealed by z-scan tests(Figure 9.5d, e) that the QDs possess superior SA properties than nanosheets, which exhibited a modulation depth of about 36% and a saturable intensity of approximately 3.3 GW cm^{-2} [35]. By utilizing the BP QDs as a saturable absorber, a mode-locked fiber laser based on the ultrashort pulse was generated with a pulse duration of about 1.08 ps centered at 1567.5 nm. These excellent broadband SA properties and ultrafast carrier relaxation dynamics make BP an ideal candidate to generate an ultrafast pulse. However, the instability of BP may be a serious obstacle to its application. Several strategies have been developed to improve the stability of BP. Xu et al. presented an approach to stabilize BP QDs by dispersing them in PMMA via an electrospinning technique (Figure 9.5g) [37]. No significant degradation of NLO property was observed after the films were stored for 3 months compared to the bare BP. Hu et al. reported the fabrication of ionic liquid (IL) polymer-assisted surface passivated few-layered BP (Figure 9.5h) with no obvious deterioration in 120 h [38]. Meanwhile, according to an open-aperture z-scan study on its SA properties, the BP erbium-based fiber microfiber device generated a mode lock pulse of 940 fs ranging from 1532 to 1570 nm [43]. The chemical stability of a MoS_2-passivated BP was investigated by Son et al., and they found that the MoS_2 protection was effective in the ambient environment for up to 3 weeks [44]. Overall, BP is a promising 2D SA material due to its broadband SA behavior and ultrafast recovery time, however, it is still a great challenge to improve the stability of BP in practical applications.

Antimonene Inspired by studies of BP, antimonene, another 2D material composed of VA group elements, has recently been extensively investigated [45–47]. Many antimonene allotropes exist, in which α-phase and β-phase have received the most attention (Figure 9.6). Theoretical studies have shown that as the thinning of antimonene from bulk to monolayer, the band structure will convert from a semimetal to an indirect bandgap semiconductor. Unlike BP, theoretical calculations show that antimonene has significant stability, which gives it more potential for practical applications. The bandgaps of α-phase and β-phase antimonene are 0.28 eV and 0.76 eV, respectively, suggesting their broadband nonlinear optical response [45]. The typical atomic structure of β-phase antimonene is shown in Figure 9.6a. Zhang et al. reported the NLO response of β-phase antimonene synthesized by the LPE method [53]. The antimonene nanosheets exhibited strong SA behavior under relatively low excitation energy and reached the threshold of nonlinear absorption (NLA). With a further increase of the excitation energy, the SA behavior was maximized, and the emerging reverse saturable absorption (RSA) behavior dominated the NLA behavior. Zhang et al. used gradient centrifugation to separate antimonene of different sizes [54]. They found that large Sb nanosheets exhibited a significant SA behavior while the nanosheets below 50 nm exhibited a strong RSA behavior. A similar phenomenon also appears in the study of Wu et al. (Figure 9.6b), which can be interpreted as a change from single-photon absorption to two-photon absorption (Figure 9.6c) [55]. Wang et al. achieved the first passive Q-switched operation of a solid-state laser based on the SA properties of antimonene [56]. Fluorinated antimonene was combined with an Nd: LuAG laser to passive Q-switched pulses, demonstrating a pulse width of 326.7 ns and a repetition rate of 733.1 kHz. Excited by 532 and 1064 nm incident lasers, the saturation intensity reached 0.43 and 0.53 $MWcm^{-2}$, and the modulation depths of 8.0% and 4.9% were achieved, respectively. The highest Q-switched response of an antimonene-based solid-state laser occurred at 1342 nm, giving a maximum pulse energy of 1.36 μJ, a minimum pulse width of 48.33 ns, and a maximum peak power of 28.17 W [55]. In conclusion, the antimonene will find a variety of uses in optics and optoelectronics benefiting from its high stability and broadband SA capabilities.

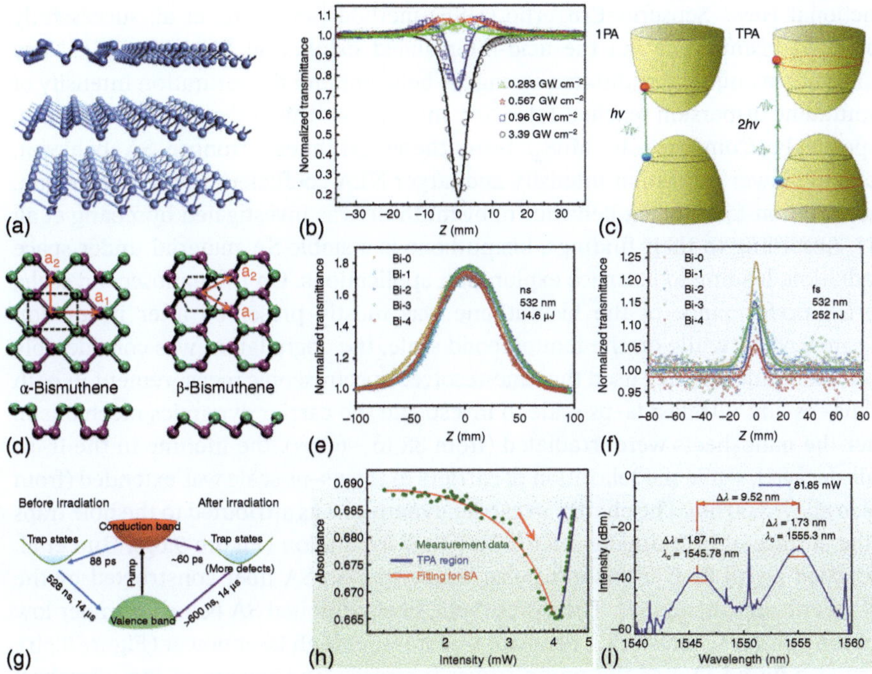

Figure 9.6 (a) Schematic diagram of the crystalline structure of antimonene. Source: Ji et al. [48]/Springer Nature/CC BY 4.0. (b) *z*-Scan results of antimonene nanosheet dispersion with different excitation intensities at 532 nm. (c) Schematic diagram of nonlinear absorption in antimonene nanosheet dispersion. Behavior of SA and TPA was exhibited at low and high intensity, respectively. Source: (b) and (c) were Wu et al. [49]/with permission of AIP Publishing LLC. (d) Schematic diagram of the crystalline structure of α- and β-bismuthene. Source: Adapted from Jain and McGaughey [50] Copyright © 2015 Springer Nature Limited. The Nano- (e) and femtosecond (f) *z*-scan curve for Bi-0, Bi-1, Bi-2, Bi-3, and Bi-4 nanosheet films with a similar linear transmittance of ~0.50 (Bi-0, Bi-1, Bi-2, and Bi-3 represent pristine Bismuthene and Bismuthene Irradiated at 5.0×10^4, 1.0×10^5, and 1.0×10^6 Gy^{60}Co γ-rays, respectively). (g) Schematic diagram of carrier dynamics of Bi nanosheets before and after irradiation. Source: (e–g) were adapted with permission Zhang et al. [51]/with permission of American Chemical Society. (h) Nonlinear optical transmittance of the bismuthene SA. (i) Spectrum of the dual-wavelength laser pulse. Source: Guo et al. [52]/with permission of American Chemical Society.

Bismuthene The final member of group VA, bismuth, is a transition metal. Similar to phosphorus and antimony, it has a layered structure that allows for exfoliation into nanosheets. Numerous crystalline forms of few-layer bismuth, commonly known as bismuthene, including α-bismuthene and β-bismuthene (Figure 9.6d). Among them, β-bismuthene is the most stable form. Based on first-principles phonon and finite temperature molecular dynamics calculations, including spin-orbit coupling (SOC), Aktürk et al. estimated the stability and band structure of single layer and bilayer bismuthene [57]. The bandgap of single layer and bilayer β-bismuthene was predicted to be 0.55 and 0.23 eV, respectively. Additionally, Zhang et al. determined that the monolayer bismuthene bandgap was 0.99 eV and that the carrier mobility was several thousand $cm^2\ V^{-1}\ s^{-1}$ based on a hybrid functional Heyd–Scuseria–Ernzerho (HSE) method [46]. Yang et al. successfully fabricated bismuthene via the acid-intercalated exfoliation technique [5]. They found that bismuthene exhibited strong SA behavior and the saturation intensity of bismuthene dispersion reached 1.4×10^{12} and 7.3×10^{11} W m^{-2} at 532 and 1064 nm, respectively. Compared to MoS_2, bismuthene exhibited stronger SA behavior, including lower saturation intensity and larger NLA coefficient in the NIR region. The radiation-tolerant SA behavior of bismuthine was investigated by Zhang et al. [51]. According to their findings, bismuthene is a stable SA material under space irradiation, leading to its space exploration applications. On the nanosecond scale, the SA performance of the bismuthene was mostly preserved after irradiation (Figure 9.6d,e), while on the femtosecond scale, the degradation was considerably more pronounced. They used the time-resolved spectroscopic measurement in both the ns–μs and ultrafast fs–ps scale to investigate the carrier dynamics mechanism. After the nanosheets were irradiated (from 88 to ~60 ps), the lifetime in the fs–ps scale decayed, while the relaxation of carriers in the ns–μs scale was extended (from 528 to above 600 ns). The change in carrier dynamics was attributed to the hole traps at the surface of bismuthene after ^{60}Co γ-rays irradiation (Figure 9.6g). Guo et al. fabricated an all-fiber ultrashort pulse laser using an SA fiber constructed by the few-layer bismuthene [52]. The bismuthene fiber exhibited SA behavior under low initiating power whereas TPA behavior under rather high laser power (Figure 9.6h). They also found that as the pump power increased, the spectra of the ultrashort laser pulse was split into two peaks (Figure 9.6i), renders its potential applications in pump–probe experiments and tunable terahertz irradiation generation.

9.3.2 Dual-Elemental 2D Materials

9.3.2.1 TMDs

TMDs are layered materials consisting of a transition metal of groups IV to X and a chalcogen. Each layer of TMDs is covalently bonded to a hexagonally packed layer of metal atoms sandwiched between two layers of chalcogen atoms (Figure 9.7a). The intralayer bonds are predominantly covalent in nature, while the adjacent planes are held together by weak van der Waals interactions [20]. The band structure of TMDs is strongly dependent on the number of layers. For instance, monolayered WS_2 is a direct bandgap semiconductor, while the thicker ones are turned to be indirect bandgap [63]. Many families of TMDs, e.g. MoS_2, WS_2, $MoSe_2$, and WSe_2, have been reported to possess strong SA response and ultrashort response time, especially in monolayer and few-layer nanosheets (Figure 9.7b, c) [58, 59, 64]. Bulk MoS_2 and WS_2 are indirect semiconductors, in which the bandgap is 1.29 and 1.35 eV, respectively [59]. However, monolayer MoS_2 and WS_2 exhibit directed bandgaps of 1.9 and 2.0 eV, respectively, due to the thickness-dependent effect [65, 66]. Zhang et al. investigated the NLO of MoS_2/WS_2 with different layers under different wavelengths of the laser [67]. It was found that the NLO phenomenon turned from TPA to SA with the blueshift of the incident laser, and the switching occurs at longer wavelengths in thicker MoS_2/WS_2, owing to the layer-dependent band gap. Noting that monolayer MoS_2 and WS_2 can also exhibit SA in the intermediate infrared (MIR) region, although the photon energy was lower than the bandgap. Several explanations have been proposed for such sub-band gaps SA phenomenon, but no conclusion has been made yet [59]. Wang et al. constructed a broadband saturable absorber based on MoS_2 with sulfur vacancy defects, showing the SA property at NIR laser [7]. Zhang et al. studied both 1T (metallic) and 2H (semiconducting) phases of the LPE MoS_2 nanosheets and found the metallic phase doped by impurities leads to the sub-bandgap SA [41]. Woodward et al. also found that MoS_2–PVA composite could be used as a saturable absorber in Q-switched fiber laser. They attributed the phenomenon to the presence of edge states within the material bandgap brought by the boundaries of a finite crystal structure [68]. Wang et al. used a femtosecond transient absorption spectroscopy to detect the ultrafast carrier dynamics process in suspended monolayer MoS_2, showing a different relaxation process in monolayer compared to the thick ones (Figure 9.7f) [60]. Wang et al. found that the liquid-phase exfoliated few-layer MoS_2 dispersion exhibited significant SA in response to the femtosecond laser [69]. Similar to MoS_2 and WS_2 in molecular structure, $MoSe_2$ and WSe_2 exhibited lower bandgap due to the heavy atom effect of Se chalcogen atoms [12]. The bulk $MoSe_2$ and WSe_2 have shown an indirect bandgap of 1.09 and 1.20 eV [59], while monolayer $MoSe_2$ and WSe_2 exhibited a direct bandgap of 1.57 and 1.67 eV, respectively [70]. Compared to MoS_2 and WS_2, the narrower bandgap and lower optical absorption coefficient could make $MoSe_2$ and WSe_2 more applicable in broadband SA for passively mode-locked lasers [71]. Mao et al. found that liquid-phase exfoliated $MoSe_2/WSe_2$ nanosheets exhibited SA at 1.55 μm (0.8 eV), although the photon energy is lower than the bandgap [59].

9.3.2.2 Topological Insulators (TIs)

TIs are a special class of materials that behave as insulators in their interior but have conducting states on the surface or edge, giving small bandgaps (0.2–0.3 eV) and high surface conductivity [13, 72]. The band structure of its surface state is similar to that of graphene, exhibiting a Dirac-like linear band dispersion [9]. Zhang et al. presented that the layered Sb_2Te_3, Bi_2Te_3, and Bi_2Se_3 are TIs via first-principles calculations [9]. Hasan et al. experimentally confirmed the single Dirac cone in TIs surface state [72]. The small bandgaps on its surface or edge states make them promising materials for SAs, especially in the range of IR region [4]. According to a study on the ultrafast surface carrier dynamics in Bi_2Te_3 carried out by Hailaoui et al., the thermalization of the surface states was initially determined by interband scattering from the bulk conduction band, lasting for about 0.5 ps [73]. Subsequently, it took a few picoseconds for the Dirac cone nonequilibrium electrons to recover a Fermi–Dirac distribution, while the relaxation extended more than 10 ps. Jung et al. reported a SA behavior of layered Bi_2Te_3 excited by 1935 nm laser [74]. Then, Yu et al. showed a SA behavior of LPE few-layered Bi_2Se_3 nanosheet, with a modulation depth of 30% [75]. Ganeev et al. studied the NLO properties of the LPE-exfoliated Bi_2Te_3 1064m and 532 nm via z-scan technique, and both RSA and TPA behavior were found at 532 nm [76]. To obtain better SA properties in high-intensity illumination, van der Waals (vdW) heterostructure could be a potential solution. Lan et al. fabricated a graphene/Bi_2Te_3 vdW heterojunction and found the enhanced optical nonlinearity in high-intensity illumination induced

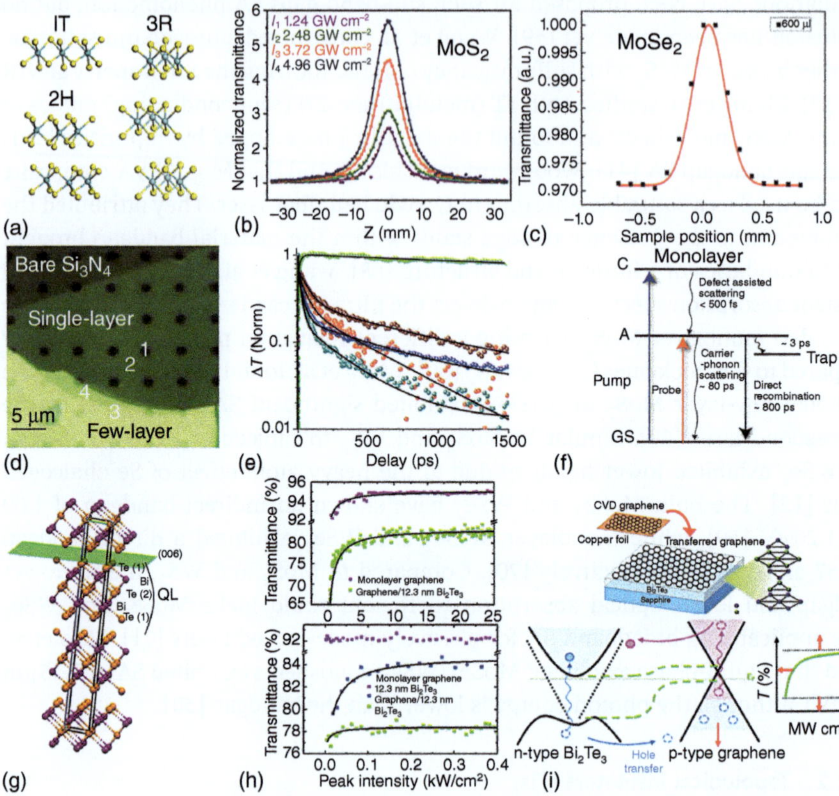

Figure 9.7 (a) Schematic diagram of the atomic structure of the three main phases of TMDs: the 1T, 2H, and 3R phases. (b, c) *z*-scan curves of (b) MoS_2 (c) $MoSe_2$. Source: (b) Durairaj and Sabari Girisun [58]/Reproduced with permission of Elsevier. (c) Reprinted with permission from Kaur et al. [59] Copyright 2020, Springer Science Business Media, LLC, part of Springer Nature. (d) Optical micrograph of single-/few-layer MoS_2 on Si_3N_4. (e) Transient absorption dynamics with a pump wavelength of 387 nm and a probe wavelength of 670 nm for positions 1, 2, 3, and 4. (f) Schematic diagram of relaxation processes in monolayer MoS_2 crystal. Source: (d), (e), and (f) were reproduced with permission from Shi et al. [60]. © 2013, American Chemical Society. (g) Schematic diagram of the atomic structure of TIs. Source: Adapted from Kong et al. [61]. (h) Nonlinear transmittances and fitting curves of monolayer graphene and a heterojunction of graphene with 12.3 nm Bi_2Te_3 at high and low optical intensity (i) Carrier dynamics and relative transmittance variations under high optical intensity. Source: Reprinted with permission from Lan et al. [62]/with permission of Royal Society of Chemistry.

by the carrier transfer [62]. They reported two SA behaviors in different intensity ranges. As shown in Figure 9.7h, compared to Bi_2Te_3, the heterojunction displayed relatively lesser modulation depth and lower saturable intensity under low-intensity light. However, the higher intensity illumination induced lower modulation depth and saturable intensity in comparison to that of graphene. Such phenomenon can be explained by the carrier transfer in the heterojunction (Figure 9.7i): under low illumination, the photoexcited electrons from the Bi_2Te_3 valence band are accumulated in the bulk conduction band. Meanwhile, some excited electrons in graphene tend to migrate into the conduction band of Bi_2Te_3, suppressing the modulation depth. Under intense illumination, the optical nonlinearity is switched on, and conduction band is saturated, and some photoexcited holes in Bi_2Te_3 migrate to the graphene valence band. Consequently, more holes and electrons were separated, resulting in a greater modulation depth [62].

9.3.3 Multi-Elemental 2D Materials

Along with the development of 2D nanomaterial chemistry, more advanced 2D materials such as MOFs, MXenes, and 2D perovskites also exhibit significant SA properties.

9.3.3.1 MOFs

MOFs are a class of porous crystalline hybrid materials interconnected by coordination between transition metal centers and organic linkers (Figure 9.8a) [77]. MOFs have been studied in many applications, including gas storage, supercapacitors, lithium batteries, and chemical sensing, due to their porous crystalline structure. However, the investigation of MOFs in NLO is just emerging owing to their strong intramolecular dipolar moment. Hao et al. investigated the NLO responses of a Ni-MOF by the open-aperture *z*-scan technique at 800, 1060, 1300, 1550, and 1800 nm and found broadband SA properties [82]. Li et al. fabricated a porphyrin-based MOF (PIZA-1) thin films by the layer-by-layer self-assembly [77], and a third-order NLO response was discovered by the *z*-scan technique (Figure 9.8b). An RSA-SA switching was found along with the increasing incident intensities or PIZA-1 film thickness, owing to the exited-state absorption (ESA) process of PIZA-1 film and the saturable ESA process occurring in relatively higher incident intensities. Liang et al. successfully synthesized MOFs denoted by Cu1 and Cu2 (Figure 9.8a) with different coordination numbers by adding trifluoroacetic acid (TFA) [83]. They used the *z*-scan method to find third-order NLO properties of these MOFs with different coordination numbers (Figure 9.8c). Due to its low coordination number, Cu1 displayed a conventional SA response, while Cu2 displayed an NLO reaction from SA to RSA. Liang's study revealed the correlation between coordination numbers and NLO properties in MOFs and proposed a general method to convert the NLO response of MOFs by changing their coordination numbers. Additionally, several recent studies have also reported the successful generation of Q-switched and mode-locked lasers using SA devices based on MOF materials [84–86].

9.3.3.2 MXenes

MXenes are a group of 2D transition metal carbides, carbonitrides, or nitrides with a chemical formula of $M_{n+1}X_nT_x$ ($n = 1$–3), whereas M is a transition metal (Ti, Cr, Sc, Zr, Hf, V, Nb, Ta, Mo, etc.), X is carbide or nitride, and T is surface terminations, which can be F, O, OH, etc. (Figure 9.8f). Numerous studies have focused on the uses of MXenes in electrochemical capacitors, water purification, biosensors, chemical catalysts, etc. However, their ultrafast photonics applications have yet been fully reported. Jon et al. studied the SA properties of a side-polished fiber (SPF) decorated by a Ti_3CNT_x monolayer film at 1560 nm (Figure 9.8e) [78]. The as-prepared SA

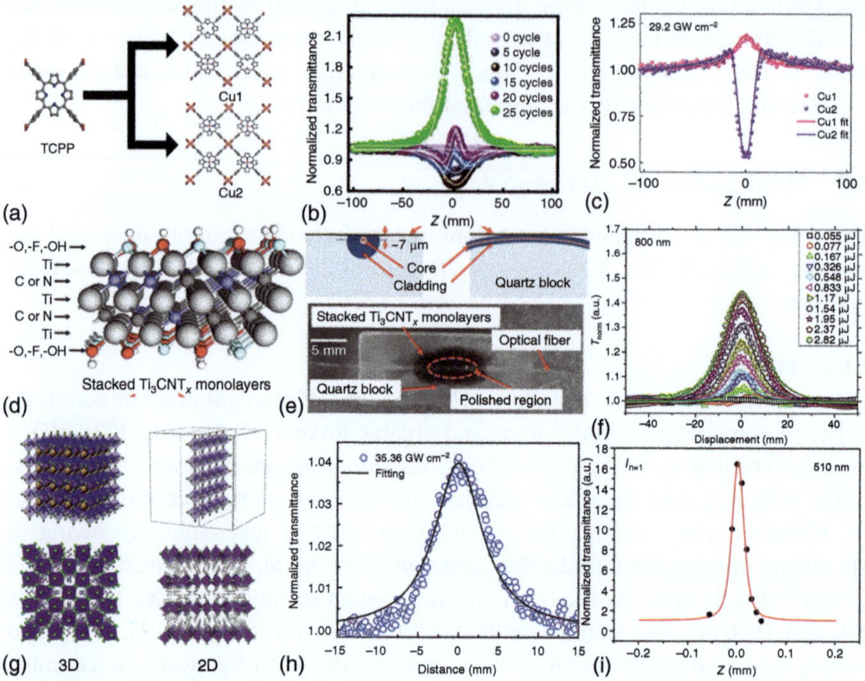

Figure 9.8 (a) Simulated molecular structure of Cu1- and Cu2-MOF. (b) Thickness-dependent NLO behavior of the PIZA-1 thin films at an incident pulse energy of 100 μJ. Source: Li et al. [77]/with permission of Royal Society of Chemistry. (c) *z*-Scan curves of Cu1 and Cu2. Source: (a) and (c) were adapted from Li et al. [77] Copyright 2020, John Wiley & Sons, Inc. (d) Structure of Ti_3CNT_x (Tx: O, F, OH). (e) Schematic diagram and SEM image of the prepared side-polished fiber deposited with the stacked Ti_3CNT_x monolayer film. Source: (d) and (e) were reproduced with permission Jhon et al. [78]. © 2017, John Wiley & Sons, Inc. (f) *z*-scan measurements of multilayered $Ti_3C_2T_x$ at a wavelength of 800 nm. Source: Jiang et al. [8]/with permission of John Wiley & Sons. (g) Schematic representation of 3D bulk, 2D perovskite materials in "structure-level" and "material-level." Source: Wang et al. [79]/with permission of John Wiley & Sons. (h) *z*-Scan curve of $(PEA)_2Cs_4Pb_5Br_{16}$ nanosheets measured at 800 nm. Source: Mushtaq et al. [80]/Royal Society of Chemistry/CC BY 3.0. (i) *z*-Scan curve of the $(BA)_2Pb_1X_4$ at plasma resonance (510 nm). Source: Adapted from Abdelwahab et al. [81] Copyright © 2019 John Wiley & Sons, Inc.

fiber device can be used in a passively mode-locked fiber laser. According to their measurements, the modulation depth and saturation power reached 1.7% and 45 W, respectively. Jiang et al. investigated the NLO response of few-layer $Ti_3C_2T_x$ over a broadband spectral range from mid-IR to visible (Figure 9.8f) and found a large SA response in $Ti_3C_2T_x$ film, two orders of magnitude greater than most of the other lossy NLA processes [8]. They also investigated the broadband nonlinear optical response from 800 to 1800 nm., finding that the $Ti_3C_2T_x$ sample exhibited a giant NLA coefficient. According to the above pioneering studies, MXene can be a promising broadband SA material. So far, the most present research has focused on the SA behavior of $Ti_3C_2T_x$, while more MXene families will be studied owing to their variable electronic and optical behavior.

9.3.3.3 2D Perovskites

Perovskites are a class of minerals having a structural formula of ABX_3 that was initially used to describe the mineral $CaTiO_3$ that was found by Russian scientist Lev Perovski [87]. Usually, A represents a small organic or inorganic cation, B is a transition metal cation (e.g. Pb^+, Sn^{2+}, or Ag^+), and the term "X" refers to a halogen anion (Cl^-, Br^-, or I^-) that shares the top angle with $[BX_6]$ octahedra. Stabilized by organic cations (R), 2D Perovskites with a molecular formula of $R_2A_{n-1}B_nX_{3n+1}$ can be considered as substructures exfoliated from the standard three-dimensional chalcogenides (Figure 9.8g). 2D Perovskites can be mainly classified into R–P (Ruddlesden–Popper) phase, D–J (Dion–Jacobson) phase, and alternating cation in the interlayer space (ACI) phase, whereas the single quantum well structure in the R–P phase is generally considered favorable for enhancing third-order NLO. Mushtaq et al. presented a compendious sketch of third-order nonlinear optical properties of $(PEA)_2Cs_4Pb_5Br_{16}$, a 2D perovskite using femtosecond Z-scan technique at 800 nm (Figure 9.8h) [80]. Li et al. investigated the nonlinear optical properties of CVD-growing $CH_3NH_3PbI_3$ nanosheets by the z-scan technique [88]. The results indicated that this 2D perovskite has strong SA with large modulation depth and low saturable intensity compared with the perovskite bulk film. The SA behavior of $CsPbBr_3$ QDs has also been studied by a femtosecond z-scan measurement [89]. The $CsPbBr_3$ QDs exhibited a NLA coefficient as high as 10^{-2} cm GW^{-1} under 515 nm laser irradiation, showing great potential in the fabrication of SA devices. Abdelwahab et al. found the broadband SA response to the femtosecond laser in $[(BA)_2(MA)_{n-1}Pb_nX_{3n+1}]$ (BA = butylammonium ($C_4H_9NH_3$), MA = methylammonium (CH_3NH_3), X = I^- or Br^-, and n is the number of inorganic layers within each quantum well) (Figure 9.8i) [81], and the giant nonlinear response can be attributed to the presence of insulator (exciton)–metal (plasma) Mott transition near the exciton resonance of the thinnest quantum well. With the rapid development of perovskites, we believe more extraordinary SA can be found in the monolayer or few-layer 2D perovskites.

9.4 Applications

Benefiting from the wide-spectral NLO response, novel excited ultrafast carrier dynamics, large modulation depth, and low saturable intensity, 2D materials have already been used in generating ultrafast pulsed lasers. By incorporating the 2D SAs into the laser cavity, the intracavity loss could be modulated by the laser intensity. And then, the ultrafast laser pulse will be generated. Therefore, the 2D SAs has great potential in the area of ultrafast optics.

9.4.1 Fabrication of SA Device Based on 2D Materials

In order to generate passive Q-switched or mode-locked lasers, many SA devices have been designed and fabricated through various techniques. Hereby, we will introduce several techniques to fabricate SA devices based on 2D materials.

9.4.1.1 Scotch Tape Exfoliation

As one of the most famous "top-down" methods, Scotch tape exfoliation has been widely used since the beginning of the 2D material adventure [90]. A single layer of 2D materials can be occasionally obtained by attaching sticky tape to bulk-layered layers and peeling it off (Figure 9.9a). It provides a simple and direct way to transfer the exfoliated nanosheet onto a substrate, e.g. a fiber pigtail to fabricate the prototype of 2D SA devices. Beyond graphene, many 2D materials, including MoS_2, WS_2, WSe_2, and BP, have been obtained by Scotch tape or thermal release tape and transferred onto the fiber laser cavity in order to fabricate prototypes of the Q-switched laser [28, 91, 94, 95]. However, the Scotch tape exfoliation is not applicable for scalable fabrications, and more approaches have been developed to manufacture the 2D SA devices.

9.4.1.2 CVD, PVD

CVD is an efficient bottom-up method to grow high-quality 2D materials with determined size and thickness on various substrates. It involves the decomposition or reaction of one or more precursor vapors on the catalytic substrate to grow 2D material films. By transferring the CVD-grown 2D materials to a desired substrate, a SA device can be obtained (Figure 9.9b). Bao et al. used the CVD method to grow graphene on Ni–SiO_2/Si substrate, and then $FeCl_3$ solution was applied to etch the Ni sacrificing layer, making the graphene floating on the solution surface. After a wet transferring onto a fiber pigtail cross section, a barricaded SA fiber generated mode-locked laser was successfully fabricated [2]. With a similar method, 2D materials, such as germanene, ReS_2, TiS_2, and WS_2, SA devices were also fabricated to generate ultrafast lasers [32, 96–98].

Physical vapor deposition (PVD), also named as physical vapor transport (PVT), is a general term for various physical process-induced vapor deposition methods, such as vacuum deposition, pulsed laser deposition (PLD), magnetron sputtering, and molecular beam epitaxial (MBE) growth, Mu et al. fabricated a Germanene–silver SA mirror (SAM) by vacuum deposition in an ultrahigh vacuum preparation chamber and simultaneously monitored the growing process by UHV-STM [28]. By simply incorporating the germanene saturable device into the fiber laser cavity (Figure 9.9d), they successfully generated a mode-locked pulse with ultrafast duration. PLD is another bottom-up PVD method assisted by the pulse laser to create 2D materials (Figure 9.9c). A MoS_2 SA device has been manufactured by the PLD-growth on the quartz glass wafer [7]. Moreover, the S defects introduced in the PLD growth can reduce the band gap of MoS_2 thin films from 1.08 to 0.08 eV, making it applicable for SA devices, e.g. Q-switched laser [99].

Overall, CVD and PVD are the efficient bottom-up methods to fabricate monolayer or few-layer homogeneous or heterostructure films with a scalable size, controllable thickness, and excellent nonlinear optical properties, opening an avenue toward the practical SA device with high linear transmission, large modulation depth, and small non-saturable losses.

9.4.1.3 LPE

LPEs, including intercalation, ion exchange, oxidation, and ultrasonic exfoliation, have been extensively used in the synthesis of 2D nanosheets. By breaking the

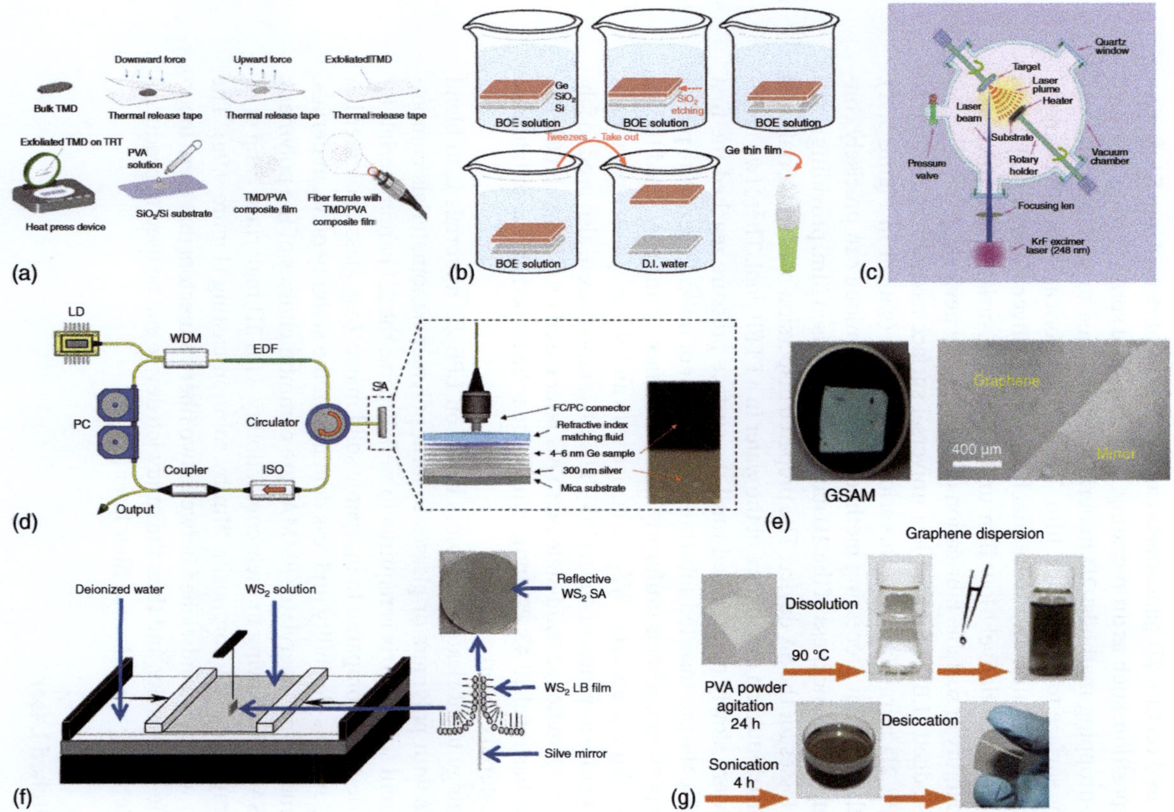

Figure 9.9 (a) The Scotch tape exfoliation process for fabricating a TMD-based SA device. Source: Reproduced with permission Ooi and Ahmad [91], © 2022, Elsevier. (b) The transfer process of the Ge thin film to fabricate the SA device. Source: Adapted with permission Lin et al. [32]/with permission of MDPI/CC BY 4.0. (c) Schematic diagram of PLD technique for preparation of MoS_2. Source: Adapted from Wang et al. [7] Copyright © 2014 John Wiley & Sons, Inc. (d) Schematic of the fiber laser experimental setup and the structure of PVD Ge saturable absorber device. Source: Reproduced with permission Mu et al. [28], Copyright © 2021 John Wiley & Sons, Inc. (e) Graphene saturable mirror (GSAM) fabricated by vacuum filtration. Source: Reproduced with permission Wang et al. [92]. 2017, © John Wiley & Sons, Inc. (f) The fabrication of the reflective WS2 saturable absorber (SA) by the Langmuir–Blodgett (LB) system. Source: Reproduced from Wang et al. [93], © 2019, MDPI/CC BY 4.0. (g) Graphene–PVA composite films proceed at different stages by the solution-cast method. Source: Reproduced with permission Feng et al. [16]. © 2015, Optica Publishing Group.

interlayer interactions in layered crystals through mechanical, thermal, or chemical procedures, ultrathin nanosheet dispersion can be fabricated in high yield together at a relatively low cost. By further processing, such as drop-casting, dip-coating, spin-coating, spray-coating, Langmuir–Blodgett assembly, and blending with polymer or glass matrix, the SA devices based on LPE 2D materials can be fabricated.

SA devices based on LPE 2D nanosheets can be directly fabricated into SA devices via solution deposition, such as drop-casting, spin-, dip- and spray-coating. Through these solution-dropping methods, 2D materials are attached to a transparent substrate or mirror surface to form a thin film. Wang et al. fabricated graphene membranes by vacuum filtration and then attached them to silver mirrors [92]. Main disadvantages include relatively low film uniformity and variable film thickness, limiting its capability of scalable fabrication of high-quality devices.

Langmuir–Blodgett (LB) technique is a universal surface assembling method to immobilize LPE 2D ultrathin films on various substrates (shown in Figure 9.9f). Compared to the solution processing method, the LB technique can precisely control the uniformity and thickness of the 2D nanomaterial-based film, providing small non-saturable losses for the SA device. The prepared dispersion of 2D materials was dropped and distributed on the surface of the water in a PTFE pool. Then, two mobilized barriers were slowly compressed until the surface pressure reached the first peak value. Then, the pre-inserted substrates, e.g. mirror or optical fiber, were pulled up slowly to deposit nanosheets uniformly on their surface. Wang et al. used the LB technique to combine a reflective WS_2 SA device with an Nd:YVO_4 laser operating at 1340 nm with a pulse duration of 409 ns [93]. The LB methods were also used to fabricate a passively Q-switched solid-state Nd: YAG laser based on GO SAs [100]. The slope efficiency has reached 43.7%, indicating LB may be a good choice to fabricate practical 2D materials absorbers.

Another strategy to fabricate SA devices based on LPE 2D materials is to blend the nanosheets with polymers or glasses to form polymer-nanomaterials composites. A transparent solid matrix is required to well disperse the 2D nanomaterials for ideal solid SA devices. Compared to inorganic solid matrix (e.g. glass), polymer is a better choice due to its flexibility and less demand on the stability of SA materials. PVA and polymethyl methacrylate (PMMA) are common choices of the polymeric matrix in the SA device, owing to their compatibility with 2D materials. Feng et al. fabricated a graphene–PVA composite film by solution casting (Figure 9.9g). The composite film was then sandwiched between two fiber connectors that were incorporated into the laser cavity, and then the sandwich device can be used to produce a passive Q-switched or mode-locked fiber laser [16].

9.4.2 Q-Switched Laser

The ratio of the total energy retained in the laser resonant cavity to the energy lost per unit of time in the cavity is known as the *Q*-factor (quality factor), which serves as a metric for the effectiveness of the optical resonant cavity in lasers. Q-switched is a technique that uses careful control of the intracavity losses to produce extremely intense pulses from a laser. One of the specific modulation methods is to employ a photoelectric crystal or acousto-optic modulator as Q-switch to control the *Q*-factor by an external drive source, known as active Q-switch. Meanwhile, the passive Q modulation employs a saturable absorber to control the intracavity losses so that the *Q*-factor of the laser spontaneously changes periodically. Since the passively Q-switched lasers are more popular used in ultrafast optics, we primarily concentrate on the application of 2D SA materials in this section.

Saturable absorbers have large linear absorption resulting in high energy dissipated per oscillation cycle, therefore the gain medium can store more energy before it reaches the threshold (Figure 9.10a). With the accumulation of inversion concentration in the gain medium, the spontaneous radiation increased gradually. While it is comparable to the I_s of the saturable absorber, its absorption decreased, and the Q factor raised significantly. Then, the laser was generated, leading to a continuous decrease in the absorption coefficient of the saturable absorber with an avalanche of stimulated radiation. Finally, the inversion concentration in the gain medium decreased to a small level as the absorption coefficient of the SA materials increased and the stimulated radiation terminated. Through the process, the passive Q-switched technology generated a giant pulse (nJ–mJ level) within several cycles. The pulse time was the same order of magnitude as L(cavity length)/c(ns–μs), with a kHz repetition rate.

According to the above discussion, if the recovery time of SA is taken into account, a relatively longer recovery time would be taken than the pulse to avoid possible energy loss in the pulse energy. In order to prevent premature recovery of the gain medium from creating stray lasers, the recovery period should also be less than the lifespan of the gain medium in its upper state. Another important factor affecting the Q-switched laser pulse is the correlation between the saturation absorption energies of SA and the gain media. Since SA should only absorb a small portion of the energy of the laser pulse, the saturation absorption energy of SA should be less than that of the gain material. When the pump power is relatively low, the generated laser intensity cannot exceed the saturation intensity of SA, producing a continuous wave (CW) laser. Whereas when the pump power exceeds the threshold, SA starts to intervene, and the mode of the cavity switches from CW to Q-switched. As the pump power further increases, the pulse width decreases, and the peak increases due to the increase in photon flow rate, resulting in an acceleration of SA action time. Figure 9.10b shows a typical Q-switched operating in the time against the wavelength domain. However, another negative effect is that significant non-saturable losses are frequently encountered in saturable absorbers, resulting in lower laser efficiency. Prevented by the presence of non-saturation losses, the pump power cannot increase infinitely, showing a reduction of the power efficiency as discussed previously [11]. Consequently, a good SA for passive q modulation requires a suitable recovery time, low saturation intensity, low non-saturable loss, and high modulation depth.

9.4.2.1 Q-Switched Fiber Laser

Fiber lasers have been widely used in scientific, medical, and industrial applications. The flexibility, reliability, and compact nature of such light sources enable them to outperform bulk lasers in many areas, offering alignment-free and turn-key

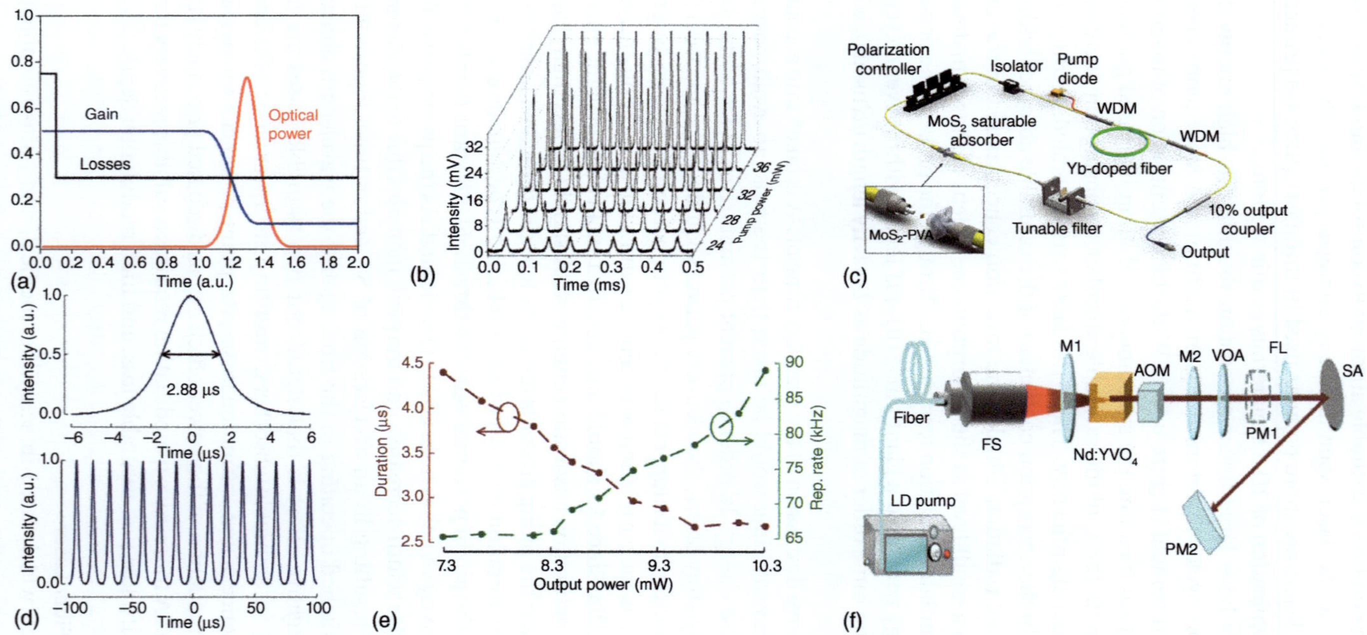

Figure 9.10 (a) Schematic diagram of the passive Q-switched process. (b) Typical characteristics of Q-switched lasers in the time and pump power domain. Source: Adapted from Tiu et al. [11]/with permission of IOP Publishing. (c) Schematic diagram of the Q-switched fiber laser setup. (d) Single pulse profile and pulse train of the Q-switched laser. (e) Variation of the repetition frequency and pulse duration with output power. Source: (c) and (e) were Woodward et al. [68]/with permission of The Optical Society. (f) Schematic diagram of the Q-switched Cr: ZnSe laser setup based on BP-SAM. Source: Wang et al. [101]/MDPI/CC BY 4.0.

operation for customers. Many designs have been developed to incorporate 2D materials into the fiber laser cavity.

The most popular fiber is based on a sandwich structure, as shown in Figure 9.10c. The 2D nanomaterial can be transferred onto the fiber laser cavity, acting as a passively Q-switched or passively mode-locked fiber laser. A simple way to process LPE 2D nanomaterials is to disperse them in a polymer film, which can be PVA, PMMA, or other materials. Then, the composite film is sandwiched between two connectors to form the fiber laser cavity. The first Q-switched laser with the 2D materials (graphene) as SA was based on such a sandwich architecture reported by Popa et al. in 2011 [102]. They mixed dispersion of monolayer (SLG) and few-layer graphene (FLG) with an aqueous solution of PVA. An approximately 50 m thick graphene–PVA composite was produced after drying in the atmosphere. In order to create a fiber-compatible SA, the film was then sandwiched between two fiber optic connectors. The sandwich structure produced a two second pulse laser with a tunable wavelength between 1522 and 1555 nm and energy of up to 40 nJ. Woodward et al. fabricated MoS_2–PVA composite by mixing LPE graphene with PVA followed by water evaporation. The generated pulse interval is 13.4 μs (Figure 9.10d), corresponding to a repetition rate of 74 kHz, and the full-width half-maximum (FWHM) pulse duration is 2.88 μs (Figure 9.10e) [65]. They also investigated the pump power-dependent Q-switched laser properties. It was observed that the pulse duration decreased from 4.40 to 2.68 μs and the repetition rate increased from 65.3 to 89.0 kHz as the output power increased (Figure 9.10e).

A saturable absorber mirror (SAM) is another popular device using 2D SA nanomaterials to produce Q-switched lasers. Benefiting from the merits of wide-spectral NLO response, novel excited carrier dynamics, and low cost of fabrication, 2D nanomaterials have the potential to be used in both commercial and scientific applications to create SAMs that can produce laser pulses with high repetition rate, higher power, and short duration over a broad wavelength range. (Figure 9.10f) Wang et al. fabricated a SAM based on LPE-prepared black phosphorene by a simple drop-casting method [103]. They inserted the BP-based SAM in a Cr: SnSe laser cavity and achieved a Q-switched MIR laser at 2.4 μm. Kong et al. also used the drop coating technique to combine the BP SAM with an Er: Y_2O_3 ceramic laser to generate a variety of Q-switched lasers from NIR (1 μm) to MIR (2.7 μm) [6]. Chen et al. used the magnetron sputtering technique to deposit WS_2 on the tip of Panda-style polarization-maintaining fiber (PMF) [104]. Subsequently, a gold film to avoid the oxidation of WS_2 was deposited on it to fabricate a WS_2-SAM. The mirror was then inserted into the Er-doped fiber cavity to generate a 1560 nm Q-switched laser with a pulse duration ranging from 1.269 μs to 154.9 ns and a repetition rate ranging from 29.5 to 367.8 kHz. Wang et al. fabricated a graphene-saturable absorber mirror (GSAM) by vacuum filtration of an LPE graphene nanosheet. The LPE-prepared graphene membrane was then peeled from the nitrocellulose porous substrate and attached to the silver mirror by a wet transferring technique. The GSAM exhibits extraordinary NLO responses, including 13% modulation depth, 9.9% non-saturable loss, and 1.2 ps carrier relaxation time at NIR (2 μm) laser [92, 100].

9.4.3 Mode-Locked Laser

To generate narrower pulses, laser beams can be specially modulated by applying mode-locked techniques. By forcing the phases of distinct modes to "lock" each other, ultrafast (fs–ps) pulsed radiation can be generated. Fixing the phase relationship of the modes in the laser cavity allows the periodic constructive interference to line up to repeatedly enhance a single laser pulse, showing the "locking" process. The mode-locked laser can also be interpreted as the Fourier component of periodic function when the phase relationship is fixed with each other and reaches a stable state. To active the mode-locked laser based on acousto-optic or the electro-optic effect, an external signal can be applied to the optical loss modulator. With the external signal, the phase relationship can be fixed to generate an ultrafast laser pulse, usually within the picosecond duration. Generally, the passively mode-locked laser with a saturable absorber can modulate the resonator losses much faster than an electronic modulator. At a steady state, the magnitude of the laser gain is just enough to compensate for the loss of the cyclic pulse, and yet much lower than the loss in the rest times when the SA device does not reach the initiating power (Figure 9.11a). Thus, the absorber can suppress any additional weaker pulses, generating a mode-locked ultrafast pulse [14].

Fiber mode-locked lasers have been extensively studied recently. Figure 9.11b exhibits a typical mode-locked fiber laser with a ring cavity. It was constructed by a pump source, a gain medium EDF, a polarization controller, an optical coupler, an independent polarization isolator, a wavelength division multiplexer (WDM), a mode-locker, and single-mode fiber (SMF). Such fiber laser generated an ideally passive mode-locked output laser pulse pattern with ultrafast pulse duration from Femto-to picosecond (Figure 9.11c) [2]. A sandwich construction can also be applied to attach 2D nanomaterials onto a fiber laser cavity and generate mode-locked lasers (Figure 9.11d). Ge et al. deposited TiS_2 nanosheets on a fiber tip using this method and fabricated an Er-doped mode-locked fiber laser [108]. Bao et al. used CVD and wet transfer process to fabricate a sandwich SA device in a ring-cavity type mode-locked fiber laser to generate an ultrafast laser pulse [2].

However, if the average power or the peak power is too high, the SA could be damaged. Therefore, sandwiched devices can only be used in passively mode-locked lasers with relatively low power. To generate mode-locked lasers with a higher power, a widely used fiber called photonic crystal fiber (PCF) may be an ideal substrate material for attaching 2D materials. Figure 9.11e illustrates a typical structure of PCF. 2D SA materials were filled into the hollow of PCF, which provides various total reflection interfaces between the 2D SA material and the high-refractive core. The evanescent-wave interaction could improve the nonlinear interaction length and increase the tolerance of high intracavity power. Lin et al. obtained a stable ultrafast mode-locked laser by dispersing electrochemical exfoliated few-layer graphene in PCF and attaching them to an EDFL system [109]. Due to the linear influence of group delay dispersion (GDD) and the nonlinear effect of self-phase modulation, the pulse width decreased from 668 to 650 fs, and the FWHM expanded from 3.75 to 3.92 nm. Yan et al. used a similar method to generate a mode-locked

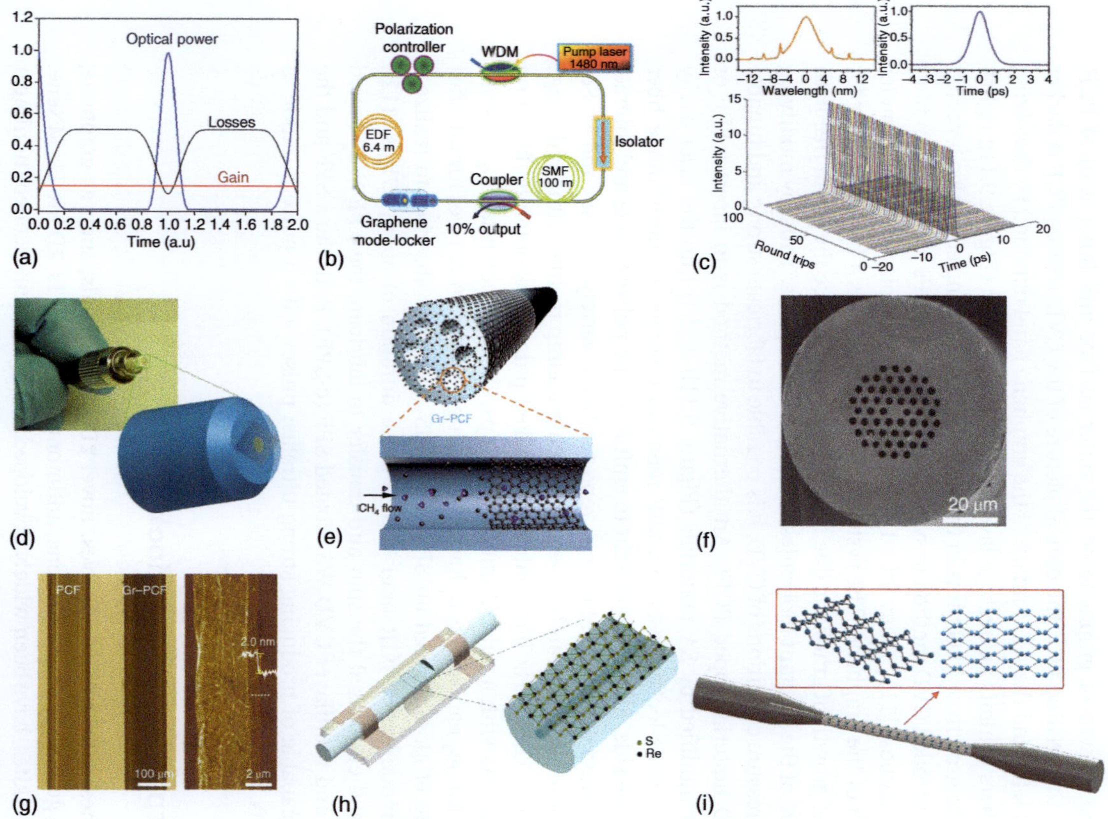

Figure 9.11 (a) Temporal evolution of optical power and losses in a passively mode-locked laser with a fast saturable absorber. (b) Laser configuration constituting a ring cavity. (c) Numerically calculated pulse evolution with mode-locked cavity round trips. (d) Photograph of fiber pigtail coated with graphene film and schematic of the end of optical fiber pigtail with a graphene film coating on a pinhole (yellow). Source: (b~d) were reproduced with permission Bao et al. [2]. 2009, © John Wiley & Sons. (e) Schematics of graphene PCF grown by the CVD method. (f) Cross-sectional SEM image of the G-PCF. (g) Optical images of bare PCF and graphene PCF (left), and AFM image of collapsed graphene ribbon with wrinkles after etching the fiber silica (right). A height of ~2.0 nm is measured along the white dashed line. Source: (e–g) were reproduced with permission Chen et al. [105], © 2019, Springer Nature. (h) Schematic diagrams of ReS_2-covered D-shaped fiber saturable absorber. Source: Cui et al. [106]/Springer Nature/CC BY 4.0. (i) Schematic model of the BP-based nonlinear device and the crystal structure of single-layer BP. Source: Li et al. [107]/with permission of IOP Publishing.

laser by dispersing TIs in PCF as the saturable absorber [110]. In contrast to Lin's work [109], they chosed to keep the TI nanosheets as a solution filled in PCF rather than drying them, which method also generated an intense mode-locked laser. However, owing to the existence of residual solvent in the cavity, large insertion losses were found. To fabricate more homogeneous 2D material-doped PCFs, Ke et al. used the CVD method to fabricate graphene PCFs [105]. Figure 9.11e illustrated the growth of graphene on the outer surface and inner wall of PCF. Compared to the pristine PCF, the optical picture of the CVD-treated PCF is darker (Figure 9.11g left), and the AFM image of the graphene band obtained by dissolving the silicon fibers exhibits an average layer thickness of ~2 nm (Figure 9.11g right), confirming the growth of FLG films in the PCFs hollows. Cheng et al. also reported a pressure-controlled CVD method to synthesize graphene PCFs with uniform fiber length and thickness [111]. They found that system pressure plays a key parameter in the growth of the graphene. The system pressure represents a key parameter for regulating the growth of graphene due to its ability to influence the mean free path and fluid state of the reactant molecules within the micropores. Thus, by modifying the system pressure conditions of CVD, it is possible to fabricate strong and tunable nonlinear 2D material-doped PCFs. An alternative method is to insert a SPF or tapered fiber modified by SA materials (Figure 9.11h and i) into the laser cavity to generate a mode-locked laser. An SPF refers to a single- or multi-mode fiber polished on one side, allowing the light to emit from the polished side and interact with the SA material coating. Mao et al. reported that a variety of WSe_2 and $MoSe_2$ nanosheet-based SPF-SAs can be operated at a high pump regime without damage [59]. A set of 3252.65 MHz harmonic mode-locked pulses was obtained based on the SPF-WSe_2 SAs with a pulse duration of 1.31 ps, showing potential applications of WSe_2 and $MoSe_2$ nanosheets in low-cost and high-power SA devices for ultrafast optics. Cheng et al. fabricated an SPF-based PdS_2 saturable absorber to realize a high stable mode-locked NIR laser with a pulse duration of about 375 ps [112]. Khazaeinezhad et al. used the spin and transfer to fabricate the LPE WS_2-coated SPF (s-SPF) and multilayer CVD WS_2-coated SPF (c-SPF). Both the s-SPF and the c-SPF were capable of producing strong ultrafast pulses with a pulse duration of 467 and 332 fs, respectively [98].

9.5 Perspectives and Outlook

As the synthesis technologies evolves, novel 2D SA materials, such as graphene, bismuthene, germanene, phosphorene, antimonene, TMDs, TIs, 2D MOFs, MXene, and 2D perovskites, have been widely developed, owing to their excellent nonlinear optical properties, e.g. wide working spectral range, ultrafast carrier dynamics, large NLA coefficient, and low saturable intensity. As a rising star in the area of nonlinear optics, 2D SA materials have become more competitive than traditional semiconductors in the field of ultrafast lasers. Moreover, the layer-dependent and defect-sensitive SA behavior in 2D materials provide convenient approaches to control their nonlinear optical performance, including chemical and physical methods, e.g. CVD, PVT, PLD, etc. In addition, their SA properties can be physically manipulated by applying an external electric field to adjust the band structure of the 2D materials. By taking advantage of the tunable SA performance, 2D material-based cavities and thus mode-locked lasers with pulse duration ranging from sub-200 fs to nanosecond scale have been successfully generated. Therefore, the 2D SA materials will play a crucial role in nonlinear optics.

However, there are still challenges to overcome: first, precise control of defects is demanded to improve the consistency of the 2D SA materials. For instance, the SA performance of TMD monolayer is significantly affected by vacancy, topological defect, orientation, edge, or doping from substrate impurities. Second, it is challenging to develop a scalable method to fabricate high-quality SA devices based on the 2D materials. An ideal fabrication technique should provide a uniform atomic layer with excellent crystallization quality at a low cost. However, the present CVD method is energy intensive and the synthesis process is difficult to be optimized. In addition, defects and aggregation of the 2D SA materials may also be induced by the chemical process. At last but not at least, it is also challenging to acquire a high-throughput prediction of the SA performance of the rapidly developed 2D materials due to the limited traditional characterization techniques. Lack of the prediction of the defects or impurities may also cause some misunderstanding of the real SA properties. As studies on 2D SA materials have increased, introducing database analysis and machine learning have become possible strategies to accelerate the rational design of 2D SA materials.

Overall, the 2D material-based SA devices have shown great promise for practical industry and commercial use. By solving the problems in quality control and scalable fabrication, it is believed that 2D SA materials will become indispensable in the area of ultrafast optics.

Acknowledgment

This work was supported by the National Science Foundation of China [No. 21972105, 2020; No. 22278302, 2023], National Key Research and Development Program (2021YFB3802500), Haihe Laboratory of Sustainable Chemical Transformations, and Key Laboratory of Special Functional Materials and Structural Design of Ministry of Education (Class B), Lanzhou University [No. lzujbky-2021-kb06].

References

1 Haus, H.A. (2000). Mode-locking of lasers. *IEEE Journal of Selected Topics in Quantum Electronics* 6 (6): 1173–1185.

2 Bao, Q., Zhang, H., Wang, Y. et al. (2009). Atomic-layer graphene as a saturable absorber for ultrafast pulsed lasers. *Advanced Functional Materials* 19 (19): 3077–3083.

3 Luo, H., Tian, X., Gao, Y. et al. (2018). Antimonene: a long-term stable two-dimensional saturable absorption material under ambient conditions for the mid-infrared spectral region. *Photonics Research* 6 (9): 900–907.

4 Yin, P., Jiang, X., Huang, R. et al. (2021). 2D materials for nonlinear photonics and electro-optical applications. *Advanced Materials Interfaces* 8 (14): 2100367.

5 Yang, Q.-Q., Liu, R.-T., Huang, C. et al. (2018). 2D bismuthene fabricated via acid-intercalated exfoliation showing strong nonlinear near-infrared responses for mode-locking lasers. *Nanoscale* 10 (45): 21106–21115.

6 Kong, L., Qin, Z., Xie, G. et al. (2016). Black phosphorus as broadband saturable absorber for pulsed lasers from 1 μm to 2.7 μm wavelength. *Laser Physics Letters* 13 (4): 045801.

7 Wang, S., Yu, H., Zhang, H. et al. (2014). Broadband few-layer MoS_2 saturable absorbers. *Advanced Materials* 26 (21): 3538–3544.

8 Jiang, X., Liu, S., Liang, W. et al. (2018). Broadband nonlinear photonics in few-layer MXene $Ti_3C_2T_x$ (T = F, O, or OH). *Laser & Photonics Reviews* 12 (2): 1700229.

9 Zhang, H., Liu, C.-X., Qi, X.-L. et al. (2009). Topological insulators in Bi_2Se_3, Bi_2Te_3 and Sb_2Te_3 with a single Dirac cone on the surface. *Nature Physics* 5 (6): 438–442.

10 Zhou, K.-G. and Zhang, H.-L. (2015). Lighten the olympia of the flatland: probing and manipulating the photonic properties of 2D transition-metal dichalcogenides. *Small* 11 (27): 3206–3220.

11 Tiu, Z.C., Ooi, S.I., Guo, J. et al. (2019). Review: application of transition metal dichalcogenide in pulsed fiber laser system. *Materials Research Express* 6 (8): 082004.

12 Lau, K.Y. and Hou, D. (2021). Recent research and advances of material-based saturable absorber in mode-locked fiber laser. *Optics and Laser Technology* 137: 106826.

13 He, J., Tao, L., Zhang, H. et al. (2019). Emerging 2D materials beyond graphene for ultrashort pulse generation in fiber lasers. *Nanoscale* 11 (6): 2577–2593.

14 Wang, G., Baker-Murray, A.A., and Blau, W.J. (2019). Saturable absorption in 2D nanomaterials and related photonic devices. *Laser & Photonics Reviews* 13 (7): 1800282.

15 Burshtein, Z., Blau, P., Kalisky, Y. et al. (1998). Excited-state absorption studies of Cr^{4+} ions in several garnet host crystals. *IEEE Journal of Quantum Electronics* 34 (2): 292–299.

16 Feng, Y., Dong, N., Wang, G. et al. (2015). Saturable absorption behavior of free-standing graphene polymer composite films over broad wavelength and time ranges. *Optics Express* 23 (1): 559–569.

17 Maiman, T.H. (1960). Stimulated optical radiation in ruby. *Nature* 187 (4736): 493–494.

18 Soffer, B.H. (1964). Giant pulse laser operation by a passive, reversibly bleachable absorber. *Journal of Applied Physics* 35 (8): 2551–2551.

19 Bret, G. and Gires, F. (1964). Giant-pulse laser and light amplifier using variable transmission coefficient glasses as light switches. *Applied Physics Letters* 4 (10): 175–176.

20 Novoselov, K.S., Jiang, D., Schedin, F. et al. (2005). Two-dimensional atomic crystals. *Proceedings. National Academy of Sciences. United States of America* 102 (30): 10451–10453.

21 Dawlaty, J.M., Shivaraman, S., Chandrashekhar, M. et al. (2008). Measurement of ultrafast carrier dynamics in epitaxial graphene. *Applied Physics Letters* 92 (4): 042116.

22 Zheng, Z., Zhao, C., Lu, S. et al. (2012). Microwave and optical saturable absorption in graphene. *Optics Express* 20 (21): 23201–23214.

23 Sun, D., Wu, Z.-K., Divin, C. et al. (2008). Ultrafast relaxation of excited dirac fermions in epitaxial graphene using optical differential transmission spectroscopy. *Physical Review Letters* 101 (15): 157402.

24 Zhang, S., Li, Y., Zhang, X. et al. (2016). Slow and fast absorption saturation of black phosphorus: experiment and modelling. *Nanoscale* 8 (39): 17374–17382.

25 Wang, K., Szydłowska, B.M., Wang, G. et al. (2016). Ultrafast nonlinear excitation dynamics of black phosphorus nanosheets from visible to mid-infrared. *ACS Nano* 10 (7): 6923–6932.

26 Zhang, F., Wang, Z., Wang, D. et al. (2016). Nonlinear optical effects in nitrogen-doped graphene. *RSC Advances* 6 (5): 3526–3531.

27 Wang, Q., Kang, J., Wang, P. et al. (2022). Broadband saturable absorption in germanene for mode-locked Yb, Er, and Tm fiber lasers. *Nano* 11 (13): 3127–3137.

28 Mu, H., Liu, Y., Bongu, S.R. et al. (2021). Germanium nanosheets with dirac characteristics as a saturable absorber for ultrafast pulse generation. *Advanced Materials* 33 (32): 2101042.

29 Zhuang, J., Liu, C., Zhou, Z. et al. (2018). Dirac signature in germanene on semiconducting substrate. *Advancement of Science* 5 (7): 1800207.

30 Matthes, L., Pulci, O., and Bechstedt, F. (2014). Optical properties of two-dimensional honeycomb crystals graphene, silicene, germanene, and tinene from first principles. *New Journal of Physics* 16 (10): 105007.

31 Claussen, S.A., Tasyurek, E., Roth, J.E., and Miller, D.A.B. (2010). Measurement and modeling of ultrafast carrier dynamics and transport in germanium/silicon-germanium quantum wells. *Optics Express* 18 (25): 25596–25607.

32 Lin, C.-Y., Cheng, C.-H., Chi, Y.-C. et al. (2022). Low-temperature PECVD growth of germanium for mode-locking of Er-doped fiber laser. *Nanomaterials* 12 (7): 1197.

33 Zhou, L., Wang, Y., Wang, Y. et al. (2021). Saturable absorption and self-defocusing response of 2D monoelemental germanium nanosheets in broadband spectra. *Optics Express* 29 (24): 39115–39124.

34 Castellanos-Gomez, A., Vicarelli, L., Prada, E. et al. (2014). Isolation and characterization of few-layer black phosphorus. *2D Materials* 1: 025001.

35 Xu, Y., Wang, Z., Guo, Z. et al. (2016). Solvothermal synthesis and ultrafast photonics of black phosphorus quantum dots. *Advanced Optical Materials* 4 (8): 1223–1229.

36 Wang, Y., Huang, G., Mu, H. et al. (2015). Ultrafast recovery time and broadband saturable absorption properties of black phosphorus suspension. *Applied Physics Letters* 107 (9): 091905.

37 Xu, Y., Wang, W., Ge, Y. et al. (2017). Stabilization of black phosphorous quantum dots in PMMA nanofiber film and broadband nonlinear optics and ultrafast photonics application. *Advanced Functional Materials* 27 (32): 1702437.

38 Hu, C.-X., Xiao, Q., Ren, Y.-Y. et al. (2018). Polymer ionic liquid stabilized black phosphorus for environmental robust flexible optoelectronics. *Advanced Functional Materials* 28 (51): 1805311.

39 Li, J., Luo, H., Zhai, B. et al. (2016). Black phosphorus: a two-dimension saturable absorption material for mid-infrared Q-switched and mode-locked fiber lasers. *Scientific Reports* 6 (1): 30361.

40 Zheng, J., Tang, X., Yang, Z. et al. (2017). Few-layer phosphorene-decorated microfiber for all-optical thresholding and optical modulation. *Advanced Optical Materials* 5 (9): 1700026.

41 Zhang, H., Lu, S.B., Zheng, J. et al. (2014). Molybdenum disulfide (MoS_2) as a broadband saturable absorber for ultra-fast photonics. *Optics Express* 22 (6): 7249–7260.

42 Lu, S.B., Miao, L.L., Guo, Z.N. et al. (2015). Broadband nonlinear optical response in multi-layer black phosphorus: an emerging infrared and mid-infrared optical material. *Optics Express* 23 (9): 11183–11194.

43 Sotor, J., Sobon, G., Macherzynski, W. et al. (2015). Black phosphorus saturable absorber for ultrashort pulse generation. *Applied Physics Letters* 107 (5): 051108.

44 Son, Y., Kozawa, D., Liu, A.T. et al. (2017). A study of bilayer phosphorene stability under MoS_2-passivation. *2D Materials* 4 (2): 025091.

45 Zhang, S., Yan, Z., Li, Y. et al. (2015). Atomically thin arsenene and antimonene: semimetal–semiconductor and indirect–direct band-gap transitions. *Angewandte Chemie, International Edition* 54 (10): 3112–3115.

46 Zhang, S., Xie, M., Li, F. et al. (2016). Semiconducting group 15 monolayers: a broad range of band gaps and high carrier mobilities. *Angewandte Chemie, International Edition* 128 (5): 1698–1701.

47 Aktürk, O.Ü., Özçelik, V.O., and Ciraci, S. (2015). Single-layer crystalline phases of antimony: antimonenes. *Physical Review B* 91 (23): 235446.

48 Ji, J., Song, X., Liu, J. et al. (2016). Two-dimensional antimonene single crystals grown by van der Waals epitaxy. *Nature Communications* 7 (1): 13352.

49 Wu, J., Wei, Y., Shen, W. et al. (2020). Antimonene nanosheets fabricated by laser irradiation technique with outstanding nonlinear absorption responses. *Applied Physics Letters* 116 (26): 261903.

50 Jain, A. and McGaughey, A.J.H. (2015). Strongly anisotropic in-plane thermal transport in single-layer black phosphorene. *Scientific Reports* 5 (1): 8501.

51 Zhang, Z.-T., Yang, Q.-Q., Zhen, X.-J. et al. (2021). Two-dimensional bismuthene showing radiation-tolerant third-order optical nonlinearities. *ACS Applied Materials & Interfaces* 13 (18): 21626–21634.

52 Guo, P., Li, X., Feng, T. et al. (2020). Few-layer bismuthene for coexistence of harmonic and dual wavelength in a mode-locked fiber laser. *ACS Applied Materials & Interfaces* 12 (28): 31757–31763.

53 Zhang, F., Wang, M., Wang, Z. et al. (2018). Excellent nonlinear absorption properties of β-antimonene nanosheets. *Journal of Materials Chemistry C* 6 (11): 2848–2853.

54 Zhang, L., Fahad, S., Wu, H.-R. et al. (2020). Tunable nonlinear optical responses and carrier dynamics of two-dimensional antimonene nanosheets. *Nanoscale Horizons* 5 (10): 1420–1429.

55 Wang, M., Zhang, F., Wang, Z. et al. (2018). Passively Q-switched Nd^{3+} solid-state lasers with antimonene as saturable absorber. *Optics Express* 26 (4): 4085–4095.

56 Zhang, G., Tang, X., Fu, X. et al. (2019). 2D group-VA fluorinated antimonene: synthesis and saturable absorption. *Nanoscale* 11 (4): 1762–1769.

57 Aktürk, E., Aktürk, O.Ü., and Ciraci, S. (2016). Single and bilayer bismuthene: stability at high temperature and mechanical and electronic properties. *Physical Review B* 94 (1): 014115.

58 Durairaj, M. and Sabari Girisun, T.C. (2023). Nanobelt and nanoplatelet structured molybdenum disulfide thin film as a saturable absorber under nanopulsed green laser excitation. *Journal of Solid State Chemistry* 320: 123865.

59 Kaur, R., Singh, K.P., and Tripathi, S.K. (2020). Study of linear and non-linear optical responses of MoSe2–PMMA nanocomposites. *Journal of Materials Science: Materials in Electronics* 31 (22): 19974–19988.

60 Shi, H., Yan, R., Bertolazzi, S. et al. (2013). Exciton dynamics in suspended monolayer and few-layer MoS_2 2D crystals. *ACS Nano* 7 (2): 1072–1080.

61 Kong, D., Dang, W., Cha, J.J. et al. (2010). Few-layer nanoplates of Bi_2Se_3 and Bi_2Te_3 with highly tunable chemical potential. *Nano Letters* 10 (6): 2245–2250.

62 Lan, J.-C., Qiao, J., Sung, W.-H. et al. (2020). Role of carrier-transfer in the optical nonlinearity of graphene/Bi_2Te_3 heterojunctions. *Nanoscale* 12 (32): 16956–16966.

63 Mao, D., Wang, Y., Ma, C. et al. (2015). WS_2 mode-locked ultrafast fiber laser. *Scientific Reports* 5 (1): 7965.

64 Bikorimana, S., Lama, P., Walser, A. et al. (2016). Nonlinear optical responses in two-dimensional transition metal dichalcogenide multilayer: WS_2, WSe_2, MoS_2 and $Mo_{0.5}W_{0.5}S_2$. *Optics Express* 24 (18): 20685–20695.

65 Jiang, Y., Miao, L., Jiang, G. et al. (2015). Broadband and enhanced nonlinear optical response of MoS_2/graphene nanocomposites for ultrafast photonics applications. *Scientific Reports* 5 (1): 16372.

66 Gutiérrez, H.R., Perea-López, N., Elías, A.L. et al. (2013). Extraordinary room-temperature photoluminescence in triangular WS_2 monolayers. *Nano Letters* 13 (8): 3447–3454.

67 Zhang, S., Dong, N., McEvoy, N. et al. (2015). Direct observation of degenerate two-photon absorption and its saturation in WS_2 and MoS_2 monolayer and few-layer films. *ACS Nano* 9 (7): 7142–7150.

68 Woodward, R.I., Kelleher, E.J.R., Howe, R.C.T. et al. (2014). Tunable Q-switched fiber laser based on saturable edge-state absorption in few-layer molybdenum disulfide (MoS_2). *Optics Express* 22 (25): 31113–31122.

69 Wang, K., Wang, J., Fan, J. et al. (2013). Ultrafast saturable absorption of two-dimensional MoS_2 nanosheets. *ACS Nano* 7 (10): 9260–9267.

70 Gusakova, J., Wang, X., Shiau, L.L. et al. (2017). Electronic properties of bulk and monolayer TMDs: theoretical study within DFT framework (GVJ-2e method). *Physica Status Solidi A: Applications and Materials Science* 214 (12): 1700218.

71 Luo, Z., Li, Y., Zhong, M. et al. (2015). Nonlinear optical absorption of few-layer molybdenum diselenide ($MoSe_2$) for passively mode-locked soliton fiber laser. *Photonic Research* 3 (3): A79–A86.

72 Hasan, M.Z. and Kane, C.L. (2010). Colloquium: topological insulators. *Reviews of Modern Physics* 82 (4): 3045–3067.

73 Hajlaoui, M., Papalazarou, E., Mauchain, J. et al. (2012). Ultrafast surface carrier dynamics in the topological insulator Bi_2Te_3. *Nano Letters* 12 (7): 3532–3536.

74 Jung, M., Lee, J., Park, J. et al. (2015). Mode-locked, 1.94 μm, all-fiberized laser using WS_2-based evanescent field interaction. *Optics Express* 23 (15): 19996–20006.

75 Yu, H., Zhang, H., Wang, Y. et al. (2013). Topological insulator as an optical modulator for pulsed solid-state lasers. *Laser & Photonics Reviews* 7 (6): L77–L83.

76 Ganeev, R.A., Popov, V.S., Zvyagin, A.I. et al. (2021). Exfoliated Bi_2Te_3 nanoparticle suspensions and films: morphological and nonlinear optical characterization. *Nano* 10 (15): 3857–3870.

77 Li, D.-J., Gu, Z.-G., and Zhang, J. (2020). Auto-controlled fabrication of a metal-porphyrin framework thin film with tunable optical limiting effects. *Chemical Science* 11 (7): 1935–1942.

78 Jhon, Y.I., Koo, J., Anasori, B. et al. (2017). Metallic MXene saturable absorber for femtosecond mode-locked lasers. *Advanced Materials* 29 (40): 1702496.

79 Wang, G., Mei, S., Liao, J. et al. (2021). Advances of nonlinear photonics in low-dimensional halide perovskites. *Small* 17 (43): 2100809.

80 Mushtaq, A., Pradhan, B., Kushavah, D. et al. (2022). Femtosecond induced third-order optical nonlinearity in quasi 2D Ruddlesden–Popper perovskite film deciphered using Z-scan. *Advanced Materials* 3 (22): 8211–8219.

81 Abdelwahab, I., Dichtl, P., Grinblat, G. et al. (2019). Giant and tunable optical nonlinearity in single-crystalline 2D perovskites due to excitonic and plasma effects. *Advanced Materials* 31 (29): 1902685.

82 Hao, G.-P., Tang, C., Zhang, E. et al. (2017). Thermal exfoliation of layered metal–organic frameworks into ultrahydrophilic graphene stacks and their applications in Li–S batteries. *Advanced Materials* 29 (37): 1702829.

83 Liang, Y., Hu, W., Yuan, X. et al. Switchable nonlinear optical absorption of metal-organic frameworks. *Advanced Optical Materials* 18: 2200779.

84 Jiang, X., Zhang, L., Liu, S. et al. (2018). Ultrathin metal-organic framework: an emerging broadband nonlinear optical material for ultrafast photonics. *Advanced Optical Materials* 6 (16): 1800561.

85 Zhang, Q., Jiang, X., Zhang, M. et al. (2020). Wideband saturable absorption in metal-organic frameworks (MOFs) for mode-locking Er- and Tm-doped fiber lasers. *Nanoscale* 12 (7): 4586–4590.

86 Huang, J., Qiao, W., Li, G. et al. (2020). Optical modulation of magnesium 2,5-dihydroxyterephthalate saturable absorber for passively Q-switched 2 μm solid-state laser. *Applied Physics Express* 13 (11): 115003.

87 Galasso, F.S. (1969). Chapter 2 - Structure of perovskite-type compounds. In: *Structure, Properties and Preparation of Perovskite-Type Compounds* (ed. F.S. Galasso), 3–49. Pergamon: Elsevier.

88 Li, P., Chen, Y., Yang, T. et al. (2017). Two-dimensional $CH_3NH_3PbI_3$ perovskite nanosheets for ultrafast pulsed fiber lasers. *ACS Applied Materials & Interfaces* 9 (14): 12759–12765.

89 Li, J., Dong, H., Xu, B. et al. (2017). $CsPbBr_3$ perovskite quantum dots: saturable absorption properties and passively Q-switched visible lasers. *Photonic Research* 5 (5): 457–460.

90 Novoselov, K.S., Geim, A.K., Morozov, S.V. et al. (2004). Electric field effect in atomically thin carbon films. *Science* 306 (5696): 666–669.

91 Ooi, S.I. and Ahmad, H. (2022). Thermal release tape assisted mechanical exfoliation of pristine TMD and the performance of the exfoliated TMD saturable absorbers for Q-switched laser generation. *Optical Materials* 128: 112363.

92 Wang, G., Wang, K., Szydłowska, B.M. et al. (2017). Ultrafast nonlinear optical properties of a graphene saturable mirror in the 2 μm wavelength region. *Laser & Photonics Reviews* 11 (5): 1700166.

93 Wang, T., Wang, Y., Wang, J. et al. (2019). 1.34 μm Q-switched Nd:YVO_4 laser with a reflective WS_2 saturable absorber. *Nanomaterials* 9 (9): 1200.

94 Sobon, G., Sotor, J., and Abramski, K.M. (2012). Passive harmonic mode-locking in Er-doped fiber laser based on graphene saturable absorber with repetition rates scalable to 2.22 GHz. *Applied Physics Letters* 100 (16): 161109.

95 Al-Masoodi, A.H.H., Ahmed, M.H.M., Latiff, A.A. et al. (2016). Q-switched ytterbium-doped fiber laser using black phosphorus as saturable absorber. *Chinese Physics Letters* 33 (5): 054206.

96 Shang, X., Guo, L., Zhang, H. et al. (2020). Titanium disulfide based saturable absorber for generating passively mode-locked and Q-switched ultra-fast fiber lasers. *Nanomaterials* 10 (10): 1922.

97 Liu, L., Chu, H., Zhang, X. et al. (2019). Heterostructure ReS_2/GaAs saturable absorber passively Q-switched Nd:YVO_4 laser. *Nanoscale Research Letters* 14 (1): 112.

98 Khazaeinezhad, R., Hosseinzadeh Kassani, S., Jeong, H. et al. (2015). Femtosecond soliton pulse generation using evanescent field interaction through tungsten disulfide (WS_2) film. *Journal of Lightwave Technology* 33 (17): 3550–3557.

99 Yang, Z. and Hao, J. (2016). Progress in pulsed laser deposited two-dimensional layered materials for device applications. *Journal of Materials Chemistry C* 4 (38): 8859–8878.

100 Wang, J., Wang, Y., Wang, T. et al. (2019). Nonlinear optical response of graphene oxide langmuir-blodgett film as saturable absorbers. *Nanomaterials* 9 (4): 640.

101 Wang, J., Wang, Y., Liu, S. et al. (2020). Nonlinear optical response of reflective MXene molybdenum carbide films as saturable absorbers. *Nanomaterials* 10 (12): 2391.

102 Popa, D., Sun, Z., Hasan, T. et al. (2011). Graphene Q-switched, tunable fiber laser. *Applied Physics Letters* 98 (7): 073106.

103 Wang, Z., Zhao, R., He, J. et al. (2016). Multi-layered black phosphorus as saturable absorber for pulsed Cr:ZnSe laser at 2.4 μm. *Optics Express* 24 (2): 1598–1603.

104 Chen, H., Chen, Y., Yin, J. et al. (2016). High-damage-resistant tungsten disulfide saturable absorber mirror for passively Q-switched fiber laser. *Optics Express* 24 (15): 16287–16296.

105 Chen, K., Zhou, X., Cheng, X. et al. (2019). Graphene photonic crystal fibre with strong and tunable light–matter interaction. *Nature Photonics* 13 (11): 754–759.

106 Cui, Y., Lu, F., and Liu, X. (2017). Nonlinear saturable and polarization-induced absorption of rhenium disulfide. *Scientific Reports* 7 (1): 40080.

107 Li, Y., He, Y., Cai, Y. et al. (2018). Black phosphorus: broadband nonlinear optical absorption and application. *Laser Physics Letters* 15 (2): 025301.

108 Ge, Y., Zhu, Z., Xu, Y. et al. (2018). Broadband nonlinear photoresponse of 2D TiS_2 for ultrashort pulse generation and all-optical thresholding devices. *Advanced Optical Materials* 6 (4): 1701166.

109 Lin, Y.-H., Yang, C.-Y., Liou, J.-H. et al. (2013). Using graphene nano-particle embedded in photonic crystal fiber for evanescent wave mode-locking of fiber laser. *Optics Express* 21 (14): 16763–16776.

110 Yan, P., Lin, R., Chen, H. et al. (2015). Topological insulator solution filled in photonic crystal fiber for passive mode-locked fiber laser. *IEEE Photonics Technology Letters* 27 (3): 264–267.

111 Cheng, Y., Yu, W., Xie, J. et al. (2022). Controllable growth of graphene photonic crystal fibers with tunable optical nonlinearity. *ACS Photonics* 9 (3): 961–968.

112 Cheng, P.K., Liu, S., Ahmed, S. et al. (2020). Ultrafast Yb-doped fiber laser using few layers of PdS_2 saturable absorber. *Nanomaterials* 10 (12): 2441.

10

Second-Harmonic and Third-Harmonic Generations in 2D Layered Materials

Xudong Jin and Min Zhao

Taiyuan University of Technology, Advanced Materials of Ministry of Education, Key Laboratory of Interface Science and Engineering, No. 79 West Yingze Street, 030024, Taiyuan, Shanxi, China

10.1 Introduction

Optical harmonic generation (HG) has attracted increasing attention with the invention of the laser in 1961 [1]. In general, nonlinear optical (NLO) processes are divided into two categories: the real part of the susceptibility dominates HG while the imaginary part is responsible for saturable absorption (SA) [2–4], multiphoton absorption [5, 6], and optical limiting [7–9]. The HG processes can provide space for micro-frequency manipulation due to the "split" or "combined" frequency. In addition, HG can be a good readout of microstructural change because of its sensitivity to the crystal structural symmetry [10–12]. Considering the high-harmonic generation (HHG) is usually quite weak, second-harmonic generation (SHG) and third-harmonic generation (THG) are the most commonly studied in HG processes.

SHG and THG has long been used independently for frequency conversion [13, 14], imaging [15], modulators [16], and ultrafast lasers [17]. Current SHG and THG devices are normally based on bulk crystals (e.g. beta barium borate (BBO), silica optical fiber, and lithium niobate ($LiNbO_3$)) [18–20] or traditional semiconductor materials [21]. Indeed, the commercial success of bulk BBO and $LiNbO_3$ crystals have been achieved due to their HG effects. In addition, a large number of organometallic and organic compounds with superior HG performance have also been discovered in the past two decades, such as fullerene derivatives [22], phthalocyanines [23], and metalloporphyrins [24].

However, these traditional materials with relatively low second-order or third-order susceptibility [13, 14] cannot fully satisfy the current technical requirement of nonlinear optics including quantum nanophotonics [25], on-chip nanophotonics [26], and nonlinear plasmonics [27]. Currently, the researches on SHG and THG have two major trends: integration and miniaturization. There are substantial challenges in integrating other material with traditional SHG and THG materials to form hybrid structures for on-chip applications. Therefore, it is significantly important to discover new materials with large SHG and THG responses

Two-Dimensional Materials for Nonlinear Optics: Fundamentals, Preparation Methods, and Applications, First Edition. Edited by Qiang Wang and Hao-Li Zhang.

that can be chip-integrated to tackle the upcoming challenges of nanophotonics for future photonics and optoelectronics.

Since the groundbreaking work on graphene, intensive research efforts on 2D layered materials (2DLMs) have been a hot topic for more than a decade due to their unique material properties [28]. 2DLMs are currently the main focus of significant research efforts due to their ultrathin scale and extraordinary photoelectric performance. In principle, the interest research toward graphene and other 2DLMs has been focusing not only on their unique electrical, mechanical, and thermal properties but also on their fascinating optical properties [29, 30]. Indeed, these 2DLMs have exhibited remarkable large optical nonlinearities [27, 28], ultrafast broadband optical response [9, 31, 32], and strong excitonic effects [33]. The special NLO properties of 2DLMs have already enabled new-type conceptual chip-integrated photonic devices fundamentally different from those based on traditional bulk materials [33, 34].

Here, we present recent advances in 2DLM-based SHG, THG, and future perspectives. Firstly, we discuss the fundamentals of different NLO processes including SHG, THG, and HHG. This is then followed by a survey of the current state of the art of 2DLM-based SHG and THG. At the end, we conclude with application perspectives on 2DLM-based nonlinear optics. These new two-dimensional materials are anticipated to provide innovative approaches to enable novel NLO devices with new functionalities and high performance.

10.1.1 Fundamentals of SHG and THG

When impinged by an optical field, the optical response of a material to an applied optical field can be expressed by the following equation [13, 14]:

$$P(t) = \chi^{(1)}E(t) + \chi^{(2)}E(t)^2 + \chi^{(3)}E(t)^3 + \dots \tag{10.1}$$

where $P(t)$, $E(t)$, and $\chi^{(n)}$ are the polarization, optical field strength, and n-th-order NLO susceptibility. The relation between E and P is linear when the electric field E is small. However, when the electric field of the input light is in close proximity to the interatomic electric field (105–108 V m^{-1}), the contribution of the nonlinear terms in Eq. (10.1) is nonnegligible [35]. The nonlinear susceptibility $\chi^{(n)}$, where n represents the order of the NLO process, is usually used to characterize the ability of materials to excite nonlinear processes. The $\chi^{(l)}$ represents the linear optical susceptibility that governs linear optical effects (e.g. absorption and refraction). However, if the NLO susceptibility is large or the optical field is intense enough, the higher-order ($n \geq 2$) terms can give rise to radiation at sum frequencies of the input light signals, which are different from the incident light frequency.

The $\chi^{(2)}$ term in Eq. (10.1) represents NLO susceptibility of the second-order NLO effects and is responsible for three-wave-mixing processes. The SHG (Figure 10.1a), optical rectification, sum- and difference-frequency generations, and the Pockels effect are typically described by the second-order NLO susceptibility $\chi^{(2)}$. The term $\chi^{(3)}$ in Eq. (10.1) describes the nonlinear susceptibility of the third-order NLO effects that is responsible for the optical Kerr effects, THG (Figure 10.1b),

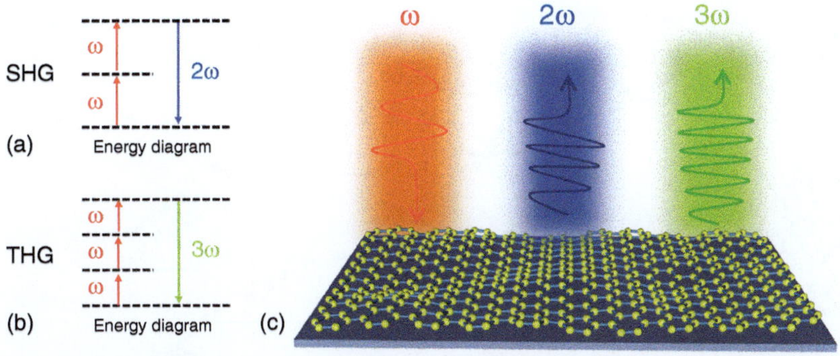

Figure 10.1 Different nonlinear frequency conversion processes of SHG (a) and THG (b). The energy diagrams represent how the input photons at a frequency "ω" add together to produce higher-frequency waves (2ω for SHG and 3ω for THG). (c) Schematic diagram illustrating the principle of SHG and THG.

four wave mixing, self-phase modulation, (SA), and self-focusing [28]. On the other hand, high-order multiphoton luminescence/absorption/scattering and HHG are described by the n-th-order NLO susceptibility $\chi^{(n)}$. Because the interaction strength of the NLO processes typically decreases with n, the SHG and THG are the most commonly investigated HG processes [13, 14].

In general, HG is utilized to convert an optical signal to higher frequencies. As discussed, SHG and THG are the most frequently used NLO processes for various frequency conversion applications. The SHG and THG processes are commonly explained with photon diagrams (Figure 10.1). As shown in Figure 10.1a, two incident photons at the frequency ω create light radiation at the frequency 2ω in SHG, while three incident photons create light radiation at the frequency 3ω in THG. Note that SHG and THG processes are typically phase sensitive; therefore efficient NLO frequency conversion requires a proper phase matching.

10.1.2 2DLM-Based SHG

Pioneering breakthrough on the micromechanical exfoliation of graphite to obtain sp^2 bonded atomic sheet of graphene by Geim and Novosolev in 2004 led to the 2010 Nobel Prize in Physics [28]. Apart from the single-crystal nature at atomic thickness, graphene exhibited world record electronic mobility, high surface area, superior Young's modulus, and excellent thermal conductivity. However, the zero bandgap of graphene hampers its prospect as an ideal candidate active layer for photoelectric chips due to the lack of sufficient charge carriers. To overcome these limitations, graphene hybridization with various low-dimensional materials and doping of graphene lattice [36, 37] have been designed. Parallelly, transition metal dichalcogenides (TMDCs) (e.g. tungsten disulfide and molybdenum disulfide) and boron nitride (BN) were developed [38]. Apart from TMDCs, new 2D materials such as elemental atomic sheets (e.g. phosphorene, stanene, germanene, and borophene) were developed in the following years [39–42]. 2D versions of MXenes (metal nitrides, metal carbides, and carbonitrides), metal phosphides [42], and metal oxides [43, 44]

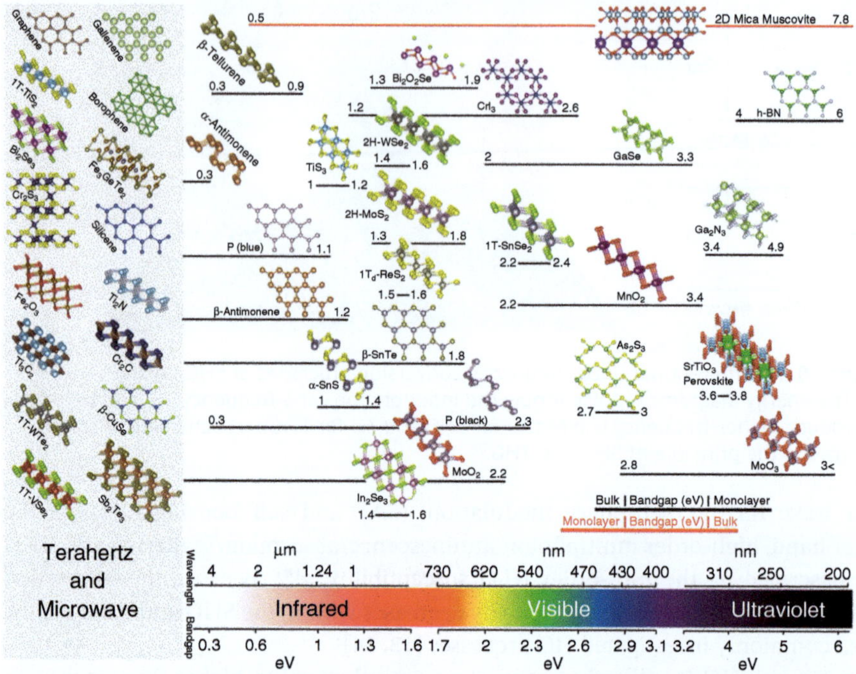

Figure 10.2 Selected family of 2D materials and their bandgaps. 2D materials are chosen for their experimental significance and demonstration, with depictions of a perspective view of their crystal structures. Arrangement is in accordance with their bandgap, guided by the bottom wavelength/bandgap scale, whereas the bar beneath each structure indicates bandgap range from bulk to monolayer. Typically, the bulk bandgap is smaller than that of its monolayer (black bars), but there are exceptions (red bars). 2D materials on the far left, indicated by a gray box, are zero or near-zero bandgap, metallic, or semimetallic. Source: [47]/Springer Nature/CC BY 4.0.

were discovered thereafter [45, 46]. A list of the 2D material family is exhibited in Figure 10.2 [47]. In this chapter, we present the latest progress on the SHG and THG of 2DLM and their heterostructures.

10.1.3 Graphene

As graphene is centrosymmetric, the electric dipolar SHG is forbidden in monolayer graphene. However, it is still possible to generate unusual SHG if the assumption can be extended from the dipole to multipolar approximation. It is found that the SHG exhibits the nature of electric quadrupole response arising from breaking the inversion symmetry by optical dressing and oblique incidence [48]. Moreover, several researchers theoretically investigated the HG response of the graphene nanostructures [49–52].

Regardless of the inversion symmetry in graphene, SHG has been observed from the interface in different hybrid graphene devices (Table 10.1). As shown in Figure 10.3a, the SHG signal is observed when the multilayer graphitic films and

Table 10.1 Recent SHG and THG studies in graphene.

NLO order	Emission wavelength (nm)	Fabrication of 2D materials	Susceptibility (m^2/V^2)	Published year
SHG	800	—	N/A	2009 [53]
	1305	—	N/A	2019 [48]
	730–830	CVD	N/A	2012 [54]
	740	CVD	N/A	2014 [55]
	740	Epitaxy	N/A	2014 [55]
	740	CVD	N/A	2013 [56]
	800	Exfoliation	N/A	2010 [57]
THG	1720	Exfoliation	0.4 E^{-16}	2013 [58]
	789	CVD	1.0 E^{-19}	2013 [59]
	1560	CVD	1.5 E^{-19}	2017 [60]
	1296	CVD	2.2 E^{-19}	2018 [61]
	1560	Exfoliation	4.0 E^{-15}	2018 [4]

exfoliated graphene are mounted on an oxidized silicon substrate. The isotropic SHG is observed as a graphene layer is rotated about the surface normal, while a threefold symmetry characteristic of significant SHG is observed in the multilayer graphitic films. The SHG signal from a bilayer graphene (BLG) sample shows a qualitatively different azimuthal pattern than that from graphene on SiO_2/(001)Si (Figure 10.3b). It also indicates the normalized SHG intensity of graphene system is slightly higher than that of the bare substrate [53].

The microscopic theory of the field-induced SHG for BLG was developed in Figure 10.3c,d. Once an in-plane electric field is applied, the AB-stacked BLG can exhibit a giant and tunable second-order nonlinear susceptibility $\chi^{(2)}$, which can be electrically tuned from 0 to ~105 pm/V. Thanks to the unique electronic spectrum of BLG, the unusually large $\chi^{(2)}$ arises from two different quantum-enhanced two-photon processes [62].

Furthermore, the SHG performance of graphene can be tuned by applied electrical field. As shown in Figure 10.4a, we find that optical SHG from the graphene/substrate system can be enhanced by the flow of direct electric current through graphene on substrate. The electric current can enhance surface SHG by about 25% for an epitaxial four-layer graphene film on SiC (0001) substrate and about 300% for a chemical-vapor-deposition (CVD) graphene monolayer on a SiO_2/Si (001) substrate. The vertical electrical field can be introduced by the induced field of trapped charges at the graphene/SiO_2 interface and the externally applied field on graphene (Figure 10.4b) [50]. A doping-induced SHG from graphene, comparable with the electric-dipole-allowed SHG in non-centrosymmetric 2D materials, is observed in Figure 10.4c,d. More remarkably, the SHG is widely tuned by chemical potential or carrier doping, which also arises in graphene-like Dirac materials [48].

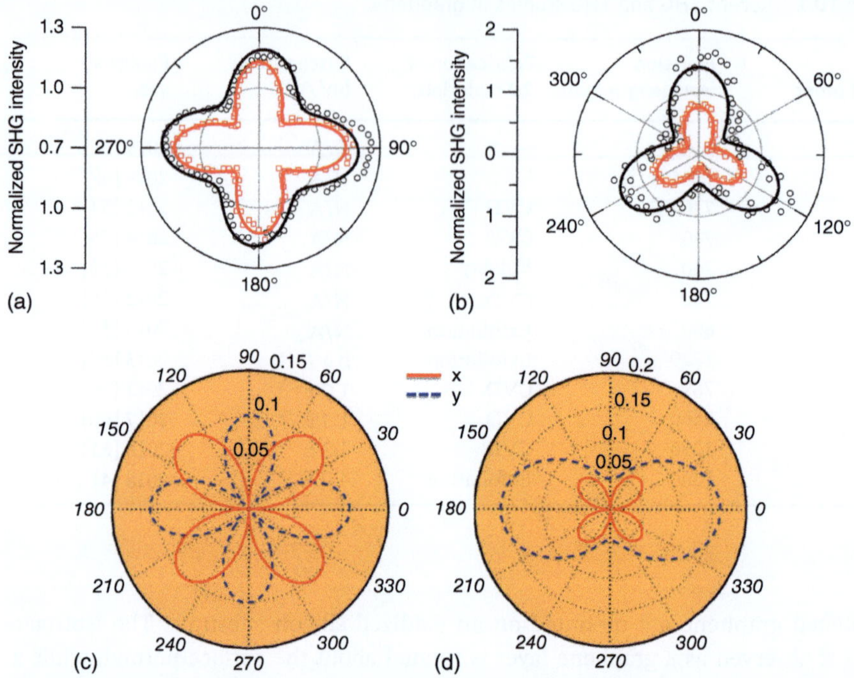

Figure 10.3 SHG signal from monolayer graphene (a) and bilayer graphene (b) on SiO_2/Si and pure SiO_2/Si as a function of azimuthal angle. Source: Reproduced with permission from J.J. Dean and H.M. van Driel [53]/American Chemical Society. (c, d) Second-order optical response of AB-stacked bilayer graphene (BLG) under linearly polarized incident laser beam. Source: Reproduced with permission from Wu et al. [62]/American Chemical Society.

In particular, SHG in graphene waveguides, graphene nanoparticle, and other nanostructures can be found [49–52]. More information about the linear optical and NLO responses of the graphene metasurfaces can be found in a recently published review [63].

10.1.4 Transition Metal Dichalcogenides

TMDC monolayers are known to have strong SHG signal under intense optical pump because of their intrinsically broken inversion symmetry (Table 10.2). The $\chi^{(2)}$ values of SHG in monolayer MoS_2 is orders of magnitude larger than that of bilayer or bulk MoS_2 with ~800-nm excitation wavelength [64, 65, 96]. Note that SHG signal in even-number-layered MoS_2 [97], spiral WS_2 nanosheets [98], and WSe_2 [99] flakes has been observed.

In addition, another interesting wavelength range is ~1550 nm that is commonly used for optical telecommunication-related applications. The reported $\chi^{(2)}$ values for MoS_2 [60, 67, 69] and $MoSe_2$ [70, 84]) with the fundamental wavelength close to 1550 nm are clearly lower than that measured at 800-nm lasers. The different results measured at 800 and 1550 nm are attributed to resonant enhancement of SHG [70]. SHG and THG images of the prepared MoS_2 sample with 1560-nm excitation wavelength are shown in Figure 10.5a. Because HG can occur when the harmonic

Table 10.2 Recent SHG studies in transition metal dichalcogenides.

Materials	NLO order	Emission wavelength (nm)	Fabrication of 2D materials	Susceptibility (m/V)	Published year
MoS_2	SHG	810	Exfoliation/CVD	$1.0E^{-7}$	2013 [64]
		680–1080	Exfoliation	$\sim 3.6E^{-10}$	2013 [65]
		1100–2000	CVD	$4.3E^{-10}$	2014 [66]
		1064–1600	CVD	$\sim 4.5E^{-11}$	2015 [67]
		800	CVT	$1.4E^{-9}$	2015 [68]
		1560	CVD	$2.9\ E^{-10}$	2017 [60]
		1560	Exfoliation	$2.2\ E^{-10}$	2017 [69]
		1000–1800	CVD	$4.0E^{-11}$	2017 [70]
		500–1600	CVT	$3.0E^{-11}$	2017 [71]
		1560	Exfoliation	$5.4E^{-12}$	2018 [72]
		900–1400	Exfoliation	$5.5E^{-10}$	2019 [73]
		750–950	CVD	N/A	2020 [74]
		900	Exfoliation	N/A	2020 [75]
WS_2	SHG	832	CVD	$4.5E^{-9}$	2014 [76]
		1064	Liquid	$4.6E^{-10}$	2017 [77]
		900 nm	Exfoliation	N/A	2017 [78]
		1560	Exfoliation/CVD	$1.6E^{-11}$	2018 [72]
		1030	Exfoliation	N/A	2018 [79]
		750–1040	CVD	N/A	2019 [80]
		800	Exfoliation	N/A	2019 [81]
		900–1400	CVD	N/A	2019 [73]
		1050	Exfoliation	$1.3E^{-9}$	2019 [82]
$MoSe_2$	SHG	752–855	Exfoliation	N/A	2015 [83]
		810	Exfoliation	$5.0E^{-11}$	2016 [84]
		820	CVD	N/A	2017 [85]
		1560	CVD	$3.7E^{-11}$	2018 [72]
		780	Exfoliation	N/A	2020 [86]
		1240–2480	Exfoliation	$6.0E^{-11}$	2015 [87]
WSe_2	SHG	1033–1416	Exfoliation	N/A	2015 [88]
		816	Exfoliation	$5.0\ E^{-9}$	2015 [89]
		1200–1420	Exfoliation	N/A	2017 [90]
		1560	CVD	$1.7E^{-11}$	2018 [72]
		800	Exfoliation	N/A	2019 [91]
		1050	CVD	$9.0E^{-10}$	2019 [82]
		800	Exfoliation	N/A	2019 [92]
$MoTe_2$	SHG	992	PVD	$3.2E^{-10}$	2016 [93]
		1040	CVT	N/A	2018 [94]
		1550	CVD	$2.5E^{-9}$	2018 [95]

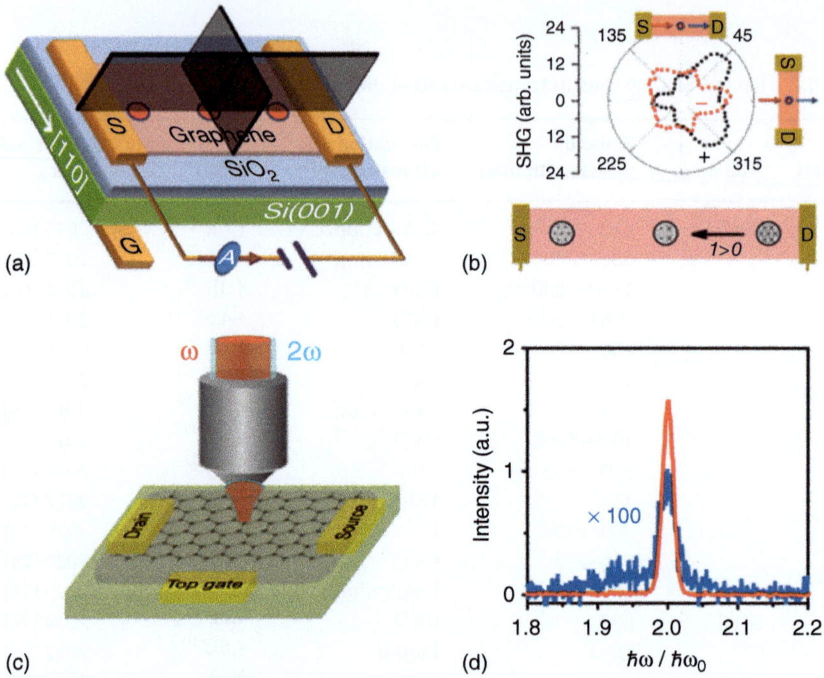

Figure 10.4 (a) Schematic illustration of the graphene/SiO_2/Si (001) sample with patterned electrodes. (b) Schematic illustration of the lateral distribution of local electric field across the SiO_2/Si interface. Source: Adapted from Majerus et al. [50]. (c) Comparison of SHG between normal and oblique incidences. (d) SHG spectra from heavily doped graphene at the same incident fluence. Source: Reproduced with permission from Zhang et al. [48]/American Physical Society.

energy is below the bandgap, the SHG signal is generated in 1L-MoS_2, while 2L-MoS_2 appears dark. As discussed above, the second-order nonlinear response is present in non-centrosymmetric 1L-MoS_2. However, strong THG is detected compared with SHG in mirrored 2L-MoS_2 and 7L-MoS_2 flakes. The output spectrum further confirms that we observe both SHG and THG. Peaks for THG and SHG at 520 and 780 nm can be seen in 1L-MoS_2, where the intensity of SHG is two orders of magnitude smaller than that of THG [69].

In Figure 10.5b, the whole area was measured with the same excitation laser using the same microscope. The exposed area is not observable, indicating that the SHG signal is not affected by the high fluence laser. In the THG image, a clear difference can be observed when the areas were exposed to higher-fluence laser or not. The THG signal on the GBs is identical to that on grains without chemical treatment. However, the GBs become clearly visible in the THG images after IPA treatment [100]. The SHG and THG for the four materials (MoS_2, $MoSe_2$, WS_2, and WSe_2) (shown in Figure 10.5c) were collected using a mode-locked fiber laser with 1560-nm center wavelength. The SHG and strong THG signals from all four TMDs were obviously observed. The small distance between the locations of the exfoliated materials allows us to easily compare the optical nonlinearities of the different materials [72].

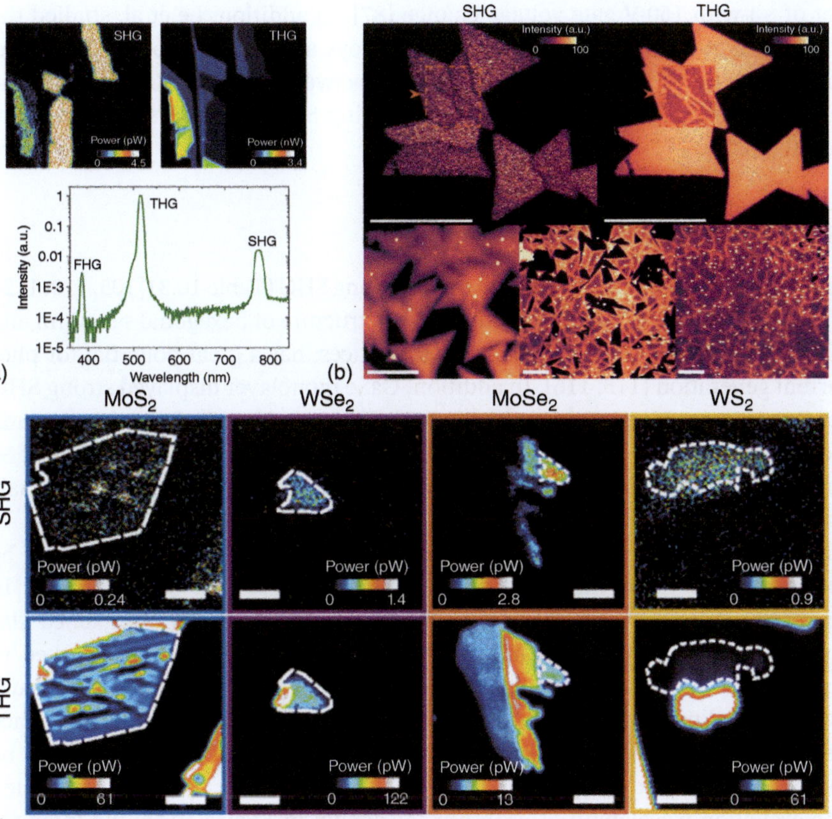

Figure 10.5 (a) SHG and THG maps of the MoS_2 flakes. Optical spectrum of the nonlinear signal from 1L-MoS_2 with a peak irradiance of ~30 GWcm^{-2}. Source: Saynatjoki et al. [69]/Springer Nature/CC BY 4.0. (b) Multiphoton imaging of CVD-grown large-area MoS_2 flakes. Source: Karvonen et al. [100]/Springer Nature/CC BY 4.0. (c) SHG (top) and THG (bottom) images from the areas marked with colored dashed rectangles in the optical image. Blue, MoS_2; purple, WSe_2; orange, $MoSe_2$; yellow, WS_2. The scale bar for MoS_2, WSe_2, and $MoSe_2$ is 10 μm, and that for WS_2 is 5 μm. Source: Reproduced with permission from Autere et al. [72]/American Physical Society.

Malard et al. [77] study the wavelength dependence of the SHG in monolayer MoS_2 by tuning the excitation wavelength from ≈680 to 1080 nm. It was found that the second-order susceptibility value of ~8 × 104 pm^2 V^{-1} was reported, when the SHG wavelength overlaps with the C exciton (~2.8 eV) [101].

For comparison purposes, an effective bulk-like second-order susceptibility of ~0.1 nm V^{-1} is typically obtained by dividing the sheet susceptibility with the thickness of the monolayer (e.g. 0.65 nm for MoS_2) [65]. Roughly, similar SHG values were also reported for WS_2 [76] and WSe_2 [89] and for MoS_2 in other studies [96] at the same wavelength range. The tunability of SHG response in monolayer WSe_2 has been demonstrated by using electrostatic doping, as the magnitude of the second-order NLO susceptibility of TMDs is largely related to the exciton resonance condition [87]. It was shown that the SHG from WSe_2 is enhanced by a factor of ∼4 with 160-V gate voltage change [87]. In addition, Le et al. studied the SHG dispersion in monolayer $MoS_{2(1-x)}Se_{2x}$ alloys as a function of Se doping (x) by measuring the SHG in a wavelength range between 500 and 900 nm [70]. The results demonstrate the possibility of engineering SHG of 2DLMs with material composition.

10.1.5 Other 2DLMs

There are many other 2DLMs that can offer a strong SHG (Table 10.3) [105, 110, 112–114]. Gallium selenide (GaSe), with the crystal structure of hexagonal selenium and gallium atoms separately occupying two sublattices, has a great potential for photocurrent generation [115, 116]. In addition, GaSe monolayer displayed strong SHG according to the nonlinear spectral [103, 117]. The monolayer GaSe is a honeycomb structure in which two layers of Ga atoms are sandwiched between two layers of Se atoms. In multilayer GaSe, the adjacent monolayers are coupled by van der Waals interactions [102].

Under excitation at 976 nm, a strong and eminent emission (488 nm) can be observed on the GaSe flakes with different thicknesses (7, 11, and 15 L) in Figure 10.6a. The different SHG responses are attributed to the weakened stability of non-centrosymmetric GaSe in the atomically thin flakes. As marked in Figure 10.6b, The SHG image with an emission filter from 450 to 500 nm clearly shows the thickness-dependent optical contrast. In addition, photoluminescence emission originating from two-photon excited fluorescence (TPEF) can also be obtained (Figure 10.6c). In Figure 10.6d, SHG signal of InSe layers was presented at the excitation wavelength of ∼800 nm. As shown in Figure 10.6e, the intensity of SHG in InSe at 400 nm increased with the increased laser excitation power (1 to 8 mW). Consistent with theory calculation, the measured intensity of SHG changes with different laser excitation power, and the square fitted result is shown

Table 10.3 Recent SHG studies in other 2DLMs.

Materials	NLO order	Emission wavelength (nm)	Fabrication of 2D materials	Susceptibility (m/V)	Published year
GaSe	SHG	1600	CVD	$0.7E^{-9}$	2015 [102]
GaSe		800	Exfoliation	$3.0\ E^{-11}$	2015 [103]
GaSe		1560	Exfoliation	$2.6\ E^{-11}$	2015 [104]
GaTe		1560	Exfoliation	$1.15\ E^{-12}$	2016 [105]
InSe		800	PVD	$6.39\ E^{-9}$	2018 [106]
InSe		800	Exfoliation	$1.3\ E^{-11}$	2019 [107]
hBN		810	Exfoliation	$1.0\ E^{-11}$	2013 [96]
hBN		810	Exfoliation	$4.2\ E^{-11}$	2019 [108]
ReS_2		1560	Exfoliation	$9.0\ E^{-10}$	2018 [109]
CrI_3		900	Exfoliation	$2.0\ E^{-9}$	2019 [110]
$AgInP_2S_6$		800	Exfoliation	$10^{-2}\chi^{(2)}\ MoS_2$	2016 [111]

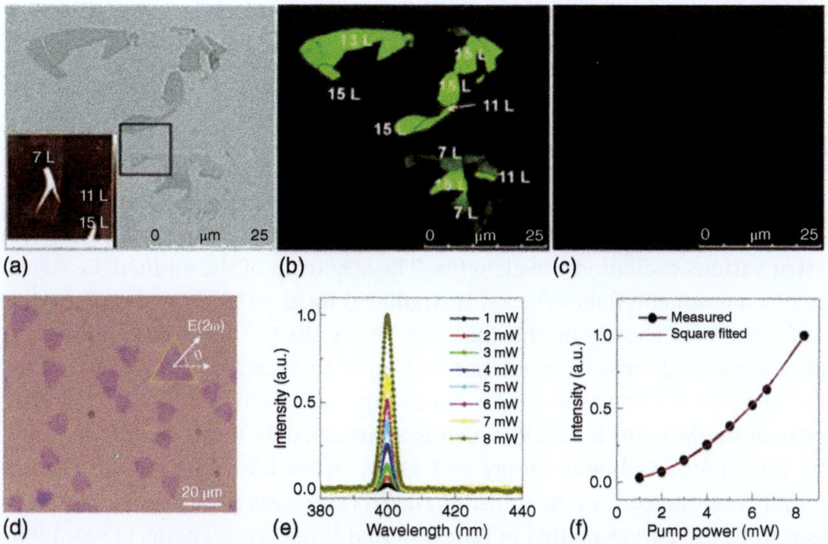

Figure 10.6 (a) Confocal laser scanning microscope image of layered GaSe flakes; (b) The SHG image with emission filter ranging from 450 to 500 nm excited by 976 nm fs laser light; (c) The fluorescent image with emission filter ranging from 600 to 700 nm. Source: Jie et al. [103]/John Wiley & Sons. (d) Optical image of monolayer InSe. (e) SHG intensity dependency on the emission power. (f) The measured and square fitted of the intensity of SHG dependence of pump power. Source: Reproduced with permission from Zhou et al. [106]/IOP Publishing.

in Figure 10.6f. The optical second-order susceptibility ($\chi^{(2)}$) of monolayer InSe is about 6.39×10^{-9}, which is around 2 orders of magnitude higher than that of monolayer MoS_2 [106].

Furthermore, the thickness-dependent SHG of hBN has been investigated, but its $\chi^{(2)}$ value is not large compared with the other 2DLMs [118, 119]. The SHG magnitude was recorded for the odd number of hBN layers, while the even number of hBN layers did not show any measurable SHG signals due to the inversion symmetry in the case of an even number of layers [96]. $AgInP_2S_6$, a new quaternary 2DLM, can be synthesized with the physical vapor transport (PVT) method and exfoliated into the atomically thin sheets [111]. An inversion symmetry breaking with odd layers like the 2H-MoS_2 was demonstrated by applying the SHG technique. Moreover, threefold symmetry was found in the monolayer $AgInP_2S_6$ in plane, and sixfold rotational symmetric response was achieved while rotating the sample normal to its surface [111].

Moreover, Glinka et al. has reported the SHG in topological insulator "Bi_2Se_3" and found that the SHG signal can be enhanced by decreasing the thickness of the Bi_2Se_3. It was also found that a 100-fold SHG was recorded while decreasing the thickness of Bi_2Se_3 from 40 to 6 nm [120]. By polarization-dependent SHG measurement, it was shown that the Weyl semimetal is a new polar material with high nonlinear second-order susceptibility $\chi^{(2)}$ [121].

10.1.6 2DLM-Based THG

10.1.6.1 Graphene

In contrast to SHG, THG can be possibly generated in materials regardless of whether they are centrosymmetric or not (Table 10.1) [13, 14]. For instance, the $\chi^{(3)}$ value of graphene is up to ~10^{-19} m^2 V^{-2}, which is nearly three orders of magnitude larger than that of fused silica (~ 2.5×10^{-22} m^2 V^{-2}) [122]. Surprisingly, the reported $\chi^{(3)}$ values of graphene range from ~10^{-15} m^2 V^{-2} [122, 123] to ~10^{-19} m^2 V^{-2} [60, 124] at various excitation wavelengths. The spectrum of the emitted 3ω signal produced by a graphene flake when it is irradiated by an ω beam with an average power of 10 mW is shown in the inset of Figure 10.7a. For convenience, the corresponding ω peak on-axis irradiance and pulse fluence are also shown on the top axis. Figure 10.7b shows the energy-band diagram, in which the exciton-shifted energy gap at the M point is 4.6 eV, which is approximately three-photon resonant with the fundamental photon energy of 1.57 eV. In addition, the graphene NLO susceptibility is enhanced via the transition matrix elements near the M point [59].

As seen from the curves plotted at four different input wavelengths (1300, 1400, 1566, and 1650 nm), the THG peak was readily observable at all μ in Figure 10.7c. Moreover, these shoulder-like features arise from stepwise switching off of resonant transitions in graphene when $|\mu|$ increases. Its THG intensity exhibits a maximum strength of ~30 times compared with that of $\mu = 0$ [125]. The standard optical multiphoton-absorption-excited fluorescence (MAEF) and THG images of a single-layer graphene are shown in Figure 10.7(d–h). The fluorescence and THG signals are both clearly distinguishable from the background, and the THG signal is nearly constant over the whole single-layer graphene flake with a length of more than 50 μm [126]. On the other hand, the large deviation of the measured $\chi^{(3)}$ values in graphene and other 2DLMs can be very sensitive to various parameters (e.g. substrates [127, 128], defects [129], doping [127], strain [85], and chemical treatment [100]).

10.1.7 Transition Metal Dichalcogenides

In the past few years, a growing number of studies on the THG of layered TMDCs have been published (Table 10.4). Because THG does not require the inversion symmetry to be broken, it can be observed in both odd and even layers of TMDCs. The THG polarization output signal does not depend on the orientation of those monolayer TMDCs and is always the same as the polarization of the incoming laser beam [69]. It is shown that the estimated $\chi^{(3)}$ value of the monolayer TMDCs is around 10^{-19} m^2 V^{-2} that is about three times smaller than that of monolayer graphene. The $\chi^{(3)}$ value of 7-layer-thick MoS_2 flake is also calculated by solving the coupled-wave equation [130]. It is clear that THG can be used as a large-area characterization method for determining the crystallographic orientations of layered TMDCs. AFM (Figures 10.8a) and Raman microscopy (Figures 10.8b) are used to characterize monolayer MoS_2, showing the successful preparation of monolayer MoS_2 with the thickness of ~0.7 nm and separation of ~19.4 cm^{-1} between $E^1{}_{2g}$ and

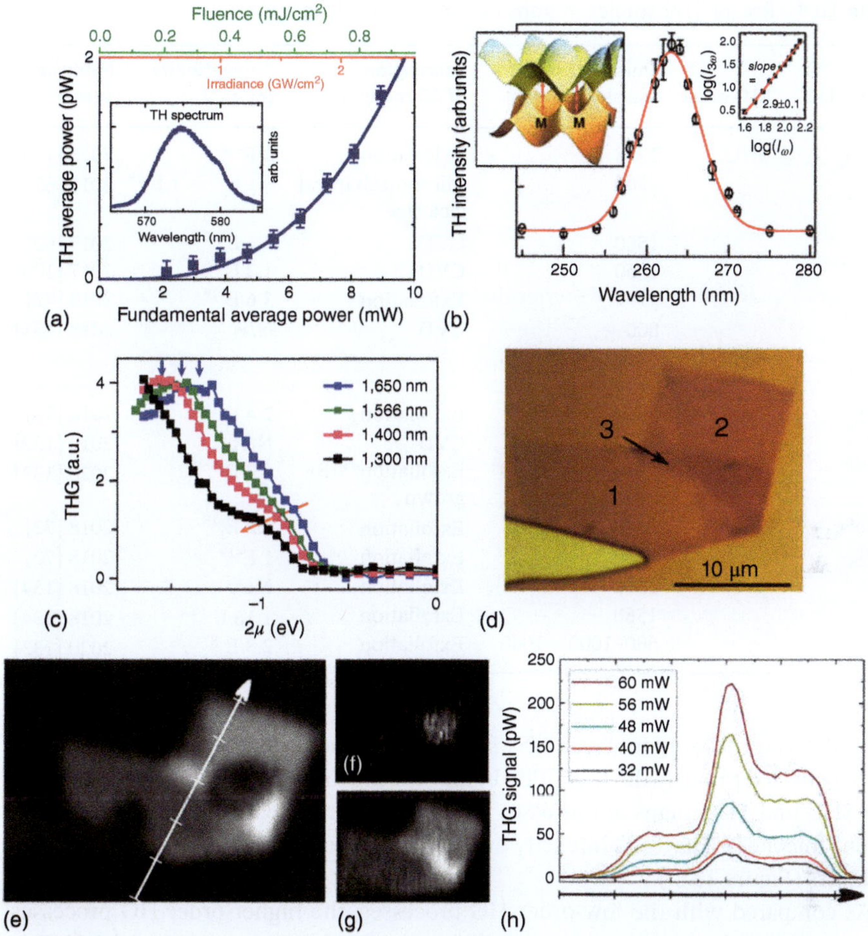

Figure 10.7 (a) The average power of the third harmonic as a function of the average power. Source: Reproduced with permission from Kumar et al. [28]/American Physical Society. (b) THG spectrum along with a right inset log–log plot showing a cubic power dependence of the TH signal intensity. Source: Reproduced with permission from Hong et al. [59]/American Physical Society/CC BY 4.0. (c) THG signal as a function of 2μ generated by different input wavelengths: 1300, 1400, 1566, and 1650 nm. Reproduced with permission from Jiang et al. [125]/Springer Nature. (d), THG (e), SHG (f), and MAEF (g) images of few-layer exfoliated graphene. (h) The THG signal along the arrow with different excitation powers. Source: Reproduced with permission from Antti et al. [126]/American Chemical Society.

$A^1{}_g$ Raman modes. Obviously, a higher contrast image recording the THG intensity with the linear optical microscopy part of the setup is produced (Figure 10.8c) [60].

Furthermore, SHG signal intensity is particularly sensitive to the strain in TMDCs. In Figure 10.8(d,e), the SHG (644 nm) and THG (429 nm) signals from exfoliated few-layer WS_2 flakes could be collected under excitation of a linearly polarized ultrafast laser (1288 nm, 100 fs). However, the THG signals in WS_2 (from one to four layers) without symmetric constraint is obviously strong. These comparison results show that THG has the potential to measure the strain in TMDCs. In Figure 10.8f, the

Table 10.4 Recent THG studies in transition metal dichalcogenides.

Materials	NLO order	Emission wavelength (nm)	Fabrication of 2D materials	Susceptibility (m^2/V^2)	Published year
MoS_2	THG	1758	Exfoliation	2 E^{-19}	2013 [130]
		1560	Micromechanical cleavage	~1 E^{-17} – 1 E^{-19}	2017 [69]
		1560	CVD	2.4 E^{-10}	2017 [60]
		1560	CVD	1.2 E^{-10}	2017 [100]
		1560	Exfoliation	3.6 E^{-19}	2018 [72]
		800 + supercontinuum generator	CVD	N/A	2018 [131]
WS_2	THG	1560	Exfoliation	2.4 E^{-19}	2018 [72]
		1288	CVD	N/A	2019 [132]
		1040 + 680–1000	Exfoliation/MBE grown	6.7 E^{-9}	2020 [133]
$MoSe_2$	THG	1560	Exfoliation	2.2 E^{-19}	2018 [72]
WSe_2	THG	1560	Exfoliation	1 E^{-19}	2018 [72]
		1546	Exfoliation	N/A	2018 [134]
		1560	Exfoliation	1.16 E^{-19}	2018 [134]
		680–1000 + 1040	Exfoliation	2.3 E^{-8}	2020 [133]

optical spectra from monolayer WS_2 flakes under a series of uniaxial tensile strain were collected to verify the feasibility in monitoring the strain. Figure 10.8g,h shows the SHG and THG maps of the WSe_2 flakes, exhibiting the distinct signal with different thickness. The THG intensity over regions clearly scales up with the number of layers (Figure 10.8i).

As compared with the low-order HG processes, the higher-order HG processes are usually weak. However, some literature suggested that the signal of the THG is greater than the SHG in the case of monolayer MoS_2. THG can also be a useful probe for recognizing the boundaries and grains, which is better than PL imaging, Raman imaging, and conventional SHG imaging [100].

10.1.7.1 Other 2DLMs

Black phosphorus (BP), in which each phosphorus atom is bonded with three neighboring ones via sp^3 hybridized orbitals, is a layered material with an orthorhombic structure [135]. As a recently discovered two-dimensional semiconductor, BP exhibits a tunable bandgap determined by the number of layers from 0.3 (bulk) to 2.0 eV (single layer) [136]. In the past few years, there has been increased interest in the NLO response of BP due to its exceptional anisotropy [137–139] and layer-dependent bandgap [140].

It is demonstrated that THG in few-layer BP can vary significantly depending on sample thickness and light polarization (Table 10.5). The increased THG signals are observed from 6- to 15-nm thickness of BP flakes, and the $\chi^{(3)}$ value was recorded as 1.64×10^{-19} m^2 V^{-2} [142]. The strong THG anisotropic results delivered by BP can be shown on the different BP flakes by the photocurrent measurement [143].

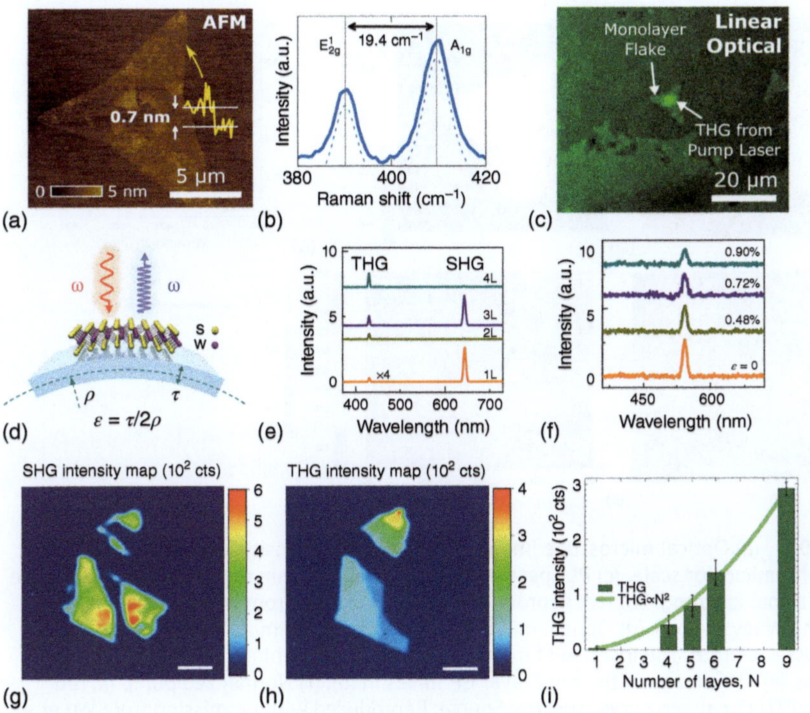

Figure 10.8 (a) AFM image and height profile inset, (b) Raman spectrum, and (c) optical image of monolayer MoS_2 flake on Si/SiO_2 substrate. Source: Woodward et al. [60]/IOP Pubilishing/Public Domain: THG response of monolayer WS_2 under strain. (d) Schematic diagram of two-point bending configuration and THG process in monolayer WS_2. The supporting substrate is optical transparent flexible Acrylic. The left red and right violet waves indicate excitation beam (ω) and generated THG signal (3ω), respectively. (e) SHG and THG spectra of mono- and few-layer WS_2. (f) The THG spectral evolution of monolayer WS_2 under uniaxial tensile strain. Source: Reproduced with permission from Liang et al. [132]/John Wiley & Sons. Optical image of the sample with labeled number of layers (N). (g) Spatial SHG and (h) THG intensity mappings across the WSe_2 sample. (i) Spatial average THG intensities as a function of N. Source: Rosa et al. [134]/Springer Nature/Public Domain.

Table 10.5 Recent THG studies in other 2DLMs.

Materials	NLO order	Emission wavelength	Fabrication of 2D materials	Susceptibility (m^2/V^2)	Published year
BP	THG	1560	Exfoliation	$10 \times \chi^{(3)}$graphene	2016 [141]
		1560	Exfoliation	1.4 E^{-19}	2017 [142]
		1560	Exfoliation	1.64 E^{-19}	2017 [143]
GaSe	THG	1560	Exfoliation	1.6 E^{-16}	2017 [4, 104]
GaTe	THG	1560	Exfoliation	2.0 E^{-16}	2016 [105]
$SnSe_2$	THG	1560	Exfoliation	4.14 E^{-19}	2019 [114]
Cd_3As_2	THG	1560	Exfoliation	2.3 E^{-19}	2020 [144]
hBN	THG	1560	Exfoliation	0.84 E^{-20}	2021 [145]

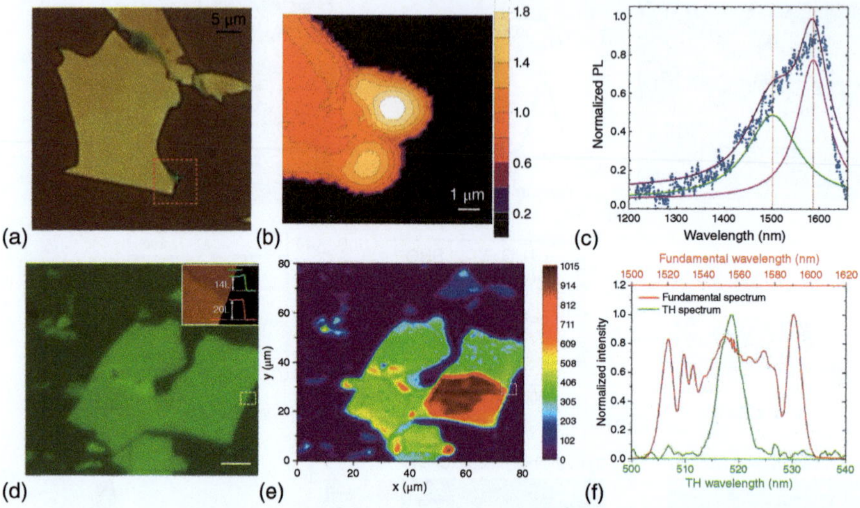

Figure 10.9 (a) Optical microscope image of BP flake. (b) Intensity map of the THG signal in a logarithmic color scale. (c) PL spectrum and analysis in a thin region of the flake where the THG signal is strong. Source: Reproduced with permission from Rodrigues et al. [141]/John Wiley & Sons. (d) Optical microscopy photograph of multilayer BP flakes on a cover glass substrate. The inset: AFM micrograph of the region highlighted by a dashed white box. (e) THG image of the multilayer BP flakes in (d). (f) Normalized pump (x, red curve) and TH ($3x$, green curve) spectra. Source: Reproduced with permission from Wu et al. [146]/AIP Publishing LLC.

As shown in Figure 10.9a–c, we found that few-layer phosphorene yields enhanced THG intensity. Figure 10.9b shows the THG intensity image obtained by collecting the reflected signal in the CCD camera. It indicates that the rightmost BP border presents a stronger signal than that of BP regions removed from the edges. This presented PL of same border at around 1501 and 1586 nm is consistent with four- to five-layer black phosphorous (Figure 10.9c) [141]. As expected, the THG signal at 519 nm is one-third of the fundamental wavelength of the pump laser at 1557 nm [142].

The optical microscopy photograph of BP flakes is shown in Figure 10.9d. The inset of Figure 10.9d is the AFM micrograph indicating the bottom side (10.5 nm; 20 L) that is thicker than the top side (7.4 nm; 14 L). It is clear that the THG signal in Figure 10.9e is nearly centered at one-third of the fundamental wavelength (1560 nm). Figure 10.9f displays the cubic power dependence of the THG signal on the pump intensity, plotted on a log–log plot (yielding a slope of 2.96). The cubic power dependence and frequency-tripled signal fully confirm the generated third-harmonic radiation from BP flake [146].

In addition, 3D Dirac semimetal (3D DSM), a graphene 3D analog, has similar optical and electronic properties that of graphene [147–150]. Surprisingly, the Fermi level and permittivity of the 3D DSM can be efficiently controlled [118, 151–153]. Moreover, the 3D DSM is fairly stable against environmental temperature [154]. 3D DSM-based linear and nonlinear optical properties have been reported by many

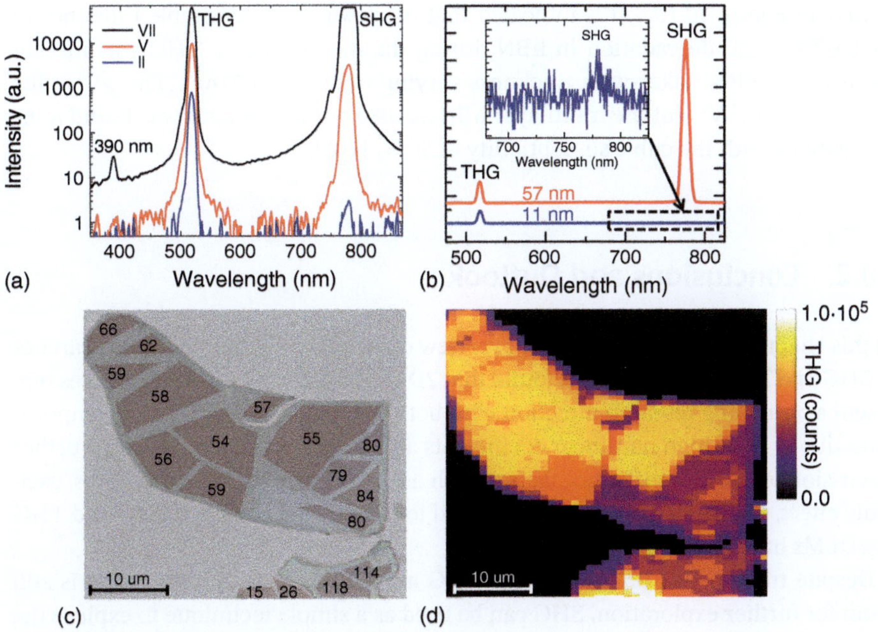

Figure 10.10 (a) Measured spectra of the generated light from three different positions on the GaSe flake. Source: Karvonen et al [104]/Springer Nature/Public Domain. (b) Measured spectra of the nonlinear signals from two different positions with different thicknesses (11 and 57 nm) on the GaTe flake. Source: Reproduced with permission from Susoma et al. [105]/AIP Publishing. (c) Optical image of the hBN flakes with a thickness map overlaid on top. (d) THG intensity map of the hBN flake. Source: Reproduced with permission from Popkova et al. [145]/American Chemical Society.

researchers in the infrared and terahertz regions [154–157]. Among the 3D DSMs, Cd_3As_2 with a large linear energy momentum space is the most widely studied material [158]. A comparison of THG signal and third-order nonlinear susceptibility $\chi^{(3)}$ was made between the Cd_3As_2 on the mica substrate, glass, graphene, and MoS_2. It was found that the $\chi^{(3)}$ value of the Cd_3As_2 is 2.3×10^{-19} m^2 V^{-2} that is 1 000 times greater than the $\chi^{(3)}$ value of glass and 1.3 and 1.5 times greater than that of the graphene and MoS_2, respectively [144].

GaSe and gallium tellurium (GaTe), layered III–VI semiconductors, have a great potential for THG (Table 10.5). The spectra of the generated light measured from the three different areas of GaSe flakes are presented in Figure 10.10a. Strong peaks at the wavelengths of ~780 and ~520 nm were observed for the flakes with ~12 layers (area V) and 40 layers (area VII) [104]. Meanwhile, Figure 10.10b present the spectra of generated light from GaTe flakes at two different positions with different thickness (11 and 57 nm). The positions and narrowness of the peaks at the wavelength of 520 and 780 nm further verify that the detected signals are second and third harmonics at the excitation wavelength of 1560 nm [105]. Furthermore, the optical image of the hBN flakes with its thickness map superimposed on top is shown in Figure 10.10c. Then, the THG intensity maps of the flakes were measured at the pump power of

30 mW in Figure 10.10d. It is indicated that there was no undesirable influence of edge effects and degradation in hBN during the measurement. THG was investigated in the hBN flakes with thickness varying from 5 to 170 nm. The $\chi^{(3)}$ value (0.84×10^{-20} m^2 V^{-2}) of the multilayer hBN at 1080 nm wavelength was found to be comparable with the cubic susceptibility of Si_3N_4 [145].

10.2 Conclusions and Outlook

In this chapter, we have provided an overview of the previous and present researches in SHG and THG of 2DLMs. It is found that 2DLM-based SHG and THG devices represent a promising research direction with both fundamental and practical implications due to their high nonlinear coefficients and large electrical tunability. Further discussion on various influence factors such as interlayer twisting, edge states, excitonic effect, strain, charge, and electric field to tune the efficiency of SHG and THG in 2DLMs have also been provided.

Despite tremendous progress in the SHG and THG of the 2DLMs, there is still room for further exploration. SHG can be used as a simple technique to explore the magnetic order in 2D magnetic materials. It is possible to unveil the dynamic influence of various optical excitations on the NLO processes in 2DLMs by combining the ultrafast transient spectroscopy with SHG and THG. In addition, 2DLM-based HG shows great potential in noninvasive bioimaging applications. It is believed that 2DLM-based SHG and THG will pave the way for a wide variety of novel applications (nonlinear integrated photonics, surface science, active nanodevice, and biosensing).

References

1 Franken, P.A., Hill, A.E., Peters, C.W., and Weinreich, G. (1961). *Physical Review Letters* 7: 118–119.
2 Kangpeng Wang, J.W., Fan, J., Lotya, M. et al. (2013). *ACS Nano* 7: 9260–9267.
3 He, M., Quan, C., He, C. et al. (2017). *The Journal of Physical Chemistry C* 121: 27147–27153.
4 Autere, A., Jussila, H., Dai, Y. et al. (2018). *Advanced Materials* 30: 1705963–1705939.
5 Richard, F.L.B., Farrer, A., Chen, V.W., and Fourkas, J.T. (2005). *Nano Letters* 5: 1139–1142.
6 Guang, L.-S.T., He, S., Zheng, Q., and Prasad, P.N. (2008). *Chemical Reviews* 108: 1245–1330.
7 Zhao, M., Chang, M.J., Wang, Q. et al. (2015). *Chemical Communication* 51: 12262–12265.
8 Zhao, M., Liu, K., Zhang, Y.-D. et al. (2015). *Materials Horizons* 2: 619–624.
9 Zhao, M., Peng, R., Zheng, Q. et al. (2015). *Nanoscale* 7: 9268–9274.

10 Zhao, M., Ye, Z., Suzuki, R. et al. (2016). *Light-Science & Applications* 5: 16131–16136.

11 Liang, J., Tu, T., Chen, G. et al. (2020). *Advanced Materials* 32: 2002831–2002837.

12 Ye, K., Liu, L., Liu, Y. et al. (2019). *Advanced Optical Materials* 7: 1900815–1900822.

13 Shen, Y.R. (1984). *The Principles of Nonlinear Optics*. New York: Wiley Press.

14 Boyd, R.W. (2007). *Nonlinear Optics*. New York: Academic Press.

15 Zipfel, W.R., Williams, R.M., and Webb, W.W. (2003). *Nature Biotechnology* 21: 1369–1377.

16 Ed, K.M.K., Wooten, L., Yi-Yan, A. et al. (2000). *IEEE Journal of Selected Topics in Quantum Electronics* 6: 69–82.

17 Keller, U. (2003). *Nature* 424: 831–838.

18 Asano, M., Komori, S., Ikuta, R. et al. (2016). *Optics Letters* 41: 5793–5796.

19 Liu, S.J., Zheng, Y.L., and Chen, X.F. (2017). *Optics Letters* 42: 3626–3629.

20 Walsh, B.M. (2017). *Optical Materials* 65: 2–7.

21 Guo, Z., Li, J., Liang, J. et al. (2022). *Nano Letters* 22: 846–852.

22 Venugopal Rao, D.N.R.S., Akkara, J.A., DeCristofano, B.S., and Rao, D.V.G.L.N. (1998). *Chemical Physics Letters* 297: 491–498.

23 Perry, K.M.J.W., Lee, I.-Y.S., Wu, X.-L. et al. (1996). *Science* 273: 1533–1536.

24 Xu, Y., Liu, Z., Zhang, X. et al. (2009). *Advanced Materials* 21: 1275–1279.

25 Chang, D.E., Vuletic, V., and Lukin, M.D. (2014). *Nature Photonics* 8: 685–694.

26 Leuthold, J., Koos, C., and Freude, W. (2010). *Nature Photonics* 4: 535–544.

27 Kauranen, M. and Zayats, A.V. (2012). *Nature Photonics* 6: 737–748.

28 Novoselov, A.K.G.K.S., Morozov, S.V., Jiang, D. et al. (2004). *Science* 306: 666–668.

29 Castro Neto, A.H., Guinea, F., Peres, N.M.R. et al. (2009). *Reviews of Modern Physics* 81: 109–162.

30 Novoselov, K.S., Fal'ko, V.I., Colombo, L. et al. (2012). *Nature* 490: 192–200.

31 Bao, Q., Zhang, H., Wang, Y. et al. (2009). *Advanced Functional Materials* 19: 3077–3083.

32 Zhipei Sun, T.H., Bonaccorso, F., Torrisi, F. et al. (2010). *ACS Nano* 4: 803–810.

33 Low, T., Chaves, A., Caldwell, J.D. et al. (2017). *Nature Materials* 16: 182–194.

34 Sun, Z., Martinez, A., and Wang, F. (2016). *Nature Photonics* 10: 227–238.

35 Wen, X., Gong, Z., and Li, D. (2019). *InfoMat* 1: 317–337.

36 Katoh, T., Imamura, G., Obata, S., and Saiki, K. (2016). *RSC Advances* 6: 13392–13398.

37 Bhushan, R., Kumar, P., and Thakur, A.K. (2020). *Solid State Ionics* 353: 115371–115376.

38 Manzeli, S., Ovchinnikov, D., Pasquier, D. et al. (2017). *Nature Reviews Materials* 2: 1–15.

39 Molle, A., Goldberger, J., Houssa, M. et al. (2017). *Nature Materials* 16: 163–169.

40 Wang, Z., Zhang, R., Zhao, M. et al. (2018). *Journal of Materials Chemistry A* 6: 23773–23779.

41 Wang, Z., Gao, L., Wei, X. et al. (2020). *Journal of Materials Chemistry C* 8: 1796–1802.

42 Mannix, A.J., Zhang, Z.H., Guisinger, N.P. et al. (2018). *Nature Nanotechnology* 13: 444–450.

43 Kumar, P., Liu, J., Ranjan, P. et al. (2018). *Small* 14: 1703346.

44 Chahal, S., Kauzlarich, S.M., and Kumar, P. (2021). *ACS Materials Letters* 3: 631–640.

45 Shekhirev, M., Shuck, C.E., Sarycheva, A., and Gogotsi, Y. (2021). *Progress in Materials Science* 120: 100757–100787.

46 Anasori, B., Lukatskaya, M.R., and Gogotsi, Y. (2017). *Nature Reviews Materials* 2: 16098–16115.

47 Chaves, A., Azadani, J.G., Alsalman, H. et al. (2020). *npj 2D Materials and Applications* 4: 29–50.

48 Zhang, Y., Huang, D., Shan, Y. et al. (2019). *Physical Reviews Letters* 122: 047401–047406.

49 Smirnova, D. and Kivshar, Y.S. (2014). *Physical Review B* 90: 165433–165437.

50 Majerus, B., Butet, J., Bernasconi, G.D. et al. (2017). *Optics Express* 25: 27015–27027.

51 Zhou, R., Yang, S., Liu, D., and Cao, G. (2017). *Optics Express* 25: 31478–31491.

52 You, J.W. and Panoiu, N.C. (2020). *Physical Review B* 102: 121403–121408.

53 Dean, J.J. and van Driel, H.M. (2009). *Applied Physics Letters* 95: 261910–261913.

54 Bykov, A.Y., Murzina, T.V., Rybin, M.G., and Obraztsova, E.D. (2012). *Physical Review B* 85: 121413–121416.

55 An, Y.Q., Rowe, J.E., Dougherty, D.B. et al. (2014). *Physical Review B* 89: 115310–115322.

56 An, Y.Q., Nelson, F., Lee, J.U., and Diebold, A.C. (2013). *Nano Letters* 13: 2104–2109.

57 Dean, J.J. and van Driel, H.M. (2010). *Physical Review B* 82: 125411–125420.

58 Kumar, N., Kumar, J., Gerstenkorn, C. et al. (2013). *Physical Review B* 87: 121406–121410.

59 Hong, S.-Y., Dadap, J.I., Petrone, N. et al. (2013). *Physical Review X* 3: 021014–021023.

60 Woodward, R.I., Murray, R.T., Phelan, C.F. et al. (2016). *2D Materials* 4: 011006–011014.

61 Soavi, G., Wang, G., Rostami, H. et al. (2018). *Nature Nanotechnology* 13: 583–588.

62 Wu, S., Mao, L., Jones, A.M. et al. (2012). *Nano Letters* 12: 2032–2036.

63 Lin, H., Lin, K.-T., Yang, T., and Jia, B. (2021). *Advanced Materials Technologies* 6: 2000963–2000988.

64 Kumar, N., Najmaei, S., Cui, Q. et al. (2013). *Physical Review B* 87: 161403–161408.

65 Malard, L.M., Alencar, T.V., Barboza, A.P.M. et al. (2013). *Physical Review B* 87: 201401–201405.

66 Clark, D.J., Senthilkumar, V., Le, C.T. et al. (2014). *Physical Review B* 90: 21409–121413.
67 Clark, D.J., Le, C.T., Senthilkumar, V. et al. (2015). *Applied Physics Letters* 107: 131113–131117.
68 Mishina, E., Sherstyuk, N., Lavrov, S. et al. (2015). *Applied Physics Letters* 106: 131901–131906.
69 Saynatjoki, A., Karvonen, L., Rostami, H. et al. (2017). *Nature Communication* 8: 893–900.
70 Le, C.T., Clark, D.J., Ullah, F. et al. (2016). *ACS Photonics* 4: 38–44.
71 Shi, J., Yu, P., Liu, F. et al. (2017). *Advanced Materials* 29: 1701486–1701494.
72 Autere, A., Jussila, H., Marini, A. et al. (2018). *Physical Review B* 98: 115426–115432.
73 Stiehm, T., Schneider, R., Kern, J. et al. (2019). *The Review of Scientific Instruments* 90: 083102–083109.
74 Carvalho, B.R., Wang, Y., Fujisawa, K. et al. (2020). *Nano Letters* 20: 284–291.
75 Khan, A.R., Liu, B., Zhang, L. et al. (2020). *Advanced Optical Materials* 8: 2000441–2000448.
76 Janisch, C., Wang, Y., Ma, D. et al. (2014). *Scientific Reports* 4: 5530–5534.
77 Forcherio, G.T., Riporto, J., Dunklin, J.R. et al. (2017). *Optics Letters* 42: 5018–5021.
78 Liu, J., Lo, T.W., Sun, J. et al. (2017). *Journal of Materials Chemistry C* 5: 11239–11245.
79 Psilodimitrakopoulos, S., Mouchliadis, L., Paradisanos, I. et al. (2018). *Light-Science & Applications* 7: 18005–18014.
80 Kikuchi, Y., Miyauchi, Y., Takaoka, R. et al. (2019). *Physical Review B* 100: 75301–075307.
81 Liu, Z., Murphy, A.W.A., Kuppe, C. et al. (2019). *ACS Nano* 13: 3896–3909.
82 Mokim, M., Card, A., and Ganikhanov, F. (2019). *Optical Materials* 88: 30–38.
83 Kim, D.H. and Lim, D. (2015). *Journal of the Korean Physical Society* 66: 816–820.
84 Le, C.T., Clark, D.J., Ullah, F. et al. (2016). *Annalen der Physik* 528: 551–559.
85 Liang, J., Zhang, J., Li, Z. et al. (2017). *Nano Letters* 17: 7539–7543.
86 Ho, Y.W., Rosa, H.G., Verzhbitskiy, I. et al. (2020). *ACS Photonics* 7: 925–931.
87 Seyler, K.L., Schaibley, J.R., Gong, P. et al. (2015). *Nature Nanotechnology* 10: 407–411.
88 Wang, G., Marie, X., Gerber, I. et al. (2015). *Physcial Review Letter* 114: 097403–097408.
89 Ribeiro-Soares, J., Janisch, C., Liu, Z. et al. (2015). *2D Materials* 2: 045015–045021.
90 Brekhov, K.A., Grishunin, K.A., Ilyin, N.A. et al. (2018). *Technical Physics Letters* 43: 1112–1114.
91 Han, A., Aljarb, A., Liu, S. et al. (2019). *Nanoscale Horizons* 4: 1434–1442.
92 Zeng, Z., Sun, X., Zhang, D. et al. (2019). *Advanced Functional Materials* 29: 1806874–1806881.
93 Beams, R., Cancado, L.G., Krylyuk, S. et al. (2016). *ACS Nano* 10: 9626–9636.

94 Hoang, A.T., Shinde, S.M., Katiyar, A.K. et al. (2018). *Nanoscale* 10: 21978–21984.

95 Song, Y., Tian, R., Yang, J. et al. (2018). *Advanced Optical Materials* 6: 1701334–1701339.

96 Li, Y., Rao, Y., Mak, K.F. et al. (2013). *Nano Letters* 13: 3329–3333.

97 Klein, J., Wierzbowski, J., Steinhoff, A. et al. (2017). *Nano Letters* 17: 392–398.

98 Fan, X., Jiang, Y., Zhuang, X. et al. (2017). *ACS Nano* 11: 4892–4898.

99 Yu, H., Talukdar, D., Xu, W. et al. (2015). *Nano Letters* 15: 5653–5657.

100 Karvonen, L., Saynatjoki, A., Huttunen, M.J. et al. (2017). *Nature Communication* 8: 15714–15421.

101 Wei-Ting Hsu, Z.-A.Z., Li, L.-J., Chen, C.-H. et al. (2014). *ACS Nano* 8: 2951–2958.

102 Zhou, X., Cheng, J., Zhou, Y. et al. (2015). *Journal of the American Chemical Society* 137: 7994–7997.

103 Jie, W., Chen, X., Li, D. et al. (2015). *Angewandte Chemie International Edition* 54: 1185–1189.

104 Karvonen, L., Saynatjoki, A., Mehravar, S. et al. (2015). *Scientific Reports* 5: 10334.

105 Susoma, J., Karvonen, L., Säynätjoki, A. et al. (2016). *Applied Physics Letters* 108: 073103–073108.

106 Zhou, J., Shi, J., Zeng, Q. et al. (2018). *2D Materials* 5: 025019–025028.

107 Hao, Q., Yi, H., Su, H. et al. (2019). *Nano Letters* 19: 2634–2640.

108 Sejeong Kim, J.E.F., Gardner, A., Li, C. et al. (2019). *Optics Letters* 44: 5792–5795.

109 Song, Y., Hu, S., Lin, M.-L. et al. (2018). *ACS Photonics* 5: 3485–3491.

110 Sun, Z., Yi, Y., Song, T. et al. (2019). *Nature* 572: 497–501.

111 Wang, X., Du, K., Liu, W. et al. (2016). *Applied Physics Letters* 109: 123103–123108.

112 Huttunen, M.J., Herranen, O., Johansson, A. et al. (2013). *New Journal of Physics* 15: 083043–083057.

113 Cappello, V., Marchetti, L., Parlanti, P. et al. (2016). *Scientific Reports* 6: 1–15.

114 Biswas, R., Dandu, M., Menon, S. et al. (2019). *Optics Express* 27: 28855–28865.

115 Late, D.J., Liu, B., Luo, J. et al. (2012). *Advanced Materials* 24: 3549–3554.

116 PingAn Hu, Z.W., Wang, L., Tan, P., and Xiao, K. (2012). *ACS Nano* 7: 5988–5994.

117 Li, X., Lin, M.W., Puretzky, A.A. et al. (2014). *Scientific Reports* 4: 5497–5505.

118 Ooi, K.J.A., Ang, Y.S., Zhai, Q. et al. (2019). *APL Photonics* 4: 034402–034409.

119 Cunha, R., Cadore, A., Ramos, S. et al. (2020). *Journal of Physics: Condensed Matter* 32: 19LT01–19LT06.

120 Glinka, Y.D., Babakiray, S., Johnson, T.A. et al. (2015). *Physical Review B* 91: 195307–195321.

121 Wu, L., Patankar, S., Morimoto, T. et al. (2016). *Nature Physics* 13: 350–355.

122 Saynatjoki, L.K.A., Riikonen, J., Kim, W. et al. (2013). *ACS Nano* 10: 8441–8446.

123 Hendry, E., Hale, P.J., Moger, J. et al. (2010). *Physics Review Letters* 105: 097401–097404.

124 Cheng, J.L., Vermeulen, N., and Sipe, J.E. (2014). *New Journal of Physics* 18: 029501–029518.

125 Jiang, T., Huang, D., Cheng, J. et al. (2018). *Nature Photonics* 12: 430–436.

126 Saynatjoki, L.K.A., Riikonen, J., Kim, W. et al. (2013). *ACS Nano* 7: 8441–8446.

127 Shi, H., Wang, C., Sun, Z. et al. (2014). *Optics Express* 22: 19375–19385.

128 Zeng, J., Yuan, M., Yuan, W. et al. (2015). *Nanoscale* 7: 13547–13553.

129 Nan, H., Wang, Z., Wang, W. et al. (2014). *ACS Nano* 8: 5738–5745.

130 Wang, R., Chien, H.C., Kumar, J. et al. (2014). *ACS Applied Materials and Interfaces* 6: 314–318.

131 Li, D., Xiao, Z., Mu, S. et al. (2018). *Nano Letters* 18: 2021–2032.

132 Liang, J., Wang, J., Zhang, Z. et al. (2019). *Advanced Materials* 31: 1808160–1808167.

133 Kim, Y., Kim, H., Jang, H. et al. (2020). *Nano Letters* 20: 4530–4536.

134 Rosa, H.G., Ho, Y.W., Verzhbitskiy, I. et al. (2018). *Scientific Reports* 8: 10035–10042.

135 Wang, X.M. and Lan, S.F. (2016). *Advances in Optics and Photonics* 8: 618–655.

136 Guowei Zhang, A.C., Huang, S., Wang, F. et al. (2018). *Science Advances* 4: 9977–9983.

137 Lee, I.-H., Martin-Moreno, L., Mohr, D.A. et al. (2018). *ACS Photonics* 5: 2208–2216.

138 Song, X., Liu, Z., Xiang, Y., and Aydin, K. (2018). *Optics Express* 26: 5469–5477.

139 Fo, Q.Q., Pan, L., Chen, X.Y. et al. (2018). *IEEE Photonics Journal* 10: 4500709–4500718.

140 Tian, R., Fei, R., Hu, S. et al. (2020). *Physical Review B* 101: 235407–235413.

141 Rodrigues, M.J., de Matos, C.J., Ho, Y.W. et al. (2016). *Advanced Materials* 28: 10693–10700.

142 Youngblood, N., Peng, R., Nemilentsau, A. et al. (2016). *ACS Photonics* 4: 8–14.

143 Autere, A., Ryder, C.R., Saynatjoki, A. et al. (2017). *Journal of Physical Chemistry Letters* 8: 1343–1350.

144 Ullah, K., Meng, Y., Sun, Y. et al. (2020). *Applied Physics Letters* 117: 011102–011107.

145 Popkova, A.A., Antropov, I.M., Fröch, J.E. et al. (2021). *ACS Photonics* 8: 824–831.

146 Wu, H.-Y., Yen, Y., and Liu, C.-H. (2016). *Applied Physics Letters* 109: 261902–261907.

147 Borisenko, S., Gibson, Q., Evtushinsky, D. et al. (2014). *Physical Reviews Letters* 113: 027603–027606.

148 Uchida, M., Nakazawa, Y., Nishihaya, S. et al. (2017). *Nature Communication* 8: 2274–2280.

149 Mosca Conte, A., Pulci, O., and Bechstedt, F. (2017). *Scientific Reports* 7: 45500–45512.

150 Yang, Y.-K., Xiu, F.-X., Wang, F.-Q. et al. (2019). *Chinese Physics B* 28: 107502–107510.

151 Liu, Z.K., Jiang, J., Zhou, B. et al. (2014). *Nature Materials* 13: 677–681.

152 Meng, H., Shang, X., Xue, X. et al. (2019). *Optics Express* 27: 31062–31074.

153 Xiong, H., Ji, Q., Bashir, T., and Yang, F. (2020). *Optics Express* 28: 13884–13894.

154 Neupane, M., Xu, S.Y., Sankar, R. et al. (2014). *Nature Communication* 5: 3786–3793.

155 Lu, W., Ge, S., Liu, X. et al. (2017). *Physical Review B* 95: 024303.

156 Sun, Y., Meng, Y., Dai, R. et al. (2019). *Optics Letters* 44: 4103–4106.

157 Sun, Y., Meng, Y., Jiang, H. et al. (2019). *Optics Letters* 44: 582–585.

158 Zhang, W., Yang, Y., Suo, P. et al. (2019). *Applied Physics Letters* 114: 221102–221107.

11

2D Perovskites for Nanolasers

Wei Yuan[1,2] *and Chuanjiang Qin*[1,2]

[1] *Chinese Academy of Sciences, Changchun Institute of Applied Chemistry, State Key Laboratory of Polymer Physics and Chemistry, No. 5625, Renmin Street, Changchun, Jilin Province 130022, China*
[2] *University of Science and Technology of China, School of Applied Chemistry and Engineering, No. 96 Jinzhai Road, Hefei, Anhui 230026, China*

11.1 Introduction

Perovskite materials have emerged as promising candidates in various optoelectronic fields, such as solar cells, light-emitting diodes, and lasers, due to their tunable photoelectric properties, facile fabrication technique, and low cost [1–5]. Owing to their large optical absorption coefficient, high luminance efficiency, and balanced ambipolar charge transport characteristics, perovskites show vast potential as gain mediums to realize electrically pumped lasers.

For typical 3D metal halide perovskite with structure of ABX_3, A refers to monovalence amine cations and fills in the voids of BX_6 octahedral network (Figure 11.1a,b). In practice, substantial defects form during the 3D perovskite crystallization process, which not only leads to severe non-radiative energy loss but also adversely impacts the operation and environment stability. One of the effective strategies to address this issue is the introduction of large organic amine cations to form low-dimensional perovskites. In formula $(A')_mA_{n-1}B_nX_{3n+1}$(m is 1 or 2) (A': large cations), $n = 1$ and $n \geq 2$ represent pure 2D and quasi-2D perovskites, respectively (Figure 11.1c). Owing to the coordination effect with metal cations and hydrophobicity of organic ligand, 2D and quasi-2D perovskites were reported to possess the capabilities of both defect passivation and excellent moisture stability. Besides, because of the enhanced quantum and dielectric confinement effect, they have larger exciton binding energy than their 3D counterparts, which contributes to radiative recombination [6–13].

Several works have revealed that actions related to lasing can be affected by organic cations. For instance, Qin et al. presented that organic cation with lower triplet energy level than that of the perovskite framework could act as the triplet quencher to prevent the accumulation of triplet excitons that hinders laser oscillation [14]. Furthermore, Liang et al. reported that the quantum confinement effect grew with decreasing the n value, resulting in stronger Auger recombination effect and exciton–phonon coupling. Consequently, it is difficult for 2D perovskite

Two-Dimensional Materials for Nonlinear Optics: Fundamentals, Preparation Methods, and Applications, First Edition. Edited by Qiang Wang and Hao-Li Zhang.

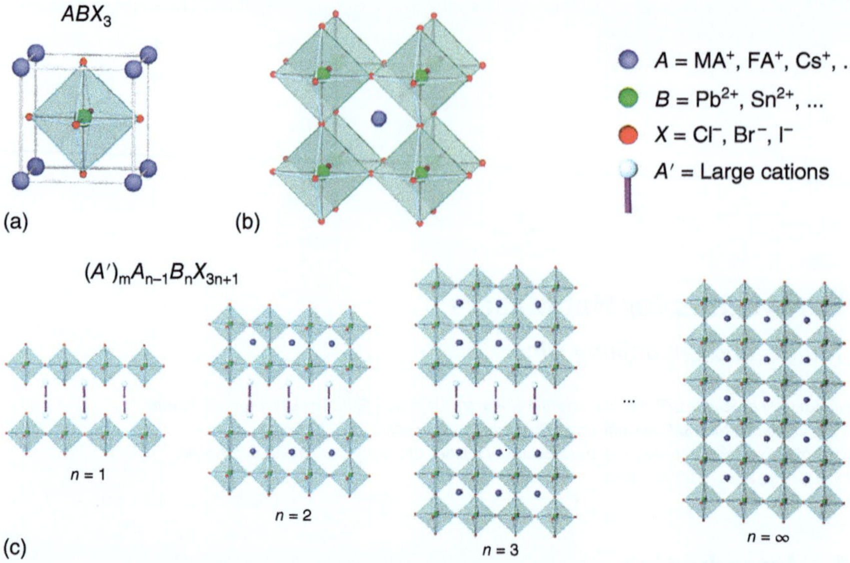

Figure 11.1 Perovskite crystal structure. (a) 3D perovskite unit cell. (b) 3D perovskite crystal packing. (c) Schematic diagram of pure 2D (n = 1) and quasi-2D RP perovskites with different numbers of perovskite layers ($n \geq 2$). Source: Reproduced with the permission from Dong et al., Chem. Soc. Rev. 49, 951 (2020). Copyright 2020 Royal Society of Chemistry.

of $n \leq 2$ to produce laser light because of the increase in laser threshold [15]. Zhang et al. indicated that ultrafast energy transfer from small-n-value domains with wide bandgap to large-n-value domains with narrow bandgap was conducive to rapidly accumulate exciton on the latter, which facilitates population inversion and reduction of the amplified spontaneous emission (ASE) and laser thresholds [16]. In addition, Lei et al. illuminated that highly oriented morphology was of importance to realize efficient energy transfer for qusai-2D perovskite films, thus achieving low ASE threshold through solvent engineering [17].

At present, low-threshold optically pumped nanolasers have been realized after coupling 3D, quasi-2D, 2D perovskite microplates, nanowires, microrings, and so on with different kinds of optical cavities (DBR, DFB, WGM, etc.) [18–23]. The choice of different solvents, antisolvents, and preparation conditions affects the crystalline orientation, morphology, and quality of perovskites and also the performance of nanolaser devices on micro- and nanoscales [17, 24]. High-quality homologous single-crystal and multi-quantum well (QW) structure polycrystalline perovskites have realized low threshold laser at room temperature [19].

This chapter is divided into three sections. Initially, the basic theories of ASE, population inversion, performance parameters, and optical cavity of perovskite nanolasers were introduced. Then, we focused on the application of 2D perovskites in diverse nanolasers. How perovskite interacts with optical resonators as a gain medium of laser devices was analyzed, and the performance of laser devices using various 2D perovskites coupled with different optical cavities was compared. Finally, we summarized the advantages and disadvantages of nanolasers prepared from 2D perovskites and analyzed the factors hindering their development and application. How to further improve the performance of 2D perovskite nanolaser devices and realize electrically pumped lasers was also presented.

11.2 Laser Formation Principle and Performance Parameters

11.2.1 Basic Optical Physical Process

There are three basic physical processes of laser, including stimulated absorption, stimulated radiation, and spontaneous radiation. These processes in which atoms, molecules, or ions absorb and radiate light are linked to the transition between different energy levels (Figure 11.2).

When external photons meet the requirement that the energy is equal to the difference between high and low energy levels, i.e. $\varepsilon = h\nu = \Delta E = E_h - E_l$, they can be absorbed to excite particles at low energy level E_l to convert into high energy level E_h, which is called stimulated absorption (Figure 11.2a). Since particles at excite states are unstable, spontaneous radiation corresponds to the process that particles at E_h spontaneously relax to E_l and emit photons with energy of $h\nu$ (Figure 11.2b). In addition, particles at E_h can also experience a relaxation process under the excitation of external source with energy of $h\nu$ (Figure 11.2c). In this case, the absorption of one photon is companied with the emission of two coherent phonons. Then, stimulated radiation appears and the light is amplified.

11.2.2 Laser Formation Condition

The occurrence of laser behavior needs to be done with the cooperation of excitation source, gain medium, and optical resonator, which are also the three primary elements for generating lasers. First of all, the excitation energy enables atom to absorb energy and transition from lower energy level to upper energy level. Then, the amplification effect of the gain medium allows stimulated radiation to prevail over absorption so that the particle number in the upper energy level is larger than that in the lower energy level. In the meantime, the optical resonator makes the effect of stimulated radiation greater than spontaneous radiation. Thus, the laser is produced.

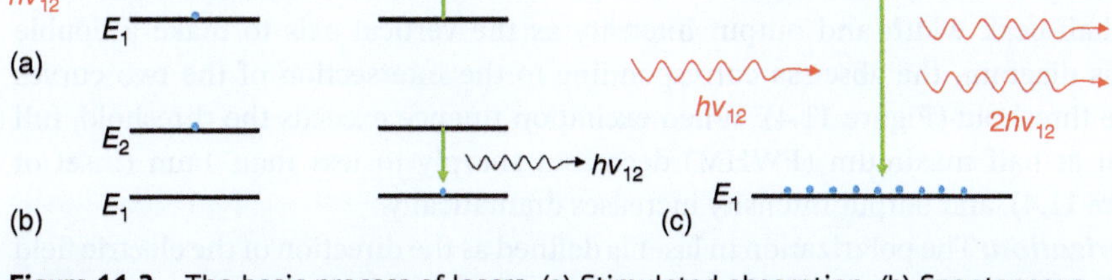

Figure 11.2 The basic process of lasers. (a) Stimulated absorption. (b) Spontaneous radiation. (c) Stimulated radiation.

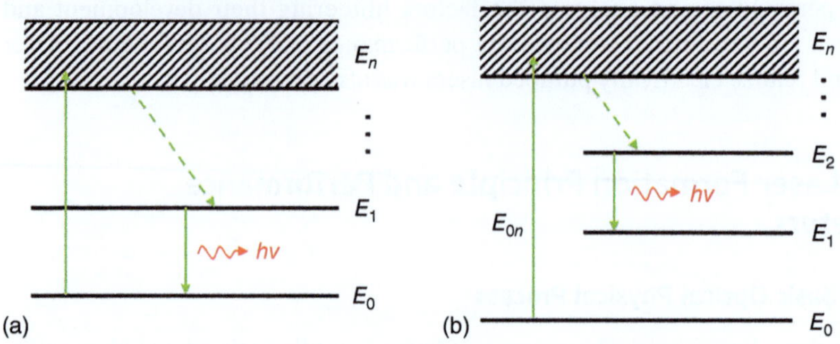

Figure 11.3 The physical process of (a) three-level system and (b) four-level system.

The energy level system to achieve particle number reversal through pump excitation can be divided into two types, namely, the three-level system and the four-level system.

Three-level system involves three energy levels in the laser generation process, in which E_0, E_1, and E_n correspond to ground state, metastable excited state, and high-energy excited state, respectively (Figure 11.3a). In fact, E_n represents a series of energy levels rather than a single energy level. After pumped by the excitation source, the particles can be excited from E_0 to E_n. Due to the short lifetime at E_n, they will quickly relax to E_1 that has relatively long lifetime and stability. When more than half of the particles at E_0 are pumped to E_1, the number of particles can be inverted between E_1 and E_0.

In terms of four-level system (Figure 11.3b), E_1 that generates laser is not a ground state, but an excited state. When particles at E_0 are pumped to E_n, they experience a quick relaxation to upper metastable E_2 through non-radiative transition. The lifetime of particles at E_1 (basically empty at room temperature) is extremely short, contributing to invert the number of particles between E_2 and E_1. Therefore, the energy of excitation source required for four-level system to generate laser is much smaller than that of three-level system, which indicates that the laser generation in four-level system should be much easier than that in three-level system.

11.2.3 Performance Parameters

Threshold: The threshold conditions are the conditions that must be met for the photon to oscillate in the resonant cavity to begin to balance with the loss, resulting in a net gain and a laser output. By taking excitation fluence as the horizontal axis and half-peak width and output intensity as the vertical axis to make a double Y-axis diagram, the abscissa corresponding to the intersection of the two curves is the threshold (Figure 11.4). When excitation fluence exceeds the threshold, full width at half maximum (FWHM) decreases sharply to less than 1 nm (inset of Figure 11.4), and output intensity increases dramatically.

Polarization: The polarization in laser is defined as the direction of the electric field oscillation of a laser beam. Since stimulated emission photons (waves) and excited photons (waves) are in the same frequency, direction, phase, and polarization, lasers

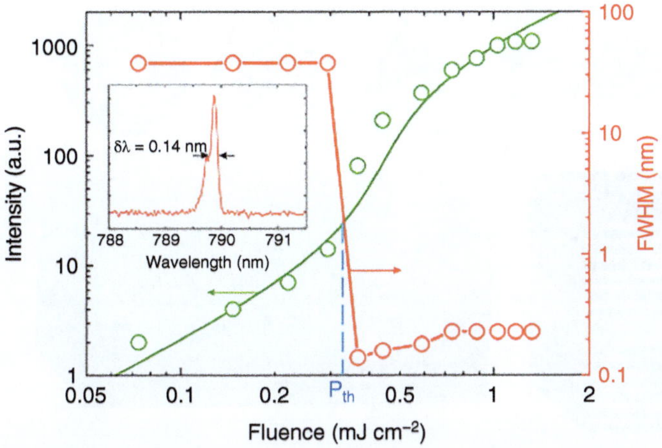

Figure 11.4 Intensity and full width at half maximum (FWHM) of the emission spectrum of the $MAPbI_3$ microdisk at the lasing wavelength. Source: Reproduced with the permission from Zhizhchenko et al., ACS Nano. 13, 4140 (2019). Copyright 2019 American Chemical Society.

generally have polarization characteristic. In exceptional cases, radially polarized beams can be generated, where the direction of polarization within the beam profile is radially oriented. Typically, radially polarized beams are generated by linearly polarized beams with certain optical elements, but radially polarized emission can also be obtained directly from the laser.

Q (Quality Factor): The quality factor Q is a parameter that characterizes the quality of the laser resonator, which is independent of the gain coefficient of the medium in the cavity and inversely proportional to the loss of the laser resonator. The higher the Q value is, the easier laser oscillation produces. The purpose of Q adjustment is to make the laser resonator in a low Q state when the laser starts to work, and the working materials continue to accumulate particles. When the number of particles accumulates to the maximum value, the Q value suddenly rises, and the laser resonator immediately establishes a strong laser oscillation and outputs a laser giant pulse in ultrashort time.

11.2.4 Optical Cavities

Optical resonator is one of the necessary components of lasers. Continuous photon oscillation in the cavity raises stimulated radiation in the gain medium, which eventually produces a high-intensity laser. Several common resonators used in perovskite lasers are Fabry–Perot (F–P) cavities, whispering-gallery-mode (WGM) cavities, distributed Bragg reflector (DBR), and distributed feedback (DFB) cavities.

In addition, perovskite crystals including nanowires, nanoplates, and microrings with smooth facets can also act as mirrors to form F–P and WGM cavities. For instance, photons propagate horizontally and oscillate under the reflection of the upper and lower crystal planes in nanowires to produce F–P lasers (Figure 11.5a). WGM cavities are usually composed of perovskite nanoplates or nanosphere

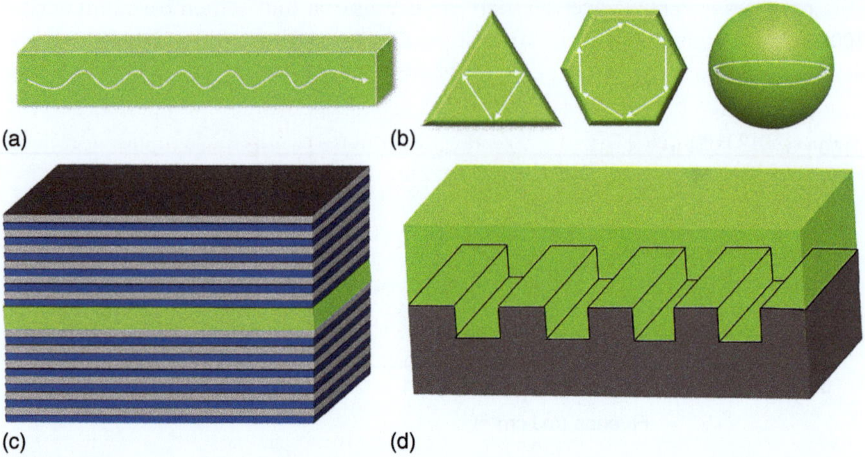

Figure 11.5 Different types of cavities. (a) A nanowire laser serving as Fabry–Pérot (F–P) laser. (b) Lasers based on the WGM cavities formed by triangular/hexagonal nanoplates or spheres. (c,d) Schematic of DBR cavity and DFB cavity, respectively.

(Figure 11.5b), which provide total internal reflection creating superior light confinement. In WGM cavities, light propagates at a specific wavelength within a plate or sphere of ultrahigh refractive index gain material, providing a relatively higher Q than F–P cavities. Unlike F–P and WGM cavities, DBR and DFB are used with gain media as external optical cavities. DBR laser devices are made up of a bottom DBR, a top DBR, and a semiconductor active layer sandwiched between two DBRs (Figure 11.5c). The spacer layer is designed so that the active layer is in the belly of the standing wave field formed by the DBR resonator. Consequently, the photons generated in the active layer have the greatest interaction with the standing wave field in the cavity and receive the strongest gain effect. The photon resonance direction in DBR is perpendicular to the semiconductor epitaxial layer, and laser is the output from the surface. In DFB cavities, the Bragg grating is distributed throughout the resonator, contacting with the gain medium directly (Figure 11.5d). Photons can achieve optical feedback and wavelength selection in a grating, after which photons with the same frequency will be emitted from the side of the laser to form a stable single-mode output.

11.3 The Application of 2D Perovskites in Nanolasers

The combination of a high-quality optical gain material with an appropriate optical cavity can generate lasers. Several materials have been studied for optical gain materials in the past decade, such as GaAs [25], TMDs [26], and MoS_2 [26]. With the use of traditional semiconductor QWs or quantum dots (QDs) as gain media, a high-quality and steady optical gain can be established [27]. For nanolasers, choosing suitable micro- and nanocavities is extremely critical to achieve high cavity reflectivity and low cavity losses [27]. The selection of 2D perovskites as gain media for nanolasers has received increasing attention due to their excellent optical properties and convenient fabrication. To decrease the threshold of nanolasers, the size and structure of gain media have strict restrictions. In this section, we introduce the application of 2D perovskites in nanolasers with different microstructures, including microplates [18], microrods [19], microcrystals [20], nanowires [16], microrings [21], thin films [17, 22, 28–30], and thin flakes [15, 23].

11.3.1 Microplates

As gain media, a large number of microcavity structures have been researched for achieving perovskite lasing. Through the use of these various optical cavities, two modes of laser were mainly realized, i.e. F–P and WGM, but both modes were implemented in different structures. In 2020, Liu et al. demonstrated two lasing modes in all-inorganic 2D perovskite $CsPb_2Br_5$ microplates with subwavelength thickness and uniform square shape at room temperature. Remarkably, upconversion lasing based on two kinds of microcavity effects, F–P and WGM modes, from microplates is successfully realized with a low threshold operating in multimode or single mode, respectively [18].

The two lasing modes were established by stimulating different areas in a $CsPb_2Br_5$ microplate under two-photon excitation. First, when they focused the excitation laser beam on the edge of a single microplate, the F–P mode lasing was realized (Figure 11.6a). In Figure 11.6b,c, the FWHM of the lasing decreased to 0.15 nm at the lasing threshold of 230 $\mu J\,cm^{-2}$. With increasing excitation fluence, the lasing peak position displayed a slight blueshift (less than 0.1 nm). At the same time, the extremely narrow FWHM also led to an incredibly high Q of 3551.

Second, significantly strong WGM lasing was achieved by focusing the excitation beam on the corner of the microplate, as shown in the inset of Figure 11.6d. When the pulsed excitation intensity increased to 180 $\mu J\,cm^{-2}$, there was an obvious transformation of the emission spectra from spontaneous emission to lasing action (Figure 11.6e). In this case, a steady single-mode laser of a $CsPb_2Br_5$ microplate was established at room temperature. The lasing peak was located at approximately 540 nm with an extremely narrow FWHM of ~0.16 nm, as shown in Figure 11.6f. To evaluate the gain effect of the material, Q was calculated as 3374. The lasing peak position occurs a slight redshift of 0.15 nm, which was attributed to the mode selection and intrinsic self-absorption arising from different laser cavities [31].

11.3.2 Microrods

Although nanolasers have been realized in 2D perovskite microplates, the lack of uniformity, defects in the grain boundary/interfacial regions, and poor stability are major challenges for achieving highly stable nanolaser devices with efficient performances. To address these issues, the slow evaporation at constant temperature (SECT) growth method has been used, which availably controls the rapid nucleation and reduces the multi-crystallinity of the perovskites, to grow millimeter- to centimeter-sized single-crystal microrods. Compared with polycrystalline thin films,

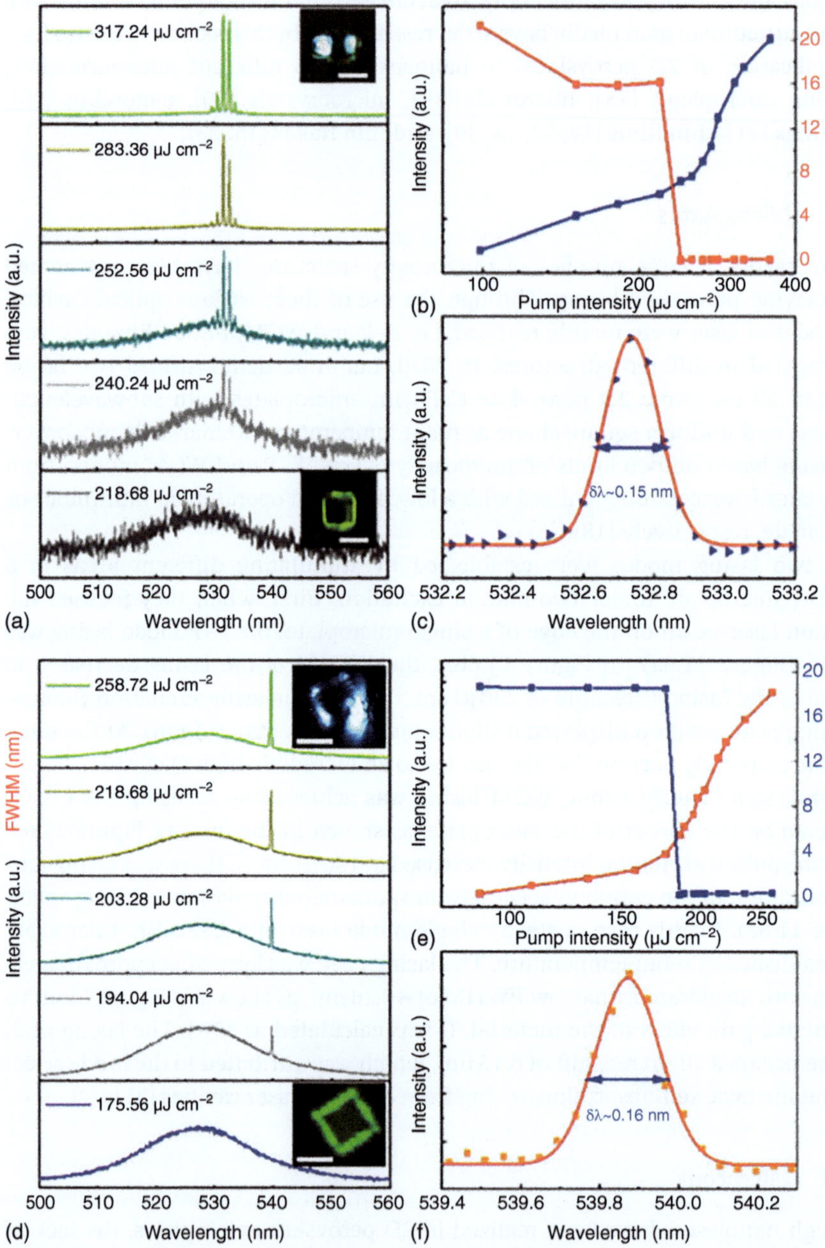

Figure 11.6 Lasing characterization of a $CsPb_2Br_5$ microplate in F–P mode (a–c) and WGM mode (d–f). (a, d) Pump-intensity-dependent emission spectra under two-photon excitation. (b, e) Output intensity (blue) and FWHM (red) of laser with increasing pump fluence. (c, f) Fitting of the lasing oscillation mode. Source: Reproduced with the permission from Liu et al. [18], © 2020, Optica Publishing Group.

perovskite microrods in single-crystal form have advantages of exceptionally long carrier diffusion length attributed to lower trap densities, charge transport, high absorption, and reduced defects due to their large grain boundaries. By utilizing the high-quality single-crystal perovskite microrods, the resultant lasers can exhibit low threshold with extremely high *Q*.

In 2020, Pradip Kumar Roy et al. studied random lasing in a high-quality long-chain organic diammonium spacer-assisted 2D hybrid perovskite FA-(N-MPDA)$PbBr_4$(N-MPDA = *N*-methylpropane-1,3-diammonium) [19]. As shown in Figure 11.7a, millimeter-sized single-crystalline perovskite microrods were fabricated by the SECT technique. Figure 11.7b showed an optical image and a top-view SEM image of the microrod, which indicated that the length and diameter of the microrod were 3 mm and 2 μm, respectively. The perovskite microrod performed a high-quality crystal structure and purity through the XRD and TEM characterization results, respectively, as shown in Figure 11.7e,f. Additionally, the spacing between the lattice fringes in FA-(N-MPDA)$PbBr_4$ microrods was approximately 0.25 nm in the HR-TEM image (Figure 11.7g).

The emission peak of the FA-(N-MPDA)$PbBr_4$ microrod located at 535 nm is shown in Figure 11.7c. Figure 11.7d presents the photomultiplier tube (PMT) mapping image of a microrod. The different colors in the emission peak plot of the spectral map represent the peak center values of the mapped region as shown in Figure 11.7h. The statistics of the emission peak (Figure 11.7i) showed that the peak was mainly concentrated at about 535 nm. All characterization results of Figure 11.7c–i indicated that the prepared microrods had excellent uniformity.

Although the prepared films display great uniformity, they still inevitably have defects that can act as scattering centers. When excitation by a pulsed laser interacted with defects, spontaneously emitted photons were scattered in all directions. After multiple scattering, closed loops were formed within the optical gain medium, and coherent feedback was provided to the emission through constructive interference, resulting in random lasers [32]. When the excitation power exceeded the threshold, there were multiple narrow peaks in the emission spectrum, as displayed in Figure 11.7k. The FWHM of every single lasing peak was at least 0.1 nm; thus, *Q* reached approximately 5350. The threshold of the random lasing was 0.5 $\mu J\ cm^{-2}$, which can be obtained from the plot of the FWHM and intensity versus the pumping energy density in Figure 11.7n. They believed that the low threshold of the laser was due to the strong quantum confinement effect of 2D perovskite single crystals, long mean free path, low nonradiative recombination rates, and large exciton binding energy caused by the introduction of long-chain organic diammonium N-MPDA. Furthermore, the consistent time of random lasing in 2D hybrid perovskite microrods could reach at least two hours (Figure 11.7i–m). Overall, a stable, extremely low threshold, high-quality random laser was achieved in 2D hybrid perovskite FA-(N-MPDA)$PbBr_4$ microrods at room temperature.

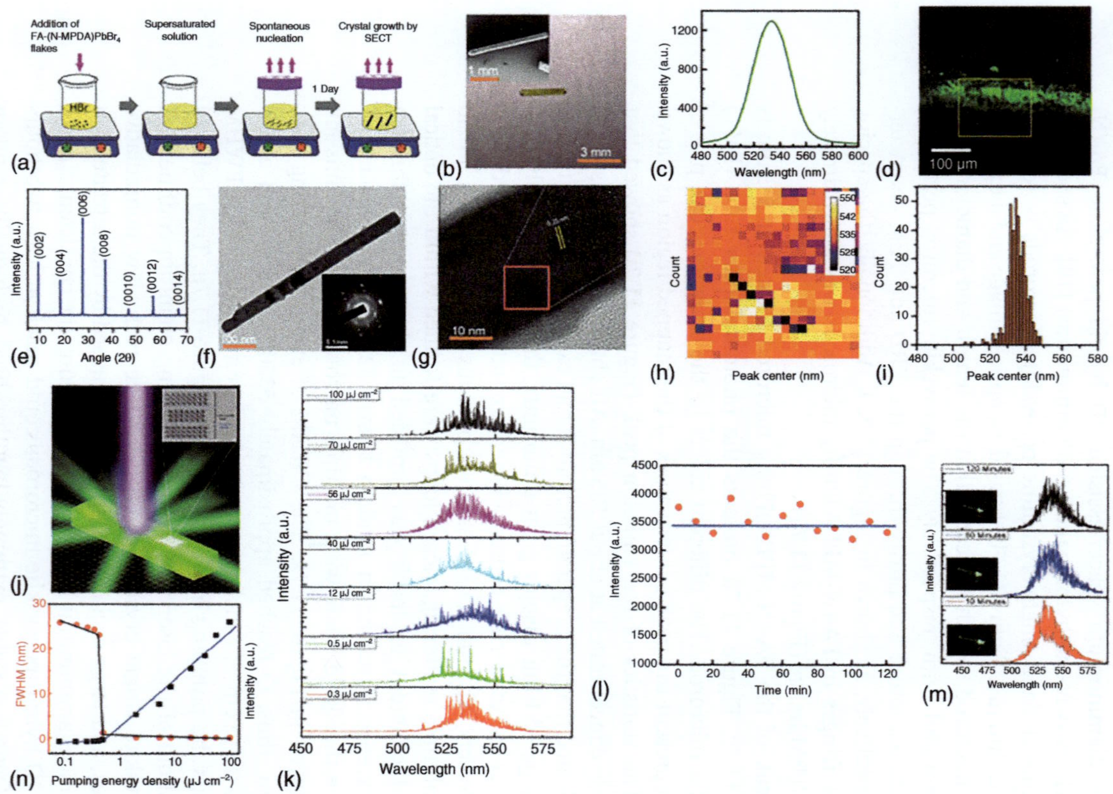

Figure 11.7 (a) Schematic illustration of the synthesis of FA-(N-MPDA)$PbBr_4$ microrods. (b) Optical image of an as-synthesized microrod ~3 mm in length (inset: top-view SEM image of the microrod). (c) PL spectrum. (d) PMT mapping of a microrod with the yellow rectangle showing the area used for spectral mapping. (e) XRD pattern of 2D microrods. (f) TEM image of an FA-(N-MPDA)$PbBr_4$ microrod at lower magnification with the SAED pattern (inset). (g) An HR-TEM image of an FA-(N-MPDA)$PbBr_4$ microrod displayed with lattice fringes of about 0.25 nm. The mapping confirms the microrod uniformity. (h and i) Peak center mapping and the corresponding histogram distribution of the yellow rectangular area shown in (d). (j) Schematic illustration of random lasers in a single-crystalline 2D perovskite microrod. The enlarged picture illustrates the multiple quantum wall structure formed by the organic and inorganic molecules in the 2D perovskite. (k) Random lasing from a 2D perovskite single-crystal microrod on a semi-log scale. (n) Integrated intensity and emission line width as a function of pump fluence. (l) Lasing (centered at ~535 nm) under pumping at 374 nm with a fluence of 12 μJ cm^{-2} with time. (m) Random lasing spectra and microscopy images recorded at a constant pump fluence of 12 μJ cm^{-2} as a function of time. Source: Reproduced with the permission from Roy et al. [19]. © AIP Publishing.

11.3.3 Microcrystals

The efficient laser emission in perovskite microcrystals is attributed to the energy/charge transfer processes in the high-quality microcrystals. The suitable refraction index and well-faceted 2D perovskite microcrystals with certain morphology symmetries could be naturally and efficiently utilized as the laser resonant cavities without any other fabrications. In 2020, Zhai et al. achieved two-mode lasing based on quasi-2D $(PEA)_2Cs_{n-1}Pb_nBr_{3n+1}$ perovskite microcrystals in a vertical single-mode optical microcavity [20]. They prepared the perovskite via spin-coating technique. The sandwich structure of the lasing is shown in Figure 11.8. When a second quasi-2D perovskite was spin-coated on PMMA, star-shaped self-organizing segregation patterns appeared in the perovskite film consisting of mixed domains with different n values, as demonstrated in Figure 11.8a–e. UV–vis absorption and PL spectra are illustrated in Figure 11.8c. The absorption peaks at 402, 433, and 509 nm represented the $n = 1$, 2, and large n phases of the 2D perovskite, respectively. The PL peak was at 519 nm. In Figure 11.8f, the spatially resolved μ-PL spectra showed that the emission peaks located at 483–496 originated from small n phase were blueshifted, markedly compared with those of large n phase, and their PL intensities were significantly reduced, proving that phase separation occurs in the quasi-2D perovskite film.

The lasing device was fabricated by quasi-2D $(PEA)_2Cs_{n-1}Pb_nBr_{3n+1}$ perovskite microcrystals combined with a single-mode vertical cavity that contained a bottom DBR and a top Ag-film (reflector) for a surface-emitting laser. The sandwich structure is shown in Figure 11.8g. Under the nanosecond pulsed excitation, lasing occurred at a peak of ~532 nm, and the FWHM of the single mode lasing was ~0.8 nm (Figure 11.8h). In Figure 11.8j, the junction of the sharp narrowing of the FWHM and the sharp increase in the integrated intensity resulted in a laser threshold of ~500 μJ cm^{-2}. The Q of the cavity was estimated as ~665. The emission of VCSEL device typically exhibited linear polarization of 94%. In addition, when the pulsed laser excited at some place of the segregated pattern, two-mode lasing (single mode or dual mode) was achieved with the threshold of ~500 μJ cm^{-2} in Figure 11.8i–l. The peaks of two modes were located at ~537 and 539 nm, respectively. However, the thresholds of two kinds of lasers were relatively high, because the gain media layer of perovskites may have a large number of defects.

11.3.4 Nanowires

Nanowire lasers with 3D perovskites exhibit extremely excellent gain properties while room-temperature optical gain and nanowire lasers with 2D perovskites have rarely been reported. In 2018, Zhang et al. achieved room-temperature nanowire lasing with multi-QWs of 2D Ruddlesden–Popper perovskite $(BA)_2(FA)_{n-1}Pb_nBr_{3n+1}$ [16]. High-density, large-area, stable NW laser arrays were fabricated by using a facile polydimethylsiloxane (PDMS) template-confined solution-growth method. The width, length, and height of the nanowire were 2 μm, 17 μm, and 420 nm,

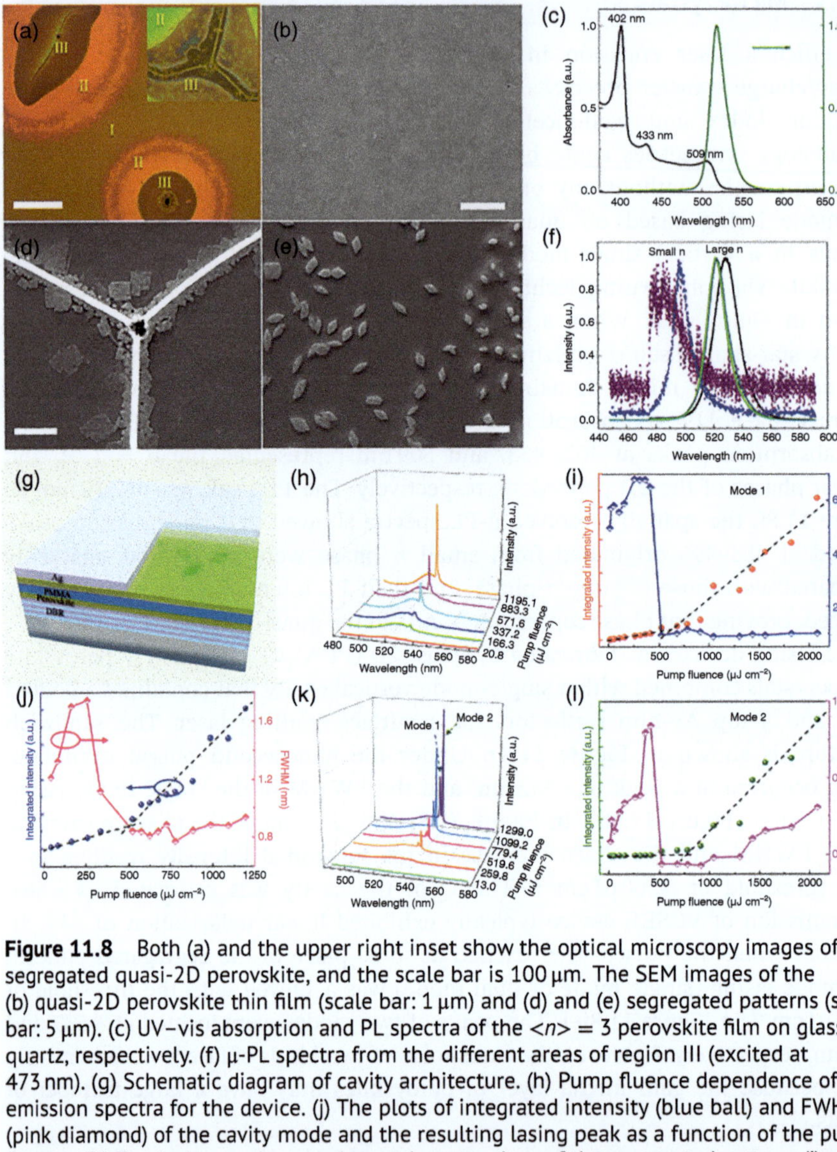

Figure 11.8 Both (a) and the upper right inset show the optical microscopy images of segregated quasi-2D perovskite, and the scale bar is 100 μm. The SEM images of the (b) quasi-2D perovskite thin film (scale bar: 1 μm) and (d) and (e) segregated patterns (scale bar: 5 μm). (c) UV–vis absorption and PL spectra of the $<n> = 3$ perovskite film on glass and quartz, respectively. (f) μ-PL spectra from the different areas of region III (excited at 473 nm). (g) Schematic diagram of cavity architecture. (h) Pump fluence dependence of the emission spectra for the device. (j) The plots of integrated intensity (blue ball) and FWHM (pink diamond) of the cavity mode and the resulting lasing peak as a function of the pump fluence. (k) Two lasing modes are observed at one place of the segregated pattern. (i) and (l) are the plots of integrated intensity and FWHM as a function of pump fluence for mode 1 and mode 2, respectively. Source: Reproduced with the permission from Zhai et al. [20].

respectively. Although the perovskite was formulated according to $n = 3$, it was actually a heterojunction structure composed of multiple different n values. The energy was transferred from QWs with small n-values to QWs with large n values to achieve population inversion, which was attributed to a decreased threshold. The PL peak of the uniform nanowire arrays consisted of $<n> = 3$ $(BA)_2(FA)_{n-1}Pb_nBr_{3n+1}$ was located at 538 nm.

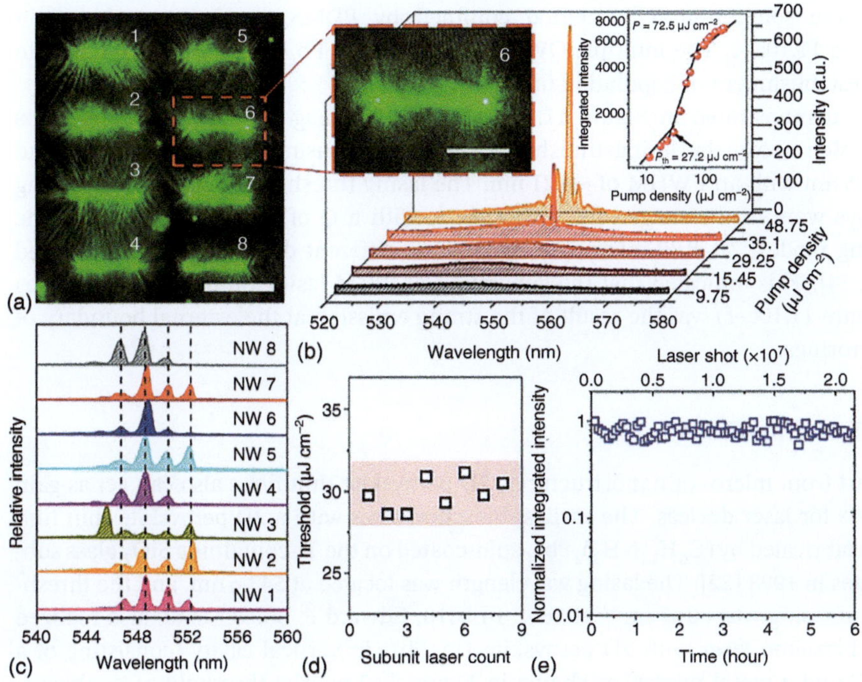

Figure 11.9 (a) μ-PL images of nanowire arrays recorded above lasing threshold. (b) Spatially resolved μ-PL spectra recorded from nanowire 6 labeled in (a) as a function of P. Inset: The plot of integrated PL intensities versus P. (c) Lasing spectra and (d) lasing threshold recorded from nanowires 1–8 labeled in (a). (e) Laser-integrated intensity of 2D nanowire under 400-nm laser pump with energy of 80 μJ cm^{-2} in different time intervals. Source: Reproduced with the permission from Zhang et al. [16], © 2018 John Wiley and Sons.

Eight nanowires were prepared (Figure 11.9a), and every single nanowire acted as a high-quality F–P laser due to its well-controlled dimensions and uniform geometries. When excitation fluence exceeded the laser threshold (27–31 μJ cm^{-2}) as shown in Figure 11.9b, all eight nanowires underwent laser action, the laser peak with a FWHM of ~0.3 nm appeared at 548 nm, and the calculated Q was 1800 (Figure 11.9c). The lifetime of the lasing could last up to six hours under ambient conditions (25 °C and 40% relative humidity) (Figure 11.9d,e).

11.3.5 Microrings

Among many resonator types, WGM microring resonators have excellent optical confinement capabilities and decrease optical losses, as a result of successive total reflection inside their inner surfaces. The microrings functioned as superb optical resonators possess an ideal circular boundary, which provide the continuous total internal reflection for optical oscillations with low scattering density and high Q. In 2018, Zhang et al. achieved lasing in 2D RP perovskites of $(BA)_2(MA)_{n-1}Pb_nBr_{3n+1}$ microring arrays at room temperature [21]. The preparation method of microring was the solution-growth method confined by PDMS template, as shown in Figure 11.10a–c. The multiple QWs in $(BA)_2(MA)_{n-1}Pb_nBr_{3n+1}$ were conducive to the establishment of population inversion.

As demonstrated in Figure 11.10d–h, the μ-PL image of nine microrings was recorded above the lasing threshold. The peak of lasing was located at around 543.5 nm with an FWHM of ~0.21 nm. The lasing thresholds of all nine microring arrays were in the range of 12–15 μJ cm^{-2}, with a Q of ~2600. To modulate the lasing modes, 2D RPP microring arrays with different diameters were fabricated [33, 34]. This indicated that the generation of WGM lasing on these microcavities (Figure 11.10e–f) was the result of the strong emission at the external boundary of microrings.

11.3.6 Thin Films

Apart from micro- or nanostructures, 2D perovskite thin films also can act as gain layers for laser devices. The earliest biexciton laser with a 2D perovskite thin film was fabricated by $(C_6H_{13}NH_3)_2PbI_4$ spin-coated on the Ti-containing SiO_2 glass substrates in 1998 [22]. The lasing wavelength was located at 543.6 nm, and the threshold of lasing was 60 μJ cm^{-2} at 16 K. In 2018, Edward P. Booker et al. also realized the biexciton laser with 2D perovskite DA_2PbI_4 in vertical cavity (consisting of a DBR and a metal mirror) as shown in Figure 11.11a. The threshold of the biexciton laser was 8.2 μJ cm^{-2} at 125 K, which was similar to the lasing action at 75 K in Figure 11.11b. The demonstration of biexciton lasing above liquid nitrogen temperatures is a crucial step for layered perovskites in photonic applications [29].

Lei et al. prepared high-quality quasi-2D perovskite $(PEA)_2(FA)_3Pb_4Br_{13}$ films with highly efficient Förster resonance energy transfer by solution engineering [17]. They used *N*-methylpyrrolidone (NMP) and DMSO solvent to prepare a precursor solution and spin-coated it to obtain a $(PEA)_2(FA)_3Pb_4Br_{13}$ 2D perovskite film to explore the influence of different solvents on the material's ability to produce lasers. They found that the use of NMP solvent can obtain a 2D perovskite film with a highly consistent crystal orientation by using GIWAXS. Compared with the 2D perovskite film with different crystal direction when using DMSO solvent, the energy transfer from the high-energy bandgap to the low-energy bandgap can be carried out faster and more completely, resulting in a lower ASE threshold. They achieved lasing under pulsed optical pumping at room temperature by spin-coating 100-nm perovskite on a DFB cavity. The SEM cross section of the DFB laser was shown in Figure 11.11c. A clear emission power threshold at 10 μJ cm^{-2} was realized with an FWHM of 1.7 nm, as shown in Figure 11.11d,e. At the inset of Figure 11.11e, the lasing emission demonstrated strong transverse electric wave polarization.

Although the room-temperature pulsed lasing has already been achieved in 2D perovskite thin films, stable room-temperature continuous-wave (CW) lasing in 2D perovskites has rarely been reported. In 2020, Qin et al. realized room-temperature CW lasing in 2D perovskites $(PEA)_2FA_{n-1}Pb_nBr_{3n+1}$ (P2F8) and $(NMA)_2FA_{n-1}Pb_nBr_{3n+1}$ (N2F8) with DFB cavities [14]. They studied the cause of lasing death and the influence of the singlet–triplet exciton annihilation to laser

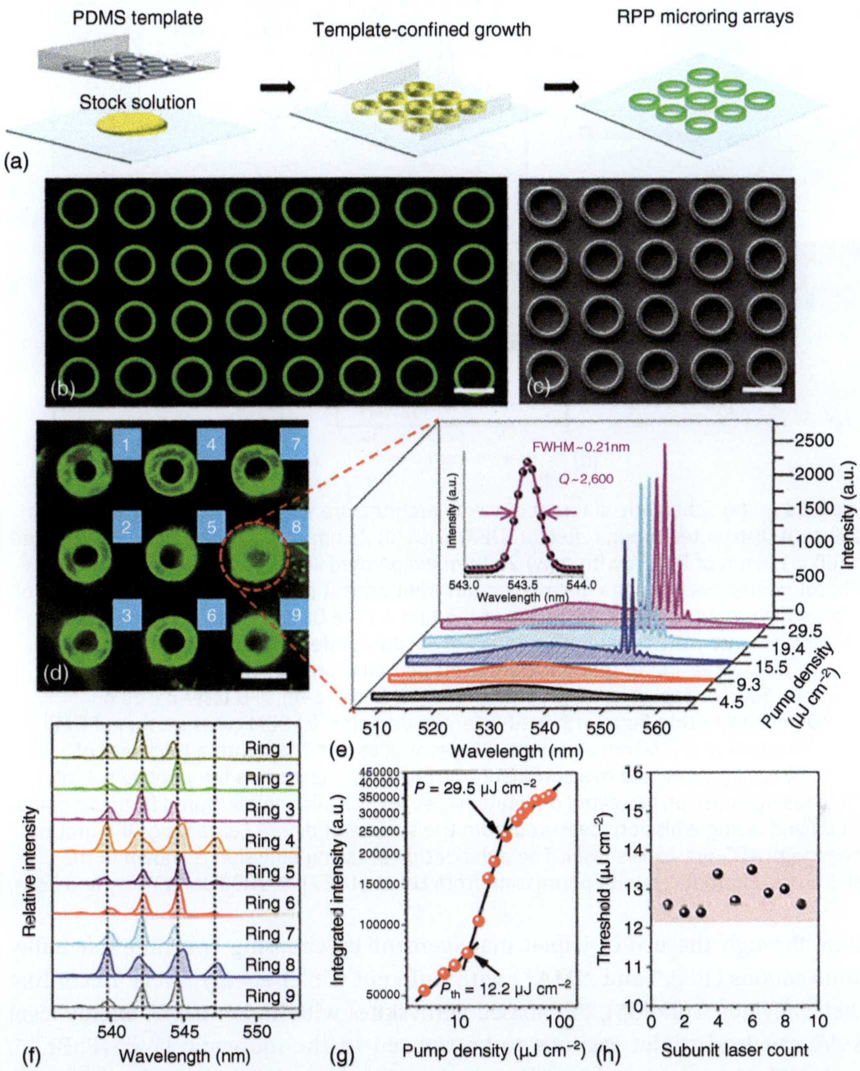

Figure 11.10 (a–c) Template-confined solution-growth of microring arrays of $\langle n \rangle = 6$ RPP. (a) Schematic illustration of the preparation procedures of microring arrays. (b, c) Low-magnification μ-PL and SEM images of $\langle n \rangle = 6$ RPP microring arrays with strict alignment and precise positioning. The scale bars in (b) and c are 20 μm. (d–h) Characterization of a 3 × 3 microring laser array. (d) μ-PL image recorded above lasing threshold. Spatial interference patterns from all nine microrings indicate coherent lasing outputs. (e) Spatially resolved μ-PL spectra recorded from microring 8 labeled in (d) as a function of P. (f) Laser spectra above the threshold with a Gaussian fitting. The FWHM is ≈0.21 nm, corresponding to a Q factor of ≈2600. (g) The plot of integrated PL intensities versus pump density, showing obvious "S" curve known for lasing oscillation. Lasing spectra (f) and lasing threshold (h) recorded from microrings 1–9 labeled in (d). Source: Reproduced with the permission from Zhang et al. [21], © 2018 John Wiley and Sons.

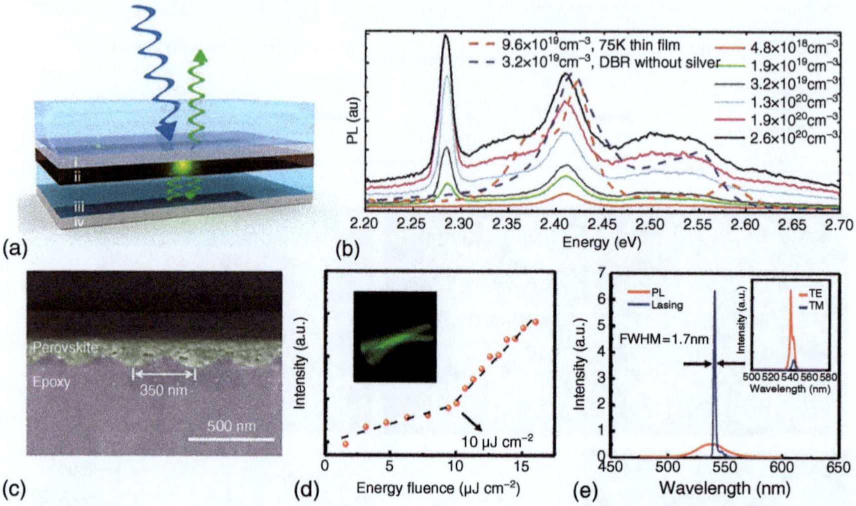

Figure 11.11 (a) Schematic diagram of cavity architecture with approximate thicknesses of layers. (i) Distributed Bragg reflector (DBR) with (ii) 25 nm of spin-coated DA_2PbI_4, coated with (iii) ≈150 nm of PMMA with a (iv) 200-nm evaporated silver mirror on top.
(b) Photoluminescence spectra of cavity at different optical pump powers, PL of thin film of DA_2PbI_4 spun on glass, and PL response of the film on the DBR without the silver mirror on top. Measurements were taken with a series of excitation densities that were not monotonically increasing to rule out significant degradation over the 4-h duration of the experiment. Source: Booker et al. [29]/John Wiley & Sons. (c–e) DFB laser based on quasi-2D perovskite thin films with well-oriented domains. (c) SEM cross section of DFB lasers with a period of 350 nm and quasi-2D perovskite thin films with a thickness of 100 nm. (d) Lasing threshold measurement of DFB laser. The inset is the photograph of far-field lasing emission projected on white paper. (e) PL emission measured from the planar thin film and lasing emission measured from the surface of device containing DFB gratings pumped with 400-nm 150-fs laser. The polarization of lasing emission is shown in the inset. Source: Reproduced with permission from Lei et al. [17], © 2020, John Wiley and Sons.

actions through the use of triplet management by choosing two different bulky organic cations (PEA^+ and NMA^+) with different triplet energy levels. According to their previous work [35], PEA-based perovskites with higher triplet energy level (3.3 eV) can lead triplet excitons to be trapped in the inorganic layer $[PbBr_6]^-$, though NMA-based perovskites that obtained a lower triplet energy level (2.6 eV) can act as triplet quencher to absorb triplets. Hence, they propounded that the long lifetime triplet excitons existed in PEA-based perovskites would hamper population inversion for ASE and lasing, causing the lasing death phenomenon. Using the oxygen-sensitive characteristic of triplet, they tested the ambient atmosphere-dependent ASE intensity and verified this assumption by the decreased ASE intensity of PEA-based perovskites in nitrogen. In consequence of NMA acting as triplet quencher, the ASE intensity of NMA-based perovskites was almost identical in oxygen and nitrogen. They also found that the laser intensity of PEA-based films declined sharply under continuous excitation in nitrogen, suggesting that singlet–triplet annihilation may be one of the intrinsic causes of lasing death. By spin-coating perovskites on the DFB cavity, both pulsed and CW lasing were

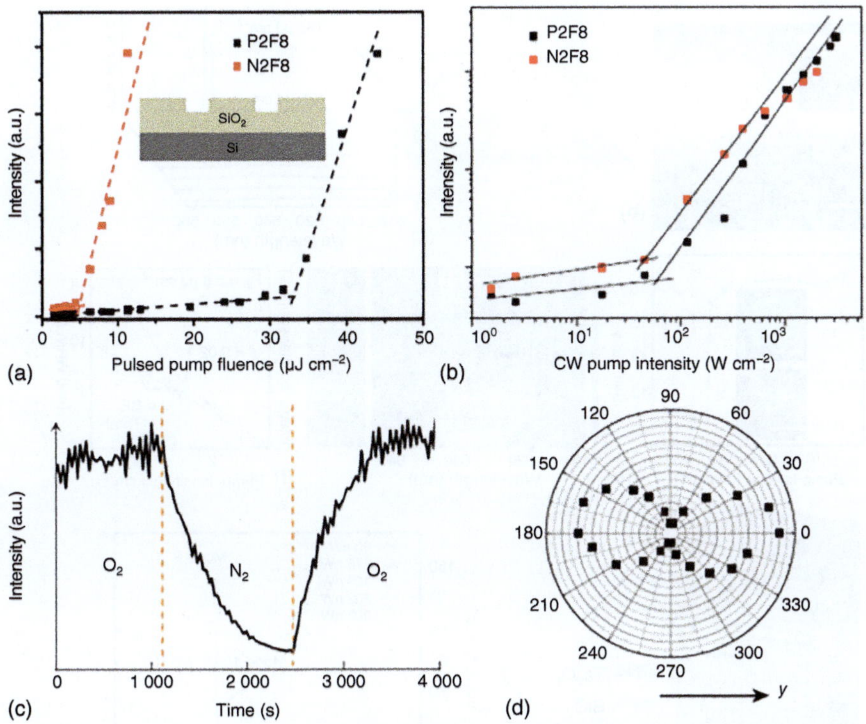

Figure 11.12 Pulsed and CW lasing properties of P2F8 and N2F8 films with DFB cavity. (a) Lasing intensity as a function of pulse energy and the structure of DFB cavity. (b) Lasing intensity as a function of CW pump intensity. (c) Lasing behavior of P2F8 in oxygen and nitrogen. (d) Polarization characteristics of CW lasing. Source: Adapted from Qin et al. [14].

observed for the P2F8 and the N2F8 at room temperature (Figure 11.12a,b). The pulsed lasing threshold of N2F8 was 4.7 μJ cm^{-2}, which was lower than that of P2F8 (32.8 μJ cm^{-2}). Under the CW pumping, a clear threshold was established to be 59 and 45 W cm^{-2} for P2F8 and N2F8, respectively. An extremely high polarization of ~83.7% was calculated in Figure 11.12d.

In 2021, Liu et al. realized subwavelength-polarized single-mode nanolasers in quasi-2D perovskite $PEA_2A_{n-1}Pb_nBr_{3n+1}$ (A: MA, Cs) with an additive of isopropylammonium (IPA) [30]. As shown in Figure 11.13a,b, there were three layers corresponding to the glass substrate, ITO, and perovskite with UV glue from bottom to top. Under the excitation of an 800-nm fs-pulse laser, the lasing action was achieved with a narrow FWHM of ~0.33 nm, and the threshold behavior at a pump fluence of ~143 μJ cm^{-2} is shown in Figure 11.13c,d. Single-mode lasing with a highly linear polarization degree of 81% and a Q of 1635 was realized. Through transient absorption tests, they found that in addition to the known rapid transfer of energy between QWs, polaron recombination also contributed to the laser generation. In addition, by incorporating the additive IPA into the precursor solution, a gain coefficient of up to 558 cm^{-1} was obtained for the 2D perovskite film, which made laser generation possible.

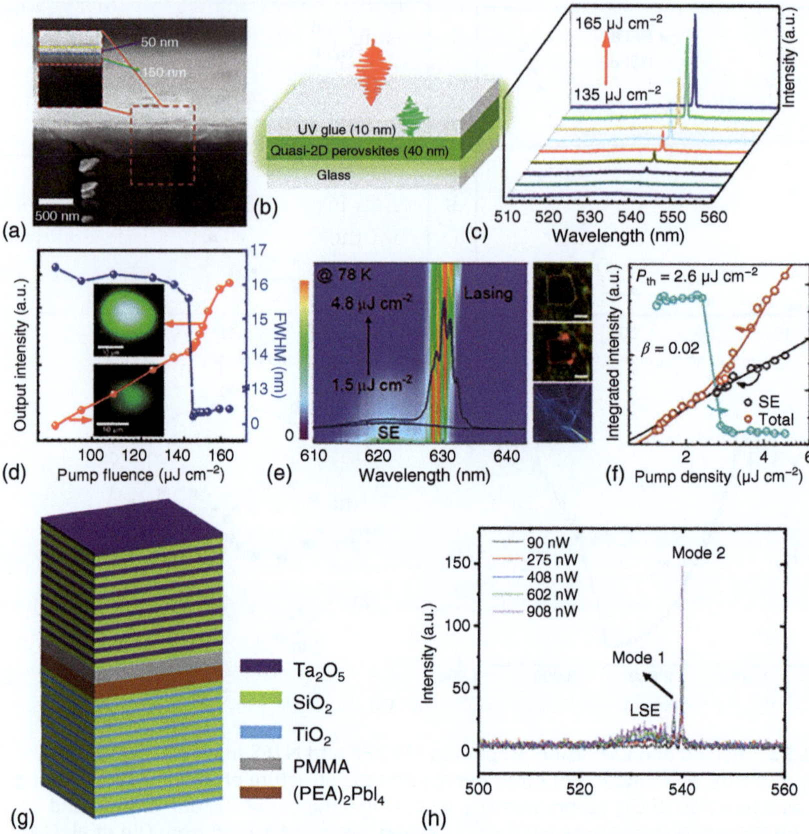

Figure 11.13 (a–d) Lasing device performance of the quasi-2D perovskites. (a) Cross-sectional SEM of the quasi-2D lasing device on the ITO. (b) Schematic of the vertical cavity under two-photon excitation with emission in green. (c) The evolution of emission spectra corresponding to the pump intensity. (d) The output intensity and FWHM of emission spectra under different pump fluence values. Insets: The optical images showing emission from the device below the threshold (bottom) and above the threshold (top). The scale bar is 10 μm. Source: Liu et al. [30]/American Chemical Society. (e–f) Lasing characteristics of mechanically exfoliated RPPs. (e) Evolution of spontaneous emission to normalized lasing spectra under various excitation fluences for $n = 3$ RPP on SiO_2/Si substrate. (f) Integrated SE intensity (dark gray), total PL intensity (red), and FWHM (blue) as a function of pump fluence exhibit a threshold behavior at $P_{th} \approx 2.6\ \mu J\ cm^{-2}$ per pulse. Source: Liang et al. [15]/John Wiley & Sons. (g) Schematic of the $(PEA)_2PbI_4$ VCSEL: green block, SiO_2 layer; orange block, $(PEA)_2PbI_4$; gray block, PMMA; purple block, Ta_2O_5; and blue block, TiO_2. (h) PL spectra collected under a variety of excitation powers. Source: Zhang et al. [23]/AIP Publishing/CC BY 4.0.

11.3.7 Thin Flakes

Indeed, lasing has been demonstrated from 2D homologous perovskite crystals, whereas the bulk size may impede its applications in nanophotonics. The understanding of the nanolasing in micro-sized thin flakes exfoliated from single crystals under high excitation conditions, which is considerably critical for low-power consumption and practical applications with size requirements, is still lacking. In 2019, Liang et al. realized multicolor lasing in mechanically exfoliated 2D $(BA)_2(MA)_{n-1}Pb_nI_{3n+1}$ RPP ($n \geq 3$) thin flakes with thicknesses from 100–300 nm [15]. Additionally, the influence of exciton–phonon interactions and Auger recombination were investigated by lasers in $(BA)_2(MA)_{n-1}Pb_nI_{3n+1}$. When the pump intensity was above the laser threshold (2.6 μJ cm^{-2}), the lasing is only emitted from the $n = 3$, 4, and 5 perovskites, as shown in Figure 11.13e,f. They found that the lasing threshold increased with n value, but there was no stimulated emission from $n = 1$ and 2 perovskites even at the ultralow temperature of 78 K. They confirmed that the lasing threshold increased with thickness, which was assigned to the enhancement of exciton–phonon interactions and Auger recombination. The FWHM and position of these oscillation peaks changed with the different values of n in $(BA)_2(MA)_{n-1}Pb_nI_{3n+1}$ at cryogenic temperatures.

Zhang et al. reported a type of room-temperature CW lasing by embedding 2D $(PEA)_2PbI_4$ perovskite in a vertical microcavity [23]. The millimeter-sized $(PEA)_2PbI_4$ single crystal exhibited superior thermal stability, high crystallization, and predominant optical properties and was fabricated by a two-step seed-growth method including two typical crystallization procedures of nucleation and growth. Using the exfoliated perovskite thin flake as the gain medium to construct the DBR, single-mode CW lasing with an ultralow threshold of 5.7 W cm^{-2} at room temperature was realized, which was attributed to the strong optical confinement in the cavity (Q ~ 1194). The emission peak of the laser was located at 537.4 nm (Figure 11.13h). The sandwich structure of the DBR and the lasing mode of the $(PEA)_2PbI_4$ thin flake are shown in Figure 11.13g.

11.4 Prospect

In this chapter, we give a detailed introduction to 2D perovskite nanolasers with different micro- and nanostructures. It is true that 2D perovskites are exceedingly promising in the preparation of gain media for low-cost, multicolor nanolasers due to their high absorption and gain coefficients, low trap state density, large exciton binding energy, and tunable wavelengths. However, there are still many unsolved problems with the use of 2D perovskites for the gain layer of lasers. (i) The quenching in 2D perovskites under high injection is serious, and the Auger recombination rate will also increase under high pumping, which hinders the development of high-injection electrical pumping perovskite lasers. (ii) The heat loss of 2D perovskite lasers severely degrades the performance of devices. The selection of device size and heat dissipation materials that are conducive to heat dissipation is also a top priority. (iii) Although the structural stability of 2D perovskites is greatly improved compared with that of 3D perovskites, in the process of preparing devices, if exposed to air, heat, and organic solvent atmospheres, 2D perovskites will also decompose quickly. Therefore, the preparation of the device should maintain a clean environment as much as possible. The device should be packed more densely. (iv) CW optically pumped lasers and electrically pumped lasers are also one of the challenges in the preparation of lasers from 2D perovskites for future applications. Although the outstanding electrical properties of 2D perovskites have been confirmed in perovskite solar cells, related research on 2D perovskite electrically pumped nanolasers is still rarely reported. (v) High thresholds are also an obstacle to the development of electrically pumped lasers.

To overcome obstacles that may be detrimental to electrically pumped lasers, the following possible ideas are proposed: the perovskite crystalline film should be continuously improved by optimizing the crystallization process and choosing different solvents during preparation, the defects that affect the high-current injection should be modified by doping or introducing additives, and by way of matching perovskites with different emission wavelengths with high *Q* optical cavities or preparing high-quality nanosized perovskite single crystals, the threshold would be as low as possible. These improved methods need to be further explored and discovered. It is believed that, in the near future, perovskite electrically pumped lasers that can meet the requirements of industrialization will become a reality.

References

1 Dong, W., Zhang, X., Yang, F. et al. (2022). *ACS Nano* 16: 9679.
2 Sun, C., Jiang, Y., Cui, M. et al. (2021). *Nature Communications* 12: 2207.
3 Wu, G., Yang, T., Li, X. et al. (2021). *Matter* 4: 582.
4 Zhang, Y. and Park, N. (2022). *ACS Energy Letters* 7: 757.
5 He, X., Gong, H., Huang, H. et al. (2022). *Advanced Optical Materials* 10: 2200238.
6 Liu, Y., Cui, J., Du, K. et al. (2019). *Nature Photonics* 13: 760.
7 Vashishtha, P., Ng, M., Shivarudraiah, S., and Halpert, J. (2019). *Chemistry of Materials* 31: 83.
8 Wang, N., Cheng, L., Ge, R. et al. (2016). *Nature Photonics* 10: 699.
9 Lee, H., Kim, H., Cho, H. et al. (2019). *Advanced Functional Materials* 29: 1901225.
10 La-Placa, M., Longo, G., Babaei, A. et al. (2017). *Chemical Communications* 53: 8707.
11 Yang, X., Zhang, X., Deng, J. et al. (2018). *Nature Communications* 9: 570.
12 Chen, Z., Zhang, C., Jiang, X. et al. (2017). *Advanced Materials* 29: 1603157.
13 Yuan, S., Wang, Z., Xiao, L. et al. (2019). *Advanced Materials* 31: 1904319.
14 Qin, C., Sandanayaka, A., Zhao, C. et al. (2020). *Nature* 585: 53.
15 Liang, Y., Shang, Q., Wei, Q. et al. (2019). *Advanced Materials* 31: 1903030.
16 Zhang, H., Wu, Y., Liao, Q. et al. (2018). *Angewandte Chemie (International Ed. in English)* 57: 7748.
17 Lei, L., Seyitliyev, D., Stuard, S. et al. (2020). *Advanced Materials* 32: 1906571.
18 Liu, Z., Wang, C., Hu, Z. et al. (2020). *Photonics Research* 8: 31.
19 Roy, P., Ulaganathan, R., Raghavan, C. et al. (2020). *Nanoscale* 12: 18269.
20 Zhai, W., Tian, C., Yuan, K. et al. (2020). *Applied Physics Letters* 114: 131107.
21 Zhang, H., Liao, Q., Wu, Y. et al. (2018). *Advanced Materials* 30: 1706186.

22 Kondo, T., Azuma, T., Yuasa, T., and Ito, R. (1998). *Solid State Communications* 102: 253.

23 Zhang, H., Hu, Y., Wen, W. et al. (2021). *APL Materials* 9: 071106.

24 Xu, L., Che, S., Huang, J. et al. (2019). *Applied Physics Letters* 115: 033101.

25 Sutherland, B. and Sargent, E. (2016). *Nature Photonics* 10: 295.

26 Reed, J., Zhu, A., Zhu, H. et al. (1967). *Nano Letters* 2015: 15.

27 Strauf, S. and Jahnke, F. (2011). *Laser & Photonics Reviews* 5: 607.

28 Dong, H., Zhang, C., Liu, X. et al. (2020). *Chemical Society Reviews* 49: 951.

29 Booker, E., Price, M., Budden, P. et al. (2018). *Advanced Optical Materials* 6: 1800616.

30 Liu, Z., Hu, M., Du, J. et al. (2021). *ACS Nano* 15: 6900.

31 Liu, X., Zhang, Q., Xiong, Q., and Sum, T. (2013). *Nano Letters* 13: 1080.

32 De Giorgi, M. and Anni, M. (2019). *Applied Sciences* 9: 4591.

33 Liao, Q., Hu, K., Zhang, H. et al. (2015). *Advanced Materials* 27: 3405.

34 Wang, X., Liao, Q., Li, H. et al. (2015). *Journal of the American Chemical Society* 137: 9289.

35 Qin, C., Matsushima, T., Potscavage, W. et al. (2019). *Nature Photonics* 14: 70.

12

2D Materials for Space Use

Shuyan Wang and Qiang Wang

Lanzhou University, State Key Laboratory of Applied Organic Chemistry (SKLAOC), Key Laboratory of Nonferrous Metal Chemistry and Resources Utilization of Gansu Province, College of Chemistry and Chemical Engineering, Key Laboratory of Special Function Materials and Structure Design, Ministry of Education, No. 222 South Tianshui Road, Lanzhou 730000, China

12.1 Introduction

Aerospace materials are the cornerstone of space exploration [1]. The ever-increasing space exploration activities demand novel radiation-tolerant materials urgently, among which NLO materials and devices are particularly scant. In the 2015 NASA technology roadmaps and 2020 NASA Technology Taxonomy, radiation-resistant pulse lasers, optical communication components, optical detectors, and other photonic technologies are among the key areas of space exploration [2]. A new-generation photonic technology based on NLO materials and devices provides core components for optical communication, photoelectric conversion, and integrated optics, such as crystals for ultrafast laser mode locking, phase modulation, photorefractive, nonlinear waveguide, all-optical switch, and optical parametric conversion [3]. Many of them are crucial technologies for space applications, including precision optical instruments and optical components of communication satellites, erbium-doped fiber amplifiers for space laser communication [4, 5], and optical limiting devices based on NLO principle for laser protection [6].

However, space exploration puts forward higher requirements for NLO materials in comparison with ground use. Various optical coatings/devices used on satellites and space telescopes need to withstand the severe test of the space environment to ensure reliability. In addition to extreme temperature, vacuum, and other factors, the space environment also has special radiation effects from high-energy particles, which determine the life and reliability of materials in space and restrict their space applications. For example, polyacrylonitrile (Pan) carbon fiber was invented as early as in 1959. But only after countless tests of space environment tolerance, has it become a mature material in today's aerospace field. At present, NLO materials

Two-Dimensional Materials for Nonlinear Optics: Fundamentals, Preparation Methods, and Applications, First Edition. Edited by Qiang Wang and Hao-Li Zhang.

suitable for the space environment are rather scant, which restricts the development of the aerospace industry. The invention of NLO materials for space applications and the study of their space environmental effects especially radiation effects are thus particularly meaningful.

Two-dimensional (2D) materials have gained an ever-increasing interest in recent years since single-atom layered graphene was reported in 2004 [7]. New graphene analogs such as the transition metal chalcogenides (TMDs) [8], graphdiyne (GD) [9], hexagonal boron nitride (h-BN) [10], and black phosphorus (BP) [11] have been incessantly fabricated. 2D materials exhibit many unique properties distinguishing from conventional materials: large surface areas [12], enhanced robustness [13], mechanical flexibility [14], superior surface sensitivity to circumstances [15], fascinating quantum effects in heterostructures [16], flexible stacking modes via van der Waals forces [17, 18], and easy manufacturing into thin films. Therefore, they have been used in a broad range of applications including nonlinear optics, optoelectronics, catalysis, and energy conversion and storage.

Particularly, 2D materials have shown outstanding advantages over traditional optical materials [19, 20]. With good NLO properties, they can be used in laser protection, wideband ultrafast laser development, and so on. As an atomic layer thickness material, they have good mechanical properties and optical transparency and are suitable for flexible micro-optoelectronic devices. In addition, the 2D material family keeps expanding and provides sufficient ammunition for space application research, such as multi-soliton NLO phenomena, current star materials of perovskites as scintillators for detecting rays, and boron nitride single-photon source for quantum communication. In particular, vibration, impact, vacuum, and thermal cycling in the space environment pose little threat to micro- and nanomaterials. Furthermore, studies have shown that 2D materials conveniently dissipate energy because of their thinness, implying the potential to resist radiation damage.

Although with the above advantages, the research on the space environment adaptability of 2D materials is still developing; especially the study on the influence of radiation effect on their NLO properties is missing. Whether the potential of 2D materials can be transformed into practical applications is also largely dependent on their adaptability in the space environment. With a deep understanding on the radiation effect and damage mechanism of the 2D materials, rational design of radiation-proof NLO 2D materials suitable for space applications is feasible.

In this chapter, we systematically review the research on the performance of 2D materials under space radiation environment, their optical nonlinearities in particular. We begin with the introduction of various radiation environments. We then comprehensively review the 2D materials' radiation effects and damage mechanism under typical irradiation of high-energy particles of electron, ion, proton, and γ-rays. Furthermore, we focus on the space use of 2D materials based on their NLO properties, including saturable absorbers, optical limiters, space X-ray detection, and integrated optoelectronic platforms. Finally, we offer our insights on the future trend to raise the technological readiness level of 2D materials for practical space use.

12.2 Space Radiation and Two-Dimensional Materials

12.2.1 Space Radiation Environment

The Earth's space can be divided into different regions: low Earth orbits (LEOs), medium Earth orbits (MEOs), geostationary orbits (GEOs), and highly elliptical orbits (HEOs) according to their altitude from the Earth. Each region poses different environmental concerns. Currently, about 50% of satellites are in LEOs, and others are mainly in the vicinity of GEOs [21]. The environment in which space vehicles work is completely different from the ground. The damage caused by space radiation is the main reason for the failure of spacecraft and the reduction of their orbital service life [21, 22]. With the continuous development of new materials and functional devices, more and more optoelectronic devices are being applied to space use. Under the influence of the radiation environment, these devices may be damaged to varying degrees, like logic circuit flipping and disservice to sensors and charge-coupled devices. It is a very challenging task for researchers to make the spacecraft system work for a long time and highly reliable in the space radiation environment.

Space radiation is very complicated, comprising various energetic particles with different fluxes dependent on the altitude, angle of inclination, etc. [23]. The three main sources, all of which are affected by solar activities, include [24] (i) Earth's radiation belts [25], which are mainly high-energy particles of protons and electrons trapped in the Earth's magnetic field; (ii) solar particle events [26], in which protons and heavy ions launched into space during solar flares; and (iii) galactic cosmic rays [27], which are mainly high-energy protons and heavy ions from outside the solar system. Among the three sources, the influence of the galactic cosmic ray is the least. Basically, the activity of the sun influences all of the above source levels. Specifically, the Earth's radiation belt is also named the Van Allen radiation belt, which is composed of the inner radiation belt and the outer radiation belt (Figure 12.1). The inner radiation belt ranges from 600 to 10 000 km and features high-energy protons, while the outer radiation belt ranges from 13 000 to 45 000 km and mainly features high-energy electrons [21].

Space radiation has a great impact on various functional systems of space vehicles, which results in performance degradation and loss of normal operation, even leads to spacecraft failure, and accounts for ~40% of anomalies induced by the space environment statistically [28, 29]. Different from plasma-induced surface charging, radiation from high-energy particles penetrate the spacecraft surface, causing internal charging, total ionizing dose effect [30], displacement damage [31], and single-event effects (SEEs) [32, 33].

Total Ionizing Dose (TID) Effect. The total dose effect refers to the generation and deposition of energy in materials such as semiconductors or insulators by energized particles through the ionization process, leading to equipment failure or biological damage to astronauts [30]. Radiation effects gradually occur and accumulate and hence cause damage to materials. For instance, ionizing radiation created electron–hole (e–h) pairs in insulators of integrated circuits. When the TID

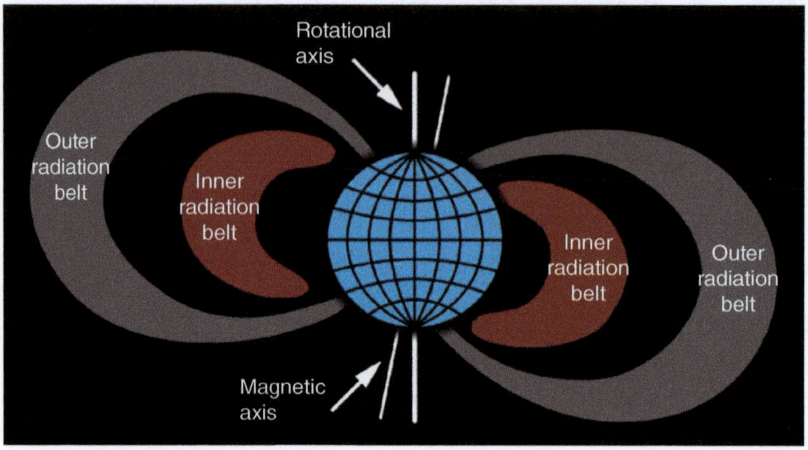

Figure 12.1 The Van Allen radiation belt that features high-energy particles of protons and electrons trapped in the Earth's magnetic field. Source: Reproduced with permission Lu et al. [21]. Copyright © The Author (s) 2019. This is an Open Access article distributed under the terms of the Creative Commons Attribution licence (https://creativecommons.org/licenses/by/4.0/), which permits unrestricted reuse, distribution, and reproduction in any medium, provided the original work is properly cited).

surpassed the tolerance limit, degradation or failure of the integrated circuits was induced [34, 35].

Displacement Damage. Displacement damage is the dislodging of atoms from their lattice sites, which imposes degradation of materials and devices [36]. Displacement damage introduces new energy levels in the bandgap and causes permanent changes in material properties like recombination lifetime. However, thermal and injection annealing may help repair the damage [36]. Basically, there are a few steps for the displacement damage [37]: defect production, defect reordering, and eventual occurrence of displacement damage effects.

Single-Event Effects (SEEs). SEEs refer to high-energy particles penetrating materials and devices (such as semiconductors and microelectronics in spacecraft) and causing dysfunction. Single-event upset (SEU), single-event functional interruption (SEFI), single-event transient (SET), and single-event latch-up (SEL) all belong to SEEs [33, 38, 39]. The SEE is usually induced by cosmic rays and high-energy protons [32, 33]. For instance, SEEs frequently posed threat to semiconductor devices by inducing a logic soft error [40].

Packages for modeling and simulating distribution and flux of high-energy particles are extremely useful in studying the radiation effects. For instance, AP-8 and ESP are strong packages to simulate high-energy protons from Van Allen radiation belt and solar activities that account for displacement damage and SEE, respectively [41, 42], and AE-8 and FLUMIC are used for modeling high-energy electrons [43, 44].

12.2.2 2D Materials Under Irradiation

The important factor that determines the long-term on-orbit service life and reliable operation performance of spacecraft is the complex space radiation environment, which causes the degradation of device and material performance. The space adaptability experiments of spacecraft functional devices and materials can be completed by on-orbit experiments and ground simulation experiments. Although the former can provide the real effect of space environmental radiation, the spacecraft usually must operate in orbit for more than ten years under a low-dose rate, which leads to a long time cycle and a high cost for the experiment [48]. Therefore, the accelerated test is often carried out in the ground-simulated space experiment to obtain the total dose effect, possessing the advantages of short experimental time cycle, low cost, and convenient operation. Furthermore, we introduce some typical radiation effects of 2D materials.

12.2.2.1 Electron Irradiation

The damage mechanism of electrons to materials is mainly caused by elastic and inelastic scatterings [45], which can induce displacement, sputtering, heating, charging, radiolysis, or contamination in the materials [49]. Meyer et al. [50] radiated graphene with 80 kV electrons, and the HRTEM images clearly displayed ejected atoms. The atomic displacement cross sections for carbon atoms were measured, which rose from zero (10^{-4} barn) at 80 keV to ~0.2 barn at 100 keV.

Algara-Siller et al. [45] raised the electron radiation resistance of single-layer MoS_2 by sandwiching it between two graphene layers, and a three orders of magnitude of resistance enhancement was observed. In Figure 12.2a, the HRTEM images corroborated that pristine MoS_2 was seriously damaged and lost long-range order after electron dose of 7×10^8 e nm^{-2}. However, for the G/MoS_2/G sample where the MoS_2 was protected by graphene, the vacancy concentration largely remained the same.

12.2.2.2 Ion Irradiation

Ion irradiation causes degrading defects to materials and can be harnessed to simulate the cosmic ray damages or that in nuclear reactors [51]. The commonly used sources include helium ions, argon ions, xenon ions, uranium ions, and fast heavy ions.

The high-angle annular dark-field (HAADF) images of MoS_2 before and after Ga^+ ion irradiation at different doses were displayed in Figure 12.2b, where the HAADF intensity changes were used to identify the elements and the atomic configuration of vacancy defects. Aberration-corrected scanning transmission election microscopy (AC-STEM) images clearly demonstrated the vacancy-defect density increased with the increase of the radiation doses [46]. Ion irradiation can also be used to tune properties of 2D materials by introducing controllable defects. Ovchinnikova and colleagues [52] used a high-dose helium ion beam to adjust the mechanical and optoelectronic properties $MoSe_2$ with nanoscale precision.

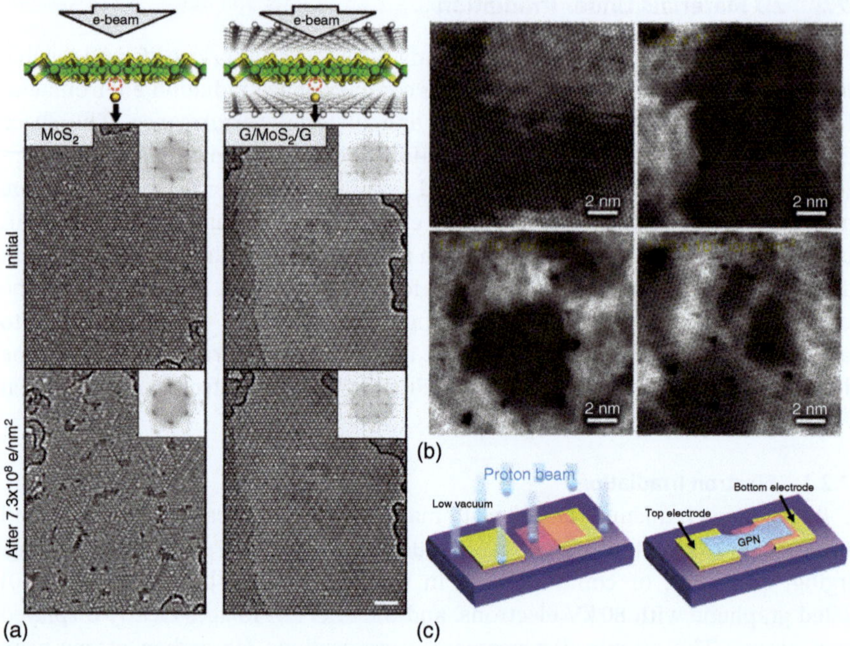

Figure 12.2 (a) MoS_2 and graphene/MoS_2/graphene heterostructure exposed to 80-keV electron irradiation. The upper and lower HRTEM images showed the samples before and after electron beam radiation, respectively. Source: Reproduced with permission Algara-Siller et al. [45]. © 2013, AIP Publishing LLC. (b) AC-STEM image of the pristine and the Ga^+ ion-irradiated MoS_2. Source: Reproduced with permission Thiruraman et al. [46]. © 2018, American Chemical Society and (c) the hexagonal boron nitride (h-BN) irradiated with high-energy protons (left) and exfoliated graphene dry-transferred onto the irradiated h-BN to form a vertical capacitor (right). Source: Adapted with permission Lee et al. [47]. Copyright © 2019 Royal Society of Chemistry.

12.2.2.3 Proton Irradiation

Lee et al. [47] exposed 2D h-BN to high-energy protons with 5- or 10-MeV proton irradiation at doses of 1×10^{13} and 5×10^{13} cm^{-2} (Figure 12.2c). After radiation, the hard breakdown field of 10.3 MV cm^{-1} of the h-BN capacitor significantly decreased. Molecular dynamics simulation indicated that the protons produced Frenkel defects within the h-BN dielectric and led to performance degradation of the device.

12.2.2.4 γ-Ray Irradiation

γ-rays with short wavelength of less than 0.01 Å easily penetrate materials, pass between atoms, and cause damage. They produce high-energy free electrons and positive ions in solids mainly through three mechanisms: photoelectric effect, Compton scattering, and pairwise generation. These free electrons and positive ions then interact with their nearby atoms in the form of ionization and remove them from their original positions, resulting in a large number of vacancy and interstitial atoms [53]. Zhang et al. [54] obtained 2D graphene flakes by γ-ray-induced graphene oxide (GO) suspension. The mechanism was that the radiation dissociation of dimethylformamide released electrons to reduce GO. Zhang et al. [55]

Table 12.1 A summary of radiation source, typical 2D materials, and radiation damage mechanism.

Radiation sources	Samples	Radiation damage mechanism	References
Electron irradiation	Graphene; Sb; TMDs (MoS_2, $MoSe_2$, WS_2); MoS_2/graphene heterostructure; Bi	Sputtering, tunneling, and vacancy defects	[28, 57–60]
Ion irradiation	Graphene; TMDs (MoS_2, $MoSe_2$, WS_2, WSe_2); h-BN; Bi_2Se_3; graphene/WSe_2 heterostructure	Doping, defects, trap states, tunneling, etc.	[59, 61–65]
Proton irradiation	Graphene; TMDs (MoS_2, $MoSe_2$, WS_2)	Displacement damage; formation of intermediate energy levels, defect energy levels, carrier traps, etc.	[28, 61, 66, 67]
γ-ray irradiation	Graphene; *r*-GO; *h*-BN; TMDs (MoS_2, $MoSe_2$, WS_2); Bi; SnS	Ionization effect; Compton effect; formation of vacancy and interfacial state defects	[28, 58, 60, 68–70]

also discovered similar effects of γ-ray-induced reduction of GO sheets in alcohol/water system.

Elafandi et al. [56] studied ^{60}Co gamma radiation effects on 2D WSe_2 and WS_2. These monolayer materials showed excellent γ radiation-resistance effects, the photoluminescence and Raman spectra of which just underwent small changes under total accumulated dosage of up to ~56 Mrad. WSe_2 monolayer was generally more stable than WS_2 samples. The result implied the inherent stability of these 2D materials under γ radiation enabled their promising space use.

Table 12.1 summarizes typical 2D materials, radiation sources, and radiation damage mechanism. Clearly, the research on the irradiation effects of 2D materials is very limited. Mainly, electrical and mechanical properties affected of the materials were investigated. The study of the optical properties of 2D materials as influenced by the radiation effects is much less, let alone their NLO properties. Moreover, new techniques and modeling/simulation methodologies on the damage mechanism are also urgently needed.

12.3 2D Materials for Space Use

12.3.1 Emerging Nonlinear Optical Applications

The NLO properties of 2D materials and their applications in mode locking, Q-switching, optical limiting, etc. have been intensively studied, which also implies promising space use. For instance, integrated photonic platforms have become indispensable for equipment in space applications [71, 72], where laser sources, optical modulators, etc. are frequently applied as key components. Lasers themselves have already been employed for free-space optical communications, space light detection and ranging (LiDAR) systems, and space laser altimeters [73, 74]. However, for practical use, large nonlinearity of the 2D materials is essential, and other features like broadband responses and high damage threshold are also required [75, 76]. Moreover, space use particularly demands excellent irradiation resistance for environmental stability. Currently researchers worldwide have endeavored to create abundant 2D materials for screening toward this purpose.

Ionization effect generates defects as much energy is deposited into electronic excitation. 2D materials possess high electron mobility and thermal conductivity, and thus, the threshold of defects generated through ionization effect is very high. Conducting 2D materials are considered more radiation-tolerant than insulating [77]. As a result, the devices based on 2D materials demonstrated high radiation tolerance. For instance, it was reported that holes in graphene appeared at ion energies in tens of MeV range [62]. In addition, the radiation hardness of 2D materials is one effective way to enhance their radiation resistance [28, 61].

At present, 2D materials continue to expand the territory. Various new materials emerge endlessly, and thousands of structures are predicted by high-throughput calculation [78]. The third-order optical nonlinearity and excited-state dynamics of some materials such as graphene and BP have been well studied [79]. However, there are few studies on their radiation damage mechanism. Research on the NLO properties of materials in the space radiation environment is particularly lacking, even for star materials such as graphene (only focusing on their mechanical and electrical properties) [20]. One of the main reasons is that the research of space radiation effect requires complex and expensive large-scale equipment simulating space environments such as accelerators. Moreover, 2D materials are a new field, and many of their properties are not fully understood, especially NLO and quantum optical properties.

12.3.1.1 Laser Protection and Mode Locking

Our group pioneers in investigating NLO properties of 2D materials, particularly in developing laser protection methods based on reversible saturation behaviors, a third-order optical nonlinear phenomenon [79, 80]. Previously, we reported composites based on 2D materials such as "PbS/graphene" [81] and "P3HT/MoS_2" [82] exhibiting excellent optical limiting performance. Recently, we prepared a series of Te nanostructures by electrochemical stripping and liquid-phase stripping methods, including one-dimensional (1D) Te nanowires (NWs), quasi-one-dimensional Te nanorods (NRs), zero-dimensional (0D) Te nanodots (NDs), and 2D Te nanosheets (NSs) [83]. Femtosecond Z-scan measurements revealed their unique size-dependent NLO properties. 1D Te NWs and quasi-1D Te NRs showed higher saturable absorption behavior than that of 0D Te nanostructures, while 2D Te NSs were highly effective optical limiters. We also developed Sb nanosheets [84] as novel laser protection materials with inherent large reverse saturation responses. Based on this, we have continued to explore the space radiation effects on the NLO properties of 2D materials toward space use.

We developed one new bottom-up venue for preparing 2D bismuthene and used them for mode-locking a fiber laser [58, 85]. Bismuthene also showed radiation-tolerant third-order optical nonlinearities under a simulated space radiation environment. When exposed to ^{60}Co γ-ray and electron irradiations, the microstructure of Bi showed negligible changes, as characterized by conventional FTIR, UV–vis, Raman, XRD patterns, and XPS measurements. However, the femtosecond saturable absorption of Bi degraded to some extent after radiation, suggesting the NLO responses were more sensitive to radiation damage. For example, the nonlinear absorption coefficient β of -8.8×10^{-12} m W^{-1} (532 nm, 370 nJ) decreased by 31.8% (Bi-1) and 39.8% (Bi-4) at the ^{60}Co γ-ray flux of 5.0×10^{4} Gy and electron radiation flux of 4.0×10^{14} e cm^{-2}, respectively (Figure 12.3a–b). Overall, the as-prepared bismuthene largely maintained saturable absorption behaviors at lower dose, indicative of suitability for mode locking in space.

Transient absorption spectroscopy offered a new method to elucidate the irradiation damage mechanism compared with conventional techniques from the perspective of the carrier dynamics. As shown in Figure 12.3c, the lifetime of carrier trapping for Bi reduced from 88 to ~60 ps, whereas the relaxation of carriers to the ground state increased significantly from 528 to >600 ns after radiation [86–88]. We contributed the observation to the fact that ^{60}Co γ-ray and electron irradiations mainly caused ionization damage and led to the increased amount of defects within the Bi nanosheets.

We then continued to investigate SnS [68], a novel two-elemental and 2D structural BP analog with enhanced stability under ambient conditions. SnS has gathered significant attention owing to its distinct optoelectronic properties and promising applications, but the radiation effect on its NLO properties is still unknown. The SnS nanoflowers (Figure 12.3d) synthesized by the solvothermal method exhibited a strong third-order NLO responses in the near-infrared region. When exposed to ^{60}Co γ-rays (Figure 12.3d–e) with a dose equal to over 45 years in a typical Earth orbit, the SnS nanosheets largely maintained good NLO properties. The β values of SnS only decreased by 5.8% under the excitation of fs pulses (120 fs, 532 nm, 0.5 μJ).

Likewise, the radiation damage mechanism was studied by ultrafast spectroscopy. The shortened carrier trapping lifetime indicated the ionization damage and defects as shown in Figure 12.3f. The carrier trapping lifetime dropped from 2.0 to 0.6 ps, but the carrier relaxation time remained on nanosecond time scale, as the γ-rays imposed ionization damage on the SnS nanosheets and produced more defects.

Kalbac et al. [63] investigated graphene under 100-keV Ar^{+} ion irradiation (Figure 12.4a). The Ar^{+} ion radiation induced defects, as proved by the increased intensity of the D peak and appearance of a new band at about 1610 cm^{-1} in Raman spectra. The Raman spectra were measured from ^{12}C and ^{13}C areas of graphene under the irradiation fluence of 0.3×10^{13} ions cm^{-2}. It was also found low-dimensional materials easily rearranged after radiation ionization and thus reduced defect damage through a self-annealing effect.

Hong et al. [57] also found that stacking of heterojunctions, i.e. inserting a graphene layer between MoS_2 and the substrate, effectively prevented the performance degradation of MoS_2 under electron beam irradiation (Figure 12.4b). With the

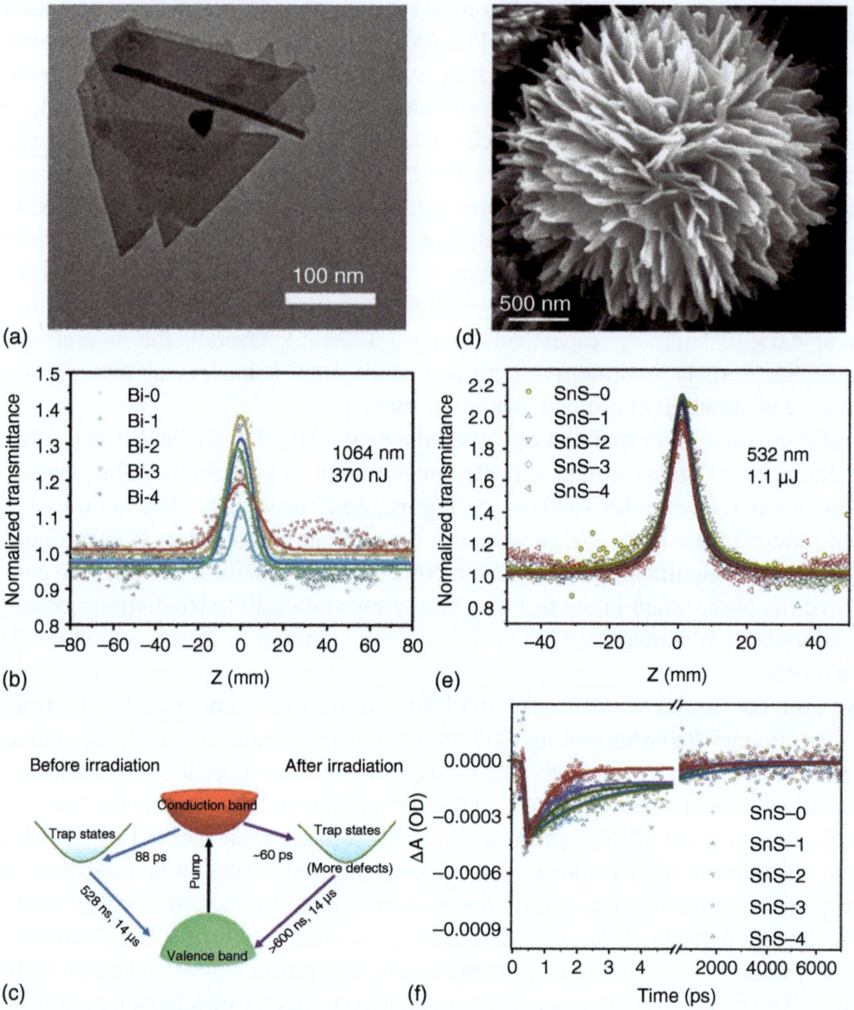

Figure 12.3 (a) The TEM images, (b) the open-aperture fs Z-scan data, and (c) the schematic carrier dynamics of bismuthene. Source: Reproduced with permission Zhang et al. [58]. © 2021, American Chemical Society. (d) The SEM images, (e) the open-aperture ns Z-scan data, and (f) the dynamic curves of 2D SnS. Source: Reproduced with permission Zhang et al. [68]. © 2022, Royal Society of Chemistry.

protection of graphene, the photoluminescence intensity of MoS_2/graphene was stronger than that of MoS_2 under the radiation dose of up to 8×10^{15} electrons cm^{-2}.

Using ion beam-modified graphene/WSe_2 heterostructure as a saturable absorber, high-performance Q-switched pulsed laser was realized in Nd: YAG waveguide cavity [64]. Carbon ion irradiation enhanced the interlayer coupling of graphene/WSe_2 heterostructure and led to variations of NLO properties (Figure 12.4c–f). The heterostructure exhibited a larger modulation depth and lower saturation intensity than individual 2D materials when graphene/WSe_2 spacing was modified.

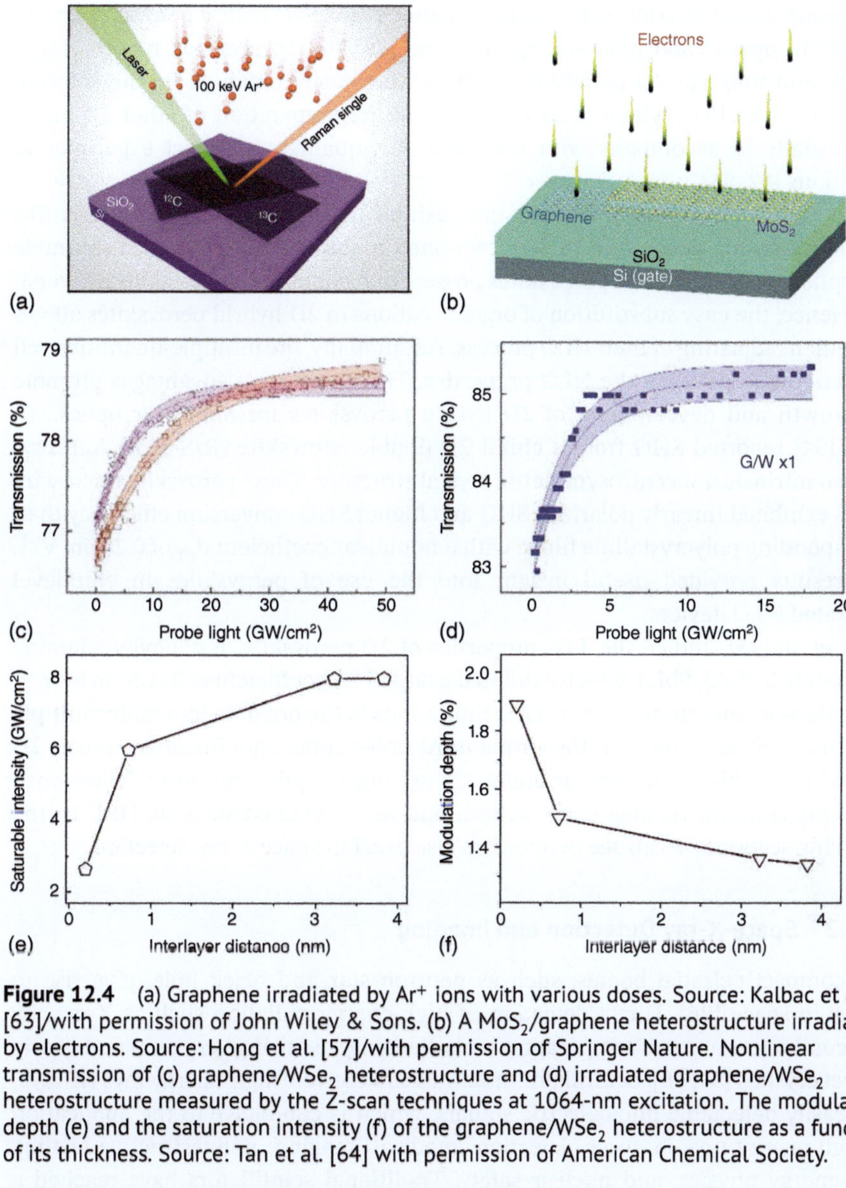

Figure 12.4 (a) Graphene irradiated by Ar^+ ions with various doses. Source: Kalbac et al. [63]/with permission of John Wiley & Sons. (b) A MoS_2/graphene heterostructure irradiated by electrons. Source: Hong et al. [57]/with permission of Springer Nature. Nonlinear transmission of (c) graphene/WSe_2 heterostructure and (d) irradiated graphene/WSe_2 heterostructure measured by the Z-scan techniques at 1064-nm excitation. The modulation depth (e) and the saturation intensity (f) of the graphene/WSe_2 heterostructure as a function of its thickness. Source: Tan et al. [64] with permission of American Chemical Society.

As a result, with the tailored graphene/WSe_2 heterostructure as a saturable absorber, the Q-switched lasers showed significantly improved lasing performance compared with that of pristine WSe_2-based system.

12.3.1.2 Other Nonlinear Phenomena

2D perovskites are also another type. 2D organic–inorganic hybrid perovskites have inherent quantum-well structures and interesting optoelectronic properties. Their applications in light emitting, sensing, modulation, and telecommunication equipment have attracted great attention. Low-cost and solution-treated fabrication and alternative organic spacer cations make 2D hybrid perovskites more tunable in optical and photonic applications [89, 90]. In previous reports, high-quality and uniform 2D perovskite crystals were easily obtained by the solution or vapor routes [91], which is advantageous to the preparation of other 2D materials usually in accompany with low-yield, low-quality, and strict experimental conditions [92, 93].

In particular, 2D perovskites can also exhibit unique NLO characteristics like second-harmonic generation (SHG), two-photon absorption (TPA), and saturable absorption (SA). 2D hybrid perovskites possess rich chemical and structural diversities. Hence, the easy substitution of organic cations in 2D hybrid perovskites allows convenient adjusting of their NLO process. Additionally, the multiple-quantum-well structure also enhances the NLO properties. These attractive advantages promote the growth and development of 2D hybrid perovskites for nonlinear optics. Yu et al. [94] reported SHG from a chiral 2D double-perovskite $(R/S\text{-}3AP)_4AgBiBr_{12}$ with an intrinsic noncentrosymmetric crystal structure. These perovskite microwire arrays exhibited linearly polarized SHG and higher SHG conversion efficiency than corresponding polycrystalline films, with a nonlinear coefficient d_{eff} of 0.28 pm V^{-1}. The results provided useful insight into the use of perovskites in chip-level integrated NLO devices.

Liu et al. [95] studied the TPA properties of 2D perovskite of phenylethylamine lead iodide ($(PEA)_2PbI_4$), which exhibited a large TPA coefficient of 211.5 cm MW^{-1}. The enhanced quantum and dielectric limitations in the organic–inorganic multiple quantum well accounted for the strong third-order optical nonlinearity. Hence, 2D layered perovskites offer new avenues for nonlinear optics due to their low cost, facile preparation, tunable construction, and large NLO coefficients [18]. In the following section, we can see perovskites also excel in space X-ray detection.

12.3.2 Space X-ray Detection and Imaging

The compact celestial bodies, such as neutron star and black hole, give rise to X-rays in space [96]. Direct detection of high-energy radiation such as X-rays by semiconductors at room temperature is challenging, requiring very pure and nearly perfect crystals [97, 98]. Scintillator materials can convert high-energy rays (X-rays) into easily detectable photoelectric signals, which is conducive to the application of high-energy rays in space detection, medical imaging, nondestructive testing, high-energy physics, and nuclear safety. Traditional scintillators have reached a limit, while the requirements for X-ray imaging and detection technology become higher. Therefore, exploring and studying new scintillators has become an urgent topic. Under this context, the emergence of metal halide perovskites has brought hope as new scintillator materials.

Metal halide perovskites of $MAPbBr_3$ have been applied as semiconductor-based X-ray detectors [101]. 3D perovskites of $CsPbBr_3$ nanocrystals [102] and nanosheets [103] were reported as scintillator screens in X-ray imaging [104]. However, compared with their 2D counterparts, they were usually less stable. In particular, 2D organic–inorganic metal halide composites are ideal for X-ray detection

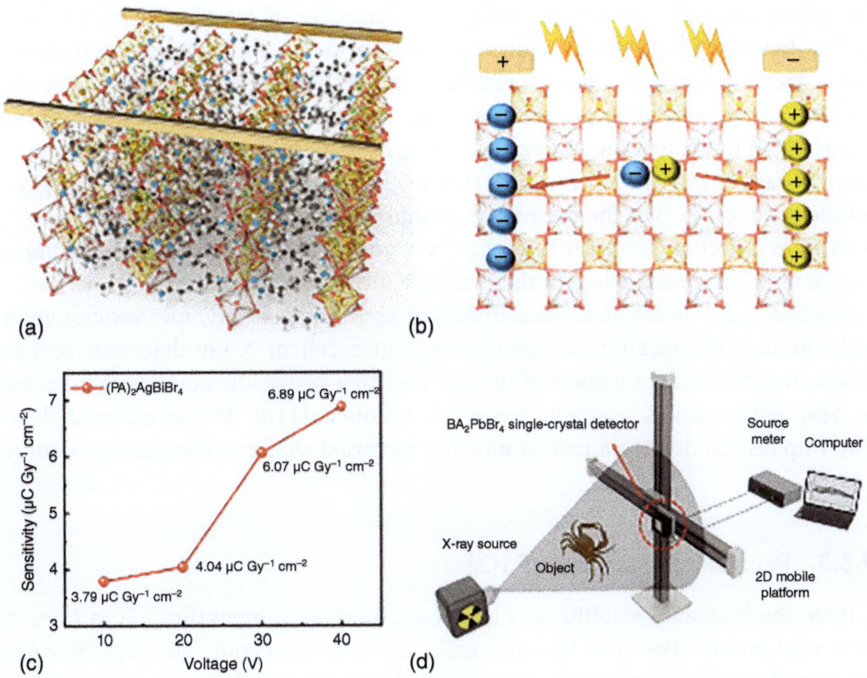

Figure 12.5 (a) Schematic diagram of the planar X-ray detector based on 2D perovskite of $(PA)_4AgBiBr_8$. (b) Carrier behavior in $(PA)_4AgBiBr_8$ under X-ray irradiation. (c) The sensitivity of $(PA)_4AgBiBr_8$ under different bias voltages. Source: Liu et al. [99]/with permission of Elsevier. (d) The plane-scan single-pixel imaging system. Source: Reproduced from Xu et al. [100], © 2022, John Wiley & Sons/CC BY 4.0.

because of their low defect density, high mobility–lifetime product (μτ) [105], high average atomic number (Z) [106], and low-cost fabrication [107]. Meng et al. [99] prepared a lead-free 2D double perovskite of $(PA)_4AgBiBr_8$ single crystal for X-ray detection (Figure 12.5a). Upon X-ray radiation, a built-in electric field was formed (Figure 12.5b). The as-prepared detector exhibited a high sensitivity of 6.89 μC Gy^{-1} cm^{-2} under 40 V bias voltage at 40 keV X-ray radiation (Figure 12.5c). Xie et al. [108] applied lithium chemical doping in a 2D $(PEA)_2PbBr_4$ crystal, which exhibited a scintillation yield of 11 000 photons per MeV under 662-keV gamma radiation.

Xu et al. [100] reported an ultrawide detectable X-ray imaging system based on 2D perovskites. Figure 12.5d displayed a setup close to a real clinical situation, where the target kept motionless but the X-ray detector moved by a 2D mobile platform. The 2D $(C_4H_9NH_3)_2PbBr_4$ single crystal provided an ultrawide detectable X-ray range of between 8.20 nG_{yair} s^{-1} (50 keV) and 15.24 mG_{yair} s^{-1} (9 MeV). Low ionic migration of the 2D perovskite gave rise to a sensitivity of 46.90 μC ${G_{yair}}^{-1}cm^{-2}$ (−1.16 V $μm^{-1}$). Furthermore, a remarkable spatial resolution of 4.3 lp mm^{-1} was obtained for the X-ray imaging.

As stated above, the new generation of metal halide perovskites has the advantages of large X-ray absorption coefficients, fast X-ray light responses, and adjustable band gaps, leading to their extraordinary X-ray detection performance. However, the toxicity of lead, the relative chemical instability of lead-based perovskites, the brittleness of high-quality single crystals, and the long growth periodicity hinder their commercialization. Flexible lead-free perovskite materials for X-ray detectors are thus becoming new trend owing to their light weight and low toxicity. Li et al. [109] fabricated metal-free PAZE-$NH_4X_3 \cdot H_2O$ perovskite for flexible transparent X-ray detection and imaging. Due to the low trap density, high mobility–lifetime products ($\mu\tau$) of the perovskite, a high sensitivity of 3708 $\mu C\,Gy_{air}^{-1}\,cm^{-2}$, an ultralow detection limit of 0.19 $\mu Gy_{air}^{-1}\,s^{-1}$, and an outstanding spatial resolution of 5.0 $lp\,mm^{-1}$ were achieved for the flexible X-ray detector.

In general, the desirable characteristics of area expansibility, low toxicity, high radiation stability, mechanical robustness, and excellent X-ray detection performance all increase the prospect of flexible lead-free perovskite composite films for the next generation of wearable electronic products [110]. Moreover, good flexibility implies the detectors can be adapted to varied shapes convenient to deploy in space.

12.3.3 Perovskite Solar Cells (PSCs)

In 1958, the first solar satellite was launched, marking a breakthrough in human space exploration. Powered by solar cells, current spacecraft, and satellites are able to go further and stay longer in space [111]. Solar panels for spacecraft face a higher standard than traditional terrestrial solar cells because they must withstand extreme conditions in space (extreme temperature cycles, high vacuum, and harsh radiation) [112], and hence, high performance is essential. Besides, they have to possess high specific power for the purpose of reducing launch load. Moreover, the difficulty of replacement and repair demands long-term stability for solar cells [96].

Besides the nonlinear optics and X-ray imaging, perovskites also show potential beyond-Earth applications for their photovoltaic properties [96, 113]. Flexible PSCs showed a higher power-to-weight ratio of 29.4 $W\,g^{-1}$ than other solar cell technologies [114], which is crucial for spacecraft payload reduction [115]. Moreover, it has been reported that perovskites exhibited better radiation resistance than glass after continuous radiation of 2.3-Mrad gamma rays for 1535 hours [96]. Due to these advantages [116], perovskite solar cells can be complementary to or even substitute conventional III–V and Si solar panels for satellites [117]. A recent report estimates a huge cost reduction of \$10–\$20 billion if perovskite solar cells replace conventional III–V photovoltaics for the LEO market [118].

Lang et al. [119] reported that perovskites of $FA_{0.75}Cs_{0.25}Sn_{0.5}Pb_{0.5}I_3$ withstood harsh radiation environments in space. Under 68-MeV proton irradiation, negligible degradation (<6%) at a dose of 10^{13} $p^+\,cm^{-2}$ was observed, whereas commercial radiation-hardened space PV degraded over 22%. The performance of perovskite ($Cs_{0.05}(MA_{0.17}FA_{0.83})_{0.95}Pb(I_{0.93}Br0_{.07})_3$) solar cells [120] under electron irradiation with 1-MeV electrons at fluences of 1×10^{14}, 1×10^{15}, and 1×10^{16} cm^{-2} was also reported. The photovoltaic characteristics of the $Cs_{0.05}(MA_{0.17}FA_{0.83})_{0.95}$

$Pb(I_{0.93}Br_{0.07})_3$ were measured about seven hours after irradiation, which implied the occurrence of annealing effect.

The PSC performance under various space irradiations have been reported, including electron [121], proton [122, 123], γ-ray irradiations [124, 125], and neutron [126]. Some pioneering experiments have been carried out under real-space conditions: in situ demonstration of PSCs' functionality and power generation by Cardinaletti et al. in 2018 [127] and Tu et al. in 2019 [128], respectively. In the latest space flight test of PSC, Reb et al. [129] reached the farthest point at the altitude of 239 km and tracked the equipment performance in the temperature range of 30–60 °C in about six minutes. These PSC space flight experiments have become valuable intermediate scientific steps on the ladder to space applications. However, these developments are limited in terms of low equipment efficiency and relatively short flight time, which is to be solved in the future.

12.3.4 2D Single-Photon Sources

Single-photon sources are indispensable components for quantum computing, quantum communication, and satellite-based quantum key distribution in space [130]. Recently, 2D materials have demonstrated advantages over traditional single-photon sources for applications in quantum technology [131]. For instance, single-photon sources are usually realized by point defects in single atoms, quantum dots, and dielectric materials [131, 132]. Point defects in 2D materials lead to efficient quantum emission, which is conveniently integrated with photonic networks [133–137].

Zhou et al. [131] displayed dislocations in TMDs as single-photon sources. Indeed, dislocation in 2D WSe_2 gives rise to emission energy at 1.29 eV, useful for optical fiber telecommunication. hBN is another popular single-photon source [138]. It has a wide bandgap of ~6 eV larger than those of most 2D materials and possesses optically active defects like nitrogen vacancy (V_N), boron vacancy (V_B), and missing atom at the nitrogen site ($N_B V_N$) [139, 140]. $N_B V_N$ defect with a peak at ~1.9 eV was highly anisotropic and the most likely defect for the quantum emitter. For instance, Tran et al. showed the emitters were polarized with record rates of $\sim 4 \times 10^6$ counts s^{-1}, indicating the potential of 2D hBN-based single emitters for quantum technologies [133].

Accompanying 2D single-photon sources is single-photon detector (SPD), which has huge potential in quantum information, LIDAR, deep space communication, etc. Niobium nitride (NbN) superconducting thin films have been used in superconducting nanowire single-photon detectors (SNSPD) due to their relatively high superconducting transition temperature. Superconducting properties of these films are associated with the film thickness. For example, a very thin (3–5 nm) NbN film is required for superconducting electrode, rapid single-flux quantum logic circuits, etc. Zhang et al. [141] applied ion irradiation control technology to modify NbN for SPD. The irradiated micron line detector exhibited higher light response sensitivity and an intrinsic detection efficiency close to saturation at 1550 nm.

12.3.5 Integrated Optoelectronic Platform

Integrated photonic and optoelectronic platforms have played indispensable roles for aerospace use, such as their broad applications in laser communication, microwave photonics, optical sensing, autonomous navigation, and astronomical imaging in spatial information systems due to the advantages of high performance and low power consumption [71, 72]. Like the microelectronic integrated circuit, the integrated optical path also utilizes lithography for fabrication, but with different devices such as lasers, detectors, modulators, and optical waveguides [142]. The development of space technology requires high-quality information systems of spacecraft and aircraft. Integrated optoelectronic platforms improve bandwidth, size, weight, power consumption, and cost of information systems that is hard to accomplish by traditional electronics and discrete optoelectronics. Hence, integrated optoelectronic platforms are expected to play crucial roles for future space business, such as microsatellite (cube satellites), deep space exploration missions, and space station construction [143]. Certainly, the radiation resistance of integrated optoelectronic devices is also a prerequisite for their space applications.

In a harsh space working environment, integrated optoelectronic devices need to cope with various forms of environmental changes, such as pressure changes, mechanical vibration, extreme temperature, space radiation, and so on. One of the most important and common is that the device is required to withstand long-term space ionizing radiation [144]. We will discuss some important components of the integrated platform.

12.3.5.1 Field-Effect Transistor and Photodetectors

Space electronics must be able to resist damage and/or failure under sustained high-throughput ionizing radiation such as particles, heavy ions, X-rays, or gamma rays. In particular, high-energy heavy ions cannot be fully shielded in space applications, which requires the discovery of antiradiation materials. These materials can also be used in other electronic systems in high-radiation environments, including high-altitude flight, military aircraft, satellites, nuclear reactors, and particle accelerators.

The evolution of the electrical characteristics of a MoS_2-based field-effect transistor (FET) under irradiation with 10-MeV proton beams was studied by Kim et al. [67]. For a low proton fluence of 10^{12} cm^{-2}, the device performance remained almost the same as that before irradiation. However, for a higher proton fluence of 10^{13} or 10^{14} cm^{-2}, a dramatic reduction in the conductance of the devices was discovered. The deterioration of the device performance was associated not with the defects produced in the 2D system, but with the presence of charge traps in the oxide substrate and their subsequent impact on the electrical characteristics of the device. In particular, Arnold et al. [61] found that ultrathin MoS_2 nanosheets can easily withstand proton and helium irradiation with fluences as high as $\sim 10^{16}$ and $\sim 10^{15}$ ions cm^{-2}, respectively, corresponding to hundreds or thousands of years of unshielded exposure to radiation in space. In the experiment reported by Lu et al. [147], MoS_2 FET was irradiated by 30-keV electrons. With the increase of

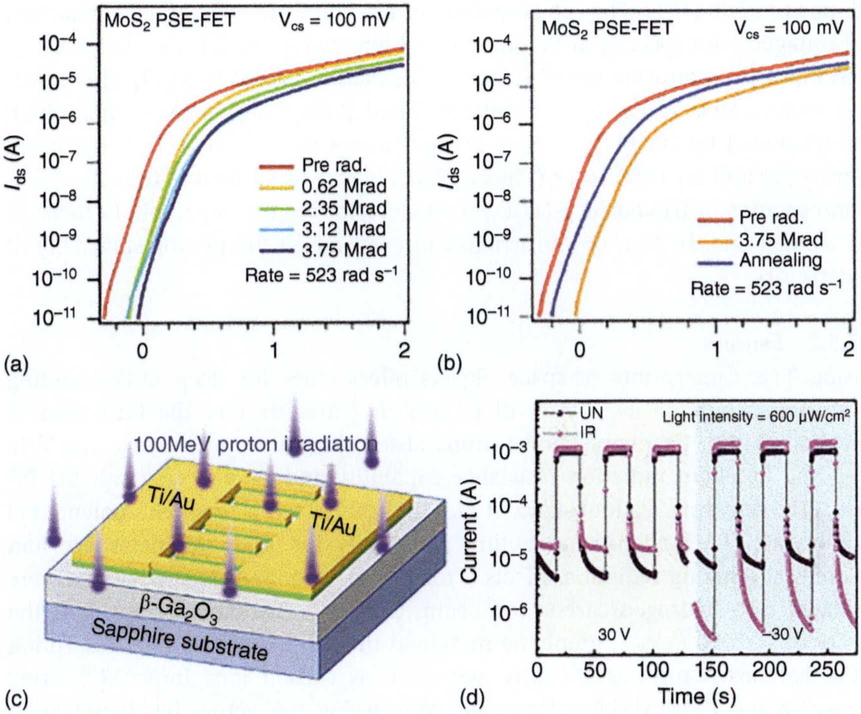

Figure 12.6 (a) Transfer characteristics of the MoS_2 PSE. (b) Annealing-induced recovery. Source: Adapted with permission Chen et al. [145]/with permission of John Wiley & Sons. (c) The schematic diagram of the β-Ga_2O_3 photodetector. (d) The I–t characteristic curves of the unirradiated (UN) and irradiated (IR) β-Ga_2O_3 photodetector. Source: Chang et al. [146]/with permission of AIP Publishing.

aging time, the transition from negative voltage offset to positive voltage offset was observed, which was explained by the dominant oxide charge after irradiation and the interface trap under high aging time.

Chen et al. [145] reported few-layer MoS_2 rad-hard FETs with polymer solid electrolyte (PSE) gate dielectrics, which demonstrated a TID tolerance of 3.75 Mrad (Si) and were repaired with thermal annealing. Figure 12.6a showed the transfer curves before and after different TIDs, and Figure 12.6b proved the recovery of the transfer characteristics. This work displayed the potential of 2D material-based devices for space use. Stanford et al. [148] reported the effect of helium ion beam irradiation on optical and electrical properties of WSe_2. In another work, Stanford and co-workers [149] reported the utilization of focused helium ion beams to bring in defects in single-layer WSe_2. Metal-like behavior of WSe_2 was induced by a dose of 1×10^{16} He^+ cm^{-2}.

β-Ga_2O_3, an emerging semiconductor material, has attracted great attention due to its extraordinary characteristics, such as an ultrawide energy bandgap of ~4.85 eV, a large breakdown electric field of ~8 MV cm^{-1}, and a high saturation electron velocity of 2×10^7 cm s^{-1} [150]. These desirable properties have allowed the application of β-Ga_2O_3 in solar-blind photodetectors, high-breakdown Schottky diodes, and high-power FETs [151]. The robust radiation resistance of wide-bandgap materials is advantageous for space applications, where high-energy particle irradiation easily deteriorates the performance of electronic devices. Besides, β-Ga_2O_3 also excels in photodetectors. Wang et al. [146] fabricated β-Ga_2O_3 photodetectors, which were irradiated by 100 MeV protons at the fluence of 10^{12} cm^{-2} with the beam intensity per unit area of 0.16 μA (Figure 12.6c). Figure 12.6d showed the solar-blind photoresponses of irradiated β-Ga_2O_3 photodetector were comparable to those of unirradiated one. In fact, proton irradiation even raised the photoresponsivity of the detectors.

12.3.5.2 Sensors

Sensing THz fingerprints of space objects offers clues for deep understanding of the universe, such as physics of pulsars and quasars and the formation of black holes [152]. Graphene shows strong absorption in the microwave and THz range and excellent radiation resistance capability and thus is very suitable for space THz detecting. Paddubskaya et al. [153] fabricated a graphene/polymethyl methacrylate (PMMA) heterostructure potentially for THz bolometer use and explored its ionizing radiation effects. Proton and secondary ion radiations were simulated by a hydrogen/carbon ion beam. Under the 290-keV ion beam at the density of 1.5×10^{12} cm^{-2}, graphene remained the same, but the THz absorption of the heterostructure significantly degraded, as carbon ions imposed heating damage to the PMMA slabs. However, considering the actual irradiation dose in GEO was significantly lower than that used in the simulated experiments, physical damage to the graphene/PMMA terahertz absorbers would be negligible in space.

Besides graphene, TMDs are emerging candidates for sensors and nanoelectronics in various industries, including nuclear power plants, LEO, space, and other harsh environments. Mondal et al. [154] fabricated a humidity sensor using 2D platinum diselenide ($PtSe_2$) and investigated γ-ray radiation effect on its properties. The result showed that the sensor was highly stable, maintaining real-time electrical output during under radiation of 10 kGy of γ-rays (^{60}Co).

12.3.6 Other Applications: Lubricants

High vacuum, as experienced by satellite components in near-Earth orbit, leads to significant tribological challenges to be solved. In recent years, there has been an increasing interest in lubrication materials capable of providing low friction and wear in vacuum conditions for implementation in the aerospace industry. The use of conventional lubricants like oils and greases involves several problems including their evaporation and outgassing [155]. To improve this situation, solid lubricants have received a good deal of attention due to the following significant advantages: (i) higher resistance to radiation than oil-based lubricants, (ii) reduced component size as no recirculation systems are required, (iii) higher resistance to extreme temperatures, (iv) lower service and maintenance required, and (v) being electrically conductive.

The advancement of aerospace endeavors, particularly the establishment of the International Space Station (ISS), has resulted in higher requirements for lubricating materials in the aerospace sector, with an ever-increasing demand for innovative lubricating materials [156]. Molybdenum disulfide (MoS_2) is one of the most widely utilized lubricating materials [157, 158]. Due to its special interlamellar hexagonal structure, MoS_2 possesses outstanding reduced friction and wear properties [159] and has been used as lubricants in a variety of modes, such as lubrication additives and physical deposition films [160]. Generally, the use of self-lubricating coatings incorporating solid lubricants has become a trend in aerospace applications [161].

12.4 Perspectives

While 2D materials offer great opportunities for space missions, their current low technological readiness level restricts deployment. In addition to further their device development, the performance of 2D materials under harsh conditions of space needs systematic and comprehensive evaluations. Space qualification studies usually consist of vacuum and thermal cycling, vibration, and shock tests and also radiation exposure. Vibration or shock usually poses no threat to nanomaterials, and vacuum and thermal cycling is routinely done in experiments. Hence, radiation effect of 2D materials is a challenging task and requires large amount of experimental data under various irradiation exposures. For future research, the following trends need particular attention:

(1) Expansion of the scope of 2D materials for space adaption evaluation. The speed of making new 2D materials has significantly outpaced their irradiation effect evaluation. For instance, although the radiation effect study on the electrical properties of graphene has been the most common, and that on TMDs has appeared, little is performed on other 2D materials, such as BP and 2D perovskites. The main reason is the measurement difficulty. The opportunities for actual space radiation environment test are extremely rare; even the simulated one on the ground is out of access to many groups and is costly. Fortunately, for its importance, more and more radiation sources for such investigations are being built worldwide, like the newly built spallation neutron source in China. Space laboratories are also available, like the earlier ISS already in orbit and the recent addition of Chinese Space Station (CSS or Tiangong) that plans to host over 1000 scientific experiments including cosmic radiation effects [162].
(2) Expansion on the radiation effects on the optical nonlinearity of 2D materials. As can be seen from the review above, current research on this topic is very rare. Nonlinear optics will be one of the most promising applications in space. For this purpose, a large quantity of work needs to be carried out to evaluate issues like whether the 2D materials are suitable for space applications and whether their unique characteristics are irreplaceable.
(3) The study of the radiation effect and damage/antiradiation mechanism of 2D materials requiring new techniques. A systematic understanding on the

mechanism transforms strenuous and aimless attempts into the rational design of 2D materials for space applications. The fs transient absorption techniques offer damage information from excited-state dynamics associated with defects caused by irradiation, which is unavailable by conventional characterizing methods like XRD and Raman techniques. Such new techniques from interdisciplinary fields, together with advanced modeling/simulating methods, become strong complementary to current methods.

(4) Expansion of nonlinear optical scope. Novel NLO phenomena of 2D materials like nonlinear quantum optics are closely related to space use. Core technologies in quantum communication include single-photon source technology, quantum coding and transmission technology, and single-photon detection technology. Quantum communication using single-photon source is considered safe and highly efficient. With quantum entanglement awarded by the 2022 Nobel Prize in Physics, the tide of the second global quantum revolution makes the development of quantum information technology in full swing. The realization of quantum entanglement to various quantum technologies such as quantum computing, encryption, communication, and sensing would profoundly change the production and lifestyle of human society. Among many methods to realize quantum entanglement, quantum photonics based on 2D layered van der Waals materials has unique advantages [163]. The wide technical universality and scalability, super high efficiency of light–matter interaction, and high flexibility of 2D materials in designing quantum states make them one of the most promising materials in the research of quantum information technology.

Acknowledgment

This work is supported by the National Natural Science Foundation of China (NSFC.22073038, 21673106), Key Research and Development Program of Gansu Province (20YF3WA013), Open Fund of National Key Laboratory of Materials Behavior and Evaluation Technology in Space Environment (6142910200106), and Supercomputing Center of Lanzhou University.

References

1 White, R.J. and Averner, M. (2001). Humans in space. *Nature* 409: 1115–1118.

2 2020 NASA Technology Taxonomy, https://www.nasa.gov/offices/oct/taxonomy/index.html.

3 Y. R. Shen, *The Principles of Nonlinear Optics*, New York (1984).

4 Li, M., Jiao, W., Song, Y. et al. (2015). Self-adaptive high anti-radiation EDFA for space optical communication systems. *Journal of Lightwave Technology* 33: 4513–4516.

5 Li, M., Ma, J., Zhang, X. et al. (2012). Investigation of black box model for erbium-doped fiber amplifiers in space radiation environment. *Journal of Lightwave Technology* 30: 3667–3671.

6 George, W.R., Abhijit, S., Salma, R. et al. (2010). Mechanistic studies for optical switching materials for space environments, nanophotonics and macrophotonics for space environments Iv. *SPIE* 7817: 22–32.

7 Novoselov, K.S., Geim, A.K., Morozov, S.V. et al. (2004). Electric field effect in atomically thin carbon films. *Science* 306: 666–669.

8 Hu, F. and Fei, Z. (2020). Recent progress on exciton polaritons in layered transition-metal dichalcogenides. *Advanced Optical Materials* 8: 1901003.

9 Wu, L., Dong, Y., Zhao, J. et al. (2019). Kerr Nonlinearity in 2D graphdiyne for passive photonic diodes. *Advanced Materials* 31: 1807981.

10 Caldwell, J.D., Aharonovich, I., Cassabois, G. et al. (2019). Photonics with hexagonal boron nitride. *Nature Reviews Materials* 4: 552–567.

11 Kanahashi, K., Pu, J., and Takenobu, T. (2020). 2D materials for large-area flexible thermoelectric devices. *Advanced Energy Materials* 10: 1902842.

12 Sun, Z., Martinez, A., and Wang, F. (2016). Optical modulators with 2D layered materials. *Nature Photonics* 10: 227–238.

13 Wang, M., Cai, S., Pan, C. et al. (2018). Robust memristors based on layered two-dimensional materials. *Nature Electronics* 1: 130–136.

14 Dong, R., Lan, C., Li, F. et al. (2019). Incorporating mixed cations in quasi-2D perovskites for high-performance and flexible photodetectors. *Nanoscale Horizons* 4: 1342–1352.

15 Anichini, C., Czepa, W., Pakulski, D. et al. (2018). Chemical sensing with 2D materials. *Chemical Society Reviews* 47: 4860–4908.

16 Novoselov, K.S., Mishchenko, A., Carvalho, A., and Castro Neto, A.H. (2016). 2D materials and van der Waals heterostructures. *Science* 353: aac9439.

17 Shiue, R.J., Efetov, D.K., Grosso, G. et al. (2017). Active 2D materials for on chip nanophotonics and quantum optics. *Nanophotonics* 6: 1329–1342.

18 Han, X., Zheng, Y., Chai, S. et al. (2020). 2D organic-inorganic hybrid perovskite materials for nonlinear optics. *Nanophotonics* 9: 1787–1810.

19 Kunitski, M., Eicke, N., Huber, P. et al. (2019). Double-slit photoelectron interference in strong-field ionization of the neon dimer. *Nature Communications* 10: 1–7.

20 Walker Ii, R.C., Shi, T., Silva, E.C. et al. (2016). Radiation effects on two-dimensional materials. *Physica Status Solidi (a)* 213: 3065–3077.

21 Lu, Y., Shao, Q., Yue, H., and Yang, F. (2019). A review of the space environment effects on spacecraft in different orbits. *IEEE Access* 7: 93473–93488.

22 Hands, A.D.P., Ryden, K.A., Meredith, N.P. et al. (2018). Radiation effects on satellites during extreme space weather events, space. *Weather* 16: 1216–1226.

23 Campajola, L. and Di Capua, F. (2016). Applications of accelerators and radiation sources in the field of space research and industry. *Topics in Current Chemistry* 374: 84.

24 Gregory, A.N. (2016). Space radiation and human exposures a primer. *Radiation Research* 185: 349–358.

25 Fok, M.C., Glocer, A., Zheng, Q. et al. (2011). Recent developments in the radiation belt environment model. *Journal of Atmospheric and Solar-Terrestrial Physics* 73: 1435–1443.

26 Coutrakon, G.B., Benton, E.R., Gridley, D.S. et al. (2007). Simulation of a 36h solar particle event at LLUMC using a proton beam scanning system. *Nuclear Instruments and Methods in Physics Research Section B: Beam Interactions with Materials and Atoms* 261: 791–794.

27 Nesterenok, A. (2013). Numerical calculations of cosmic ray cascade in the Earth's atmosphere-Results for nucleon spectra. *Nuclear Instruments and Methods in Physics Research Section B: Beam Interactions with Materials and Atoms* 295: 99–106.

28 Vogl, T., Sripathy, K., Sharma, A. et al. (2019). Radiation tolerance of two-dimensional material-based devices for space applications. *Nature Communications* 10: 1202.

29 Sen, J.S.Z. and Wang, J.L. (2010). Satellite on-board failure statistics and analysis. *Spacecraft Engineering* 19: 41.

30 Gong, H., Liao, W., Zhang, E.X. et al. (2018). Proton-induced displacement damage and total-ionizing-dose effects on silicon-based MEMS resonators. *IEEE Transactions on Nuclear Science* 65: 34–38.

31 Tang, D., Martin-Bragado, I., He, C. et al. (2016). Time dependent modeling of single particle displacement damage in silicon devices. *Microelectronics Reliability* 60: 25–32.

32 Yang, W.-T., Li, Y.-H., Guo, Y.-X. et al. (2020). Investigation of single event effect in 28-nm system-on-chip with multi patterns*. *Chinese Physics B* 29: 108504.

33 Luo, J., Liu, J., Sun, Y. et al. (2017). Influence of heavy ion flux on single event effect testing in memory devices. *Nuclear Instruments and Methods in Physics Research Section B: Beam Interactions with Materials and Atoms* 406: 431–436.

34 Dressendorfer, E.P.V. and Ma, T.P. (1989). *Ionizing Radiation Effects in MOS Devices and Circuits*. New York, NY USA: Wiley.

35 Siedle, A.H. and Adams, L. (1993). *Handbook of Radiation Effects (Oxford Science Publications)*, 2e. London, UK: Oxford Univ. Press.

36 Srour, J.R. and Palko, J.W. (2013). Displacement damage effects in irradiated semiconductor devices. *IEEE Transactions on Nuclear Science* 60: 1740–1766.

37 Srour, J.R., Marshall, C.J., and Marshall, P.W. (2003). Review of displacement damage effects in silicon devices. *IEEE Transactions on Nuclear Science* 50: 653–670.

38 Sexton, F.W. (2003). Destructive single-event effects in semiconductor devices and ICs. *IEEE Transactions on Nuclear Science* 50: 603–621.

39 Bi, J.S., Zeng, C.B., Gao, L.C. et al. (2014). Estimation of pulsed laser-induced single event transient in a partially depleted silicon-on-insulator 0.18-μm MOSFET. *Chinese Physics B* 23: 088505.

40 Crain, S.H., Mazur, J.E., Katz, R.B. et al. (2001). Analog and digital single-event effects experiments in space. *IEEE Transactions on Nuclear Science* 48: 1841–1848.

41 Sawyer, D.M. and Vette, J.I. (1976). Ap-8 trapped proton environment for solar maximum and solar minimum, NASA sti/recon technical report n 77-18983.

42 Xapsos, M.A., Summers, G.P., Barth, J.L. et al. (2000). Probability model for cumulative solar proton event fluences. *IEEE Transactions on Nuclear Science* 47: 486–490.

43 Vette, J.I. (1991). The AE-8 trapped electron model environment, National Space Science Data Center (NSSDC), World Data Center A for Rockets and Satellites (WDC-AR &S), National Aeronautics and Space Administration, Goddard Space Flight Center.

44 ESA Space Environment & Effects. FLUMIC 99% worst day electron flux, ESA Space Environment & Effects, https://nom.esa.int/models/flumic (accessed 10 October 2022).

45 Algara-Siller, G., Kurasch, S., Sedighi, M. et al. (2013). The pristine atomic structure of MoS_2 monolayer protected from electron radiation damage by graphene. *Applied Physics Letters* 103: 203107.

46 Thiruraman, J.P., Fujisawa, K., Danda, G. et al. (2018). Angstrom-size defect creation and ionic transport through pores in single-layer MoS_2. *Nano Letters* 18: 1651–1659.

47 Lee, D., Yoo, S., Bae, J. et al. (2019). High-energy proton irradiation damage on two-dimensional hexagonal boron nitride. *RSC Advances* 9: 18326–18332.

48 La Tessa, C., Di Fino, L., Larosa, M. et al. (2009). Estimate of the space station shielding thickness at a USLab site using ALTEA measurements and fragmentation cross sections. *Nuclear Instruments and Methods in Physics Research Section B: Beam Interactions with Materials and Atoms* 267: 3383–3387.

49 Jin, C., Lin, F., Suenaga, K., and Iijima, S. (2009). Fabrication of a freestanding boron nitride single layer and its defect assignments. *Physical Review Letters* 102: 195505.

50 Meyer, J.C., Eder, F., Kurasch, S. et al. (2012). Accurate measurement of electron beam induced displacement cross sections for single-layer graphene. *Physical Review Letters* 108: 196102.

51 Was, G.S., Jiao, Z., Getto, E. et al. (2014). Emulation of reactor irradiation damage using ion beams. *Scripta Materialia* 88: 33–36.

52 Iberi, V., Liang, L., Ievlev, A.V. et al. (2016). Nanoforging single layer $MoSe_2$ through defect engineering with focused helium ion beams. *Scientific Reports* 6: 30481.

53 Oser, S., Bhattacharya, D., Boone, L.M. et al. (2001). High-energy gamma-ray observations of the crab nebula and pulsar with the solar tower atmospheric cerenkov effect experiment. *The Astrophysical Journal* 547: 949–958.

54 Zhang, Y., Ma, H.-L., Zhang, Q. et al. (2012). Facile synthesis of well-dispersed graphene by γ-ray induced reduction of graphene oxide. *Journal of Materials Chemistry* 22: 13064–13069.

55 Zhang, B., Li, L., Wang, Z. et al. (2012). Radiation induced reduction: an effective and clean route to synthesize functionalized graphene. *Journal of Materials Chemistry* 22: 7775–7781.

56 Elafandi, S., Christiansen, R., Azam, N. et al. (2020). Monolayer 2D quantum materials subjected to gamma irradiation in high-vacuum for nuclear and space applications. *Applied Physics Letters* 116: 213105.

57 Hong, S., Fu, D., Hou, J. et al. (2018). Robust photoluminescence energy of MoS_2/graphene heterostructure against electron irradiation. *Science China Materials* 61: 1351–1359.

58 Zhang, Z.T., Yang, Q.Q., Zhen, X.J. et al. (2021). Two-dimensional bismuthene showing radiation-tolerant third-order optical nonlinearities. *ACS Applied Materials & Interfaces* 13: 21626–21634.

59 Krasheninnikov, A.V. and Nordlund, K. (2010). Ion and electron irradiation-induced effects in nanostructured materials. *Journal of Applied Physics* 107: 071301.

60 Wang, Y., Feng, Y., Mo, F. et al. (2014). Influence of irradiation upon few-layered graphene using electron-beams and gamma-rays. *Applied Physics Letters* 105: 023102.

61 Arnold, A.J., Shi, T., Jovanovic, I., and Das, S. (2019). Extraordinary radiation hardness of atomically thin MoS_2. *ACS Applied Materials & Interfaces* 11: 8391–8399.

62 Nebogatikova, N.A., Antonova, I.V., Erohin, S.V. et al. (2018). Nanostructuring few-layer graphene films with swift heavy ions for electronic application: tuning of electronic and transport properties. *Nanoscale* 10: 14499–14509.

63 Kalbac, M., Lehtinen, O., Krasheninnikov, A.V., and Keinonen, J. (2013). Ion-irradiation-induced defects in isotopically-labeled two layered graphene: enhanced in-situ annealing of the damage. *Advanced Materials* 25: 1004–1009.

64 Tan, Y., Liu, X., He, Z. et al. (2017). Tuning of interlayer coupling in large-area graphene/WSe_2 van der Waals Heterostructure via ion irradiation: optical evidences and photonic applications. *ACS Photonics* 4: 1531–1538.

65 Li, Z. and Chen, F. (2017). Ion beam modification of two-dimensional materials: characterization, properties, and applications. *Applied Physics Reviews* 4: 011103.

66 Mathew, S., Chan, T.K., Zhan, D. et al. (2011). The effect of layer number and substrate on the stability of graphene under MeV proton beam irradiation. *Carbon* 49: 1720–1726.

67 Kim, T.-Y., Cho, K., Park, W. et al. (2014). Irradiation effects of high-energy proton beams on MoS_2 field effect transistors. *ACS Nano* 8: 2774–2781.

68 Wang, S.Y., Zhen, X.J., Feng, Z.Z. et al. (2022). Third-order optical nonlinearities of two-dimensional SnS under irradiation: implications for space use. *Journal of Materials Chemistry C* 10: 18025–18032.

69 Bhimanapati, G.R., Wetherington, M., Mahabir, S., and Robinson, J.A. (2016). Synthesis and radiation response of BCON: a graphene oxide and hexagonal boron nitride hybrid. *2D Materials* 3: 025028.

70 Cazalas, E., Hogsed, M.R., Vangala, S. et al. (2019). Gamma-ray radiation effects in graphene-based transistors with h-BN nanometer film substrates. *Applied Physics Letters* 115: 223504.

71 Wang, X.J., Fang, H.H., Sun, F.W., and Sun, H.B. (2022). Laser writing of color centers. *Laser & Photonics Reviews* 16: 2100029.

72 Singh, A., Jo, S.S., Li, Y. et al. (2020). Refractive uses of layered and two-dimensional materials for integrated photonics. *ACS Photonics* 7: 3270–3285.

73 Cui, T., Bai, B., and Sun, H.B. (2019). Tunable metasurfaces based on active materials. *Advanced Functional Materials* 29: 1806692.

74 Zhou, H., Chen, Y., Hyyppä, J., and Li, S. (2017). An overview of the laser ranging method of space laser altimeter. *Infrared Physics & Technology* 86: 147–158.

75 Muhammad, A.R., Zakaria, R., Ahmad, M.T. et al. (2019). Pure gold saturable absorber for generating Q-switching pulses at 2 μm in Thulium-doped fiber laser cavity. *Optical Fiber Technology* 50: 23–30.

76 Shu, Y., Guo, P., Li, X. et al. (2019). Gold nanorods as saturable absorber for harmonic soliton molecules generation. *Frontiers in Chemistry* 7: 715.

77 Krasheninnikov, A.V. (2020). Are two-dimensional materials radiation tolerant? *Nanoscale Horizons* 5: 1447–1452.

78 Mounet, N., Gibertini, M., Schwaller, P. et al. (2018). Two-dimensional materials from high-throughput computational exfoliation of experimentally known compounds. *Nature Nanotechnology* 13: 246–252.

79 Li, N., Wang, Q., and Zhang, H.L. (2020). 2D materials in light: excited-state dynamics and applications. *The Chemical Record* 20: 413–428.

80 Zhai, X.P., Ma, B., Wang, Q., and Zhang, H.L. (2020). 2D materials towards ultrafast photonic applications. *Physical Chemistry Chemical Physics* 22: 22140–22156.

81 Zhao, M., Peng, R., Zheng, Q. et al. (2015). Broadband optical limiting response of a graphene-PbS nanohybrid. *Nanoscale* 7: 9268–9274.

82 Zhao, M., Chang, M.J., Wang, Q. et al. (2015). Unexpected optical limiting properties from MoS_2 nanosheets modified by a semiconductive polymer. *Chemical Communications* 51: 12262–12265.

83 Xiao, Q., Ma, B., Fei, X. et al. (2021). Unveiling the dimension-dependence of femtosecond nonlinear optical properties of tellurium nanostructures. *Nanoscale Horizons* 6: 918–927.

84 Zhang, L., Fahad, S., Wu, H.-R. et al. (2020). Tunable nonlinear optical responses and carrier dynamics of two-dimensional antimonene nanosheets. *Nanoscale Horizons* 5: 1420–1429.

85 Yang, Q.Q., Liu, R.T., Huang, C. et al. (2018). 2D bismuthene fabricated via acid-intercalated exfoliation showing strong nonlinear near-infrared responses for mode-locking lasers. *Nanoscale* 10: 21106–21115.

86 Kim, W.D., Kim, J.H., Lee, S. et al. (2016). Role of surface states in photocatalysis: study of chlorine-passivated CdSe nanocrystals for photocatalytic hydrogen generation. *Chemistry of Materials* 28: 962–968.

87 Jiao, X., Chen, Z., Li, X. et al. (2017). Defect-mediated electron-hole separation in one-unit-cell $ZnIn_2S_4$ layers for boosted solar-driven CO_2 reduction. *Journal of the American Chemical Society* 139: 7586–7594.

88 Yang, B., Hong, F., Chen, J. et al. (2019). Colloidal synthesis and charge-carrier dynamics of $Cs_2AgSb_{1-y}Bi_yX_6$ (X: Br, Cl; 0 ≤y≤1) double perovskite nanocrystals. *Angewandte Chemie International Edition* 58: 2278–2283.

89 Xu, J., Li, X., Xiong, J. et al. (2020). Halide perovskites for nonlinear optics. *Advanced Materials* 32: 1806736.

90 Ferrando, A., Martínez Pastor, J.P., and Suárez, I. (2018). Toward metal halide perovskite nonlinear photonics. *The Journal of Physical Chemistry Letters* 9: 5612–5623.

91 Huo, C., Cai, B., Yuan, Z. et al. (2017). Two-dimensional metal halide perovskites: theory. *Synthesis, and Optoelectronics, Small Methods* 1: 1600018.

92 Dong, R., Zhang, T., and Feng, X. (2018). Interface-assisted synthesis of 2D materials: trend and challenges. *Chemical Reviews* 118: 6189–6235.

93 Lim, D., Suh, H., Suryawanshi, M. et al. (2018). Kinetically controlled growth of phase-pure SnS absorbers for thin film solar cells: achieving efficiency near 3% with long-term stability using an SnS/CdS heterojunction. *Advanced Energy Materials* 8: 1702605.

94 Yu, Z., Cao, S., Zhao, Y. et al. (2022). Chiral lead-free double perovskite single-crystalline microwire arrays for anisotropic second-harmonic generation. *ACS Applied Materials & Interfaces* 14: 39451–39458.

95 Liu, W., Xing, J., Zhao, J. et al. (2017). Giant two-photon absorption and its saturation in 2D organic-inorganic perovskite. *Advanced Optical Materials* 5: 1601045.

96 Tu, Y., Wu, J., Xu, G. et al. (2021). Perovskite solar cells for space applications: progress and challenges. *Advanced Materials* 33: 2006545–2006567.

97 He, Y., Hadar, I., and Kanatzidis, M.G. (2022). Detecting ionizing radiation using halide perovskite semiconductors processed through solution and alternative methods. *Nature Photonics* 16: 14–26.

98 Zhang, H., Yang, Z., Zhou, M. et al. (2021). Reproducible X-ray imaging with a perovskite nanocrystal scintillator embedded in a transparent amorphous network structure. *Advanced Materials* 33: 2102529.

99 Liu, P., Xiao, Y., Yang, Z. et al. (2022). Two-Dimensional hybrid double perovskite $(PA)_4AgBiBr_8$ single crystals for X-ray detection. *Optical Materials* 133: 112972.

100 Xu, X., Wu, Y., Zhang, Y. et al. (2022). Two-dimensional perovskite single crystals for high-performance X-ray imaging and exploring MeV X-ray detection. *Energy & Environmental Materials* e12487.

101 Wei, H., Fang, Y., Mulligan, P. et al. (2016). Sensitive X-ray detectors made of methylammonium lead tribromide perovskite single crystals. *Nature Photonics* 10: 333–339.

102 Zhang, Y., Sun, R., Ou, X. et al. (2019). Metal halide perovskite nanosheet for X-ray high-resolution scintillation imaging screens. *ACS Nano* 13: 2520–2525.

103 Wei, H., DeSantis, D., Wei, W. et al. (2017). Dopant compensation in alloyed $CH_3NH_3PbBr_{3-x}Cl_x$ perovskite single crystals for gamma-ray spectroscopy. *Nature Materials* 16: 826–833.

104 Zhou, Y., Chen, J., Bakr, O.M., and Mohammed, O.F. (2021). Metal halide perovskites for X-ray imaging scintillators and detectors. *ACS Energy Letters* 6: 739–768.

105 Li, Z., Zhou, F., Yao, H. et al. (2021). Halide perovskites for high-performance X-ray detector. *Materials Today* 48: 155–175.

106 Xia, M., Yuan, J.H., Niu, G. et al. (2020). Unveiling the structural descriptor of $A_3B_2X_9$ perovskite derivatives toward X-ray detectors with low detection limit and high stability. *Advanced Functional Materials* 30: 1910648.

107 Li, H.Y., Song, J.M., Pan, W.T. et al. (2020). Sensitive and stable 2D perovskite single-crystal X-ray detectors enabled by a supramolecular anchor. *Advanced Materials* 32: 2003790.

108 Xie, A., Hettiarachchi, C., Maddalena, F. et al. (2020). Lithium-doped two-dimensional perovskite scintillator for wide-range radiation detection. *Communications Materials* 1: 37.

109 Li, Z., Peng, G., Chen, H. et al. (2022). Metal-free PAZE-$NH_4X_3{\cdot}H_2O$ perovskite for flexible transparent X-ray detection and imaging. *Angewandte Chemie International Edition* 61: e202207198.

110 Zhao, J., Zhao, L., Deng, Y. et al. (2020). Perovskite-filled membranes for flexible and large-area direct-conversion X-ray detector arrays. *Nature Photonics* 14: 612–617.

111 Andreev, V.M. (2003). *GaAs and High-Efficiency Space Cells, Practical Handbook of Photovoltaics*, 417–433. Elsevier Science.

112 Iles, P.A. (2001). Evolution of space solar cells. *Solar Energy Materials and Solar Cells* 68: 1–13.

113 Berry, J.J., van de Lagemaat, J., Al-Jassim, M.M. et al. (2017). Perovskite photovoltaics: the path to a printable terawatt-scale technology. *ACS Energy Letters* 2: 2540–2544.

114 Kang, S., Jeong, J., Cho, S. et al. (2019). Ultrathin, lightweight and flexible perovskite solar cells with an excellent power-per-weight performance. *Journal of Materials Chemistry A* 7: 1107–1114.

115 Min, H., Lee, D.Y., Kim, J. et al. (2021). Perovskite solar cells with atomically coherent interlayers on SnO_2 electrodes. *Nature* 598: 444–450.

116 Kaltenbrunner, M., Adam, G., Głowacki, E.D. et al. (2015). Flexible high power-per-weight perovskite solar cells with chromium oxide-metal contacts for improved stability in air. *Nature Materials* 14: 1032–1039.

117 Akyildiz, I.F. and Kak, A. (2019). The internet of space things/CubeSats. *IEEE Network* 33: 212–218.

118 Ho-Baillie, A.W.Y., Sullivan, H.G.J., Bannerman, T.A. et al. (2022). Deployment opportunities for space photovoltaics and the prospects for perovskite solar cells. *Advanced Materials Technologies* 7: 2101059.

119 Lang, F., Eperon, G.E., Frohna, K. et al. (2021). Proton-radiation tolerant all-perovskite multijunction solar cells. *Advanced Energy Materials* 11: 2102246.

120 Miyazawa, Y., Kim, G.M., Ishii, A. et al. (2021). Evaluation of damage coefficient for minority-carrier diffusion length of triple-cation perovskite solar cells under 1 MeV electron irradiation for space applications. *The Journal of Physical Chemistry C* 125: 13131–13137.

121 Song, Z., Li, C., Chen, C. et al. (2020). High remaining factors in the photovoltaic performance of perovskite solar cells after high-fluence electron beam irradiations. *The Journal of Physical Chemistry C* 124: 1330–1336.

122 Miyazawa, Y., Ikegami, M., Chen, H.-W. et al. (2018). Tolerance of perovskite solar cell to high-energy particle irradiations in space environment. *iScience* 2: 148–155.

123 Lang, F., Nickel, N.H., Bundesmann, J. et al. (2016). Radiation hardness and self-healing of perovskite solar cells. *Advanced Materials* 28: 8726–8731.

124 Yang, S., Xu, Z., Xue, S. et al. (2019). Organohalide lead perovskites: more stable than glass under gamma-ray radiation. *Advanced Materials* 31: 1805547.

125 Yang, K., Huang, K., Li, X. et al. (2019). Radiation tolerance of perovskite solar cells under gamma ray. *Organic Electronics* 71: 79–84.

126 Paternò, G.M., Robbiano, V., Santarelli, L. et al. (2019). Perovskite solar cell resilience to fast neutrons. *Sustainable Energy Fuels* 3: 2561–2566.

127 Cardinaletti, I., Vangerven, T., Nagels, S. et al. (2018). Organic and perovskite solar cells for space applications. *Solar Energy Materials and Solar Cells* 182: 121–127.

128 Tu, Y., Xu, G., Yang, X. et al. (2019). Mixed-cation perovskite solar cells in space. *Science China Physics, Mechanics & Astronomy* 62: 974221.

129 Reb, L.K., Böhmer, M., Predeschly, B. et al. (2020). Perovskite and organic solar cells on a rocket flight. *Joule* 4: 1880–1892.

130 Liao, S.K., Cai, W.Q., Liu, W.Y. et al. (2017). Satellite-to-ground quantum key distribution. *Nature* 549: 43–47.

131 Zhou, X., Zhang, Z., and Guo, W. (2020). Dislocations as single photon sources in two-dimensional semiconductors. *Nano Letters* 20: 4136–4143.

132 Srivastava, A., Sidler, M., Allain, A.V. et al. (2015). Optically active quantum dots in monolayer WSe_2. *Nature Nanotechnology* 10: 491–496.

133 Tran, T.T., Bray, K., Ford, M.J. et al. (2016). Quantum emission from hexagonal boron nitride monolayers. *Nature Nanotechnology* 11: 37–41.

134 Vogl, T., Campbell, G., Buchler, B.C. et al. (2018). Fabrication and deterministic transfer of high-quality quantum emitters in hexagonal boron nitride. *ACS Photonics* 5: 2305–2312.

135 Vogl, T., Lu, Y., and Koy Lam, P. (2017). Room temperature single photon source using fiber-integrated hexagonal boron nitride. *Journal of Physics D: Applied Physics* 50: 295101.

136 Tran, T.T., Elbadawi, C., Totonjian, D. et al. (2016). Robust multicolor single photon emission from point defects in hexagonal boron nitride. *ACS Nano* 10: 7331–7338.

137 Choi, S., Tran, T.T., Elbadawi, C. et al. (2016). Engineering and localization of quantum emitters in large hexagonal boron nitride layers. *ACS Applied Materials & Interfaces* 8: 29642–29648.

138 Song, L., Ci, L., Lu, H. et al. (2010). Large scale growth and characterization of atomic hexagonal boron nitride layers. *Nano Letters* 10: 3209–3215.

139 Xia, F., Wang, H., Xiao, D. et al. (2014). Two-dimensional material nanophotonics. *Nature Photonics* 8: 899–907.

140 Attaccalite, C., Bockstedte, M., Marini, A. et al. (2011). Coupling of excitons and defect states in boron-nitride nanostructures. *Physical Review B* 83: 144115.

141 Zhang, W., Jia, Q., You, L. et al. (2019). Saturating intrinsic detection efficiency of superconducting nanowire single-photon detectors via defect engineering. *Physical Review Applied* 12: 044040.

142 Wang, J., Paesani, S., Ding, Y. et al. (2018). Multidimensional quantum entanglement with large-scale integrated optics. *Science* 360: 285–291.

143 Guilhot, D. and Ribes-Pleguezuelo, P. (2019). Laser technology in photonic applications for space. *Instruments* 3: 50.

144 Johnston, A.H. (2013). Radiation effects in optoelectronic devices. *IEEE Transactions on Nuclear Science* 60: 2054–2073.

145 Chen, D., Li, J., Wei, Z. et al. (2022). Repairable polymer solid electrolyte gated MoS_2 field effect devices with large radiation tolerance. *Advanced Electronic Materials* 8: 2100619.

146 Chang, M.M., Guo, D.Y., Zhong, X.L. et al. (2022). Impact of 100 MeV high-energy proton irradiation on β-Ga_2O_3 solar-blind photodetector: Oxygen vacancies formation and resistance switching effect. *Journal of Applied Physics* 132: 123105.

147 Lu, M.Y., Wu, S.C., Wang, H.C., and Lu, M.P. (2018). Time-evolution of the electrical characteristics of MoS_2 field-effect transistors after electron beam irradiation. *Physical Chemistry Chemical Physics* 20: 9038–9044.

148 Stanford, M.G., Pudasaini, P.R., Belianinov, A. et al. (2016). Focused helium-ion beam irradiation effects on electrical transport properties of few-layer WSe_2: enabling nanoscale direct write homo-junctions. *Scientific Reports* 6: 27276.

149 Stanford, M.G., Pudasaini, P.R., Gallmeier, E.T. et al. (2017). High conduction hopping behavior induced in transition metal dichalcogenides by percolating defect networks: toward atomically thin circuits. *Advanced Functional Materials* 27: 1702829.

150 Mastro, M.A., Kuramata, A., Calkins, J. et al. (2017). Perspective-opportunities and future directions for Ga_2O_3. *ECS Journal of Solid State Science and Technology* 6: 356–359.

151 Oh, S., Yang, G., and Kim, J. (2016). Electrical characteristics of vertical Ni/β-Ga_2O_3 schottky barrier diodes at high temperatures. *ECS Journal of Solid State Science and Technology* 6: Q3022.

152 Huang, Z., Chen, H., Huang, Y. et al. (2018). Ultra-broadband wide-angle terahertz absorption properties of 3D graphene foam. *Advanced Functional Materials* 28: 1704363.

153 Paddubskaya, A., Batrakov, K., Khrushchinsky, A. et al. (2021). Outstanding radiation tolerance of supported graphene: towards 2D sensors for the space millimeter radioastronomy. *Nanomaterials* 11: 170.

154 Mondal, S., Min, B.K., Yi, Y., and Choi, C.G. (2022). Highly sensitive and fast responsive humidity sensor based on 2D $PtSe_2$ with gamma radiation tolerance. *Advanced Materials Technologies* 7: 2100751.

155 Briscoe, H.M. (1990). Why space tribology? *Tribology International* 23: 67–74.
156 Bai, Y., Zhang, C., Yu, Q. et al. (2021). Supramolecular PFPE gel lubricant with anti-creep capability under irradiation conditions at high vacuum. *Chemical Engineering Journal* 409: 128120.
157 Li, L., Qin, Z., Ries, L. et al. (2019). Role of sulfur vacancies and undercoordinated Mo regions in MoS_2 nanosheets toward the evolution of hydrogen. *ACS Nano* 13: 6824–6834.
158 Zhang, T., Liu, S., Zhang, X. et al. (2021). Fabrication of two-dimensional functional covalent organic frameworks via the thiol-ene "click" reaction as lubricant additives for antiwear and friction reduction. *ACS Applied Materials & Interfaces* 13: 36213–36220.
159 Liu, C., Yin, Y., Tong, B., and Zhang, G. (2021). Tribological properties of MoS_2 powder-lubricated interface. *Industrial Lubrication and Tribology* 73: 839–845.
160 Zhang, J., Chen, B., Chen, X., and Hou, X. (2021). Liquid-based adaptive structural materials. *Advanced Materials* 33: 2005664.
161 Torres, J.M. (2017). Risk taking as a new academic librarian through media outreach: "The Walton Reader's Spotlight Web Series". *Public Services Quarterly* 13: 309–314.
162 Mallapaty, S. (2021). China's space station is preparing to host 1,000 scientific experiments. *Nature* 596 (7870): 20–21.
163 Turunen, M., Brotons-Gisbert, M., Dai, Y. et al. (2022). Quantum photonics with layered 2D materials. *Nature Reviews Physics* 4: 219–236.

Index

a

ABA-stacked trilayer graphene 65
aberration-corrected scanning transmission election microscopy (AC-STEM) 307, 308
AB-stacked bilayer graphene (BLG) 261–262
aerospace materials 303
aggregation-induced emission (AIE) polymer 190
air biased coherent detection (ABCD) 139, 142, 143
air plasma 132–133, 136–139, 142, 143
all-optical modulation 41, 73–75
ammonium dihydrogen phosphate ($NH_4H_2PO_4$, ADP) 55
amplified spontaneous emission (ASE) 282, 294, 296
angle-resolved photoemission spectroscopy (ARPES) technique 228, 229
antimonene
  group VA-elemental 2D materials 232
  novel 2D materials 197
β-antimonene nanosheets 198
artificial plasmonic and metamaterial 55
atomic fabrication 190

b

balanced photodiodes (BPDs) 135–136
bandgap renormalization (BGR) 111, 113, 115, 118, 122
beta barium borate (BaB_2O_4, BBO) crystal 55, 58, 142–143
Bi_2Se_3 1, 33–34, 39, 235, 267
bismuthene 32, 40, 42, 221, 248, 311–312
  group VA-elemental 2D materials 233–234
black phosphorus (BP) 1, 29, 32, 56, 58, 104, 115–118, 124, 186, 196–197, 221, 230–232, 270, 272, 304, 310–311, 321
bottom-up method 240

c

calcium titanate ($CaTiO_3$) 200, 239
central dielectric layer 91
charge carriers 33, 116, 131, 148–149, 163, 259
chemical vapor deposition (CVD) 240
  growth of 2D materials 6–8
  growth of wafer-scale single crystal 2D materials 8
circular frequency 188
^{60}Co gamma radiation effects 309
coherent anti-Stokes Raman scattering (CARS) 29–31, 95
color-changing materials 206
complementary-metal–oxide–semiconductor (CMOS) technology 34, 56
continuous wave (CW) laser 185, 223, 243
  in 2D perovskites 294
conventional nonlinear optical 21

Two-Dimensional Materials for Nonlinear Optics: Fundamentals, Preparation Methods, and Applications, First Edition. Edited by Qiang Wang and Hao-Li Zhang.

Coulomb correlations 131, 149–150
cross-phase-modulation (XPM) effect 40, 97
$CsPb_2Br_5$ microplate 287–288

d

DAST 132–133, 135
defect-state assisted recombination 108
degenerate FWM (DFWM) 95
Dennard scaling 103
diamond anvil cell (DAC) 143–145, 147
difference frequency generation (DFG) 21–22, 57–58, 132–133, 258
Dirac fermions 109, 110, 163, 226
direct exfoliation methods 4
displacement damage 305, 306
distributed Bragg reflector (DBR) 73, 282, 285–286, 291, 294, 296, 299
distributed feedback (DFB) cavities 282, 285–286, 294, 296–297
donor-acceptor (D-A) type blends 196
dual-beam Z-scan 90
dual-elemental 2D materials
- topological insulators (TIs) 235–237
- transition metal dichalcogenides (TMDs) 234–235

e

edge states 32, 56, 65–66, 235, 274
electrical modulation 75–78
electric-field-induced second harmonics (EFISH) 75–76
electrochromic devices 206–207
electrochromism 206, 210
electron irradiation 307–309, 311, 316
electro-optic effect 21, 246
electrospinning technique 232
energy level model 186
erbium-doped fiber (EDF) 38–39, 230, 246, 303
- amplifiers 303

Er-doped fiber laser (EDFL) 40, 230, 246
E_{THz} 135, 137–139
excited state absorption (ESA) 32, 35, 37, 187–188, 200, 225, 231, 237
exciton-exciton annihilation (EEA) 111, 115
excitonic effects enhanced nonlinearity 66–69
excitons 25, 31, 33–35, 43–44, 66, 70, 111–112, 118, 131–132, 149–150, 154–155, 281, 296

f

Fabry–Pérot (F–P) cavity 73, 206, 285–286
FA-(*N*-MPDA)$PbBr_4$ microrod 289, 290
Faraday cage-assisted PECVD 179
fast saturable absorber 223–225, 247
Fengxia Geng's group 3
Fermi-Dirac distribution 226, 235
Fermi level modulation 190
field-effect transistors (FETs) 164, 198, 318–320
force separation method 190
Fourier transform 91–92, 132, 143
four-wave-mixing (FWM) microscopy 24, 26–27, 44, 55–59, 68, 70, 76, 87–98, 137–138, 164
four-wave rectification (FWR) 137–138
Frantz–Nodvik equation 224
free-carrier absorption (FCA) 27, 35, 111, 122, 186–188
FTO (fluorine-doped tin oxide) 207
full-width half-maximum (FWHM) 153, 245–246, 284–285, 287–289, 291–299

g

GaAs 59, 66, 132–135, 144–146, 165, 286
gallium selenide (GaSe) 25, 132–133, 266–267, 273
gallium tellurium (GaTe) 140–141, 273
β-Ga_2O_3 319–320
geostationary orbits (GEOs) 305, 320
germanium 25, 37, 76
- group IVA-elemental 2D materials 228–230

graphene 109, 110
  group IVA-elemental 2D materials 226–228
  2DLMs based SHG 260–262
  2DLMs based THG 268
  zero-bandgap structure of 228
graphene glass
  graphene doped glass 167
  high-order harmonic generation of 165–166
  high-temperature-resistant directly growth 168–171
  light absorption of 163
  metal-catalyst assisted growth 173–177
  on molten growth 171–173
  plasma-enhanced low-temperature growth 176, 179
  saturable absorption of 165
  spin coating 167–168
  third-order nonlinear optical properties of 164–165
  transfer method 168
graphene saturable absorber mirror (GSAM) 109–110, 241, 245
group delay dispersion (GDD) 246
group IVA-elemental 2D materials
  germanium 228–230
  graphene 226–228
group VA-elemental 2D materials
  antimonene 232
  bismuthene 233–234
  phosphorene 230–232

h

harmonic generation (HG) 22, 44, 66, 87, 179, 257
heat loss of 2D perovskite lasers 299
Hercher's rate equations 223, 225
hexagonal boron nitride (h-BN) 1, 6–7, 10, 12, 29, 58, 73, 78, 176–177, 197, 304, 308
Heyd–Scuseria–Ernzerho (HSE) method 234
high-angle annular dark-field (HAADF) 229, 307
high harmonic generation (HHG) 24–26, 59, 166, 257–259
high-intensity pulsed lasers 189
highly elliptical orbits (HEOs) 305
high-order harmonics (HHG) 57, 165–166
high resolution spatial and temporal THz spectroscopy 155
hydrogen induced exfoliated graphene (HEG) 206, 207

i

incoherent detection 138–139
inhomogeneous effects in 2D materials 88
integrated optoelectronic platform 318
  field-effect transistor 318–320
  lubricants 320–321
  sensors 320
interfacial segregation growth 174–175
International Space Station (ISS) 321
intramolecular charge transfer effect 194
ion irradiation 307–308, 311, 312, 317
irradiation 307
  electron irradiation 307
  ion irradiation 307–308
  proton irradiation 308
  γ-rays irradiation 308–309

k

Kerr effect 24, 40, 41, 44, 57, 89, 92, 258
Kirchhoff diffraction theory 89

l

Langmuir–Blodgett (LB) technique 242
large aperture photoconductive antennas (LAPCAs) 134
laser filamentation 138
laser intensity 28, 89–92, 137–139, 187, 195, 224, 228, 239, 243, 296

laser protection 185
  materials 189–208
  and mode-locking 310–313
lasers
  formation condition 283–284
  optical cavities 285–286
  performance parameters
    polarization 284
    quality factor Q 285
    threshold 284
  physical processes of 283
layer number 10, 14, 38, 41, 56, 60–64, 196
$LiNbO_3$ crystal 132, 133, 145, 257
linear absorption coefficient 27, 92, 188
liquid-phase exfoliations (LPE) 4–6, 14, 38, 109, 115, 190, 196, 198, 199, 201, 228, 230, 232, 235, 240–242, 245, 248
lithium niobate ($LiNbO_3$) 27, 43, 132, 133, 135, 136, 145, 257
localized surface plasmon resonance (LSPR) 190, 205
low Earth orbits (LEOs) 305
low-threshold optically-pumped nanolasers 282
lubricants 320–321

m
magnetical modulation 78
magnetization-induced SHG (MSHG) 78
mechanical exfoliation of two-dimensional (2D) materials 1–4
medium Earth orbits (MEOs) 305
metal-catalyst assisted graphene
  interfacial segregation growth 174–175
  metal element-containing precursors 175–177
  metal vapor catalyst 173–174
  sacrificial metal coating approach 175
metal halide perovskites of $MAPbBr_3$ 314–315
metal-organic frameworks (MOFs) 201–203, 221, 237
metal vapor catalyst 173–174
metamaterial 33, 41, 42, 55, 56, 73, 203–205
methyl methacrylate (MMA) 29, 167, 242, 320
microcrystals 287, 291
microplates 282, 287
microring 26, 282, 285, 287, 293–295
microrods 287–290
microscopic theory 261
mode-locked laser 38–40, 235, 237, 239, 246–249, 340
mode-locked pulses 230, 248
mode-locked techniques 246
molybdenum disulfide (MoS_2) 259, 321
  MoS_2-PVK 195, 196
  P3HT/MoS_2 310
molybdenum/tungsten disulfide/diselenide 111
Moore's Law 103
multi-elemental 2D materials
  metal-organic frameworks (MOFs) 237
  MXenes 237–238
  2D perovskites 238, 239
MXenes 237–238

n
nanoparticle on mirror (NPoM) system 73
nanoscale THz scanning probe microscopy 149–154
  van der Waals materials 149
nanowire lasers 286, 291, 293
NASA technology 303
Neumann principle 22
niobium nitride (NbN) superconducting thin films 317
N-methylpyrrolidone (NMP) 4, 294
non-layered 2D platinum sulfide 120–123

nonlinear absorption (NLA) 27–29, 35, 37, 40, 89–91, 144, 167, 186, 188, 189, 197–199, 203, 206, 207, 209, 210, 227, 232–234, 238, 239, 248, 311
4f nonlinear-imaging technique with phase object (NIT-PO)
  advantage 93
  CCD camera 91
  experimental arrangement of 91
  schematic diagram of 94, 95
nonlinear light-matter interactions 88
nonlinear optical (NLO) processes 55, 257
  imaging 22, 29–31
  metasurfaces 203
  susceptibility 25, 65, 73, 88
nonlinear optical limiting mechanism
  free carrier absorption (FCA) 188
  multifunctional optical limiting devices 206–208
  nonlinear refraction 188–189
  nonlinear scattering 189
  reverse saturable absorption 186–189
  two photon/multiphoton absorption 187–188
nonlinear photonic metamaterials 203–205
nonlinear refraction (NLR) 27–29, 37, 89, 91, 186, 188–189, 192
nonlinear scattering (NLS) 37, 186, 189, 194, 196
nonlinear two-photon process 21
nonparametric nonlinear optics 59–60

o

odd-order nonlinearity 35, 37, 57, 190, 199
optical cavity enhanced nonlinearity 71–73
optical limiting (OL) materials 35, 37
optical limiting effect 37, 38, 186, 188, 189, 196, 203, 206
optical modulation
  all-optical modulation 73–75
  excitonic effects enhanced nonlinearity 66–69
  optical cavity enhanced nonlinearity 71–73
  valley dependent nonlinearity 69–71
optical parametric amplifiers (OPA) 25, 132, 133
optical switch and modulation 40–41
organic materials 186

p

palladium diselenide 113, 116
Panda-style polarization maintaining fiber (PMF) 245
parametric nonlinear optics 58–60
Pauli exclusion principle 164, 221, 226
PbS/graphene 310
$(PEA)_2PbI_4$ perovskite 299
perovskite crystal structure 282
perovskite solar cells (PSCs) 300, 316–317
perovskite *via* spin-coating technique 291
phase control in CVD growth of 2D materials 12–14
phase engineering of 2D materials 13
phase transformation method 13
phase transition materials 185
phenylethylamine lead iodide($(PEA)_2PbI_4$) 38, 40, 201, 238, 239, 291, 294, 296, 298, 299, 314, 315
phosphorene, group VA-elemental 2D materials 230–234
photobleaching (PB) 113, 122
photoconductive antennas (PCAs) 132–135
photoinduced absorption (PIA) 113, 122
photomultiplier tube (PMT) 24, 29, 90, 138, 143, 289, 290
photonic crystal fiber (PCF) 39, 246, 248
physical vapor transport (PVT) 240, 249, 267

physical vapour deposition (PVD) 43, 198, 229, 240, 241
plasmon hybridization 206
plasmonic effect enhanced nonlinearity 205–206
plasmonic hot carrier 75
platinum diselenide 111, 113, 320
platinum sulfide (PtS) 120–123
polariton 33, 34, 43, 73, 124, 149–151
polarization, definition 284
polyacrylonitrile (Pan) carbon fiber 303
polydimethylsiloxane (PDMS) template 291, 294
poly(methylmethacrylate) (PMMA) 29, 37, 167, 168, 194–197, 231, 232, 245, 291, 296, 298, 320
poly (N-vinyl carbazole) (PVK) 195
potassium dihydrogen phosphate (KH_2PO_4, KDP) 55
proton irradiation 308, 316, 320
pump-probe measurement 104–106, 109, 113, 115
pump-probe techniques 22, 32–35, 44, 223

q
Q-switched and Mode-Locked Lasers 38–40, 237
Q-switched fiber laser 235, 243–245
Q-switched laser 37, 240, 242–245, 313
quantum communication 27, 304, 322
quantum NLO 27, 43, 44
quasi-2D hybrid perovskite (BPCSA 2D PVK) 38, 201
quasi-2D $(PEA)_2Cs_{n-1}Pb_nBr_{3n+1}$ perovskite microcrystals 291

r
γ-rays irradiation 234, 308
reflection Z-scan 90
reverse saturable absorption (RSA) 27–29, 32, 33, 60, 167, 186–187, 232
rhenium disulfide 113, 115

s
sacrificial metal coating approach 175, 176
saturable absorber mirror (SAM) 109, 110, 240, 245
saturable adsorption (SA)
- CVD 240
- dual-elemental 2D materials 234–237
- fast 224, 225
- liquid-phase exfoliations (LPE) 240–242
- mode-locked laser 246–248
- mono-elemental 2D materials 226–234
- multi-elemental 2D materials 237–239
- PVD 240
- Q-switched fiber laser 243–245
- Q-switched laser 242–243
- scotch tape exfoliation 240
- slow 224–225
- theory 222–223

Scotch tape exfoliation 240, 241
Scotch tape method 2, 3
second harmonic generation (SHG) 8, 21, 22, 55, 142, 164, 205, 257, 314
- mapping 8

second-order NLO effects 258
second-order nonlinearities 22
second-order nonlinear polarization 78
second-order nonlinearity counterpart 22
self-defocusing effect 189
self-phase modulation (SPM) 40, 41, 57, 59, 87, 89, 246, 259
SESAM 109, 110
shear exfoliation 4
SHG BBO 142
Shockley-Read-Hall 108
side-polished fiber (SPF) 237, 238, 248
single event effects (SEEs) 305, 306
single-mode fiber (SMF) 246
single photon sources 317
slow evaporation at constant temperature (SECT) 287, 288

slowly varying envelope approximation (SVEA) 92
slow saturable absorber 224–225
sonication-assisted exfoliation (SAE) 4, 172
space electronics 318
space environment adaptability 304
space radiation and two-dimensional materials
  electron irradiation 307
  ion irradiation 307
  irradiation 307
  proton irradiation 308
  γ-rays irradiation 308–309
  space radiation environment 305–306
space radiation environment 304, 305, 307, 310, 311, 321, 322
space X-ray detection and imaging 314–316
spin-orbit coupling (SOC) 69, 233
s-SNOM tip 151
stacking order 41, 56, 60, 64–65, 148
stimulated absorption 283
stimulated Raman scattering (SRS) 29, 95
structural engineering
  edge states 65–66
  layer number 60–62
  stacking order 64–65
  twisted angle 62–64
sum-and difference frequency generation (SFG, DFG) 21, 258
sum-frequency generation (SFG) 21, 22, 27, 44, 57, 58, 64, 70
superconducting nanowire single photon detectors (SNSPD) 317
super Rayleigh scattering 21
susceptibility dominates HG 257

t

tellurene 198–200
terahertz spectroscopy for 2D materials
  charge carriers 131
  nanoscale THz scanning probe microscopy of
    scanning near-field 150–151
    THz-STM 151–154
    van der Waals materials 149
  nonlinear optical processes
    advantages and disadvantages 133
    air plasma 136–139
    electro-optic (EO) crystals 135–136
    photoconductive antennas (PCAs) 133–135
    THz emission of 2D semiconductor materials 140–141
  photoinduced changes of the dielectric function 147–149
  time-domain spectroscopy setup 141–147
TFISH 138, 139
thermally assisted conversion (TAC) method 111, 113, 115, 120
thermally managed Z-scan 90
thickness control in CVD growth of 2D materials 10–12
thin films 39, 113, 147, 167, 198, 203, 205, 206, 238, 240, 287, 294–298, 304, 317
thin flakes 25, 266, 287, 298–299
third-harmonic generation (THG) 24, 55, 90, 143, 257–274
third-order NLO measurement technique 88, 97
third-order nonlinear optical process 76
third-order nonlinear optical susceptibility 188
third-order nonlinearity 22, 27, 57, 137, 138, 185, 207, 209
3D Dirac semimetal (3D DSM) 272
3D metal halide perovskite 281
three-photon processes 21
three-stage sonication-assisted liquid-phase exfoliation (TS-LPE) method 4, 5
tilted pulse front (TPF) rectification technology 133
time-resolved THz time-domain spectroscopy (TRTS) 145, 155

Tm-doped fiber laser (TDFL) 230
top-down method 6, 240
top-hat Z-scan 90
topological insulators (TIs) 1, 33, 141, 221, 226, 235–237
total ionizing dose (TID) effect 305
TPA-enhanced difference-frequency mixing 95
traditional laser protection equipment 185
traditional laser protection materials 189
transient absorption spectroscopy (TAS) 111, 115–118, 122, 148, 235, 311
transient electron current model 137
transient THz spectroscopy 154–155
transition-Cherenkov radiation (TCR) 138
transition metal dichalcogenides (TMDCs/TMDs) 1, 141, 149, 154, 194–196
  molybdenum/tungsten disulfide/diselenide 111
  palladium diselenide 113, 116
  platinum diselenide 111–113
  rhenium disulfide 113, 115, 116
  2DLMs based SHG 262, 264–266
  2DLMs based THG 268–270
trifluoroacetic acid (TFA) 237
twisted angle 56, 58, 62–64
2D $(BA)_2(MA)_{n-1}Pb_nI_{3n+1}$ 299
2D coordination polymers 201
two-dimensional coherent spectroscopy (2DCS) 97
2D layered materials (2DLMs) 147
  based SHG 259–260
    gallium selenide (GaSe) 266
    graphene 260–262
    TMDCs 262–266
  based THG
    black phosphorus 270–274
    graphene 268
    SHG and THG fundamentals 258–259
    TMDCs 268–270
2D materials for space use
  emerging nonlinear optical applications 309–314
  integrated optoelectronic platform 318–320
  perovskite solar cells (PSCs) 316–317
  single photon sources 317
  space X-ray detection and imaging 314–316
2D metal–organic frameworks (MOFs) 201, 203
2D nonlinear optical limiting materials 191
  atomic fabrication method 190
  Fermi level modulation 190
  graphene and analogs 190–194
    black phosphorus 196–197
    force separation method 190
    hexagonal boron nitride (h-BN) 197
    transition metal dichalcogenides (TMDs) 194–196
  materials with structures and mechanism
    nonlinear photonic metamaterials 203–205
    plasmonic effect enhanced nonlinearity 205–206
  novel 2D materials
    2D perovskites 200–201
    antimonene 197–198
    metal-organic frameworks (MOFs) 201, 203
    tellurene 198–200
  tuning bandgaps 189
2D nonlinear optical materials
  application of
    optical limiting (OL) materials 35–38
    optical switch and modulation 40–41
    plasmonics 42–43
    Q-switched and Mode-Locked Lasers 38–40
    quantum NLO 43
    ultrafast phonon dynamics 41–42

atomic thickness and dangling free surface 1
basics 57–58
chemical vapor deposition growth of 6–8
CVD growth of wafer-scale single crystal 8–10
ever-lasting expansion 44
via four-wave-mixing microscopy 26–27, 95–97
high resolution spatiotemporal characterizing techniques 44
illustration of 23
liquid-phase exfoliation (LPE) 4–6
mechanical exfoliation of 1–4
modulation in
  electrical 75–77
  magnetical 78
  optical 66–75
  structural engineering 60–66
new physics 44
NIT-PO 91–95
nonlinear optical imaging 29–31
nonparametric nonlinear optics 59–60
parametric nonlinear optics 58–59
perspectives 78, 97–98
phase control in CVD growth of 12–14
precise fabrication and functionalization of 2D materials 43
principle of
  nonlinear optical interaction 88
  self-phase modulation (SPM) 89
  third-harmonic generation (THG) 90
  two-photon induced photoluminescence (TPL) method 90
  Z-scan technique 27–29, 89–90
pump-probe techniques 32–35
SHG, THG and HHG Setups 24–26
thickness control in CVD growth of 10–12
2D organic-inorganic hybrid perovskites 40, 313
2D perovskites 118, 120, 200, 238, 239
  for nanolasers
    microcrystals 291
    microplates 287
    microring 293–294
    microrods 287, 289
    nanowire lasers 291–293
    thin films 294–298
    thin flakes 298–299
two-photon absorption (TPA/2PA) 55, 59, 90, 97, 111
two photon absorption/multiphoton absorption (TPA/MPA) 186
two-photon excited fluorescence (TPEF) 29, 31, 266
two-photon induced photoluminescence (TPL) method 27, 66, 70, 87, 90
two photon/multiphoton absorption 187

u

ultrafast phonon dynamics 23, 41
ultrafast time-resolved spectroscopy
  experiment for 107
  mechanism of 107–108
  pump-probe measurement 104–106

v

valley dependent nonlinearity 69–71
van der Waals 2D materials
  black phosphorus (BP) 115–118
  graphene 109, 110
  transition metal dichalcogenides (TMDC) 104, 109–115
van der Waals (vdW) heterostructures 147
van der Waals (vdWs) force 3, 56

w

wavelength division multiplexer (WDM) 38, 246
whispering–gallery–mode (WGM) cavities 282, 285–288, 293, 294

white light supercontinuum (WLS) 107
Wollastion prism (WP) 135

y
Yb-doped fiber laser (YDFL) 39, 40, 230
Young's modulus 259

z
Zernike phase contrast microscopy 91, 93
Z-scan technique 22, 27–29, 89–91, 167, 196, 203, 207, 230, 235, 237, 239, 313